CHIRAL DRUGS

CHIRAL DRUGS
Chemistry and Biological Action

Edited by

GUO-QIANG LIN
QI-DONG YOU
JIE-FEI CHENG

WILEY

A JOHN WILEY & SONS, INC., PUBLICATION

Published by John Wiley & Sons, Inc., Hoboken, New Jersey
Published simultaneously in Canada

For general information on our other products and services or for technical support, please contact our Customer Care Department within the United States at (800) 762-2974, outside the United States at (317) 572-3993 or fax (317) 572-4002.

Wiley also publishes its books in a variety of electronic formats. Some content that appears in print may not be available in electronic formats. For more information about Wiley products, visit our web site at www.wiley.com.

Library of Congress Cataloging-in-Publication Data:

Chiral drugs : chemistry and biological action / edited by Guo-Qiang Lin, Qi-Dong You, Jie-Fei Cheng.
    p. ; cm.
  Includes bibliographical references and index.
  ISBN 978-0-470-58720-1 (hardback)
  1. Chiral drugs. 2. Drug development. 3. Structure-activity relationships (Biochemistry) I. Lin, Guo-Qiang, 1943- II. You, Qi-Dong. III. Cheng, Jie-Fei.
  [DNLM: 1. Drug Discovery—methods. 2. Pharmaceutical Preparations—chemistry.
3. Structure-Activity Relationship.   QV 744]
  RS429.C483 2011
  615′.19—dc22

                                                              2011002203

Printed in Singapore

oBook ISBN: 978-1-118-07564-7
ePDF ISBN: 978-1-118-07562-3
ePub ISBN: 978-1-118-07563-0

10 9 8 7 6 5 4 3 2 1

CONTENTS

ABOUT THE EDITORS

Professor **Guo-Qiang Lin** received his BS degree in chemistry from Shanghai University of Science and Technology in 1964. After completion of his graduate study at the Shanghai Institute of Organic Chemistry in 1968, he remained in the same institute and worked on natural products chemistry. He was promoted to full professorship in 1991. In 2001, he was elected as an Academician of the Chinese Academy of Sciences. His research interests include the synthesis of natural products and biologically active compounds, asymmetric catalysis, and biotransformation. He is an Executive Board Member of Editors for *Tetrahedron* Publications, Vice Editor-In-Chief of *Acta Chimica Sinica,* and *Scientia Sinica Chimica*. He has served as Director of the Division of Chemical Science, National Natural Science Foundation of China since 2006.

Dr. **Qi-Dong You** is the Dean and a Professor of the School of Pharmacy, at China Pharmaceutical University. He received his BS degree in pharmacy from the China Pharmaceutical University and completed his PhD degree in medicinal chemistry at the Shanghai Institute of Pharmaceutical Industry in 1989. He then returned to CPU as a lecturer and associate director of the Department of Medicinal Chemistry. He spent one year and a half as a senior visiting scholar in the Department of Pharmaceutical Sciences, University of Strathclyde, Glasgow, UK, before he was promoted to a full professorship in 1995. He is a council member of the China Pharmaceutical Association (CPA) and the Vice-Director of the Division of Medicinal Chemistry of CPA. His research interests include the design, synthesis, and biological evaluation of new therapeutic agents for cancer and cardiovascular and infectious diseases. He is an Associate Editor of *Progress in Pharmaceutical Sciences* and serves on the Editorial Board of the *International Journal of Medicinal Chemistry* and *Acta Pharmaceutica Sinica*.

Dr. **Jie-Fei (Jay) Cheng** was born in 1964 in Jiangxi, China. He obtained his BS degree in chemistry from the Jiangxi Normal University in 1983 and continued his graduate studies at the Shanghai Institute of Organic Chemistry, Chinese Academy of Sciences, under the guidance of Professors Wei-Shan Zhou and Guo-Qiang Lin. After receiving his Master's degree in chemistry in 1986, he joined the research group of Professor Yoshimasa Hirata and Dr. Junichi Kobayashi (now a Professor at Hokkaido University) at the Mistubishi-Kasei Institute of Life Sciences, Tokyo, Japan. He then moved to Keio University to pursue his Ph.D in Professor Shosuke Yamamura's lab. Since 1993, he has been working at various pharmaceutical companies/biotechs in the United States, focusing on small-molecule drug discovery. He is currently the Director of Otsuka Shanghai Research Institute, a fully owned subsidiary of Otsuka Pharmaceutical Co. Ltd, Japan and an adjunct professor at Fudan Univeristy, China.

CONTRIBUTORS

CARL BEHRENS, Wilmington PharmaTech Company LLC, Newark, DE, USA, and University of Delaware, Newark, DE, USA

HAI-ZHI BU, 3D BioOptima Co. Ltd, Suzhou, Jiangsu, China

JIE-FEI (JAY) CHENG (EDITOR), Otsuka Maryland Medicinal Laboratories, Inc., Rockville, MD, USA, and, Otsuka Shanghai Research Institute, Shanghai, China

HANQING DONG, OSI Pharmaceuticals, A Wholly Owned Subsidiary of Astellas US, Farmingdale, NY, USA

XIAO-HUI GU, Otsuka Maryland Medicinal Laboratories, Inc., Rockville, MD, USA

XIAOCHUAN GUO, Drumetix Laboratories, LLC, Greensboro, NC, USA

ERIC HU, Gilead Sciences Inc., Foster City, CA, USA

HUI-YIN (HARRY) LI, Wilmington PharmaTech Company LLC, Newark, DE, USA, and University of Delaware, Newark, DE, USA

ZENGBIAO LI, Drumetix Laboratories, LLC, Greensboro, NC, USA

GUO-QIANG LIN (EDITOR), Key Laboratory of Synthetic Chemistry of Natural Substances, Shanghai Institute of Organic Chemistry, Chinese Academy of Sciences, Shanghai, China

DINGGUO LIU, Pfizer, San Diego, CA, USA

YONGGE LIU, Otsuka Maryland Medicinal Laboratories, Inc., Rockville, MD, USA

RUI LIU, Wilmington PharmaTech Company LLC, Newark, DE, and University of Delaware, Newark, DE, USA

WENYA LU, Department of Chemistry, Iowa State University, Ames, Iowa, USA

CHAO-YING NI, Wilmington PharmaTech Company LLC, Newark, DE, USA, and University of Delaware, Newark, DE, USA

FENG-LING QING, Key Laboratory of Organofluorine Chemistry, Shanghai Institute of Organic Chemistry, Chinese Academy of Sciences, Shanghai, China, and College of Chemistry and Chemistry Engineering, Donghua University, Shanghai, China

XIAO-LONG QIU, Key Laboratory of Organofluorine Chemistry, Shanghai Institute of Organic Chemistry, Chinese Academy of Sciences, Shanghai, China

JIANGQIN SUN, Otsuka Shanghai Research Institute, Shanghai, China

XING-WEN SUN, Department of Chemistry, Fudan University, Shanghai, China

DEPING WANG, Biogen IDEC Inc., Cambridge, MA, USA

JIANQIANG WANG, ArQule Inc., Woburn, MA, USA

ZHIMIN WANG, Sundia MedTech Company Ltd., Shanghai, China

GUANG YANG, GLAXOSMITHKLINE, R&D China, Shanghai, China

QI-DONG YOU (EDITOR), China Pharmaceutical University, Nanjing, Jiangsu, China

XUYI YUE, Key Laboratory of Organofluorine Chemistry, Shanghai Institute of Organic Chemistry, Chinese Academy of Sciences, Shanghai, China

JIAN-GE ZHANG, School of Pharmaceutical Science, Zhengzhou University, Zhengzhou, Henan, China

INTRODUCTION

The book consists of 11 chapters. The first part of the book introduces the general concept of chirality and its impact on drug discovery and development. The history and the trends of chiral drug development, the technologies for the preparation of chiral drugs, and the industrial applications of chiral technologies are discussed. This part covers three important chiral technologies, namely, asymmetric synthesis, biocatalytic process, and chiral resolution, and discusses their impact on chiral drug development. Without question, fluorine atoms play an important role in chiral drug discovery and development. The significance and the preparation of fluorine-containing chiral drugs are the topic of a separate chapter.

The second part of the book mainly deals with some unique aspects of chiral drugs in terms of pharmaceutical, pharmacological, and toxicological properties. For instance, pharmacology, pharmacokinetic properties, and toxicology of chiral drugs are discussed in comparison with racemic drugs. Additionally, computational modeling as applied to chiral drug discovery and development is discussed. This part of the book provides a general knowledge of design, synthesis, screening, and pharmacology from the preclinical point of view, hoping to raise interest from a broad range of readers.

Finally, Chapter 11 covers 25 representative chiral drugs that have been approved or are in advanced clinical trials. Some natural products are not included. The most important criteria for their selection are the involvement of chiral processes during their preparation and the significance of chirality in their development. Every entry contains the trade name, chemical name and properties, a representative synthetic pathway, pharmacological characterizations, and references.

This book is intended to introduce chemists to pharmacological aspects of drug development and to form a fruitful cooperation among academic synthetic chemists, medicinal chemists, pharmaceutical scientists, and pharmacologists from the pharmaceutical and biotechnology industries. The references after each chapter will give readers an opportunity for further reading on the topics discussed. This is the first book of its kind to combine synthetic organic chemistry, medicinal chemistry, process chemistry, and pharmacology in the context of chiral drug discovery and development.

CHAPTER 1

OVERVIEW OF CHIRALITY AND CHIRAL DRUGS

GUO-QIANG LIN
Key Laboratory of Synthetic Chemistry of Natural Substances, Shanghai Institute of Organic Chemistry, Chinese Academy of Sciences, Shanghai, China

JIAN-GE ZHANG
School of Pharmaceutical Science, Zhengzhou University, Zhengzhou, China

JIE-FEI CHENG
Otsuka Shanghai Research Institute, Pudong New District, Shanghai, China

Chiral Drugs: Chemistry and Biological Action, First Edition. Edited by Guo-Qiang Lin, Qi-Dong You and Jie-Fei Cheng.
© 2011 John Wiley & Sons, Inc. Published 2011 by John Wiley & Sons, Inc.

3

1.1 INTRODUCTION

The pharmacological activity of a drug depends mainly on its interaction with biological matrices or drug targets such as proteins, nucleic acids, and biomembranes (e.g., phospholipids and glycolipids). These biological matrices display complex three-dimensional structures that are capable of recognizing specifically a drug molecule in only one of the many possible arrangements in the three-dimensional space, thus determining the binding mode and the affinity of a drug molecule. As the drug target is made of small fragments with chirality, it is understandable that a chiral drug molecule may display biological and pharmacological activities different from its enantiomer or its racemate counterpart when interacting with a drug target. In vivo pharmacokinetic processes (ADME) may also contribute to the observed difference in in vivo pharmacological activities or toxicology profiles. One of the earliest observations on the taste differences associated with two enantiomers of asparagines was made in 1886 by Piutti [1]. Colorless crystalline asparagine is the amide form of aspartic or aminosuccinic acid and is found in the cell sap of plants in two isomeric forms, levo- and dextro-asparagin. The *l*-form exists in asparagus, beet-root, wheat, and many seeds and is tasteless, while the *d*-form is sweet. Thalidomide is another classical example. It was first synthesized as a racemate in 1953 and was widely prescribed for morning sickness from 1957 to 1962 in the European countries and Canada. This led to an estimated over 10,000 babies born with defects [2]. It was argued that if one of the enantiomers had been used instead of the racemate, the birth defects could have been avoided as the *S* isomer caused teratogenesis and induced fatal malformations or deaths in rodents while the *R* isomer exhibited the desired analgetic properties without side effects [3]. Subsequent tests with rabbits proved that both enantiomers have desirable and undesirable activities and the chiral center is easily racemized in vivo [4]. Recent identification of thalidomide's target solved the long-standing controversies [5]. The chirality story about thalidomide, although not true, has indeed had great impact on modern chiral drug discovery and development (Fig. 1.1).

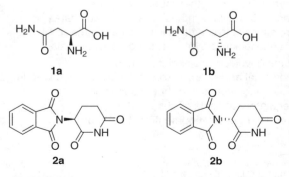

FIGURE 1.1 Asparagine (**1**) and thalidomide (**2**).

1.2 OVERVIEW OF CHIRALITY

1.2.1 Superimposability

Chirality is a fundamental property of three-dimensional objects. The word "chiral" is derived from the Greek word *cheir*, meaning hand, or "handedness" in a general sense. The left and right hands are mirror images of each other no matter how the two are arranged. A chiral molecule is the one that is not superimposable with its mirror image. Accordingly, an achiral compound has a superimposable mirror image. Two possible mirror image forms are called enantiomers and are exemplified by the right-handed and left-handed forms of lactic acids in Figure 1.2. Formally, a chiral molecule possesses either an asymmetric center (usually carbon) referred to as a chiral center or an asymmetric plane (planar chirality).

In an achiral environment, enantiomers of a chiral compound exhibit identical physical and chemical properties, but they rotate the plane of polarized light in opposite directions and react at different rates with a chiral compound or with an achiral compound in a chiral environment. A chiral drug is a chiral molecule with defined pharmaceutical/pharmacological activities and utilities. The description "chiral drug" does not indicate specifically whether a drug is racemic, single-enantiomeric, or a mixture of stereoisomers. Instead, it simply implies that the drug contains chiral centers or has other forms of chirality, and the enantiomeric composition is not specified by this terminology.

1.2.2 Stereoisomerism

In chemistry, there are two major forms of isomerism: constitutional (structural) isomerism and stereoisomerism. Isomers are chemical species (or molecular entities) that have the same stoichiometric molecular formula but different constitutional formulas or different stereochemical formulas. In structural isomers, the atoms and functional groups are joined together in different ways. On the

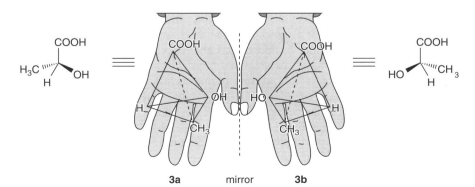

FIGURE 1.2 Mirror images of lactic acid.

other hand, stereoisomers are compounds that have the same atoms connected in the same order but differ from each other in the way that the atoms are oriented in space. They include enantiomers and diastereomers, the latter indicating compounds that contain two or more chiral centers and are not superimposable with their mirror image. Diastereomers also include the nonoptical isomers such as *cis-trans* isomers.

Many molecules, particularly many naturally derived compounds, contain more than one chiral center. In general, a compound with n chiral centers will have 2^n possible stereoisomers. Thus 2-methylamino-1-phenylpropanol with two chiral centers could have a total of four possible stereoisomers. Among these, there are two pairs of enantiomers and two pairs of diastereomers. This relationship is exemplified by ephedrines (**4a, 4b**) and pseudoephedrines (**4c, 4d**) shown in Figure 1.3. In certain cases, one of the stereoisomeric forms of a molecule containing two or more chiral centers could display a superimposable mirror image, which is referred to as a *meso* isomer.

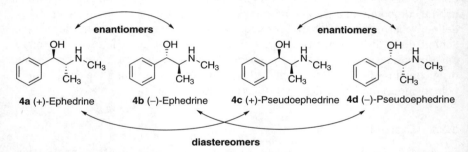

FIGURE 1.3   Enantiomers and diastereomers.

1.2.3   Absolute Configuration

It is important to define the absolute configuration of a chiral molecule in order to understand its function in a biological system. Many biological activities are exclusive to one specific absolute configuration. Without a good understanding of absolute configuration of a molecule, it often is hard to understand its chemical and biological behavior. As mentioned above, two enantiomers of a chiral compound will have identical chemical and physical properties such as the same boiling/melting points and solubility in a normal achiral environment.

The *R/S* nomenclature or Cahn–Ingold–Prelog (CIP) system for defining absolute configuration is the most widely used system in the chemistry community. The key to this system is the CIP priority rule, which defines the substituent priority based on the following criteria: 1) Higher atomic number or higher atomic mass is given higher priority; 2) when the proximate atom of two or more of the substituents are the same, the atomic number of the next atom determines the priority; 3) double bonds or triple bonds are counted as if they were split into two or three single bonds, respectively; 4) *cis* is given higher priority than *trans*;

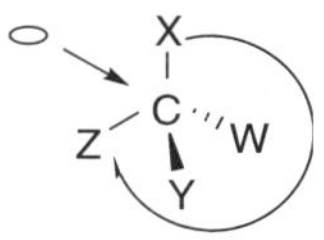

5

FIGURE 1.4 A central chiral system.

5) long pair electrons are regarded as an atom with atomic number 0; and 6) proximal groups have higher priority than distal groups.

The carbon atom in compound **5** (Fig. 1.4) is defined as a chiral center if the four substituents (X, Y, Z, and W) around the center are different. If the molecule is oriented in a way that the lowest-priority group W is pointed away from the observer and the other three groups have a priority sequence X→Y→Z in a clockwise direction, the chiral center will have a *R* configuration; otherwise it is defined as an *S* configuration. The *R/S* system can be used for other chiral molecules without a chiral center (e.g., planar chirality) as well [6].

Fischer's convention with D or L prefix (small cap) is sometimes used for the description of the absolute configuration of a molecule, particularly for carbohydrates or amino acids. For example, D-glyceraldehyde **6a** by Fischer's convention is shown in Figure 1.5 and is identical to (*R*)-glyceraldehyde according to *CIP* rules. By relating compounds to glyceraldehydes, the absolute configuration of other compounds can be defined. For example, naturally occurring alanine **6d** is designated as L-form with an *S* configuration.

Enantiomers do differ from each other in rotating the plane-polarized light, which is referred to as optical activity or optical rotation. When an enantiomer rotates the plane of polarized light clockwise (as seen by a viewer toward whom the light is traveling), it is labeled as (+). Its mirror-image enantiomer is labeled as the (−) isomer. The (+) and (−) isomers have historically been termed *d*- and *l*-, respectively, with *d* for dextrorotatory and *l* for levorotatory rotation of the lights. This *d/l* system is now obsolete, and (+/−) should be used instead to specify the optical rotation. It should also be pointed out that the optical rotation (+/−) convention has no direct relation with the *R/S* or D/L systems. It is used in most cases for description of *relative*, not *absolute* configuration. Thus compound **3b**, which rotates the plane-polarized light in a clockwise direction, is denoted as *R*-(+)-lactic acid, while the enantiomer (**3a**) is referred to *S*-(−)-lactic acid.

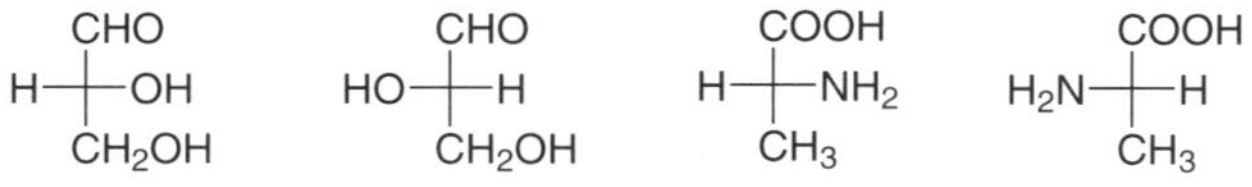

CHO	CHO	COOH	COOH
H——OH	HO——H	H——NH₂	H₂N——H
CH₂OH	CH₂OH	CH₃	CH₃
6a	**6b**	**6c**	**6d**

FIGURE 1.5 Structure of (D)- and (L)-glyceraldehyde and analogs.

Absolute configuration is most commonly determined either by X-ray crystallography or through chemical conversion to a known compound with defined stereochemistry. Other instrumental procedures for determining absolute stereochemistry without derivatization include circular dichroism (CD), vibrational circular dichroism (VCD) [7], and optical rotator dispersion (ORD) or specific optical rotation. The NMR-based method for deducing the absolute configuration of secondary carbinol (alcohol) centers using the "modified Mosher method" [8] was first described by Kakisawa and co-workers [9]. This modified Mosher ester analysis relies on the fact that the protons in diasteromeric α-methoxy-α-trifluoromethylphenylacetates display different arrays of chemical shifts in their ^{1}H NMR spectra. When correctly used and supported by appropriate data, the method can be used to determine the absolute configuration of a variety of compounds including alcohols, amines, and carboxylic acids [10]. However, it is always advisable to examine the complete molecular topology in the neighborhood of the asymmetric carbon centers and confirm with another analytical method.

1.2.4   Determination of Enantiomer Composition (ee) and Diastereomeric Ratio (dr)

It is important to measure enantiomer composition and diastereomeric ratio for a chiral molecule, in particular a chiral drug, as the biological data may closely relate to the optical purity. The enantiomer composition of a sample is described by enantiomeric excess, or *ee*%, which describes the excess of one enantiomer over the other. Correspondingly, the diastereomer composition of a diastereomer mixture is the measure of an extent of a particular diastereomer over the others. This is calculated as shown in Equations 1 and 2, respectively, for $[S] > [R]$ (Fig. 1.6).

A chiral molecule containing only one enantiomeric form is regarded as optically pure or enantiopure or enantiomerically pure. Enantiomers can be separated via a process called resolution (Chapter 4), while in most cases diastereomers can be separated through chromatographic methods. A variety of methods for determination of optical purity or *ee*/de value are available [6]. One of the widely used methods for analyzing chiral molecules is polarimetry. For any compound of which the optical rotation of the pure enantiomer is known, the *ee* can be determined simply from the observed rotation and calculated by Equations 3 and 4 (Fig. 1.7).

$$ee\% = \frac{[S] - [R]}{[S] + [R]} \times 100\% \qquad \text{eq. 1}$$

$$de\% = \frac{[S^*S] - [S^*R]}{[S^*S] + [S^*R]} \times 100\% \qquad \text{eq. 2}$$

FIGURE 1.6   Method of calculating enantiomer or diastereomer excess.

$$[\alpha]_D^{20} = \frac{[\alpha]}{L(dm) \times c(g/100mL)} \times 100\%$$ **Eq. 3**

$[\alpha]^D$ = measured rotation
L = path length of cell (dm)
c = concentration (g/100mL)
D = D line of wavelength of light used for measurement
20 = temperature

$$ee\ \% = \frac{[\alpha]_{obs}}{[\alpha]_{max}} \times 100\%$$ **Eq. 4**
(optical purity)

FIGURE 1.7   *ee* value is directly determined from the observed rotation.

Chromatography with chiral stationary columns, for example, chiral high-pressure liquid chromatography (HPLC) or chiral gas chromatography (GC), has also been utilized extensively for analyzing and determining enantiomeric composition of a chiral compound. Nuclear magnetic resonance (NMR) spectroscopy can also be used to evaluate the enantiomeric purity in the presence of chiral shift reagents [6,11] or through its diastereomer derivatives (e.g., Mosher's esters) [8].

1.3 GENERAL STRATEGIES FOR SYNTHESIS OF CHIRAL DRUGS

Asymmetric synthesis refers to the selective formation of a single stereoisomer and therefore affords superior atom economy. It has become the most powerful and commonly employed method for preparation of chiral drugs. Since the 1980s, there has been progress in many new technologies, in particular, the technology related to catalytic asymmetric synthesis, that allow the preparation of pure enatiomers in quantity. The first commercialized catalytic asymmetric synthesis, the Monsanto process of L-DOPA (**9**) (Fig. 1.8), was established in 1974 by Knowles [12], who was awarded a Nobel Prize in Chemistry in 2001 along with Noyori and Sharpless. In the key step of the synthesis of L-DOPA, a gold standard drug for Parkinson disease, enamide compound **7** is hydrogenated in the presence of a catalytic amount of [Rh(*R,R*)-DiPAMP)COD]$^+$BF$_4$ complex, affording the protected amino acid **8** in quantitative yield and in 95% *ee*. A simple acid-catalyzed hydrolysis step completes the synthesis of L-DOPA (**9**).

The discovery of an atropisomeric chiral diphosphine, BINAP, by Noyori in 1980 [13] was revolutionary in the field of catalytic asymmetric synthesis. For example, the BINAP-Ru(II) complexes exhibit an extremely high chiral recognition ability in the hydrogenation of a variety of functionalized olefins and ketones. This transition metal catalysis is clean, simple, and economical to operate and hence is capable of conducting a reaction on a milligram to kilogram scale with a very high (up to 50%) substrate concentration in organic solvents. Both

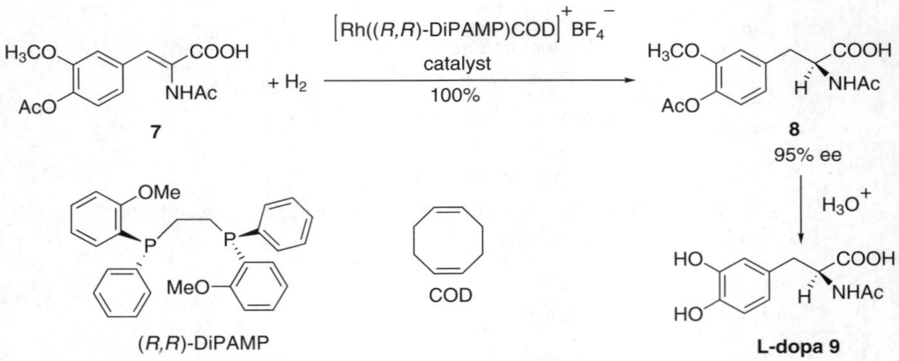

FIGURE 1.8   Monsanto process of L-dopa (**9**).

enantiomers can be synthesized with equal efficiency by choosing the appropriate enantiomers of the catalysts. It has been used in industrial production
of compounds such as (R)-1,2- propanediol, (S)-naproxen, a chiral azetidinone
intermediate for carbapenem synthesis, and a β-hydroxylcarboxylic acid intermediate for the first-generation synthesis of Januvia [14] among others. The
Sharpless–Katsuki epoxidation was also published in 1980 [15]. It has also been
used for the chiral drug synthesis on an industrial scale.

Chiral compounds can now be accessed in one of many different approaches:
1) via chiral resolution of a racemate (Chapter 4); 2) through asymmetric synthesis, either chemically or enzymatically (Chapters 2 and 3); and 3) through
manipulation of chiral starting materials (chiral-pool material). In the early 1990s,
most chiral drugs were derived from chiral-pool materials, and only 20% of all
drugs were made via purely synthetic approaches. This has now been reversed,
with only about 25% of drugs made from chiral pool and over 50% from other
chiral technologies [16]. The following is a brief account of catalytic enantioselective synthesis with commercial applications.

1.3.1   Enantioselective Synthesis via Enzymatic Catalysis

Enzyme-catalyzed reactions (biotransformation) are often highly enantioselective
and regioselective, and they can be carried out at ambient temperature, atmospheric pressure, and at or near neutral pH. Most of the enzymes used in the
asymmetric synthesis can be generated in large quantity with modern molecular biology approaches. The enzyme can be degraded biochemically, therefore
eliminating any potential hazardous caused by the catalysis, providing a superior and environmentally friendly method for making chiral drug molecules. It
is estimated that the value of pharmaceutical intermediates generated by using
enzymatic reactions was $198 million in 2006 and is expected to reach $354.4
million by 2013 [17].

(S)-6-hydroxynorleucine (**11**) is a key intermediate for the synthesis of omapatrilat (**12**), an antihypertensive drug that acts by inhibiting angiotensin-converting enzyme (ACE) and neutral endopeptidase (NEP). **11** is prepared from 2-keto-6-hydroxyhexanoic acid **10** by reductive amination using beef liver glutamate dehydrogenase at 100 g/l substrate concentration. The reaction requires ammonia and NADH. NAD produced during the reaction is recycled to NADH by the oxidation of glucose to gluconic acid with glucose dehydrogenase from *Bacillus megaterium*. The reaction is complete in about 3 h with reaction yields of 92% and >99% *ee* for (S)-6-hydroxynorleucine **11** (Fig. 1.9) [18].

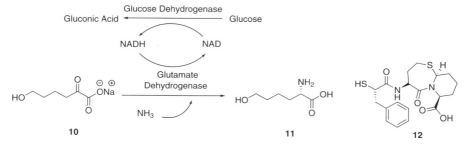

FIGURE 1.9   Enzymatic synthesis of chiral synthon (S)-6-hydroxynorleucine (**11**).

There are some exceptions and limitations to the enzymatic-catalyzed reactions. For example, the reaction type may be limited, and reactions may preferably be conducted in aqueous media and at low substrate concentration. However, a lot of new development in the technology of engineering enzymes have been witnessed recently [19]. Enzymes can be immobilized and reused in many cycles. Selective mutations of an enzyme can alter the enzyme's performance or even make the opposite enantiomer formation possible.

1.3.2   Enantioselective Synthesis via Organometallic Catalysis

In asymmetric synthesis, a chiral agent should behave as a catalyst with enzymelike selectivity and turnover rate. Transition metal-based catalysts have been prevalent in organic synthesis for many years. Since the introduction of the Monsanto process of L-DOPA and BINAP-based ligands, asymmetric hydrogenation has become one of the most important processes in the pharmaceutical industry to synthesize key intermediates or active pharmaceutical ingredients. More than 3,000 chiral diphosphine and many monophosphine ligands have been reported, and approximately 1% of those ligands are currently commercially available [20]. Besides the asymmetric hydrogenation of olefins, the ligand-mediate asymmetric hydrogenation of ketone to the corresponding alcohol [21] is becoming an indispensable alternative to other known processes such as transfer hydrogenation and biocatalytic and hydride reduction. However, a lot still remains to be improved in this field in terms of catalyst sensitivity to atmosphere, high cost, and possible toxicity.

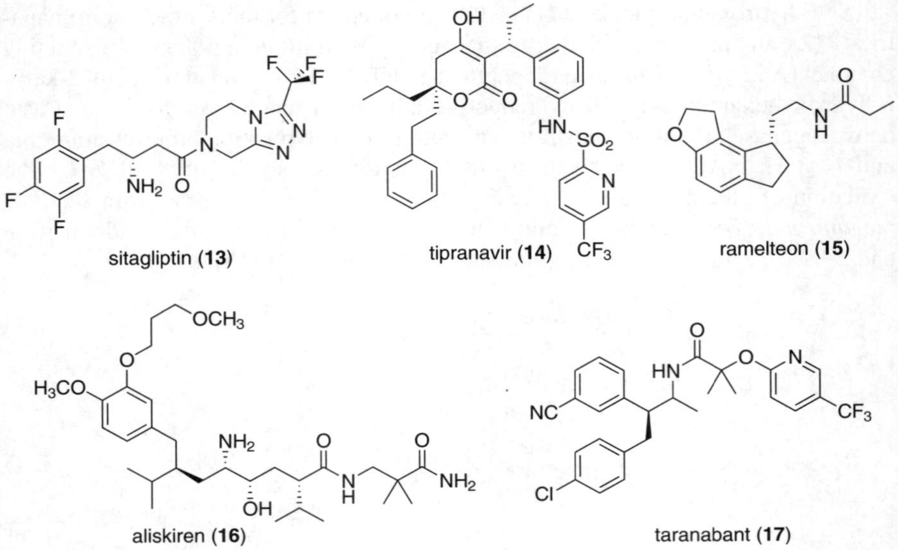

sitagliptin (**13**)   tipranavir (**14**)   ramelteon (**15**)

aliskiren (**16**)   taranabant (**17**)

FIGURE 1.10 Example compounds generated via catalytic asymmetric hydrogenation.

Compounds **13–17** are examples that were generated via catalytic asymmetric hydrogenation. According to reference [22], they are sitagliptin (**13**), an oral diabetes drug, tipranavir (**14**), an HIV protease inhibitor, ramelteon (**15**), a sleep aid, aliskiren (**16**), which is a hypertension drug, and taranabant (**17**), the antiobesity agent (Fig. 1.10).

1.3.3 Enantioselective Synthesis via Organocatalysis

Organocatalysts [23] have emerged as a powerful synthetic paradigm to complement organometallic- and enzyme-catalyzed asymmetric synthesis. Although examples of asymmetric organocatalysis appeared as early as the 1970s [24], the field was not born until the late 1990s and matured at the turn of the new century. Organocatalysis is now widely accepted as a new branch of enantioselective synthesis. A survey conducted by MacMillan [25] in 2008 showed only a few papers describing organocatalytic reactions before 2000, while the number of papers published in 2007 is close to 600. There have been a number of special issues of journals dedicated to asymmetric organocatalysis [26].

Organocatalysts are loosely defined as low-molecular-weight organic molecules having intrinsic catalytic activity. If an organocatalyst is modified to contain a chiral element, the reaction catalyzed by it could become enantioselective. Aside from being catalytically active, asymmetric organocatalysis are in general relatively inexpensive and readily available, are stable to atmospheric conditions, and have low toxicity. Many organocatalysts are simple derivatives of commonly available naturally occurring compounds. Representative examples

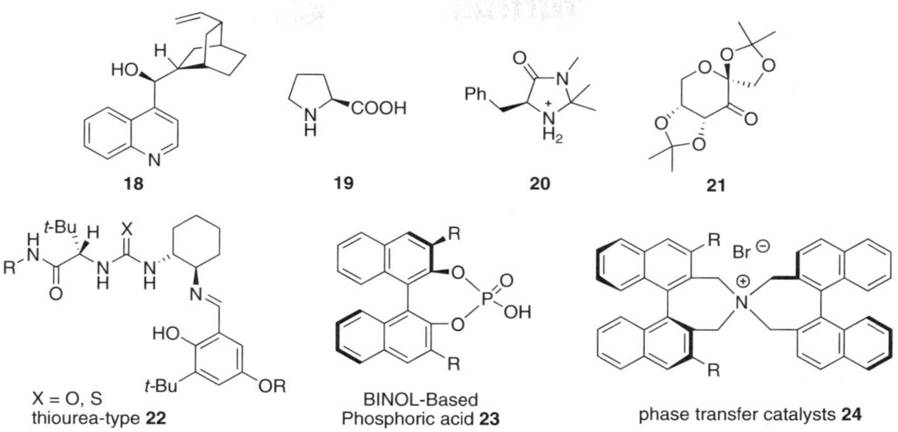

FIGURE 1.11 Representative organocatalysts.

include alkaloids and their derivatives (e.g., cinchonidine **18**) or L-proline (**19**) and other natural amino acids, which function, for example, as starting materials for MacMillan-type catalysts like **20**. The chiral ketone (**21**) generated from fructose was reported for dioxirane-mediated asymmetric epoxidation [27] (Fig. 1.11). A number of privileged organocatalysts, such as **22, 23**, and **24** have been designed, synthesized, and applied to various asymmetric reactions, which include C-C, C-heteroatom bond formation, oxidation, and reduction reactions.

The versatility of asymmetric organocatalysis is demonstrated by the practical synthesis of methyl (2*R*,3*S*)-3-(4-methoxyphenyl) glycidate [(−)-**27**], a key inter-mediate in the synthesis of diltiazem hydrochloride **28**, which has been used as a medicine for the treatment of cardiovascular diseases since the 1970s. Methyl (*E*)-4-methoxycinnamate **25** underwent asymmetric epoxidation with a chiral dioxirane, generated in situ from Yang's catalyst **26**, to provide the product (−)-**27** in both high chemical (>85%) and optical (>70%*ee*) yields (Fig. 1.12) [28].

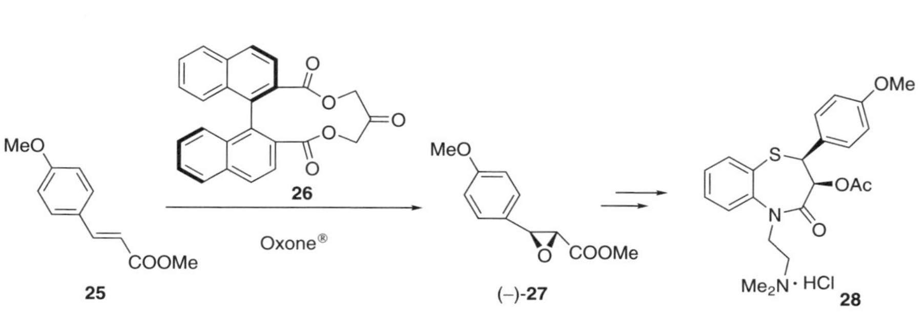

FIGURE 1.12 Asymmetric epoxidation of methyl (*E*)-4-methoxycinnamate (**25**).

1.4 TRENDS IN THE DEVELOPMENT OF CHIRAL DRUGS

1.4.1 Biological and Pharmacological Activities of Chiral Drugs

Many of the components associated with living organisms are chiral, for example, DNA, enzymes, antibodies, and hormones. The enantiomers of a chiral drug may display different biological and pharmacological behaviors in chiral living systems. This can be easily understood with the example of a drug-receptor model depicted in Figure 1.13. In possession of different spatial configurations, one active isomer may bind precisely to the target sites (α, β, γ), while an inactive isomer may have an unfavorable binding or bind to other unintended targets [29]. Pharmacological effects of enantiomeric drugs may be categorized as follows [30].

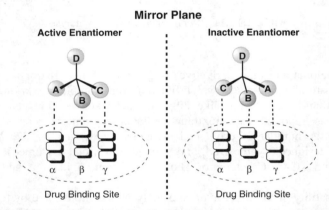

FIGURE 1.13 Stereoselective binding of enantiomers of a chiral drug.

1. *Both enantiomers act on the same biological target(s), but one isomer has higher binding affinity than the other:* For example, carvedilol (**29**) is marketed as a racemate for the treatment of hypertension and congestive heart failure [31]. It is a nonselective β- and α-adrenergic receptor blocking agent. Nonselective β-blocking activity resides mainly in the (S)-carvedilol, and the α-blocking effect is shared by both (R)- and (S)-enantiomers [32]. Sotalol (**30**) is a racemic β-adrenergic blocker. The (R)-enantiomer possesses the majority of the β-blocking activity, and the (R)- and (S)-enantiomers of sotalol share an equivalent degree of class III antiarrhythmic potency [33] (Fig. 1.14).

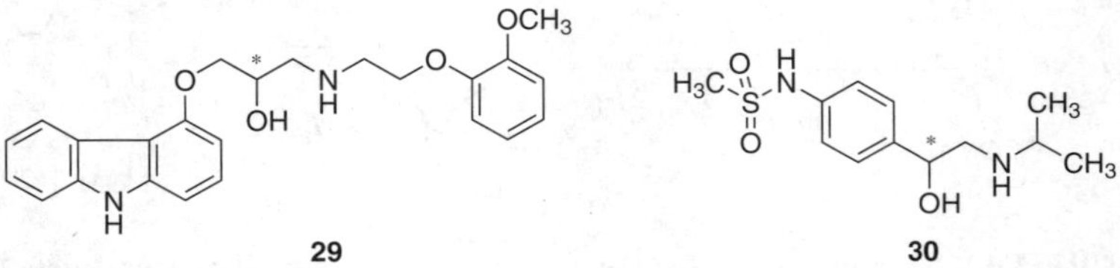

29   **30**

FIGURE 1.14 Structures of carvedilol (**29**) and sotalol (**30**).

2. *Both enantiomers act on the same biological target, but exert opposed pharmacological activities:* For example, (−)-dobutamine **31** demonstrated an agonistic activities against α-adrenoceptors, whereas its antipode (+)-dobutamine is an antagonist against the same receptors. The latter also acts as an β1-adrenoceptor agonist with a tenfold higher potency than the (−) isomer and is used to treat cardiogenic shock. The individual enantiomers of the 1,4-dihydropyridine analog Bayk8644 (**32**) have opposing effects on L-type calcium channels, with the (*S*)-enantiomer being an activator and the (*R*)-enantiomer an antagonist [34] (Fig. 1.15).

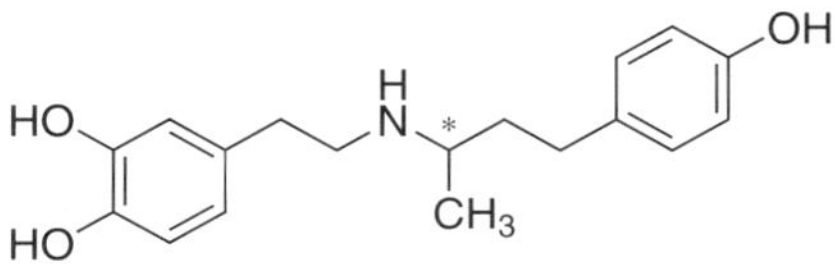

31 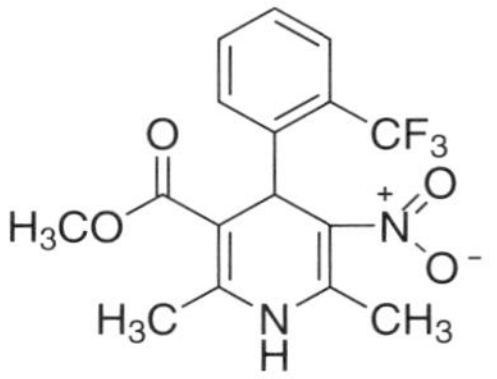   **32**

FIGURE 1.15   Structures of (−)-dobutamine (**31**) and Bayk8644 (**32**).

3. *Both enantiomers may act similarly, but they do not have a synergistic effect:* Two enantiomers of Δ-3-tetrahydrocannabinol (*S*)-**34** or (*R*)-**34** were assayed in humans for psychoactivity. The 1*S* enantiomer **34** had definite psychic actions, qualitatively similar to those of Δ-1-tetrahydrocannabinol, but quantitatively less potent (1:3 to 1:6). Adding two enantiomers together did not increase the effect, confirming that activity was solely in one enantiomer and that there was no synergistic effect between the two isomers [35] (Fig. 1.16).

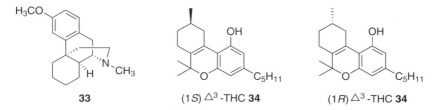

33   (1*S*) Δ³-THC **34**   (1*R*) Δ³-THC **34**

FIGURE 1.16   Structures of dextromethorphan and Δ-3-tetrahydrocannabinol (*S*)-**34** or (*R*)-**34**.

4. *Both enantiomers have independent therapeutic effects through action on different targets:* The classical example of this behavior is quinine **35** and quinidine **36** (Fig. 1.17). Quinine, which was originally obtained from the bark of cinchona trees, has been used for the treatment of malaria for

centuries. Quinidine, on the other hand, is used as a class 1A antiarrhythmic agent and acts by increasing action potential duration [36].

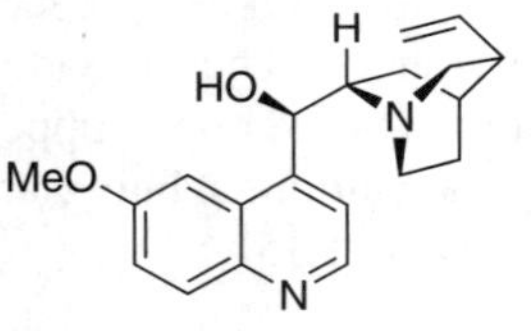

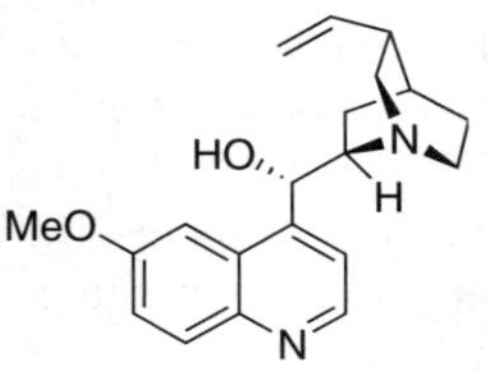

FIGURE 1.17   Structures of quinine (**35**) and quinidine (**36**).

5. *One or both enantiomers have the desired effect; at the same time, only one enantiomer can cause unwanted side effects:* Racemic dropropizine (**37**) has long been used in human therapy as an antitussive agent. Recent studies have revealed that (*S*)-dropropizine possesses the same antitussive activity as the racemic mixture, but has much lower selective activity on the CNS [37]. Therefore, particular clinical significance is attached to drugs of which one enantiomer may contribute side or toxic effects (Fig. 1.18).

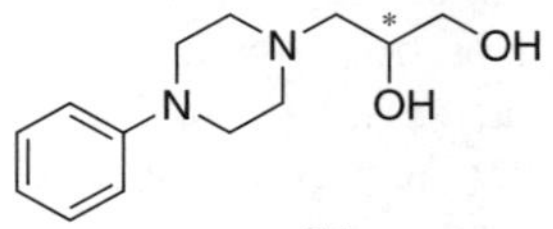

FIGURE 1.18   Structures of dropropizine (**37**).

6. *The inactive enantiomer might antagonize the side effects of the active antipode:* In such cases, taking into account both efficacy and safety aspects, the racemate seems to be superior to either enantiomer alone. For example, the opioid analgesic tramadol (**38**) is a used as a racemate and is not associated with the classical side effects of opiate drugs, such as respiratory depression, constipation, or sedation [38]. The (+)-enantiomer is a selective agonist for μ receptors with preferential inhibition of serotonin reuptake and enhances serotonin efflux in the brain, whereas the (−)-enantiomer mainly inhibits noradrenaline reuptake. The incidence of side effects, particularly opioid-mediated effects, was higher with the (+)-enantiomer than with ±-tramadol or the (−)-enantiomer. Therefore, the racemate of tramadol is superior to the enantiomers for the treatment of severe postoperative pain [39]. Albuterol (**39**), an adrenoceptor agonist bronchodilator, is the racemic form of 4-[2-(*tert*-butylamino)-1-hydroxyethyl]-2-(hydroxymethyl) phenol and can increase bronchial

airway diameter without increasing heart rate. The bronchodilator activity resides in (R)-albuterol. (S)-albuterol, however, is not inert, as it indirectly antagonizes the benefits of (R)-albuterol and may have proinflammatory effects [40] (Fig. 1.19).

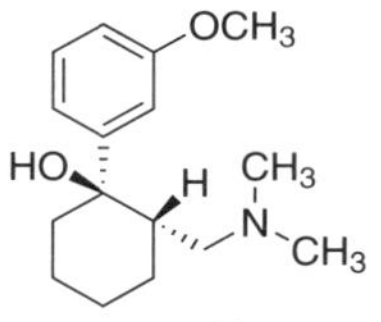

FIGURE 1.19 Structures of tramadol (**38**) and albuterol (**39**).

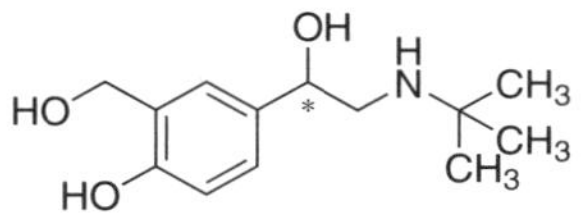

It is a difficult task to rationally predict the biological/pharmacological activity difference for two enantiomers. Fokkens and Klebe developed a simple protocol using isothermal titration calorimetry in an attempt to semiquantitatively determine the difference in binding affinity of two enantiomers to a protein without requiring prior resolution of the racemates [41]. In some cases, the affinity difference could be explained in terms of differences in the structural fit of the enantiomers into the binding pocket of the protein. [42].

Many attempts were made to develop a quantitative structure-activity relationship between the two enantiomers and a specific target or target families. The ratio of potency or affinity of two enantiomers is defined as the eudismic ratio (ER). The more potent enantiomer is generally called the eutomer, and the less potent enantiometer is the distomer. The logarithm of the eudimic ratio is regarded as the eudismic index (EI). Pfeiffer made an initial observation that the logarithm of the ratio of the activities of the optical isomers was proportional to the logarithm of the human dose. The generalization that the lower the effective dose of a drug, the greater the difference in pharmacological effect between the optical isomers is referred as Pfeiffer's rule [43]. Indeed, a linear correlation between the logarithm of the EI of 14 randomly chosen enantiomeric pairs and the logarithm of the average human dose was observed.

Eudismic analysis was made for a series of five cholinesterase inhibitors, derivatives of S-alkyl p-nitrophenyl methylphosphonothiolates (R: methyl to pentyl), a series of four derivatives of 1,3-dioxolane (R: H, Me, Et, i-propyl) active at the muscarinic receptor, and Pfeiffer's original set of 14 nonhomologous enantiomeric pairs with a computer-aided drug design method [44]. It was concluded that eudismic ratios of potent drugs belonging to homologous sets can be correlated with their chirality coefficients, which was defined as the quantitative index of the dissimilarity between the enantiomers and was calculated from a combination of data from the superimposition of computer-optimized conformations and electrostatic potential (ESP) calculations. Linear correlations were observed between the calculated chirality coefficients and experimentally determined eudismic ratios for both sets of homologous derivatives. With Pfeiffer's set (members include atropine, norepinephrine, epinephrine, and methadone) correlation was observed for the first (most potent) eight members of the series. The lack of correlation for the less potent compounds in Pfeiffer's set was explained as a function of kinetic differences becoming more influential than drug-receptor interactions [44].

On the qualitative side, 3D binding molecule modeling studies can point out some interesting binding differences for two enantiomers. Two enantiomers of citalopram were demonstrated to bind to human serotonin transporter in reversed orientation [45].

1.4.2 Pharmacokinetics and Drug Disposition

In addition to the differences in biological activities, stereoisomers may differ in their pharmacokinetic properties such as absorption, distribution, metabolism, and excretion (ADME) as a result of chiral discrimination during the pharmacokinetic processes [46]. The difference in bioavailability, rate of metabolism, metabolite formation, excretion rate, and toxicity may be further influenced by other factors such as the route of administration, the age and sex of the subjects, disease states, and genetic polymorphism in cytochrome P450 (CYP) isoenzymes involved in drug metabolism [47].

Active transport processes may discriminate between the enantiomers, with implications for bioavailability. For example, a longer plasma half-life in the rabbit and greater accumulation of propranolol in the heart and brain of the rat were found for the active (S)-$(-)$-enantiomer (**40**) as compared to the corresponding racemate.

Plasma binding capacity for two enantiomers may also be significantly different, thus influencing drug efficacy. Methadone (**41**), introduced to treat opioid dependence in 1965, has therapeutic benefits that reside in the (R)-enantiomer. Compared to the (S)-enantiomer of methadone, methadone's (R)-enantiomer shows 10-fold higher affinity for μ and κ opioid receptors and up to 50 times the antinociceptive activity in animal model and clinical studies. Methadone's enantiomers show markedly different pharmacokinetics. The (R)-enantiomer shows a significantly greater unbound fraction and total renal clearance than the (S)-enantiomer. This reflects higher plasma protein binding of the (S)-enantiomer [48] (Fig. 1.20).

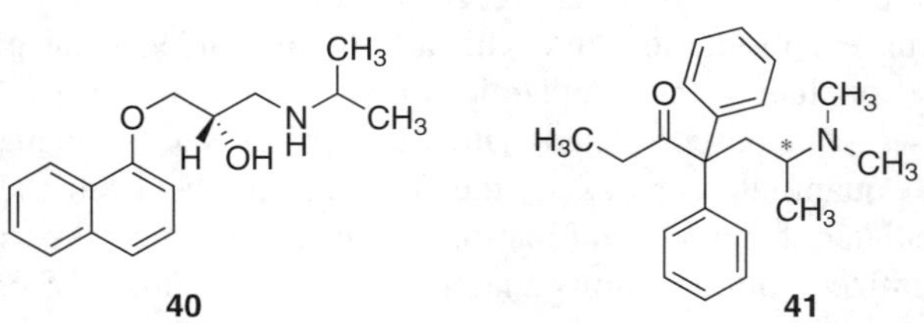

FIGURE 1.20 Structures of (S)-$(-)$-propranolol (**40**) and methadone (**41**).

Similarly, enzymes that metabolize drug molecules may also discriminate enantiomers differently. For example, esomeprazole **42** (S-isomer of omeprazole), an optical isomer proton pump inhibitor, generally provides better acid control than the current racemic proton pump inhibitors and has a favorable pharmacokinetic profile relative to omeprazole. However, the metabolic profiles of the two drugs are different, leading to different systemic exposures and thus different pharmacodynamic effects. Metabolism of the (R)-enantiomer is more dependent on CYP2C19, whereas the (S)-enantiomer can be metabolized by alternative pathways like CYP3A4 and sulfotransferases (Fig. 1.21). This results in the less active (R)-enantiomer achieving higher concentrations in poor metabolizers, which may in the long term cause adverse effects like gastric carcinoids and hyperplasia [29,49].

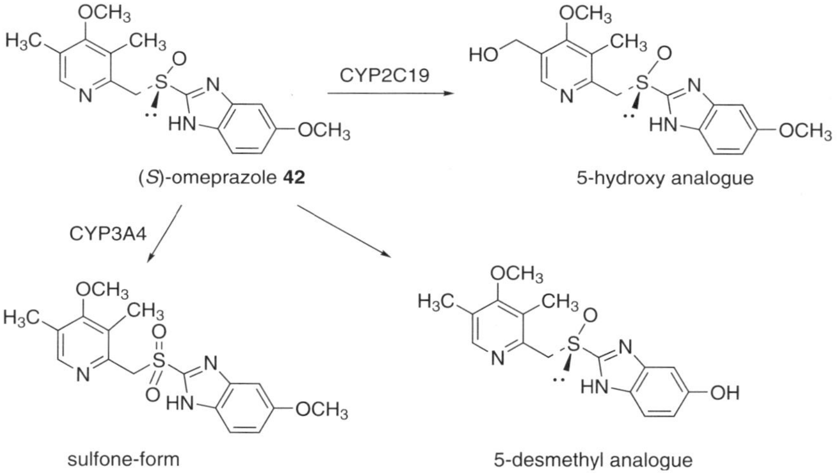

(S)-omeprazole **42**

5-hydroxy analogue

sulfone-form

5-desmethyl analogue

FIGURE 1.21 Main metabolites of (S)-omeprazole (**42**).

The more advantageous pharmacokinetics for both enantiomers is found in the metabolic distribution, clearance, and so on. Cetirizine (Zyrtec), the potent histamine H1 receptor antagonist, is a racemic mixture of (R)- and (S)-dextrocetirizine **43** (Fig. 1.22). In binding assays, levocetirizine has demonstrated a twofold higher affinity for the human H1 receptor compared to cetirizine, and an approximately 30-fold higher affinity than dextrocetirizin [50]. However, levocetirizine is rapidly and extensively absorbed and poorly metabolized and exhibits comparable pharmacokinetic profiles with the racemate. Its apparent volume of distribution is smaller than that of dextrocetirizine (0.41 l/kg vs. 0.60 l/kg). Moreover, the nonrenal (mostly hepatic) clearance of levocetirizine is also significantly lower than that of dextrocetirizine (11.8 ml/min vs. 29.2 ml/min). All evidence available indicates that levocetirizine is intrinsically more active and more efficacious than dextrocetirizine, and for a longer duration [51].

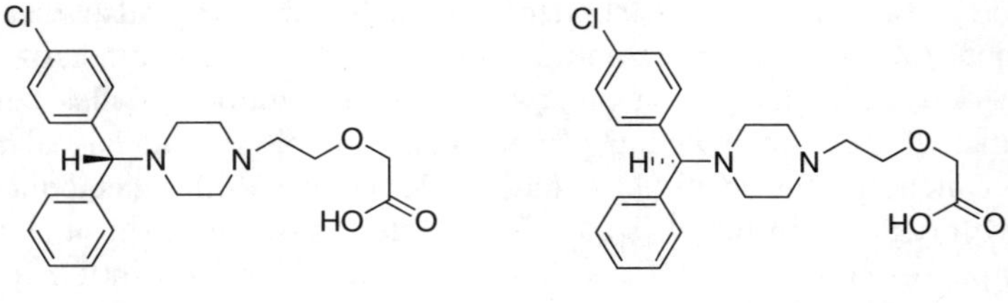

(R)-**43**   (S)-**43**

FIGURE 1.22   Structures of (R)-levocetirizine and (S)-dextrocetirizine.

Although it is well known that cytochrome P450 enzyme can accommodate a wide range of substrates, it is still possible for two isomers of a drug to induce or inhibit these enzymes differently, either with different modes of action or with different enzyme subtypes. Trimipramine (**44**) is a tricyclic antidepressant with sedative and anxiolytic properties. Desmethyl-trimipramine (**45**), 2-hydroxy-trimipramine (**46**) and 2-hydroxy-desmethyl-trimipramine (**47**) are the main metabolites of trimipramine (Fig. 1.23). However, trimipramine appears to show stereoselective metabolism with preferential N-demethylation of D-trimipramine and preferential hydroxylation of L-trimipramine, mediated by CYP2D6. CYP2C19 appears to be involved in demethylation and favors the D-enantiomer, while CYP3A4 and CYP3A5 seem to metabolize L-trimipramine to a currently undetermined metabolite [52].

The advances in technologies in chiral synthesis and stereoselective bioanalysis in the 1990s made it possible to investigate the relationship between stereochemistry and pharmacokinetics and pharmacodynamics. When drug is administered into a system, it will encounter numerous biological barriers before it reaches the target to exert pharmacological effects. Administration of a chiral drug can have potential advantage as compared to its racemate counterpart in many aspects.

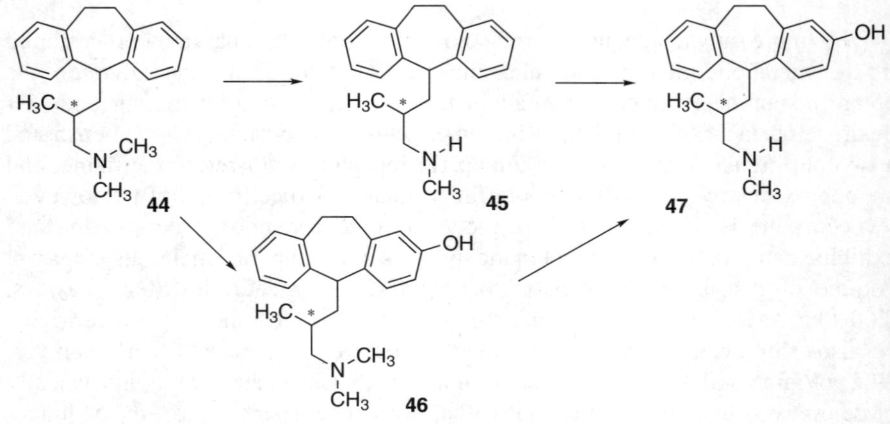

FIGURE 1.23   Metabolism of trimipramine (**44**).

The overall dose and thus the toxicity can be reduced and minimized because of the increase of potential potency. The development of a single enantiomer is particularly desirable when only one of the enantiomers has a toxic or undesirable pharmacological effect. Further investigation of the properties of the individual enantiomers and their active metabolites is warranted if unexpected toxicity or pharmacological effect occurs with clinical doses of the racemate.

1.4.3   Regulatory Aspects of Chiral Drugs

Regulatory authorities [53] issue general scientific guidance and regulatory principles for drug development. The presence of chiral center(s) adds a different challenge to these known principles, especially for drugs showing enantioselective pharmacokinetic and/or pharmacodynamic profiles. Authorities many countries began to issue regulatory guidelines on chiral drugs in the mid-1980s owing to the accessibility of enantiomerically pure drug candidates and the accumulation of knowledge on chiral drugs. Japanese regulatory authorities were among the first to respond to the emerging issue of chirality in drug development, although officially they never issued a document related specifically to chiral drug development [54]. The Ministry of Health and Welfare of Japan stated in 1986 that 1) when the drug is a racemate, it is recommended to investigate the ADME patterns of each isomer, including the possible occurrence of in vivo inversion, and 2) for a mixture of diastereoisomers, in particular, it is necessary to investigate how each isomer is metabolized and disposed, and how each isomer contributes to efficacy. The US Food and Drug Administration (FDA) officially announced a policy statement in 1992 on the development of steroisomeric drugs [55]. Guidelines on investigations of chiral active substances were issued by a commission of the European countries [56] in 1994. The Canadian government announced a Therapeutic Product Programme to address stereochemical issues in chiral drug development in 2000 [57]. All regulatory guidance emphasizes the importance of chirality of active ingredient in the testing of the bulk drug, the manufacturing of the finished product, the design of stability testing protocols, and the labeling of the drug. It strongly urges companies to evaluate racemates and enantiomers for new drugs, but not to draw definitive guidelines forbidding racemic drugs. The importance of evaluating the behavior of stereoisomers was further highlighted in an FDA regulatory document in 2005. Applicants must recognize the occurrence of chirality in new drugs, attempt to separate the stereoisomers, assess the contribution of the various stereoisomers to the activity of interest, and make a rational selection of the stereoisomeric form that is proposed for marketing [58].

1.4.4   Trends in the Development of Chiral Drugs

With the tightening of regulations, demand for chiral raw materials, intermediates, and active ingredients has grown dramatically since 1990 [58,59]. Pharmaceutical companies responded to this rising demand with honing, acquiring, development, and expanding chiral technologies [60]. Current technologies of asymmetric synthesis and chiral separation make it possible for pharmaceutical companies to develop single-enantiomer drugs. The single-enantiomer drug segment has become an important part of the overall pharmaceutical market. The unique aspects of drug discovery and development made it possible to develop a new chiral drug through a "chiral switch" from the existing racemic drugs [61].

The chiral industry can be divided into two main categories: the manufacture of chiral compounds and the analysis of chiral compounds. Chiral manufacturing will continue to dominate the market: It rose from revenues of over $1 billion in 2003 to over $1.6 billion in 2008. This category can be further subdivided into chiral synthesis and chiral separation. Compared with chiral separation, the market size of chiral synthesis is much bigger. The synthesis market was worth $986 million in 2003 and was expected to reach $1.5 billion by 2008, reflecting an average annual growth rate (AAGR) of 9.2%. Just under 98% of this market is accounted for by chiral intermediates, with the remainder accounted for by chiral chemical catalysts and other materials used in chiral synthesis, biocatalysts, and chiral auxiliaries.

Single-enantiomer therapeutics had sales of $225 billion in 2005, representing 37% of the total final formulation pharmaceutical market of $602 billion (Table 1.1) based on estimates from Technology Catalysts International and IMS Health. The compound annual growth rate for single-enantiomer products over the past 5 years is 11%, which is on par with the pharmaceutical market as a whole [62].

Chiral drugs comprise more than half the drugs approved worldwide, including many of the top-selling drugs in the world. For example, among the top 10

TABLE 1.1   Worldwide Sales of Single-Enantiomer Pharmaceutical Products Final Formulation

Therapeutic Category	2000 Sales (in $ billions)	2004 Sales (in $ billions)	2005 Sales (in $ billions)	CAGR (%)* 2000–2005
Cardiovascular	27.650	34.033	36.196	6
Antibiotics and antifungals	25.942	32.305	34.298	6
Cancer therapies	12.201	21.358	27.172	17
Hematology	11.989	20.119	22.172	13
Hormone and endocrinology	15.228	20.608	22.355	8
Central nervous system	9.322	17.106	18.551	15
Respiratory	6.506	12.827	14.708	18
Antiviral	5.890	11.654	14.683	20
Gastrointestinal	4.171	11.647	13.476	26
Ophthalmic	2.265	3.063	3.416	9
Dermatological	1.272	1.486	1.561	4
Vaccines	1.427	2.450	3.100	17
Other	7.128	10.400	13.268	13
Total	130.991	199.056	225.223	11

*CAGR is compound annual growth rate (source: Technology Catalysis International).

best-selling US prescription small-molecule pharmaceuticals in 2009, six of these are single enantiomers, two are achiral, and only two racemates (Table 1.2) [63].

An analysis of the new molecular entities (NMEs) approved by the US FDA in 2008 gave the following approximate distribution: 63% single enantiomers, 32% achiral drugs, and only 5% racemates. Overall, there is clear trend that the racemate drugs are decreasing from 1992 (about 21%) to 2008 (5%), with no racemic drug introduction at all in 2001 and 2003 [64]. The chiral drugs increased

TABLE 1.2   Top 10 Best-Selling Small-Molecule Therapeutics in US in 2009

Rank*	Product	Active Ingredient	Form of Ingredient
1	Lipitor	Atorvastatin	Single enantiomer
2	Nexium	Esomeprazole	Single enantiomer
3	Plavix	Clopidogrel	Single enantiomer
4	Advair Diskus	Fluticasone salmeterol	Single enantiomer, racemate
5	Seroquel	Quetiapine	Achiral
6	Abilify	Aripiperazol	Achiral
7	Singular	Montelukast	Single enantiomer
8	OxyContin	Oxycodone	Single enantiomer
9	Actos	Pioglitazone	Racemate
10	Prevacid	Lansoprazole	Racemate

*http://www.drugs.com/top200.html

TABLE 1.3   Annual Distribution of FDA-Approved Drugs from 1992 to 2008

Year	Racemic (%)	Chiral (%)	Achiral (%)
1992	21	44	35
1993	16	45	39
1994	38	38	24
1995	21	46	33
1996	9	41	50
1997	24	30	46
1998	15	50	35
1999	19	50	31
2000	3	67	30
2001	0	72	28
2002	6	58	36
2003	0	76	24
2004	6	76	18
2005	5	63	32
2006	10	55	35
2007	5	68	27
2008	5	63	32

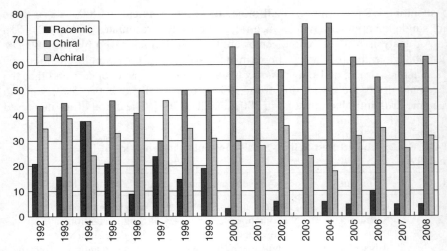

FIGURE 1.24   Annual distribution of FDA-approved drugs

from 30–40% in the 1990s to around 60% or above since 2000. A significant amount of drugs are still achiral (Table 1.3 and Fig. 1.24).

REFERENCES

1. Piutti, A. (1886). Cristallographic Chimique-Sur une nouvelle espke d'asparagine. *Compt. Rend*. **103**, 134–138.

2. Anonymous. (1962). Medicine: the thalidomide disaster. *Time*, August 10.

3. Blaschke, G., Kraft, H. P., Fickentscher, K., Kohler, F. (1979). Chromatographic separation of racemic thalidomide and teratogenic activity of its enantiomers. *Arzneimittelforschung* **29**, 1640–1642.

4. a) Fabro, S., Smith, R. L., Williams, R. T. (1967). Toxicity and teratogenicity of optical isomers of thalidomide. *Nature* **215**, 296–297;   b) Reist, M., Carrupt, P. A., Francotte, E., Testa, B. (1998). Chiral inversion and hydrolysis of thalidomide: mechanisms and catalysis by bases and serum albumin, and chiral stability of teratogenic metabolites. *Chem. Res. Toxicol*. **11**, 1521–1528;   c) Eriksson, T., Bjorkman, S., Roth, B., Fyge, A., Hoglund, P. (1995). Stereospecific determination, chiral inversion in vitro and pharmacokinetics in humans of the enantiomers of thalidomide. *Chirality* **7**, 44–52.

5. Ito, T., Ando, H., Suzuki, T., Ogura, T., Hotta, K., Imamura, T., Yamaguchi, Y., Handa, H. (2010). Identification of a primary target of thalidomide teratogenicity. *Science* **327**, 1345–1350.

6. Lin, G. Q., Li, Y. M., Chen, A. S. C. (2001). *Principles and Application of Asymmetric Synthesis*. New York: Wiley-Interscience, p. 10, 17.

7. Rodger, A., Nordén, B. (1997). *Circular Dichroism and Linear Dichroism*. Oxford, UK: Oxford University Press, ISBN 019855897X.

8. Dale, J. A., Mosher, H. S. (1973). Nuclear magnetic resonance enantiomer regents. Configurational correlations via nuclear magnetic resonance chemical shifts of diastereomeric mandelate, O-methylmandelate, and alpha-methoxy-alpha-trifluoromethylphenylacetate (MTPA) esters. *J. Am. Chem. Soc.* **95**, 512–519.

9. Ohtani, I., Kusumi, T., Kashman, Y, Kakisawa, H. (1991) High-field FT NMR application of Mosher's method. The absolute configurations of marine terpenoids. *J. Am. Chem. Soc.* **113**, 4092–4096.

10. (a) Seco, J. M., Quinoa, E., Riguera, R. (2004) The assignment of absolute configuration by NMR. *Chem. Rev.* **104**, 17–117;   (b) Hoye, T. R., Jeffrey, C. S., Shao, F. (2007) Mosher ester analysis for the determination of absolute configuration of stereogenic (chiral) carbinol carbons. *Nat. Protoc.* **2**, 2451–2458.

11. Rothchild, R. (1989). Chiral lanthanide NMR shift reagents and equilibria with substrate enantiomers: rationale for the observed signals. *J. Chem. Educ.* **66**, 814.

12. Knowles, W. S. (1983). Asymmetric hydrogenation. *Acc. Chem. Res.* **16**, 106–112.

13. Miyashita, A., Yasuda, A., Takaya, H., Toriumi, K., Ito, T., Souchi T., Noyori, R. (1980). Synthesis of 2,2′-bis(diphenylphosphino)-1,1′-binaphthyl (BINAP), an atropisomeric chiral bis(triaryl)phosphine, and its use in the rhodium (I)-catalyzed asymmetric hydrogenation of α-(acylamino)acrylic acids. *J. Am. Chem. Soc.* **102**, 7932–7934.

14. Hansen, K.B., Balsells, J., Dreher, S., Hsiao, Y., Kubryk, M., Palucki, M., Rivera, N., Steinhuebel, D., Armstrong, J. D. III, Askin, D., Grabowski, E. J. J. (2005). First generation process for the preparation of the DPP-IV inhibitor sitagliptin. *Org. Process Res. Dev.* **9**, 634–639.

15. Katsuki, T., Sharpless, K.B. (1980). The 1st practical method for asymmetric epoxidation. *J. Am. Chem. Soc.* **102**, 5974–5976.

16. Crabtree, R. H. (2009). *Handbook of Green Chemistry, vol. 3: Biocatalysis*. Weinheim: Wiley-VCH, p. 173.

17. Broussy, S. B., Cheloha, R. W., Berkowitz, D. B. (2009). Enantioselective, keto-reductase-based entry into pharmaceutical building blocks: ethanol as tunable nicotinamide reductant. *Org. Lett.* **11**, 305–308.

18. Patel, R., Hanson, R., Goswami, A., Nanduri, V., Banerjee, A., Donovan, M. J., Goldberg, S., Johnston, R., Brzozowski, D., Tully, T., Howell, J., Cazzulino, D., Ko, R. (2003). Enzymatic synthesis of chiral intermediates for pharmaceuticals. *J. Ind. Microbiol. Biotechnol.* **30**, 252–259.

19. Drauz, K., Waldmann, H. (2004). *Enzyme Catalysis in Organic Synthesis*. Weinheim: Wiley-VCH, p. 151, 164.

20. Paquette, L. A. (2003) *Handbook of Reagents for Organic Synthesis, Chiral Reagents for Asymmetric Synthesis*. New York: Wiley, ISBN: 978-0-470-85625-3.

21. Ohkuma, T., Ooka, H., Hashiguchi, S., Ikariya, T., Noyori, R. (1995). Practical enantioselective hydrogenation of aromatic ketones. *J. Am. Chem. Soc.* **117**, 2675–2676.

22. Thayer. A. M. (2007). Centering on chirality. *Chem. Eng. News*, **85**, 11–19.

23. a) List, B. (2007). Introduction: organocatalysis. *Chem. Rev.* **107**, 5413–5415;   b) MacMillan, D. W. C. (2008). The advent and development of organocatalysis. *Nature* **455**, 304–308;   c) Akiyama, T. (2007). Stronger Brønsted acids. *Chem. Rev.* **107**, 5744–5758;   d) Doyle, A. G., Jacobsen, E. N. (2007). Small-molecule H-bond donors in asymmetric catalysis. *Chem. Rev.* **107**, 5713–5743.

24. a) Hajos, Z. G., Parrish, D. R. (1971). DE 2102623; b) Hajos, Z. G., Parrish, D.R. (1974). Asymmetric synthesis of bicyclic intermediates of natural product chemistry. *J. Org. Chem.* **39**, 1615–1621;   c) Deer, U., Sauer, G., Wiechert, R. (1971). DE 2014757;   b) Eder, U., Sauer, G., Wiechert, R. (1971). New type of asymmetric cyclization to optically active steroid CD partial structures. *Angew. Chem. Int. Ed. Engl.* **10**, 496–497.

25. MacMillan, D. W. C. (2008). The advent and development of organocatalysis. *Nature* **455**, 304–308.

26. a) Houk, K. N., List, B. (2004). Asymmetric organocatalysis. *Acc. Chem. Res.* **37**, 487–621;   b) Kocovsky, P., Malkov, A. V. (2006). Organocatalysis in organic synthesis. *Tetrahedron* **62**, 255;   c) List, B. (2007). Introduction: organocatalysis. *Chem. Rev.* **107**, 5413–5883.

27. Berkessel, A., Gröger, H. G. (2005). *Asymmetric Organocatalysis: From Biomimetic Concepts to Application in Asymmetric Synthesis*. Weinheim: Wiley-VCH, p. 1, 2.

28. Seki, M. (2008). A practical synthesis of a key chiral drug intermediate via asymmetric organocatalysis. *Synlett* **2**, 164–176.

29. Patil, P. A., Kothekar, M. A. (2006). Development of safer molecules through chirality. *Indian. J. Med. Sci* **60**, 427–437.

30. McConathy J., Owens, M. J. (2003). Stereochemistry in drug action. *Primary Care Companion J Clin. Psychiatry* **5**, 70–73.

31. Tenero, D., Boike, S., Boyle, D., Ilson, B., Fesniak, H. F., Brozena, S., Jorkasky, D. (2000). Steady-state pharmacokinetics of carvedilol and its enantiomers in patients with congestive heart failure. *J. Clin. Pharmacol.* **40**, 844–853.

32. Bartsch, W., Sponer, G., Strein, K., Müller-Beckmann, B., Kling, L., Böhm, E., Martin, U., Borbe, H. O. (1990). Pharmacological characteristics of the stereoisomers of carvedilol. *Eur. J. Clin. Pharmacol.* **38**, S104–S107.

33. Mehvar, R., Brocks, D. R. (2001). Stereospecific pharmacokinetics and pharmacodynamics of beta-adrenergic blockers in humans. *J. Pharm. Pharm. Sci.* **4**, 185–200.

34. Krogsgaard-Larsen, P., Liljefors, T., Madsen, U. (2002). *Textbook of Drug Design and Discovery*. New York: Taylor and Francis, p. 56.

35. Hollister, L. E., Gillespie, H. K., Mechoulam, R., Srebnik. M. (1987). Human pharmacology of 1S and 1R enantiomers of delta-3-tetrahydrocannabinol. *Psychopharmacology* **92**, 505–507.

36. Leffingwell, J. C. (2003). Chirality & bioactivity I: pharmacology. *Leffingwell Rep.* **3**, 1–27.

37. Salunkhe, M. M., Nair, R. V. (2001). Novel route for the resolution of both enantiomers of dropropizine by using oxime esters and supported lipases of *Pseudomonas capacia*. *Enzyme. Microb. Technol.* **28**, 333–338.

38. Sacerdote, P., Bianchi, M., Gaspani, L., Panerai, A. E. (1999). Effects of tramadol and its enantiomers on Concanavalin-A induced-proliferation and NK activity of mouse splenocytes: involvement of serotonin. *Int. J. Immunopharmacol.* **21**, 727–734.

39. Rojas-Corrales, M. O., Giber-t-Rahola, J., Mica, J. A. (1998). Tramadol induces antidepressant-type effects in mice. *Life Sci.* **63**, 175–180.

40. Leffingwell, J. C. (2003). Chirality & bioactivity I: pharmacology. *Leffingwell Rep.* **3**, 1–27.

41. Fokkens, J., Klebe, G. (2006). A simple protocol to estimate differences in protein binding affinity for enantiomers without prior resolution of racemates. *Angew. Chem. Int. Ed*. **45**, 985–989.

42. Thurkauf, A., Hillery, P., Mattson, M. V., Jacobson, A. E., Rice, K. C. (1988). Synthesis, pharmacological action, and receptor binding affinity of the enantiomeric 1-(1-phenyl-3-methylcyclohexyl) piperidines. *J. Med. Chem*. **31**, 1625–1628.

43. Pfeiffer, C. C. (1956). Optical isomerism and pharmacological action, a generalization. *Science* **124**, 29–31.

44. Seri Levy, A., Richards, W. G. (1993). Chiral drug potency: Pfeiffer's rule and computed chirality coefficients. *Tetrahedron Asym*. **4**, 1917–1923.

45. Koldsø, H., Severinsen, K., Tran, T. T., Celik, L., Jensen, H. H., Wiborg, O., Schiøtt, B., Sinning, S. (2010). The two enantiomers of citalopram bind to the human serotonin transporter in reversed orientations. *J. Am. Chem. Soc*. **132**, 1311–1322.

46. Brocks, D. R. (2006). Drug disposition in three dimensions: an update on stereoselectivity in pharmacokinetics. *Biopharm. Drug Dispos*. **27**, 387–406.

47. Lane, R. M., Baker, G. B. (1999). Chirality and drugs used in psychiatry: nice to know or need to know? *Cell Mol. Neurobiol*. **19**, 355–372.

48. a) Foster, D. J. R., Somogyi, A. A., Dyer, K. R., White, J. M., Bochner. F. Br. (2000). Steady-state pharmacokinetics of (R)- and (S)-methadone in methadone maintenance patients. *J. Clin. Pharmacol*. **50**, 427–440; b) Baumann, P., Zullino, D. F., Eap, C. B. (2002). Enantiomers' potential in psychopharmacology-a critical analysis with special emphasis on the antidepressant escitalopram. *Eur. Neuropsychopharmacol*. **12**, 433–444.

49. Andersson, T., Weidolf, L. (2008). Stereoselective disposition of proton pump inhibitors. *Clin. Drug. Invest*. **28**, 263–279.

50. Gillard, M., Van Der Perren, C., Moguilevsky, N., Massingham, R., Chatelain, P. (2002). Binding characteristics of cetirizine and levocetirizine to human H1 histamine receptors: contribution of Lys191 and Thr194. *Mol. Pharmacol*. **61**, 391–399.

51. Tillementa, J. P., Testab, B., Brée. F. (2003). Compared pharmacological characteristics in humans of racemic cetirizine and levocetirizine, two histamine H1-receptor antagonists. *Biochem. Pharmacol*. **66**, 1123–1126.

52. Eap, C. B., Bender, S., Gastpar, M., Fischer, W., Haarmann, C., Powell, K., Jonzier-Perey, M., Cochard, N., Baumann, P. (2000). Steady state plasma levels of the enantiomers of trimipramine and of its metabolites in CYP2D6-, CYP2C19- and CYP3A4/5-phenotyped patients. *Ther. Drug. Monit*. **22**, 209–214.

53. Web site addresses for regulatory agencies: http://www.hc-sc.gc.ca/hpb/Canada; http://www.ifpma.org/ich1.html ICH; http://www.mhw.go.jp/english/index.html Japan; http://www.health.gov.au/tga/Australia; http://eudraportal.eudra.org/EMEA.

54. a) Shindo, H., Caldwell, J. (1995). Development of chiral drugs in Japan: an update on regulatory and industrial opinion. *Chirality* **7**, 349–352; b) Shimazawa, R., Nagai, N., Toyoshima, S., Okuda, H. (2008). Present status of new chiral drug development and review in Japan. *J. Health Sci*. **54**, 23–29.

55. FDA's policy statement for the development of new stereoisomeric drugs. (1992). *Chirality*, **4**, 338–340.

56. a) Branch, S. (2001). International regulation of chiral drugs. In: *Chiral Separation Techniques. A. Practical Approach*. 2nd Edition, Subramanian, G., editor. Weinheim: Wiley-VCH, p. 319–342; (b) Huang, Z. W. (2007). *Drug Discovery Research: New Frontiers in the Post-Genomic Era*. Hoboken, NJ: John Wiley & Sons, Inc., p. 243.

57. http://www.hc-sc.gc.ca/dhp-mps/prodpharma/applic-demande/guide-ld/chem/stereo-eng.php.

58. Caner, H., Groner, E., Levy, L., Agranat I. (2004). Trends in the development of chiral drugs. *Drug Discov. Today* **9**, 105–110.

59. Agranat, I., Caner, H., Caldwell, J. (2002). Putting chirality to work: the strategy of chiral switches. *Nat. Rev. Drug. Discov*. **1**, 753–768.

60. Maier, N. M., Franco, P., Lindner, W. (2001). Separation of enantiomers: needs, challenges, perspectives. *J. Chromatograph. A.* **906**, 3–33.

61. a) Agranat, I., Caner, H., Caldwell, J. (2002). Putting Chirality to work: the strategy of chiral switches. *Nat. Rev. Drug Discov*. **1**, 753–768;   b) Tucker, G. (2000). Chiral switches. *Lancet* **355**, 1085–1087.

62. Erb, S. (2006). Single-enantiomer drugs poised for further market growth. *Pharmaceut. Technol*. **30**, ps14–s18.

63. Van Arnum, P. (2006). Single-enantiomer drugs drive advances in asymmetric synthesis (cover story). *Pharmaceut. Technol*. **30**, 58–66.

64. Anonymous. (2009). *Annual Report in Medicinal Chemistry, 1992–2009*, Vol. 27–44.

CHAPTER 2

CHIRAL DRUGS THROUGH ASYMMETRIC SYNTHESIS

GUO-QIANG LIN

Key Laboratory of Synthetic Chemistry of Natural Substances, Shanghai Institute of Organic Chemistry, Chinese Academy of Sciences, Shanghai, China

XING-WEN SUN

Department of Chemistry, Fudan University, Shanghai, China

Chiral Drugs: Chemistry and Biological Action, First Edition. Edited by Guo-Qiang Lin, Qi-Dong You and Jie-Fei Cheng.
© 2011 John Wiley & Sons, Inc. Published 2011 by John Wiley & Sons, Inc.

2.1 CATALYTIC ASYMMETRIC SYNTHESIS AND ITS APPLICATION

Regardless of the fact that majority of therapeutics and natural products are in single enantiomeric form, for example, amino acids and carbohydrates, there are still great demands for chiral artificial products in high enantiopurities and for use in various ways, especially in the pharmaceutical industry. Asymmetric synthesis refers to the conversion of an achiral starting material to a chiral product in a chiral environment. It is presently the most powerful and most commonly used method for chiral molecule preparation. The resolution of racemates has been an important technique for obtaining enantiomerically pure compounds. Other methods involve the conversion or derivatization of readily available natural chiral compounds (chiral pools) such as amino acids, tartaric and lactic acids, terpenes, carbohydrates, and alkaloids. Biological transformations using enzymes, cell cultures, or whole microorganisms are also practical and powerful means of access to enantiomerically pure compounds from prochiral precursors, even though the scope of such reactions is limited because of the highly specific action of enzymes. Organic synthesis is characterized by generality and flexibility. During the last three decades, chemists have made tremendous progress in discovering a variety of versatile stereoselective reactions that complement biological processes. In an asymmetric reaction, substrate and reagent combine to form diastereomeric transition states. One of the two reactants must have a chiral element to induce asymmetry at the reaction site. Thus far, most of the best asymmetric syntheses have been catalyzed by enzymes, and the challenge before us today is to develop chemical systems as efficient as the enzymatic systems.

To meet the categorical requirement, the best synthetic catalysts should demonstrate a useful level of enantioselectivity for a wide range of substrates. Such generality of substrate scope is usually not observed in enzyme catalysts. So far, through great effort, certain classes of man-made ligand catalysts, the so-called "privileged structures," have been successfully applied to a wide range of asymmetric reactions. Examples of such structures are shown in Figure 2.1 [1].

The major concerns of chemical synthesis are the cost of the production and environmental issues. Therefore, a desirable asymmetric catalyst must be readily available and highly efficient as compared to the traditional and conventional technologies for producing a chiral target. Ligands are the key issue to chirality. In this chapter, the main points that will be useful for both research and development are introduced.

Asymmetric reactions have been executed in several industrial processes, such as the asymmetric synthesis of L-DOPA (**11**) (Scheme 2.1), a drug for the treatment of Parkinson disease, via a Rh(DIPAMP)-catalyzed hydrogenation of the enamide [2]. The high enantioselectivity is thought to arise from the

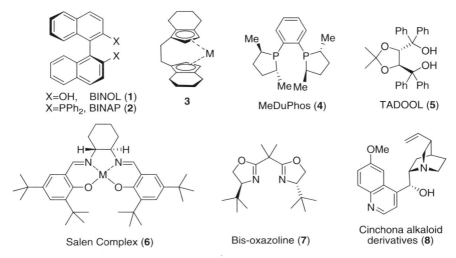

X=OH, BINOL (1)
X=PPh2, BINAP (2)
3
MeDuPhos (4)
TADOOL (5)
Salen Complex (6)
Bis-oxazoline (7)
Cinchona alkaloid derivatives (8)

FIGURE 2.1 Examples of privileged chiral ligands and catalysts.

H2, Ru/DIPAMP
H2O, IPA
9
10
H3O+
11
(R,R)-DIPAMP (12)

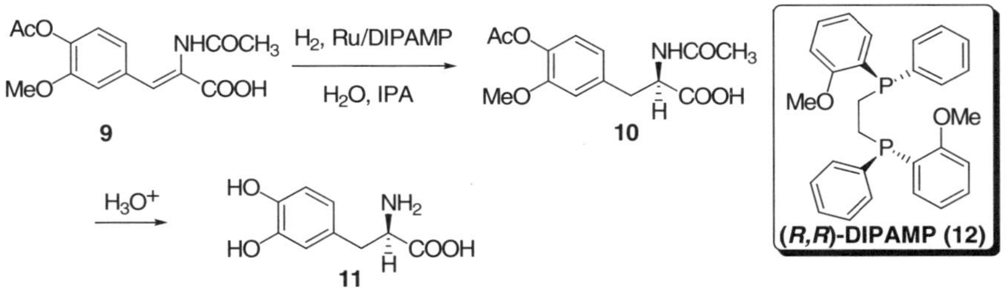

SCHEME 2.1 Preparation of L-DOPA (**11**) by Rh (DIPAMP)-catalyzed hydrogenation.

chirality residing on the P atoms which are close to the metal center during the coordination.

The industrial synthesis of (−)-menthol (**17**) and (+)-citronellal (**16**) through asymmetric isomerization of an allylic amine [3] are shown in Scheme 2.2, where (*S*)-BINAP (**2**) is used as a chiral ligand.

Currently, the requisite side chain (**22**) for the semisynthesis of Taxol® and docetaxel [4] can also be efficiently synthesized via a three-step sequence featuring a highly enantioselective epoxidation reaction catalyzed by the readily available (salen)Mn (III) complex (**19**) (Scheme 2.3).

From the beginning of this century, enantioselective synthesis with small-molecule catalysts through hydrogen-bonding donors or iminium ionic intermediates to construct C—C and C-hetero atom bonds has emerged as a frontier of organic synthesis. This topic is discussed in Section 2.5.

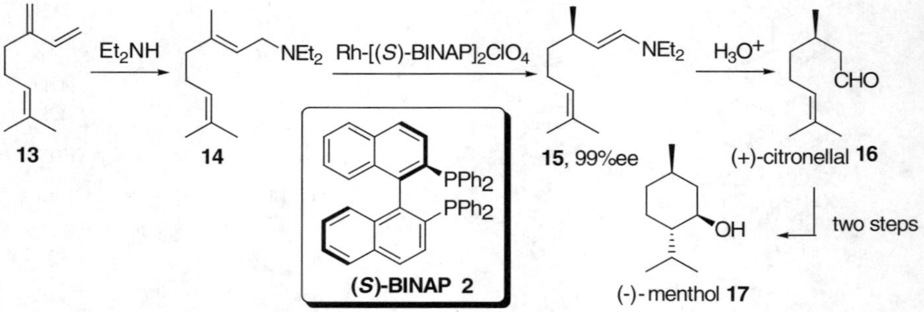

SCHEME 2.2   Synthesis of (−)-menthol (**17**) and (+)-citronellal (**16**).

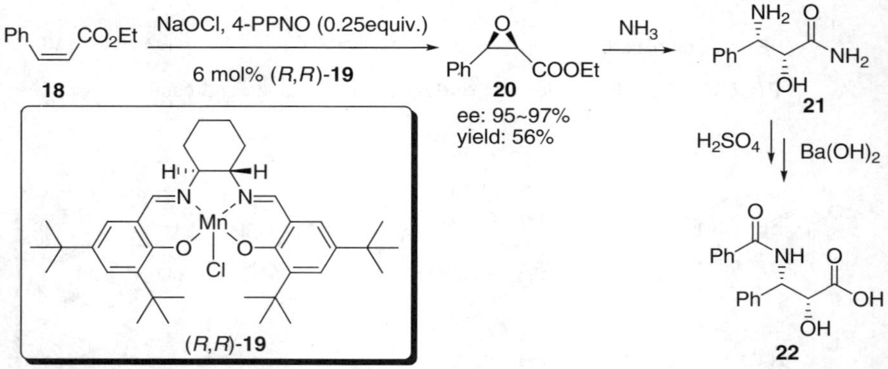

SCHEME 2.3   Synthesis of side chain (**22**) of Taxol® by asymmetric epoxidation.

2.2   ASYMMETRIC HYDROGENATION REACTION AND REDUCTION REACTION

Catalytic asymmetric hydrogenation has become one of the most powerful and reliable methods for the preparation of optically active compounds. High enantioselectivity, low catalyst loadings, essentially quantitative yields, perfect atomic economy with a broad scope of substrates, and mild conditions are the attractive features of this transformation. The success of asymmetric hydrogenation of C≡C or C≡X (X≡N, O) bond relies mainly on the proper selection of metals and ligands. Accordingly, exploration of new effective chiral ligands has played a fundamental role in asymmetric hydrogenation. As shown in Scheme 2.2, the astropisomeric biaryl ligand BINAP associated with Rh has been used on a wide range of substrates, from olefins to ketones. Moreover, Duphos and BPE designed by Burk have been found to be versatile ligands for Rh-catalyzed hydrogenation of functional olefins. The family of ferrocene-based Josiphos ligands with C_1-symmetric planar chirality provides the opportunity to adjust two chelating donors easily to fit steric and electronic requirements. The *P,N*-ligand PHOX shows the preponderance versus other ligands in the Ir-catalyzed asymmetric hydrogenation of imines. A recent conceptual breakthrough is the introduction of monodentrate ligands for hydrogenation reaction [5]. Among the ligands shown in Figure 2.1, the Josiphos ligands complexed with Rh, Ir, and Ru selectively catalyze the enantioselective hydrogenation of enamides, itaconic acid derivatives, acetoacetates, acrylic acids, and *N*-aryl imines as well as other transformations. Josiphos ligands have been utilized for the preparation of some important pharmaceutical intermediates or products as listed in Figure 2.2 [6].

2.2.1 Asymmetric Hydrogenation of C=C Double Bond

About 70% of the new small-molecule drugs approved in 2007 by FDA (US) contained at least one chiral center. Tremendous efforts have been undertaken to construct chiral scaffolds. The desire to generate enantiomers in high yield and optical purity continues to yield new catalysts and even new reactions. Hydrogen is the simplest molecule and is available in abundance at very low cost;

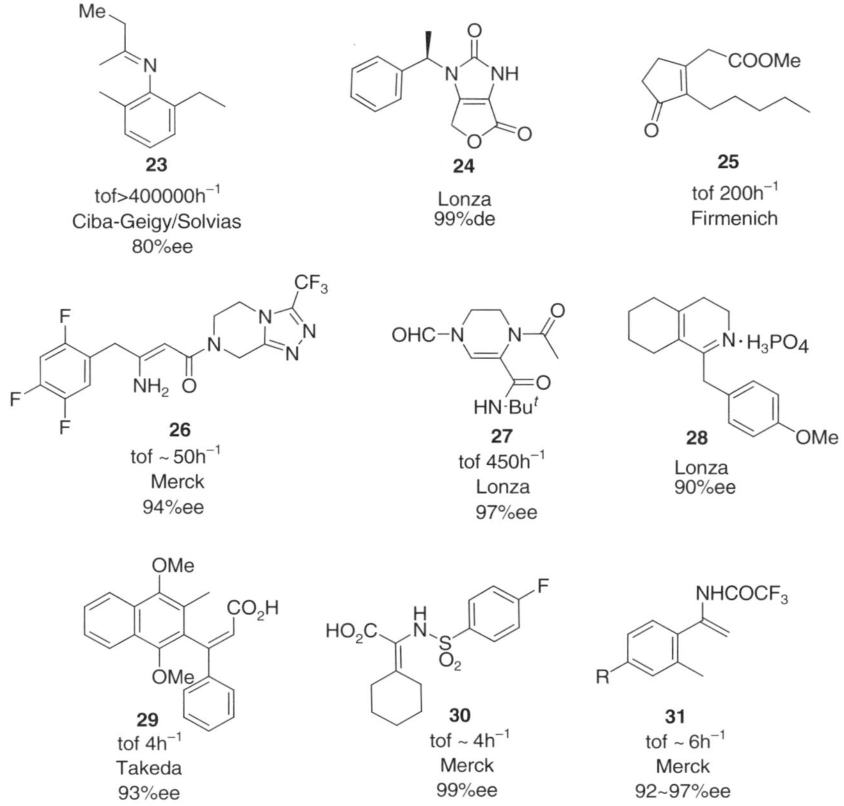

FIGURE 2.2 Josiphos ligands have been utilized for the hydrogenation of some pharmaceutically important intermediates.

therefore, asymmetric hydrogenation has become one of the core technologies in academia and industry. A number of recently launched drugs such as tipranavir (**32**), sitagliptin (**26**), aliskiren (**33**), and rozerem (**34**) [7] are all reported to employ asymmetric hydrogenation in their preparation (Fig. 2.3) [8].

Since Knowles [9] and Horner [10] reported the first chiral phosphine-based Wilkinson's catalyst [Rh(PPh$_3$)$_3$]Cl for homogeneous hydrogenation, several historically important ligands (Fig. 2.4) have been developed over the past decades, providing extremely valuable insights into ligand design.

A breakthrough in this area came when Kagan reported DIOP (**35**), a C_2-symmetric chiral diphosphine readily available from tartaric acid. DIOP-Rh(I) complex catalyzed the enantioselective hydrogenation of α-(acylamino) acrylic acids or esters to produce the corresponding amino acid derivatives with up to 80% *ee* [11]. This achievement led to the discovery of a variety of bidentate chiral diphosphines that have emerged as the most useful and versatile ligands in catalytic asymmetric hydrogenation. It has been found that many ligands with C_2 symmetry are effective in enantioselective hydrogenation reactions (Fig. 2.4). DIOP (**35**) was implemented with two sp^3 asymmetric carbons and a C_2-symmetric axis, while DIPAMP (**12**) (2.1) possesses two asymmetric phosphorus atoms. The C_2 axial chirality in BINAP (**2**) (Fig. 2.1) and its analogs has shown great industrial potential. Since Kang's discovery of a practical synthesis of an air-stable ferrocenyl bis(phosphine) and its application in the rhodium(I)-catalyzed enantioselective hydrogenation of dehydroamino acid derivatives, several ferrocenyl phosphines have found applications in the synthesis of chiral pharmaceuticals and agrochemicals. The chirality of these catalysts was derived from the dissymmetric nature of the cylopentadienyl rings attached to the metal. A process has been developed by Lonza Fine Chemicals in partnership with Ciba-Geigy for the production of (+)-biotin (**53**) by the Rh-ligand (**51**) complex-catalyzed enantioselective hydrogenation of the tetra-substituted C=C double bond [12] (Scheme 2.4).

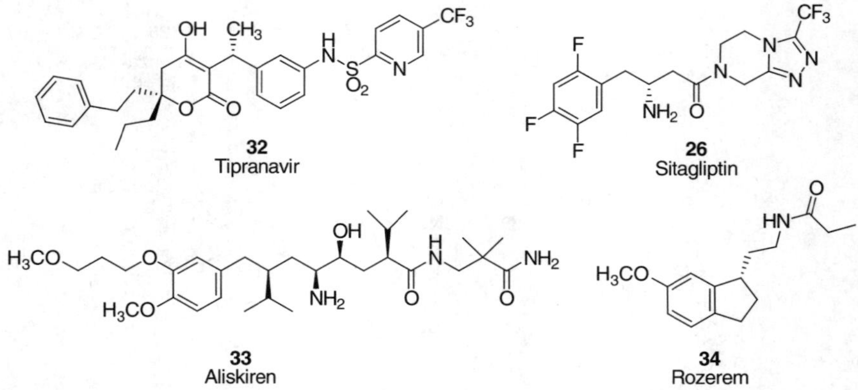

32
Tipranavir

26
Sitagliptin

33
Aliskiren

34
Rozerem

FIGURE 2.3 Structures of tipranavir (**32**), sitagliptin (**26**), aliskiren (**33**), and rozerem (**34**).

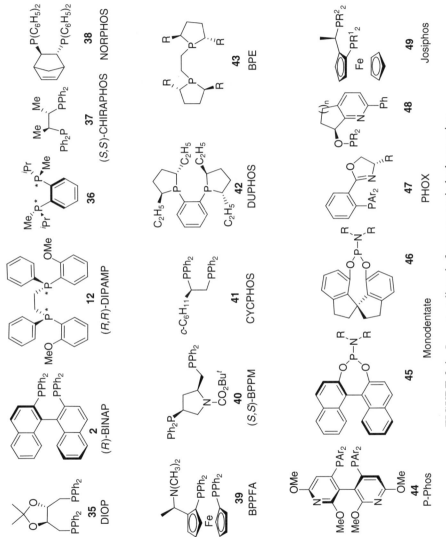

FIGURE 2.4 Important ligands for asymmetric hydrogenation.

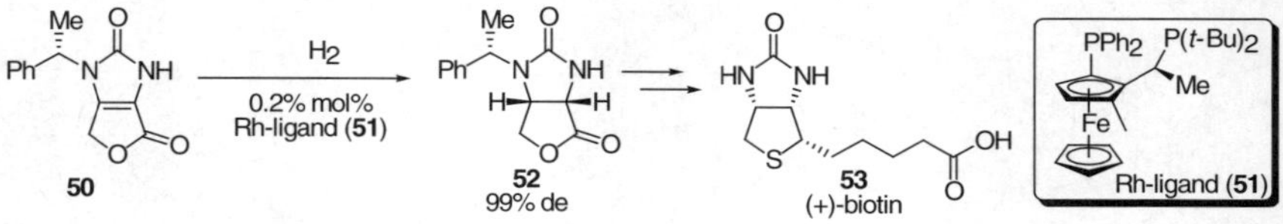

SCHEME 2.4 Asymmetric synthesis of (+)-biotin (**53**).

In practice, asymmetric hydrogenation offers an economical method for the large-scale production of chiral compounds, such as the famous L-DOPA (**11**) synthesis (Scheme 2.1), carbapenem synthesis [13] and (*S*)-metolachlor synthesis [14]. Thus far, many pharmaceutically important chiral building blocks, including amino acids, amines, alcohols, esters, acids, and amino alcohols, have been synthesized by asymmetric hydrogenation in a highly efficient manner. Recently, iridium catalysts with chiral *P,N*-ligands have emerged as a new class of efficient catalysts. A variety of substrates, for which it was difficult to find suitable catalysts, can now be hydrogenated with excellent optical purity. Iridium complexes with five- and six-member rings (**54a** and **54b**) proved to be efficient catalysts [15]. These catalysts are also efficient for enantioselective hydrogenation of alkyl-substituted olefins (Scheme 2.5).

Zhang and co-workers described an effective Ru-catalyzed asymmetric synthesis for atorvastatin calcium (Lipitor®, **58**) one of the best-selling drugs (>10 billion dollar sales per year) for the treatment of hypercholesterolemia. For Lipitor [(*S*)-ethyl 4-chloro-3-hydroxyl- butanoate (**57**)] synthesis, a key chiral intermediate was prepared *via* Ru-catalyzed asymmetric hydrogenation of ethyl 4-chloroacetoacetate (**55**). Excellent enantioselectivity (>98% *ee*) has been achieved at 45,000 turnovers with complete conversion [16] (Scheme 2.6).

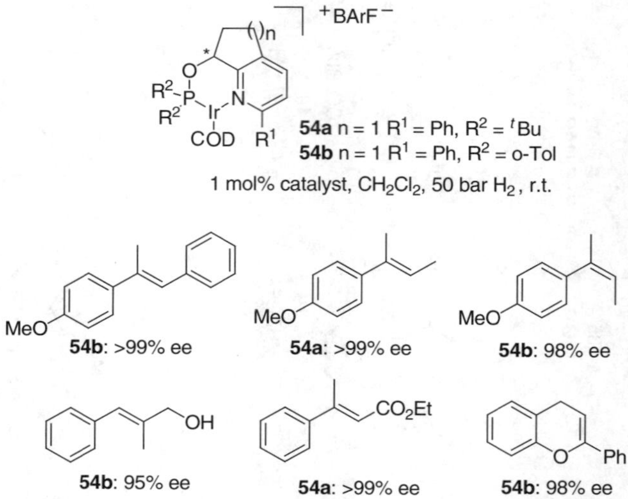

SCHEME 2.5 Hydrogenation of various substrates catalyzed by Ir complex **54a** or **54b**.

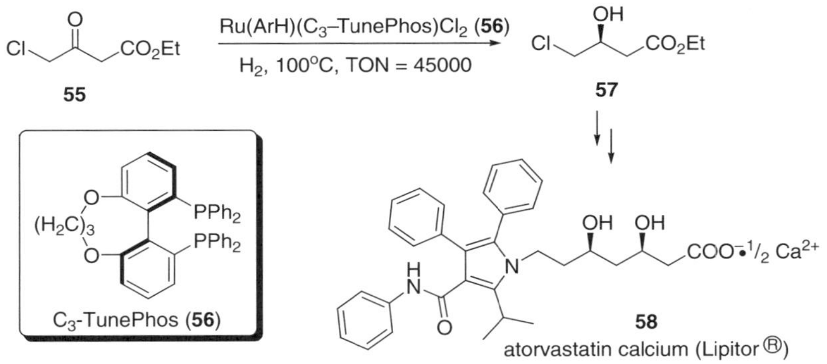

SCHEME 2.6 Asymmetric synthesis of atorvastatin calcium (Lipitor®).

2.2.2 Asymmetric Hydrogenation of C=O Bond

Asymmetric hydrogenation of ketones is one of the most efficient methods for making chiral alcohols. Ru-BINAP catalysts are highly effective in the asymmetric hydrogenation of functionalized ketones. This method has been used in the industrial production of synthetic intermediates for some important antibiotics. A particularly great successful system is the combination of Ru-complex, 1,2-diamine, and an inorganic base (Fig. 2.5). With the use of **59**, simple α,β-unsaturated ketones can be catalytically hydrogenated to give allylic alcohols in high *ee*, leaving the double bonds unchanged. The aromatic ketones are also good substrates under the same conditions [17].

Synthesis of carbapenem (**64**) and sanfetrinem (**67**) illustrates the power of asymmetric hydrogenation. By combining this asymmetric reaction with dynamic kinetic resolution (Chapter 4), an important chiral intermediate, (2*S*, 3*R*)-**62**, for the synthesis of carbapenem antibiotics [13a] was synthesized, as shown in Scheme 2.7.

In the case of the preparation of sanfetrinem (**67**), 2-methoxycyclohexanone (**65**) was hydrogenated in the presence of Ru-(*S*)-3,5-xylyl-BINAP-(*S,S*)-DPEN-KOH catalyst (**59c**) to give rise to (1*R*,2*S*)-2-methoxycyclohexanol (**66**) as the nearly sole stereoisomer [18] (Scheme 2.8).

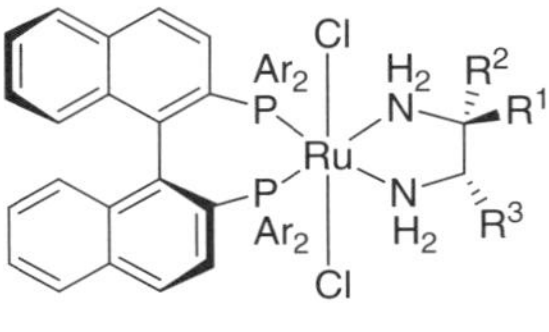

59a: Ar = 3,5-(CH$_3$)$_2$C$_6$H$_3$; R^1 = R^2 = 4-CH$_3$OC$_6$H$_4$; R^3 = CH(CH$_3$)$_2$
59b: Ar = 3,5-(CH$_3$)$_2$C$_6$H$_3$; R^1 = R^3 = (CH$_2$)$_4$; R$_2$ = H
59c: Ar = 3,5-(CH$_3$)$_2$C$_6$H$_3$; R^1 = R^3 = C$_6$H$_5$; R$_2$ = H

FIGURE 2.5 BINAP-Ru diamine complex.

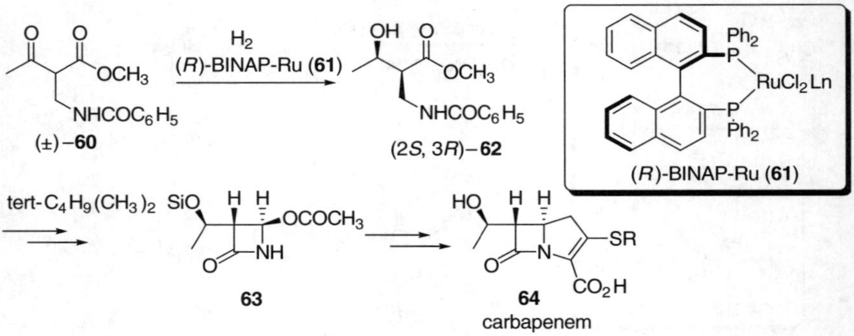

SCHEME 2.7 Synthesis of carbapenem (**64**).

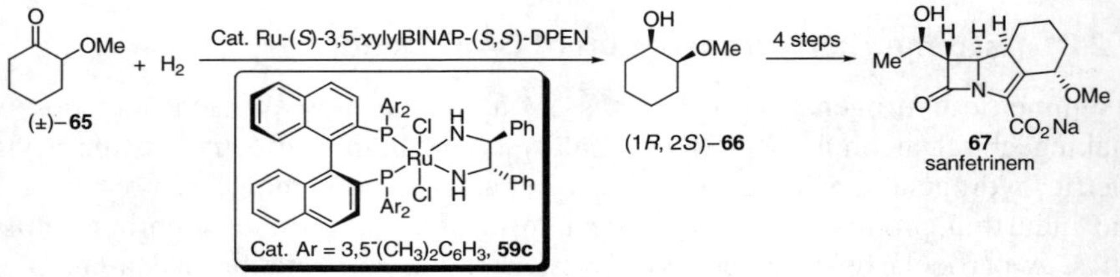

SCHEME 2.8 Synthesis of sanfetrinem (**67**).

Reduction of ketone with chiral boron complexes has recently been used for the synthesis of pharmaceuticals and other optically active products [19]. This chemistry involves the formation of a borane-tetrahydrofuran (BH_3/THF) complex in the presence of a catalytic amount of sterically hindered oxazaboraline (CBS, **71**). This method has been utilized for the preparation of a key intermediate in the synthesis of sertraline (**70**) [20], an antidepressant drug (Scheme 2.9).

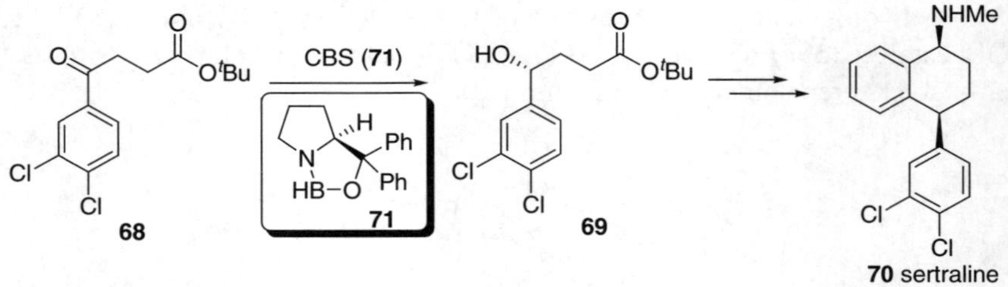

SCHEME 2.9 Synthesis of the antidepressant drug sertraline (**70**).

2.2.3 Asymmetric Hydrogenation of C=N Bond

Asymmetric hydrogenation of imines is equally important as that of ketones since corresponding chiral amines are universal synthetic intermediates in the preparation of many pharmacologically active compounds. The direct approach for the preparation of chiral amines is asymmetric hydrogenation of $C=N$-containing functional groups, mainly ketimines generated from ketone by condensation with amino-containing compounds.

Zhou and co-workers reported a convenient and economical method to synthesize optically active tetrahydroquinolines and chiral drugs via Ir [(R)-MeO-BiPhep]-catalyzed hydrogenation of the quinolines [21] (Scheme 2.10). N-methylation of hydrogenation products completed the concise syntheses of (−)-angustrureine (**75**), (S)-flumequine (**76**), (−)-galipinine (**77**), (−)-cuspareine (**78**), and (−)-galipeine (**79**) in high overall yields.

Lennon developed an asymmetric hydrogenation process for the synthesis of N-acetylcolchinol (**81**), a water-soluble phosphate pro-drug of ZD6126. Good enantiomeric excess (ee) was achieved with either (S,S)-^{i}Pr-FerroTANE Ru(methallyl)$_2$ or $[(R,R)$-^{t}Bu- FerroTANE Rh(COD)]BF$_4$ [22] (Scheme 2.11).

The antibacterial agent levofloxacin (**85**) can be prepared from intermediate (**84**), which in turn is obtained from cyclicimine (**82**) through asymmetric hydrogenation with the iridium (I) complex of $(2S, 4S)$-BPPM (**83**) in the presence of bismuth (III) iodide [23] (Scheme 2.12).

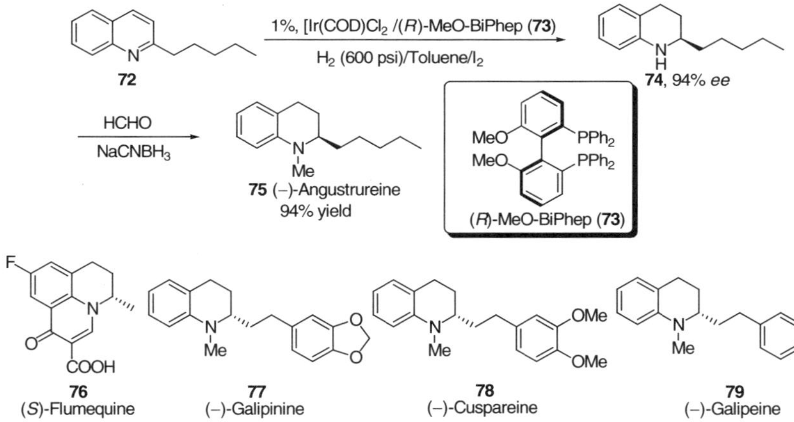

SCHEME 2.10 Iridium-catalyzed asymmetric hydrogenation of quinolines.

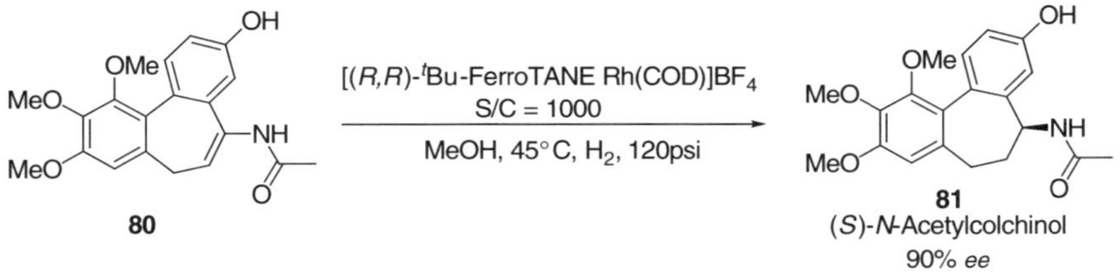

SCHEME 2.11 Asymmetric hydrogenation of a cyclic enamide.

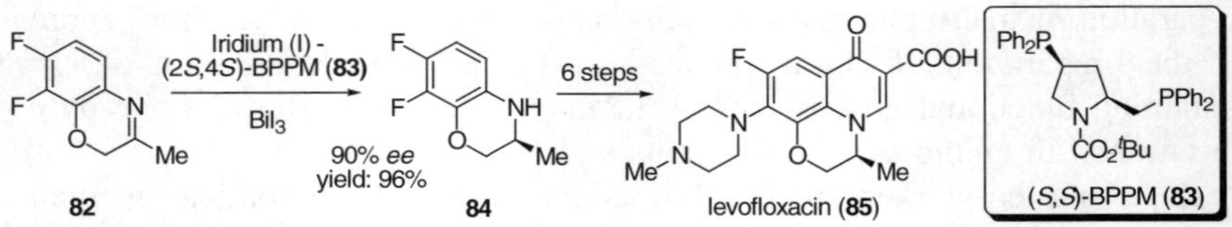

SCHEME 2.12 Synthesis of levofloxacin (**82**).

As mentioned above, metallocene-based chiral ligands have found increasing utility in asymmetric synthesis over the past decades. The synthesis of a herbicide, (S)-metolachlor (**89**), is shown in Scheme 2.13 through chloroacetylation of the chiral amine obtained through the Ir-catalyzed enantioselective hydrogenation of the requisite ketimine in the presence of ligand **87** [24].

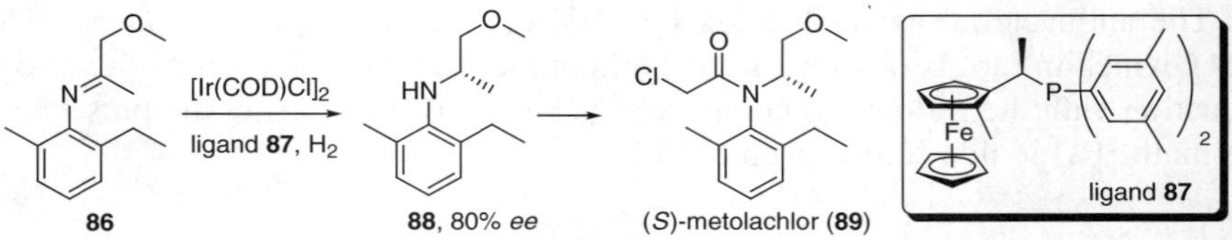

SCHEME 2.13 Synthesis of (S)-metolachlor (**89**).

Lennon reported the use of [diphosphine-RuCl$_2$-diamine] precatalysts for the asymmetric hydrogenation of imines [25]. It was applied to the asymmetric hydrogenation of an N-sulfonylimine substrate for the manufacture of S-18986 (**91**), an (R)-amino-3-hydroxy-5-methyl-4-isoxazolepropionic acid (AMPA) receptor modulator. For this substrate, S-18986 was produced in 87% ee but was readily upgraded to >99% ee by recrystallization from acetonitrile, which also reduced the content of the ruthenium metal to acceptable levels (Scheme 2.14).

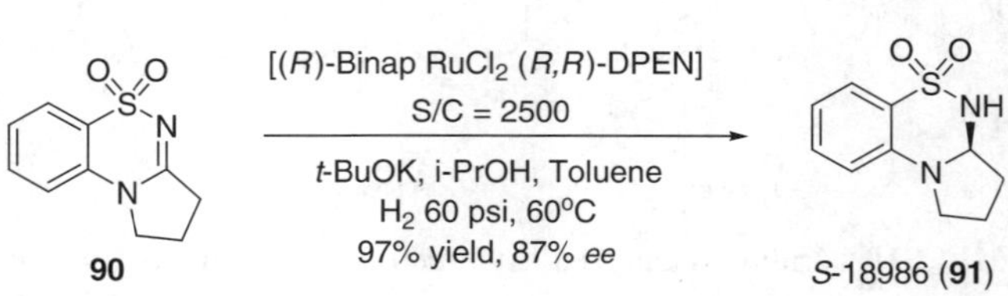

SCHEME 2.14 Synthesis of S-18986 (**91**).

2.3 ASYMMETRIC OXIDATION REACTION

The asymmetric oxidation reactions have been extensively investigated and have found widespread application for the preparation of many important intermediates for natural products and pharmaceutically significant compounds. This section describes asymmetric dihydroxylation, aminohydroxylation, and sulfoxidation reactions.

2.3.1   Asymmetric Dihydroxylation Reaction

Sharpless reported the first attempt to effect the asymmetric *cis*-dihydroxylation of olefins with osmium tetroxides in 1980 [26]. The rate of osmium ester (VI) formation can be accelerated by the presence of nucleophilic ligands. The naturally occurring cinchona alkaloids quinine and quinidine and their derivatives were found to be excellent chiral ligands (Scheme 2.15). Their derivatives have also been used successfully as ligands to afford the dihydroxylated products in 85–90% yields and excellent enantioselectivities. OsO_4 can be used catalytically in the presence of a secondary oxidant donor (e.g., H_2O_2, TBHP, N-methylolmorphaline-N-oxide, sodium periodate, O_2, sodium hypochlorite, potassium ferricyanide) [27]. Depending on the configuration of the ligands, both enantiomers can be synthesized conveniently in high yields and excellent *ee* values.

Asymmetric dihydroxylation (AD) of alkenes mediated by OsO_4-cinchona alkaloid complexes has been utilized to prepare chiral drugs or intermediates. For example, the nonsteroidal anti-inflammatory drug (NSAID) naproxen (**96**) has been synthesized in high enantiomeric excess by utilizing Sharpless asymmetric dihydroxylation of appropriate methyl styrene as a key step [28] (Scheme 2.16).

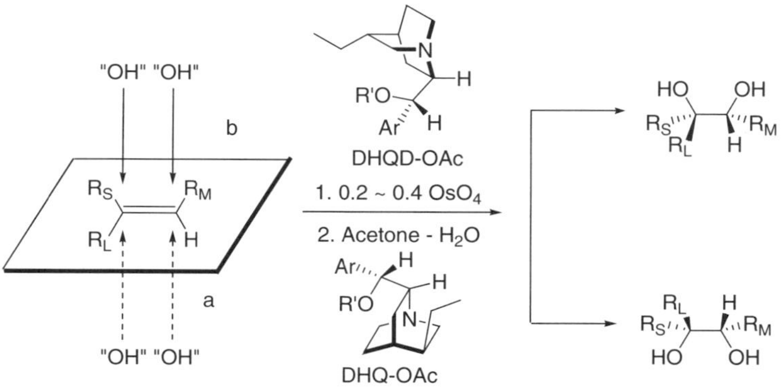

SCHEME 2.15   Sharpless dihydroxylation reaction.

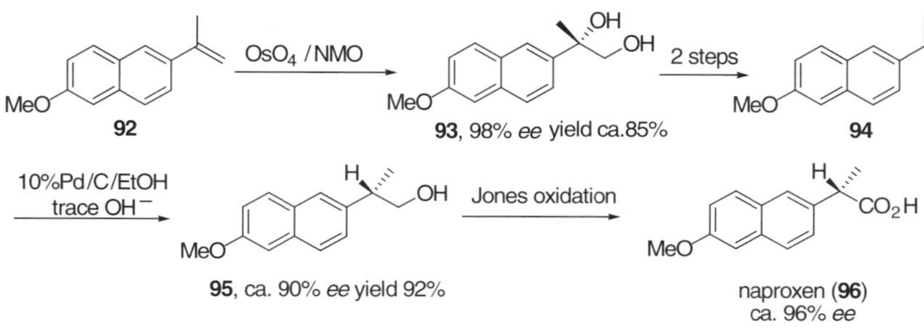

SCHEME 2.16   Synthesis of naproxen (**96**).

An AD-mix formulation was developed to simplify the handling of the asymmetric dihydroxylation on a millimole scale. In this formulation, only trace amounts of the ligand and the osmium salt are required. The components are blended into the bulk ingredients ferricyanide and carbonate (99.4% by wt), producing a convenient yellow powder. The mix is commercially available in two variations, "AD-mix-α" and "AD-mix-β," following ingredient lists published by Sharpless [29]. The mixture contains potassium osmate [$K_2OsO_2(OH)_4$] as the source of osmium tetroxide and potassium ferricyanide [$K_3Fe(CN)_6$], which is the reoxidant in the catalytic cycle, along with potassium carbonate and chiral ligands [in AD-mix α, (DHQ)$_2$PHAL is the phthalazine adduct with dihydroquinine; in AD-mix-β, (DHQD)$_2$PHAL is the phthalazine adduct with dihydroquinidine]. These AD-mixes are stable for months when protected from prolonged exposure to moisture.

Commercially available AD-mix-β was successful used in asymmetric dihydroxylation of aryl allyl ethers. The application of this sequence for synthesis of (S)-propranolol (**100**) is shown in Scheme 2.17 [30].

Asymmetric dihydroxylation can also be applied for cyclic olefins (Scheme 2.18) [31]. Fang and co-workers found that with the (DHQD)$_2$-PYR (pyrimidine) ligand compound **102** can be obtained in 94% ee from a cyclic vinyl ether starting material (**101**). Oxidation of the hemiacetal product afforded the corresponding lactone, which was then converted to camptothecin (**104**) in three steps. This intermediate has also used in the synthesis of camptothecin analog GI147211C (**105**) [32].

The Sharpless AD system has been applied to the preparation of 3,5-bis-trifluoro-methylstyrene oxide. AD-mix-α provided an 80% yield of the diol with 92% ee; recrystallization increased the optical purity to 97–99% ee. This diol was then dehydrated under Mitsunobu cyclodehydration to provide the epoxide **107**, a precursor to neurokinin-1 (NK-1) antagonist **108** (Scheme 2.19) [33].

Scientists at Merck utilized the Sharpless AD reaction with (DHQD)$_2$PHAL to prepare tertiary (R)-hydroxyketone **110** (Scheme 2.20), a precursor to COX-2 inhibitor L-784,512 (**112**). The produced diol was recrystallized to raise ee to >98% [34].

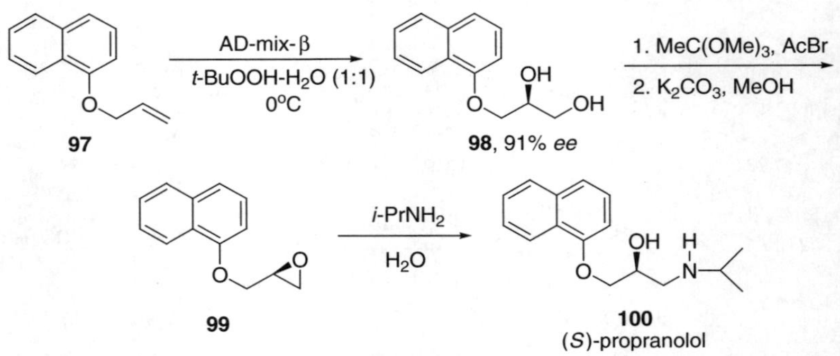

SCHEME 2.17   Synthesis of (S)-propranolol (**100**).

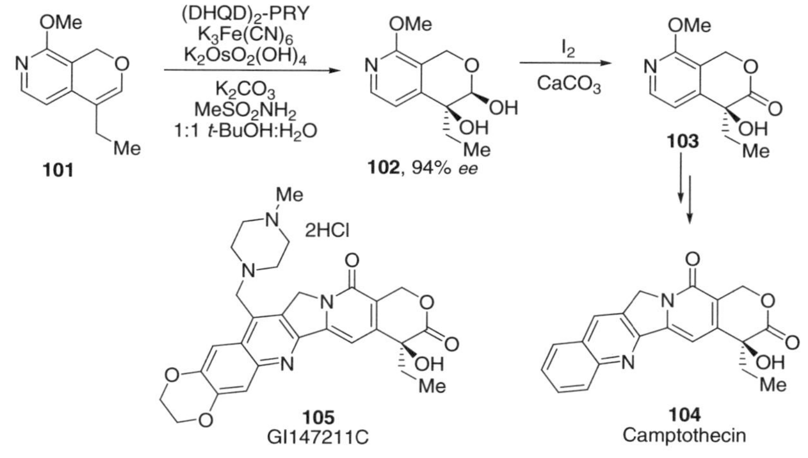

SCHEME 2.18 Dihydroxylation in the synthesis of camptothecin (**104**) and analog GI147211C (**105**).

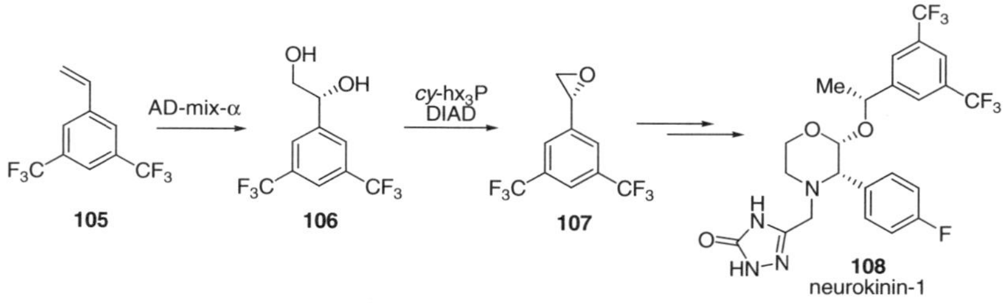

SCHEME 2.19 Preparation of NK-1 antagonist (**108**).

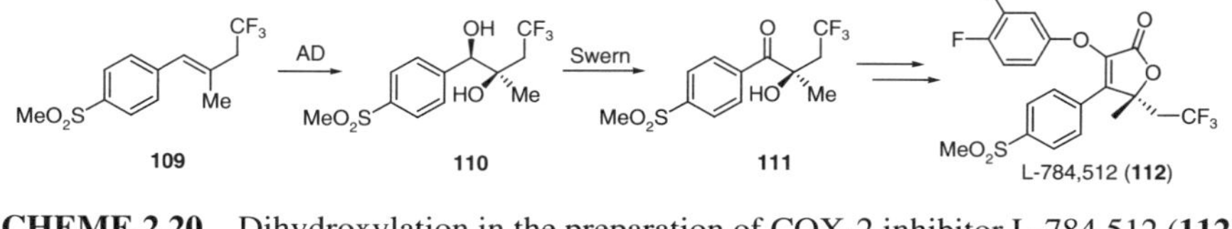

SCHEME 2.20 Dihydroxylation in the preparation of COX-2 inhibitor L-784,512 (**112**).

2.3.2 Asymmetric Aminohydroxylation Reaction

β-Amino alcohol is a key unit in many biologically important molecules. The most efficient way of creating this functionality is the direct addition of hydroxyl amino groups to an olefin. This functionality can be realized by the attack of TsNClNa (chlorament) as the nitrogen source/oxidant to olefins with excellent yields and enantioselectivities. This is also a process that greatly benefits from ligand-accelerated catalysts [35]. Through proper selection of the ligand (DHQ)$_2$PHAL (**114**) or (DHQD)$_2$PHAL (**116**), the enantio-enriched pair

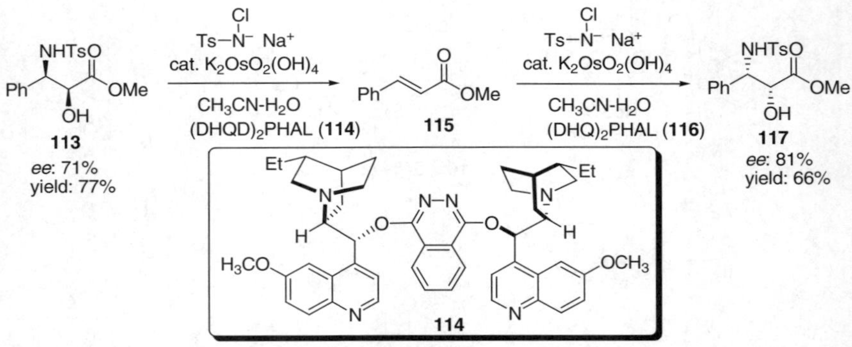

SCHEME 2.21 Sharpless aminohydroxylation.

of amino alcohols can be prepared (Scheme 2.21). Several commonly used ligands are shown in Figure 2.6.

Loracarbef (**122**) is a carbacephalosporin antibiotic with improved chemical and serum stability. It is particularly useful for the treatment of pediatric ear infections. A formal synthesis of loracarbef was accomplished with high stereoselectivity using aminohydroxylation as a key step. Stereoselective construction of

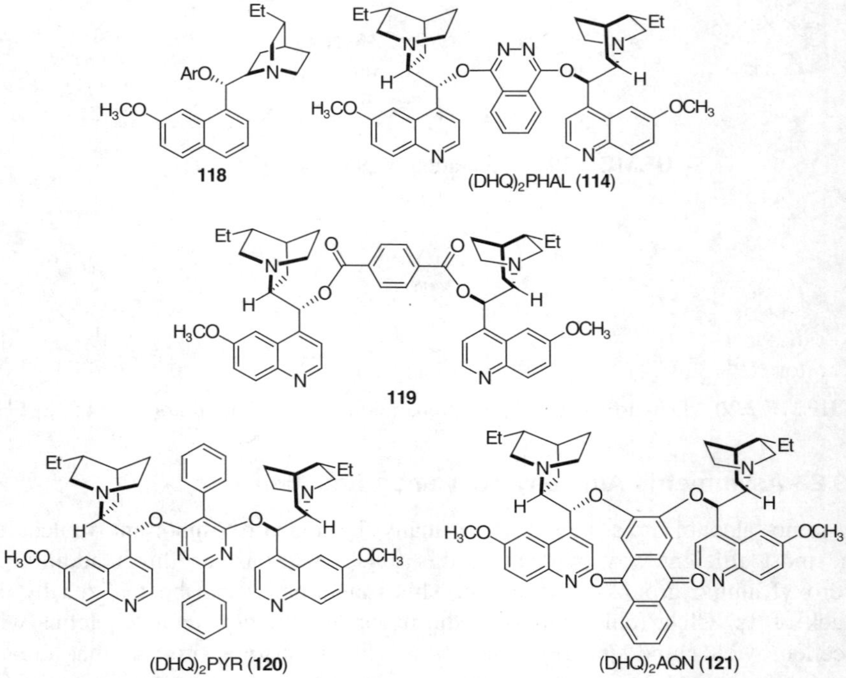

FIGURE 2.6 Ligands for Sharpless aminohydroxylation.

the key intermediate *cis*-3,4-disubstituted azetidinone (**123**) was accomplished by employing intramolecular cyclization of the *cis*-substituted azetidinone skeleton (**124**), which was obtained from Sharpless asymmetric aminohydroxylation of α, β-unsaturated ester (**125**) in good yield and with high regioselectivity (>13:1) and enantioselectivity (89% *ee*) [36] (Scheme 2.22).

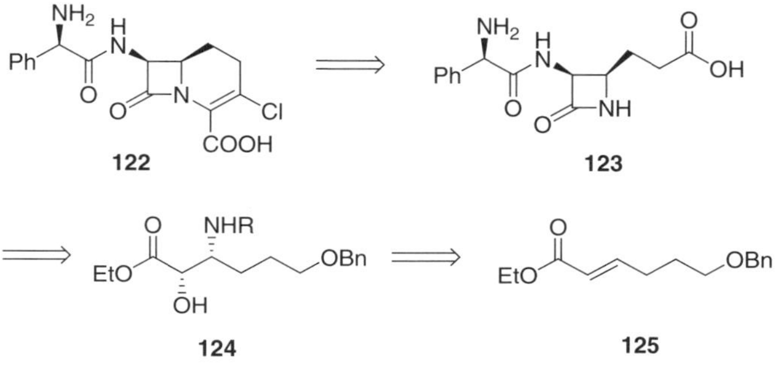

SCHEME 2.22   Retrosynthetic analysis of loracarbef (**122**).

2.3.3 Asymmetric Sulfoxidation Reaction

A number of biologically significant molecules have a stereogenic sulfinyl sulfur atom and therefore exist as a pair of enantiomers. The importance of chirality of sulfoxides in the drug molecule has not received much attention in all the many monographs dealing with asymmetric synthesis, despite the fact that the world's many billion-dollar drug molecules, such as omeprazole, contain a chiral sulfoxide moiety. Omeprazole (**127**), marketed under the names of Losec and Prilosec, is the leading gastric proton pump inhibitor (PPI) used as an antiulcer agent. Consequently, a large number of pharmaceutical companies seek to develop their own version of gastric acid secretion inhibitors based on the framework of omeprazole. In recent years great efforts have been made to synthesize optically pure (*S*)-omeprazole (esomeprazole). Other important sulfoxides include the ACAT inhibitor RP 73163 (**128**) and the calcium channel antagonists (Fig. 2.7).

The Kagan modification [37] of the Sharpless reagent has been successfully utilized for a number of substrates (Scheme 2.23). These conditions were claimed to be advantageous because the product precipitated from the reaction mixture, avoiding overoxidation to the sulfone. Alkyl peroxides appear to give the best stereoselectivity. The enantioselectivity is highest for rigid substrates or those with a large disparity in size between the two substituents on sulfur, such as substrates **131** [38], **133** [39], and **135** [40].

Esomeprazole, Nexium® (**138**), which is an active ingredient in Astra Zeneca's antiulcer drug, represents an excellent example where a Ti-catalyzed asymmetric sulfide oxidation is carried out on a scale of 100 tons per annum as shown in Scheme 2.24 [41].

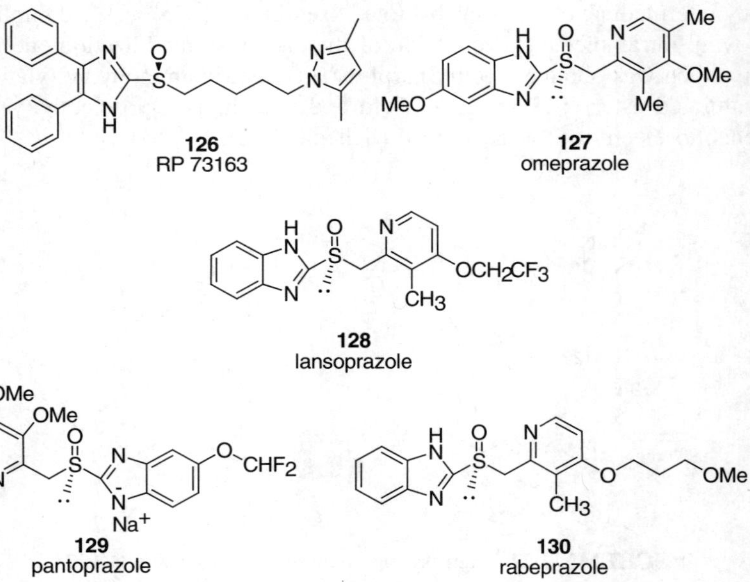

FIGURE 2.7 Some important sulfoxide drugs.

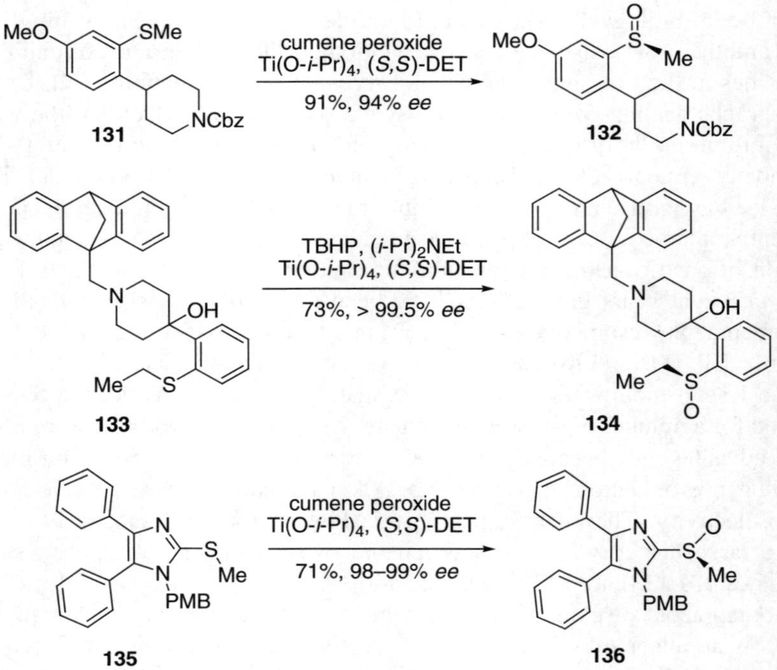

SCHEME 2.23 Stereoselective oxidation of the sulfur by Sharpless reagent.

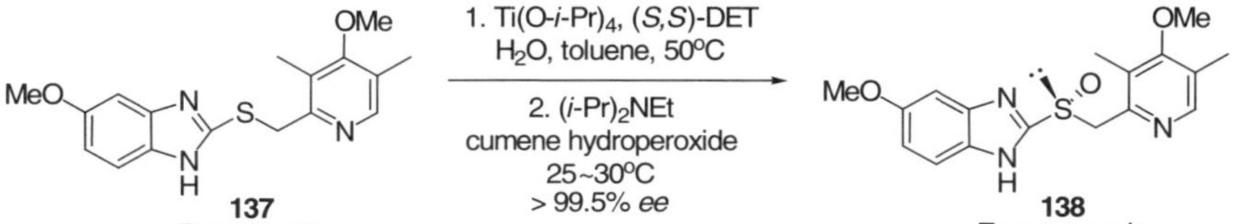

137
Pyrmetazole

138
Esomeprazole

SCHEME 2.24 Synthesis of esomeprazole (**138**) from pyrmetazole (**137**).

2.4 ASYMMETRIC C—C BOND FORMATION

2.4.1 Asymmetric Cyclopropanation Reaction

The chiral cyclopropyl group is a common structural motif seen in some biologically active molecules, such as curacin A (**139**) and plakoside A (**140**) in Figure 2.8. Curacin A (**139**) is a novel antimitotic agent recently isolated from a Caribbean cyanobacterium, *Lyngbya majuscule* [42]. Plakoside A (**140**) is a prenylated galactosphingolipid isolated from the marine sponge *Plakortis simplex* and exhibits strongly immunosuppressive activity with rather lower cytotoxicity [43].

Double-asymmetric Simmons–Smith cyclopropanation of the diene (**142**) derived from diethyl L-tartrate in four steps proceeded with excellent diasterofacial selectivity (>99% *de*) to give cyclopropane (**144**), which was converted to the desired carboxylic acid (−)-**145** in three steps [44] (Scheme 2.25). The optically pure compound (−)-**145** is the key intermediate for the synthesis of curacin A.

Cilastatin (**152**), a dehydropeptidase-I inhibitor that increases the *in vivo* stability of imipenem, has desired pharmacological properties [45]. The (*S*)-(+)-2,2-dimethylcyclopropane carboxylic acid (+)-**150**, a key building

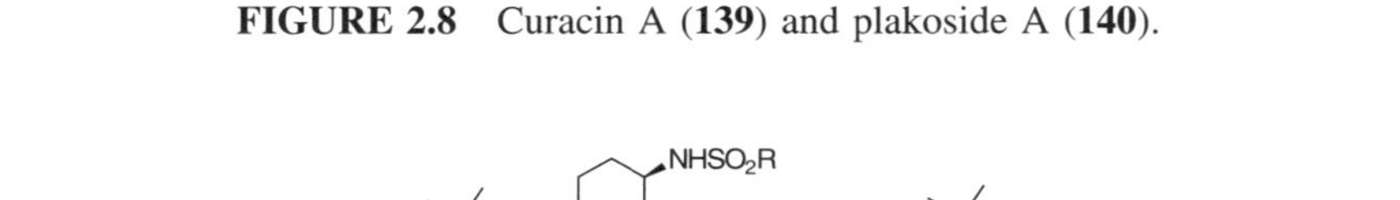

139
curacin A

140
plakoside A

FIGURE 2.8 Curacin A (**139**) and plakoside A (**140**).

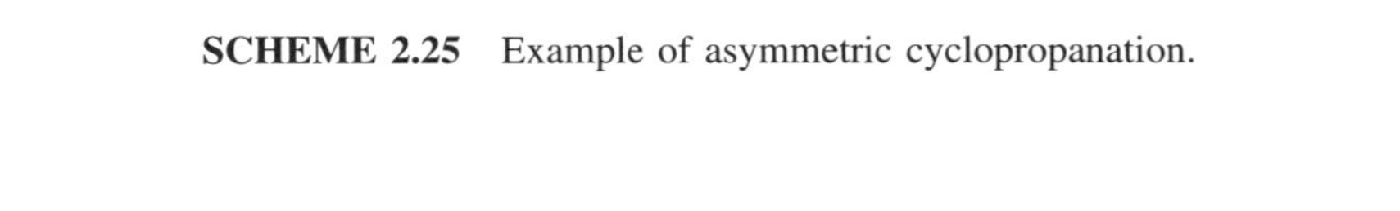

141
142
144
(−)-145

SCHEME 2.25 Example of asymmetric cyclopropanation.

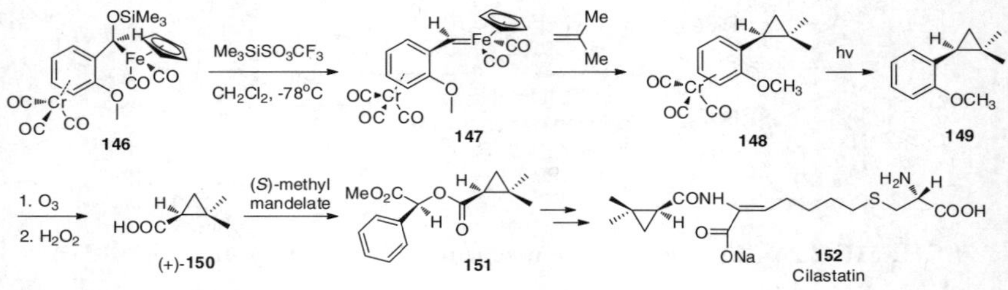

SCHEME 2.26 Synthesis of cilastatin (**152**).

block for the synthesis of cilastatin (**152**), was prepared from 2-methylpropene and chiral iron carbene complex via asymmetric cyclopropanation reaction and subsequent oxidations with up to 92% *ee* [46] (Scheme 2.26).

2.4.2 Asymmetric Hydroformylation Reaction

Hydroformylation, the conversion of olefins into aldehydes through the addition of CO and H_2, can be achieved via homogeneous catalysis by transition metals (Scheme 2.27) [47]. It is a valuable reaction in organic chemistry because it produces highly versatile aldehyde intermediates. The key features of the reaction include high atomic efficiency, tolerance of other functionalities, and the use of readily available syngas.

Hydroformylation is the largest homogeneous catalytic process in industry. Over 6 million tons of hydroformylation products are being produced worldwide per year. (*S*)-ibuprofen (**156**) and (*S*)-naproxen (**158**) can be readily synthesized through asymmetric hydroformylation on an industrial scale (for reviews, see [48]) (Scheme 2.28).

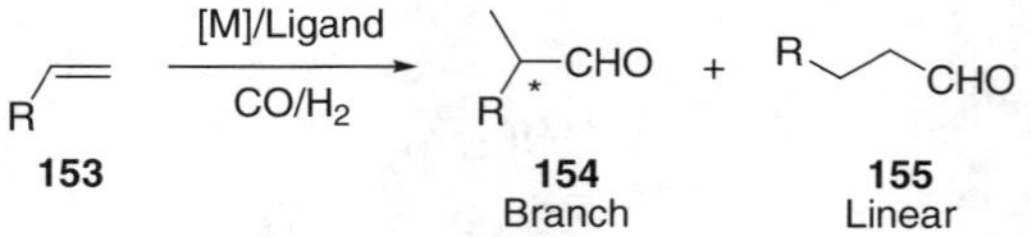

SCHEME 2.27 Hydroformylation reaction.

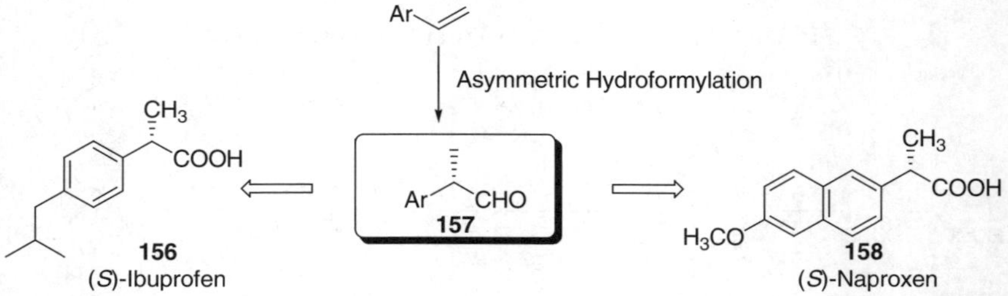

SCHEME 2.28 Preparation of pharmaceutical products by asymmetric hydroformylation.

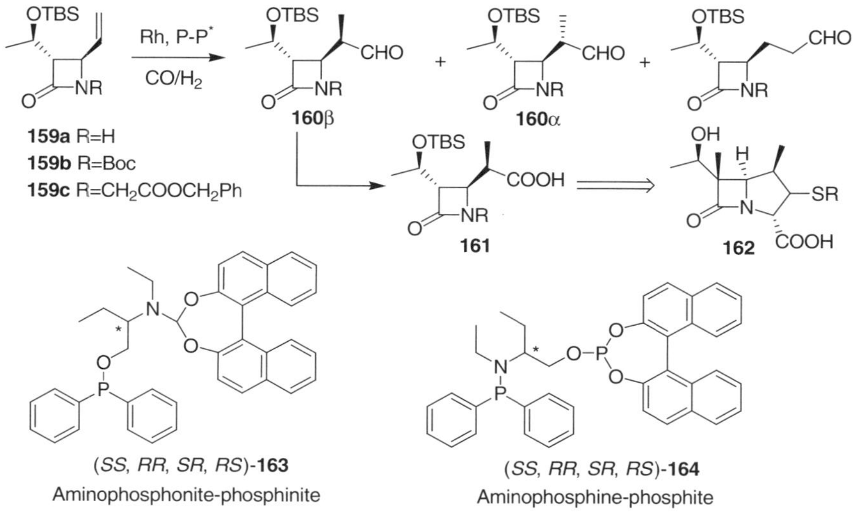

SCHEME 2.29 Synthesis of 1β-methylcarbapenem by asymmetric hydroformylation.

1β-Methylcarbapenem (**162**) is an antibiotic with an excellent antibacterial profile as well as enhanced chemical and metabolic stability. Monocyclic (*R/S*)-lactam (**161**), a key intermediate in the preparation of (**162**), was synthesized by the asymmetric hydroformylation of the 4-vinyl β-lactam (**159**) in the presence of rhodium amino phosphonite (**163**) and rhodium aminophosphinephosphite (**164**) complexes. The stereoselectivity of the reaction was controlled by the substituent at the nitrogen atom. The regioselectivity (branched/linear) but not the stereoselectivity was found to be dependent on the ratio of substrate to catalyst [49] (Scheme 2.29).

2.4.3 Asymmetric Michael Addition

Enantioselective addition of nucleophiles to the β-position of α, β-unsaturated compounds is one of the most powerful carbon-carbon bond-forming reactions for the construction of enantio-enriched synthons for biological active and natural compounds. Significant advantages of the process include high compatibility with many functional groups, low cost, and often high regio- and enantioselectivities. Generally, the asymmetric conjugate addition of a α, β-unsaturated compound is achieved by a carbon nucleophile to form a new stereogenic carbon center.

Hayashi and co-workers have developed an extremely efficient enantioselective protocol for the Rh-catalyzed addition of boronic acids to unsaturated systems. They showcased this method by applying it to the enantioselective synthesis of (*R*)-tolterodine (**168**) (Scheme 2.30). Thus, coumarin is treated with phenylboronic acid in the presence of Segphos and a Rh(I) precatalyst, to yield conjugate product in extremely high *ee* [50].

Chemists at Merck discovered that a ZnCl₂-MAEP-system is extremely useful for diastereoselective Michael addition and applied it to the synthesis of

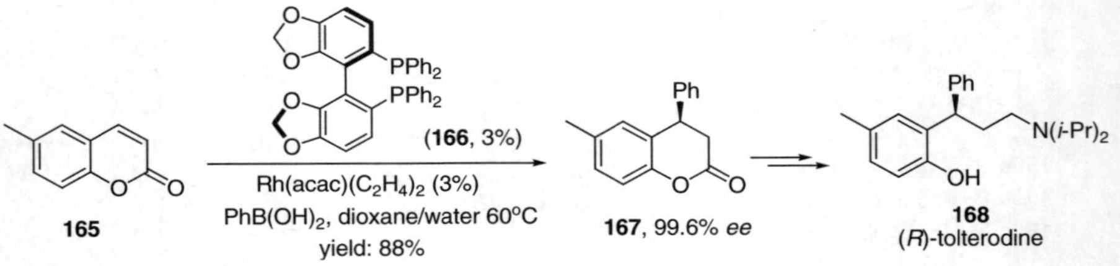

SCHEME 2.30   The synthesis of (*R*)-tolterodine (**168**).

muscarinic receptor antagonist **173** [51]. Enolization of dioxolane with LDA followed by addition to 2-cyclopenten-1-one in the presence of 15% of $ZnCl_2$-MAEP afforded the adduct in 74% yield and 99.0% *de*. It was believed that the unsymmetrical triamine ligand (MAEP) stabilizes the lithium enolate. A homogeneous reaction mixture was required before the addition of cyclopentenone for high selectivity. This synthesis has been scaled up on a multikilogram scale for the synthesis of muscarinic receptor antagonist **173** (Scheme 2.31).

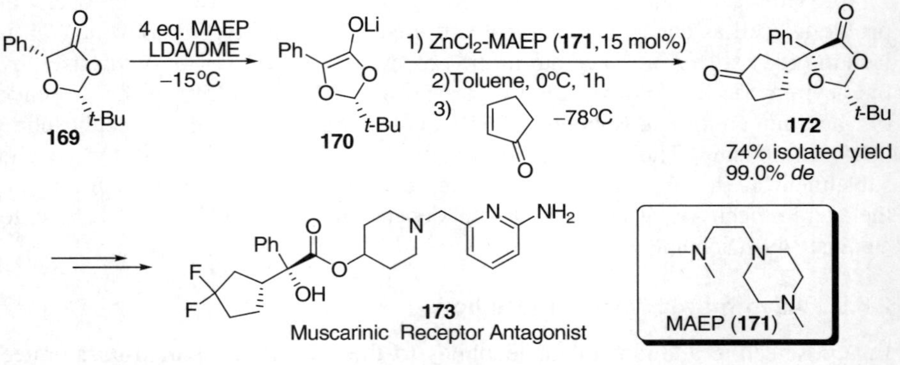

SCHEME 2.31   Synthesis of muscarinic receptor antagonist (**173**).

In the preparation of endothelin receptor antagonist (**179**), Song et al. at Merck utilized imidazolidine as an auxiliary for an enantioselective Michael addition following a sequence similar to that described above (Scheme 2.32) [52]. The process is practical and applicable to a wide variety of structurally complex chiral diaryl propanoates on a large scale.

In 2002, Barnes and co-workers reported a chiral Lewis acid-catalyzed Michael addition of diethyl malonate to the fully elaborated nitrostyrene [53]. Because of the high enantioselectivity and yield of malonates in addition to nitrostyrenes, this reaction was successfully employed in the enantioselective synthesis of (*R*)-rolipram (**185**), an inhibitor of PDE-IV for the treatment of depression (Scheme 2.33).

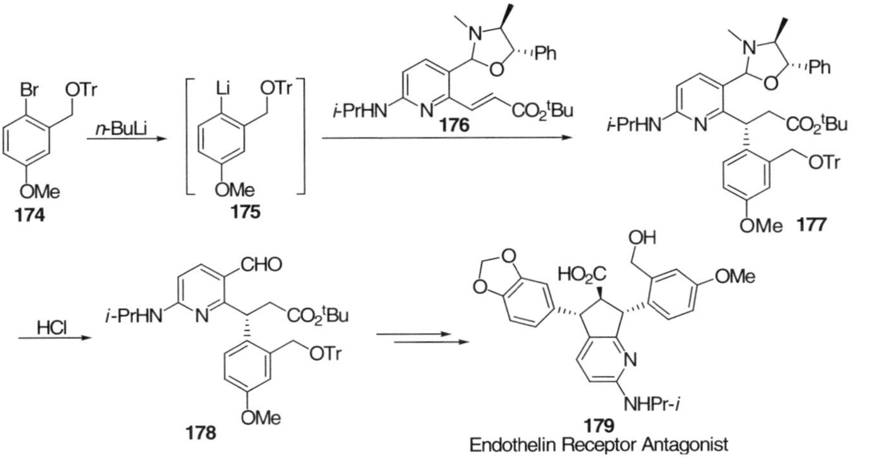

SCHEME 2.32 Synthesis of endothelin receptor antagonist **179**.

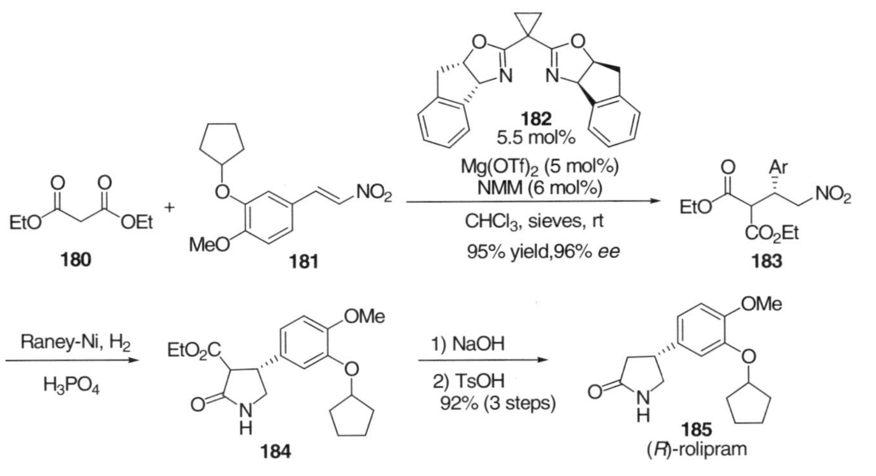

SCHEME 2.33 Synthesis of (R)-rolipram (**185**) by chiral Lewis acid-catalyzed Michael addition.

2.5 ASYMMETRIC REACTIONS VIA ORGANOCATALYSIS

One of the major concerns in the chemistry community is environmental sustainability. The concept of green chemistry and engineering is based on the premise that it is better to prevent waste than to clean it up after it is created. To this end, organocatalysis in asymmetric synthesis has received more and more attention (for reviews on organocatalysts, see [54]).

In comparison with organometallic systems, organocatalysts have obvious advantages. Organic molecules are generally insensitive to air and moisture; a wide variety of building blocks for organocatalysts, such as amino acids, carbohydrates, hydroxyl acids, and alkaloids, are readily available; small organic molecules are usually nontoxic and environmentally friendly.

Enantioselective synthesis with low-molecular-weight organic molecules with hydrogen donors has become a frontier of research on the use of asymmetric synthesis in forming C—C and C-hetero atom bonds, which has shown promising applications in the pharmaceutical industry.

Figure 2.9 shows representive organic molecules used as catalysts. Of all classes of structure of organomolecules, the cinchona alkaloids are perhaps the most remarkable. These compounds have shown excellent utility for a variety of enantioselective transformations. The presence of the tertiary quinalidine nitrogen, the basic functionality, renders them effective ligands for a variety of metal-catalyzed transformations.

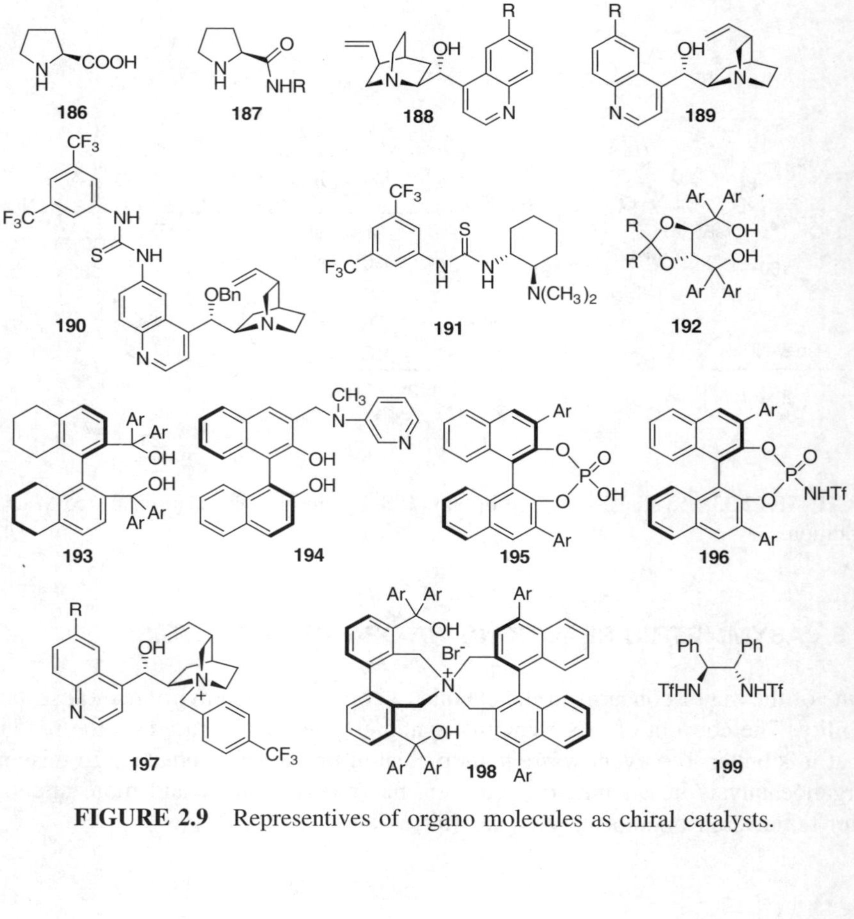

186 187 188 189 190 191 192 193 194 195 196 197 198 199

FIGURE 2.9 Representives of organo molecules as chiral catalysts.

2.5.1 Proline and Its Derivatives

Soluble chiral acids and bases have also been utilized in the synthesis of pharmaceutical and agrochemical products. The chiral acids and bases selected for these reactions typically originate from the chiral pool. Aldol reactions catalyzed by L-proline are one example of a chiral base-catalyzed process. A bicyclic intermediate for the synthesis of 19-norsteroids has been prepared in this manner. Discovered in the early 1970s, this reaction was the first example of a highly enantioselective organocatalytic process [55]. The use of proline as a chiral catalyst illustrates the importance of H-bonding in asymmetric catalysts. Recent investigation has provided the evidence that proline acts as a multifunctional catalyst as well [51]. Substantial research has been devoted to synthesis of proline derivatives bearing H-bond donors (Scheme 2.34).

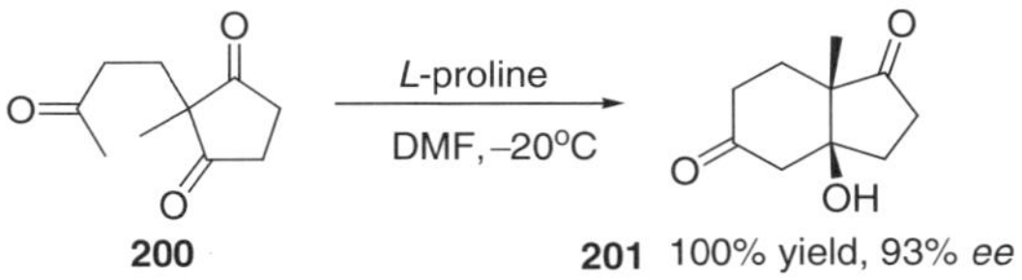

SCHEME 2.34 The first example of a highly enantioselective organocatalytic process.

Northrup and MacMillan reported a useful variant of the proline-catalyzed intermolecular aldol reaction [56]. α-Unbranched aldehydes can be used as donors in the reaction with aldehyde acceptors. Proline catalyzed the cross-aldolizations of two different aldehydes to furnish β-hydroxy aldehyde anti-aldols with excellent enantioselectivities, good yields, and diastereoselectivities. This reaction has recently been used in an extremely efficient and short synthesis of prelactone B (**206**) by Pikho and Erkkilä (Scheme 2.35) [57].

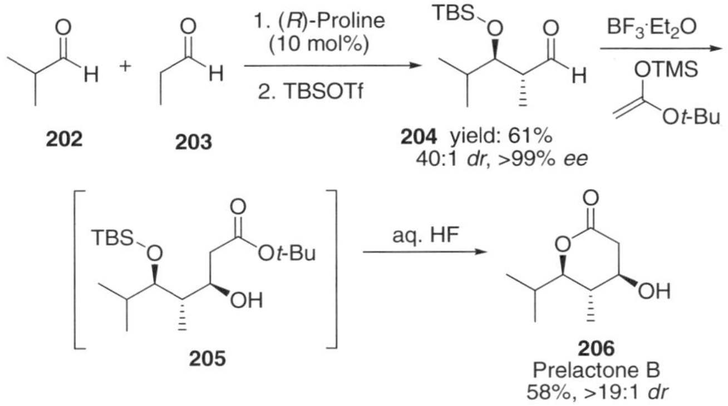

SCHEME 2.35 Preparation of prelactone B by proline-catalyzed cross-aldolizations.

In 2006, Jørgensen and co-workers [58] reported the asymmetric conjugate addition of malonate with enals. The group applied this method in the enantioselective formal total syntheses of $(-)$-paroxetine (**212**) and $(+)$-femexotine (**213**).

Addition of dibenzyl malonate to cinnamaldehyde or *p*-fluorocinnamaldehyde established one of the two required stereocenters in 80% or 72% yield and 91% or 86% *ee*, respectively (Scheme 2.36). Following the synthesis of these essential building blocks, a reductive amination-cyclization sequence yielded the respective chiral lactams that could be converted to the desired compounds through known transformations.

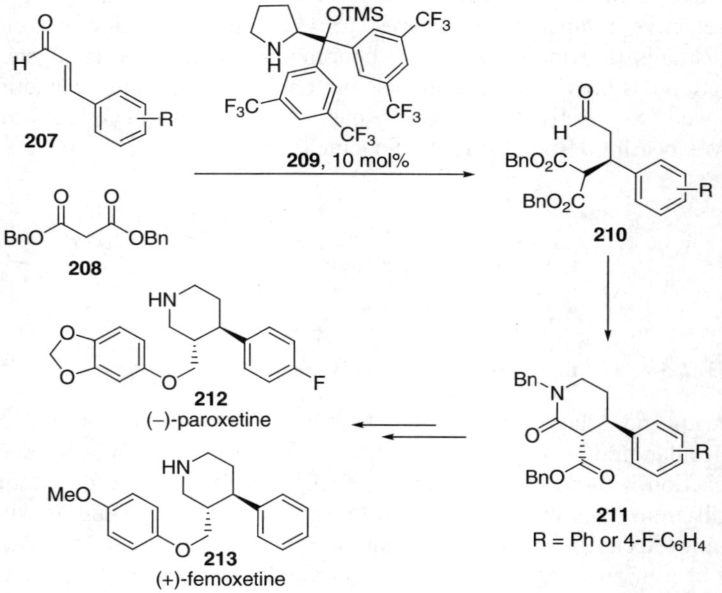

SCHEME 2.36   Synthesis of (−)-paroxetine (**212**) and (+)-femexotine (**213**).

The proline-catalyzed Mannich reaction has been applied in the enantioselective synthesis of (+)-*epi*-cytoxazone (**76**) [59] by the same authors. The reaction provided (+)-*epi*-cytoxazone with 35% overall yield and 81% *ee* in six steps starting from readily available *p*-anisaldehyde (Scheme 2.37) [60].

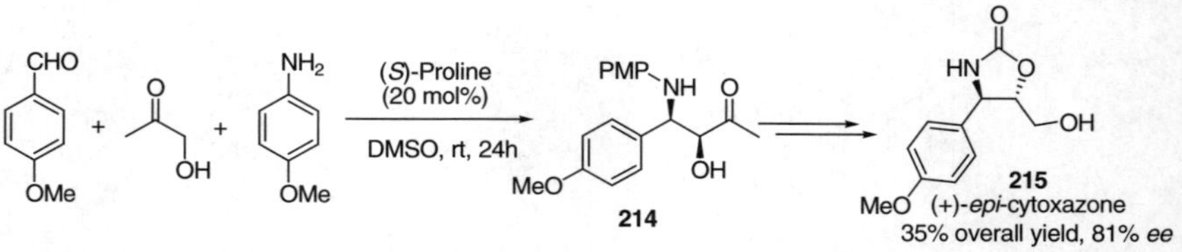

SCHEME 2.37   Synthesis of (+)-*epi*-cytoxazone (**215**) by proline-catalyzed Mannich reaction.

Itoh et al. described the asymmetric synthesis of *ent*-sedridine (**220**) by applying the proline-catalyzed Mannich reaction of 4-hydroxybutanal, *p*-anisidine, and acetone as the key step (Scheme 2.38) [61]. The resulting ketone product (**219**) was converted into *ent*-sedridine in six steps with good yield.

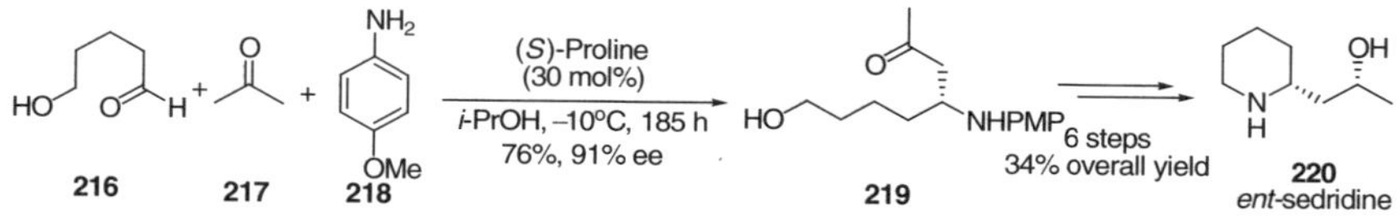

SCHEME 2.38   Synthesis of *ent*-sedridine (**220**).

Wang et al. described pyrrolidine sulfonamide as an efficient catalyst for α-aminoxylation, α-sulfenylation, α-selenenylation, aldol, Mannich, and Michael reactions [62]. For example, the pyrrolidine-containing sulfonamide catalyst was successfully employed in the synthesis of the potent H$_3$ agonist Sch50917 (**225**) [63], with the stereoselective addition of propionaldehyde to nitroolefin as the key step (Scheme 2.39).

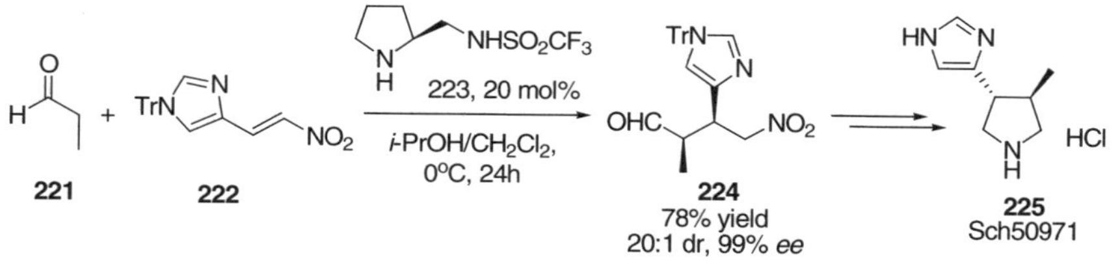

SCHEME 2.39   Synthesis of the potent H$_3$ agonist Sch50917 (**225**).

The utility of the proline-catalyzed direct asymmetric α-amination of α,α-disubstituted aldehydes was illustrated in an application to the synthesis of the metabotropic glutamate receptor ligands (*S*)-APICA (**229**) and (*S*)-AIDA (**230**) by Barbas (Scheme 2.40) [64].

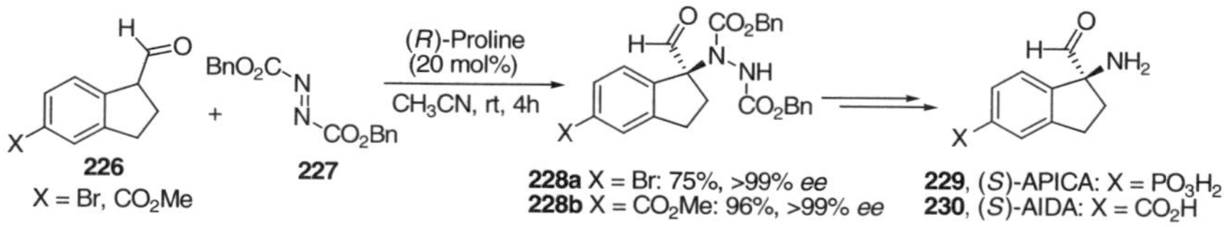

SCHEME 2.40   Synthesis of the metabotropic glutamate receptor ligands (*S*)-APICA (**229**) and (*S*)-AIDA (**230**).

Another application of the simple yet useful transformation described above for the construction of a quaternary stereocenter was developed for the total synthesis of the LFA-1 antagonist BIRT-377 (**232**) [65]. (*S*)-proline-derived tetrazole turned out to be a good choice of catalyst, affording the product in 95% yield after 3 h with 80% *ee* (Scheme 2.41). The corresponding α-aminated aldehyde was then elaborated to BIRT-377 (**233**) by standard transformations.

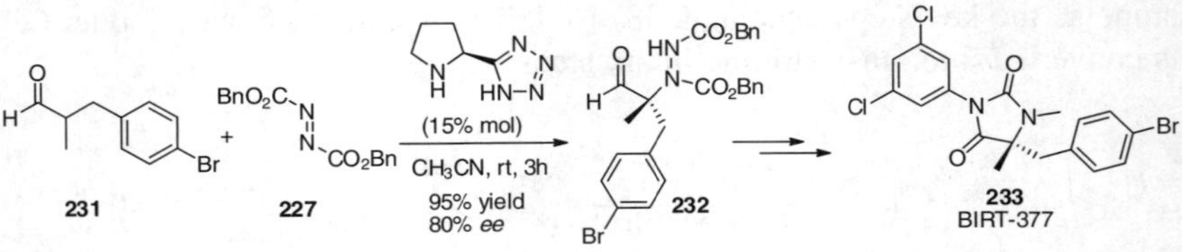

SCHEME 2.41   (*S*)-proline-derived tetrazole-catalyzed α-aminated reaction.

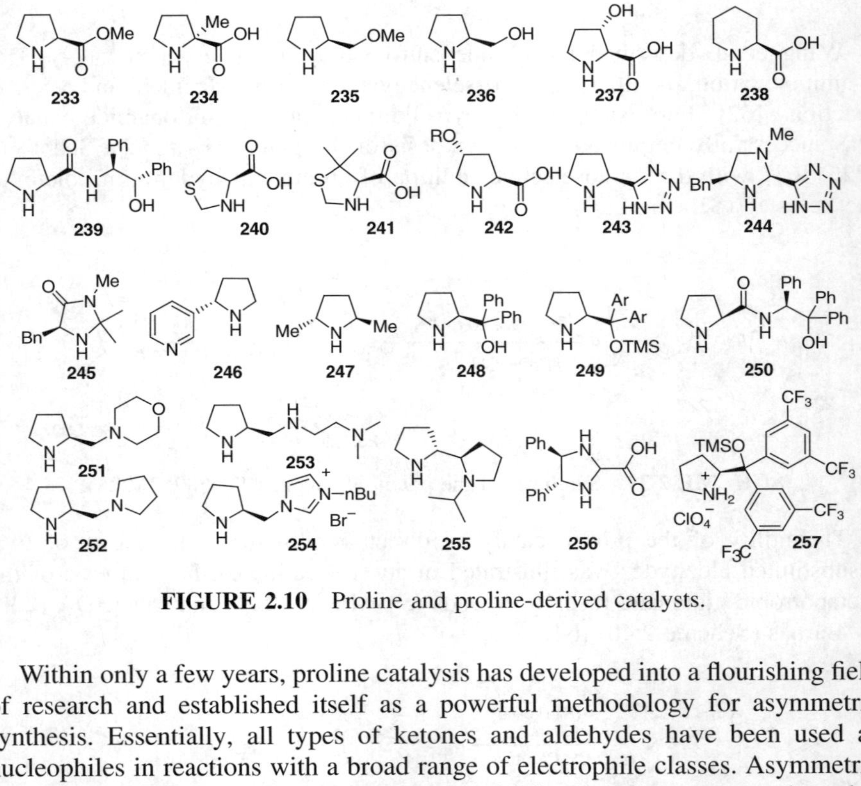

FIGURE 2.10   Proline and proline-derived catalysts.

Within only a few years, proline catalysis has developed into a flourishing field of research and established itself as a powerful methodology for asymmetric synthesis. Essentially, all types of ketones and aldehydes have been used as nucleophiles in reactions with a broad range of electrophile classes. Asymmetric enamine catalysis has delivered profoundly useful and rather unexpected results. Figure 2.10 outlines an ever-increasing number of amino catalysts.

2.5.2   Chiral Ketones

Dioxiranes represent a new generation of oxidants for olefin epoxidation. Dioxiranes are usually generated from Oxone® (potassium peroxomonosulfate) and ketones. Owing to the rapidity, mildness, and safety of the reaction, this system has been extensively investigated (Fig. 2.11).

The epoxidation can be performed with either isolated dioxiranes or dioxiranes formed in situ. Generally, when the dioxirane is formed in situ, the ketone can

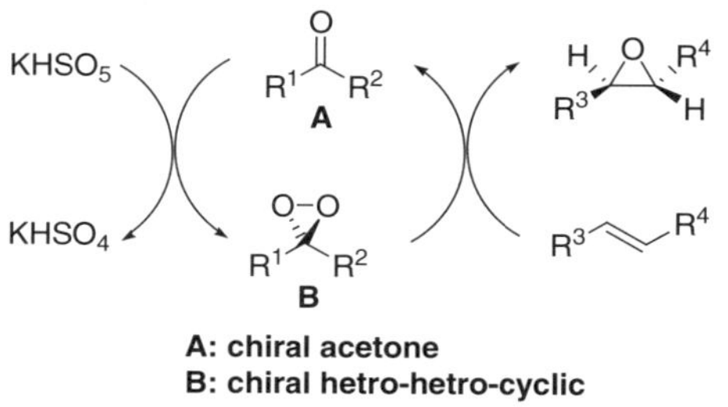

A: chiral acetone
B: chiral hetro-hetro-cyclic

FIGURE 2.11 Olefin epoxidated by dioxiranes.

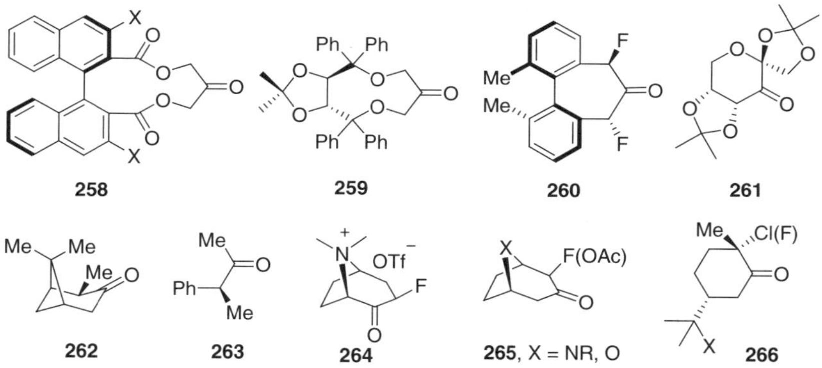

258   259   260   261

262   263   264   265, X = NR, O   266

FIGURE 2.12 Representative examples of chiral ketone catalysts reported.

be used as a catalyst since it can be regenerated upon epoxidation. The first asymmetric epoxidation using a chiral ketone was reported by Curci et al. in 1984 [66]. During the past decade, various chiral ketones have been studied and reported by a number of laboratories. Nowadays, chiral dioxiranes have been shown to be powerful agents for asymmetric epoxidation of olefins (Fig. 2.12) (for references on dioxiranes, see [67]; for reviews on chiral ketone-catalyzed asymmetric epoxidation, see [68]).

In the late 1990s, Yang et al. [70a], Shi and co-workers [72a], and Denmark et al. [69] demonstrated independently that an enantiomerically pure ketone-Oxone® system could be used to catalyze the enantioselective epoxidation of simple alkenes, via the chiral dioxirane intermediate formed in situ.

Yang and co-workers designed and synthesized C_2-symmetric cyclic chiral binaphthyl ketone catalysts [70]. The chiral binaphthyl ketone (**259**) catalysts have been applied by Tanabe Seikayu Company in Japan to a large-scale asymmetric epoxidation of methyl p-methoxycinnamate (MPC). The chiral epoxide product is an important intermediate in the synthesis of diltiazem hydrochloride (**262**), a coronary vasodilator for the treatment of angina pectoris and hypertension (Scheme 2.42) [71]. It is worth noting that the chiral ketone

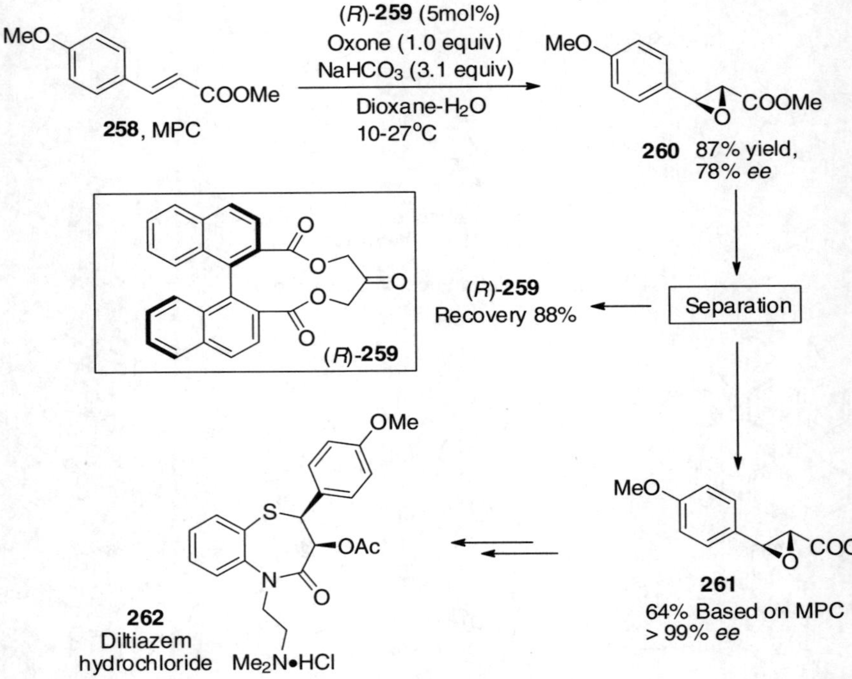

SCHEME 2.42   Asymmetric epoxidation of MPC catalyzed by chiral binaphthyl ketone catalyst.

can be readily separated and recovered in 88% yield and then reused without any loss of activity and chiral induction, while the chiral epoxide products can be isolated by recrystallization in 64% yield and >99%*ee*.

In 1996, Shi reported a highly efficient and mild asymmetric epoxidation of *trans-* and trisubstituted olefins with Oxone® and a fructose-derived ketone. It is an efficient epoxidation catalyst with high *ee* values for a variety of *trans-* and trisubstituted olefins [72]. Subsequently, they developed a nitrogen analog of the chiral ketone, which achieved generally high *ee* values for a class of acyclic and cyclic *cis*-olefins [73] (Scheme 2.43).

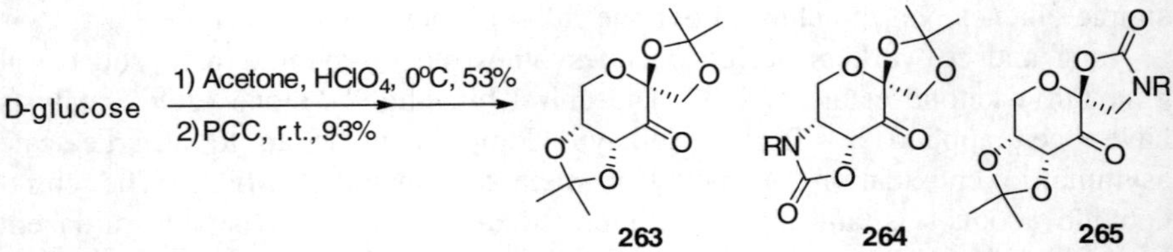

SCHEME 2.43   Shi's chiral ketones and nitrogen analogs.

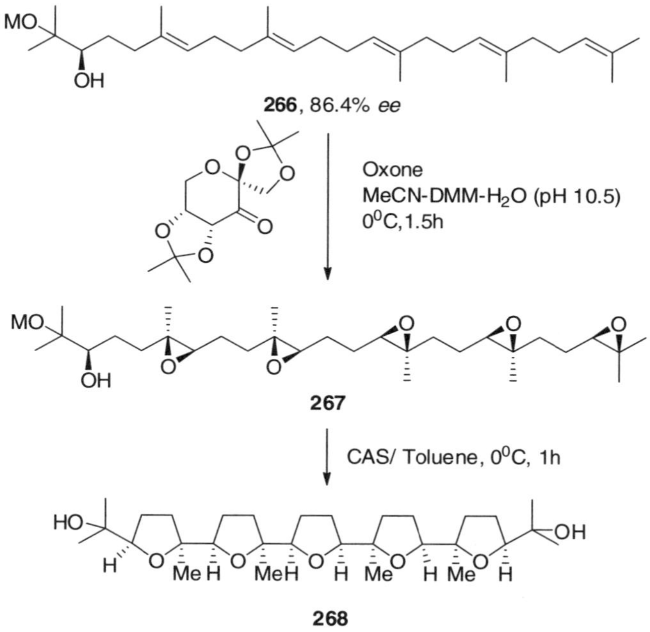

SCHEME 2.44   Applications of Shi's chiral ketone.

Shi's epoxidation reaction has been successfully applied in the asymmetric synthesis of ladder polyether natural products by a cascade of epoxide opening [74] (Scheme 2.44).

2.5.3 Chiral Brønsted Acids

Asymmetric catalysis by Brønsted acids is a recent topic in the growing field of organic catalysis. Most organocatalysis are bifunctional with a Brønsted acid and a Lewis base center.

In contrast to the Lewis acids, Brønsted acids are easy to handle and generally stable toward oxygen and moisture and can be stored for a long period of time. Chiral Brønsted acid (for reviews on chiral Brønsted-acid catalysis, see [75]) catalysts have recently been recognized as a new class of organocatalysts for a number of enantioselective carbon-carbon bond or carbon-hetero bond formation reactions. Chiral Brønsted acids activate the substrates, such as carbonyls or imines, by providing acidic proton binding to them and thereby establishing chiral surroundings (Fig. 2.13).

In 2004, the research groups of Akiyama [76] and Terada [77] independently described the use of chiral phosphoric acids that exhibited excellent catalytic activity as chiral Brønsted acid catalysts for Mannich reaction of N-aryl or N-Boc imines. It was found that chiral phosphoric acids were bifunctional catalysts

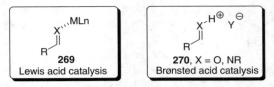

FIGURE 2.13 Brønsted acid and Lewis acid catalysis.

bearing both a Brønsted acidic site and a Lewis basic site. The another key finding was a significant effect of an aromatic substituent (X) at the 3,3′-position of one binaphthyl subunit of the catalyst on the enantiofacial discrimination (Fig. 2.14).

Following those initial findings, chiral phosphoric acids have now been recognized as novel chiral catalysts and have received great the attention from synthetic organic chemists. Several novel phosphoric acids have been reported with adjustable pH of the corresponding acidic proton (Fig. 2.15).

In 2008 Schaus reported the enantioselective synthesis of the enantio-enriched dihydropyrimidone core by the chiral phosphoric acid-catalyzed Biginelli reaction (Scheme 2.45) [78]. Dihydropyrimidone is a class of heterocyclic compounds that possesses a wide range of biological activities such as calcium channel modulators, antihypertensive agents, mitotic kinesin Eg5 inhibitors, and melanin concentrating hormone receptor (MCH1-R) antagonists [79].

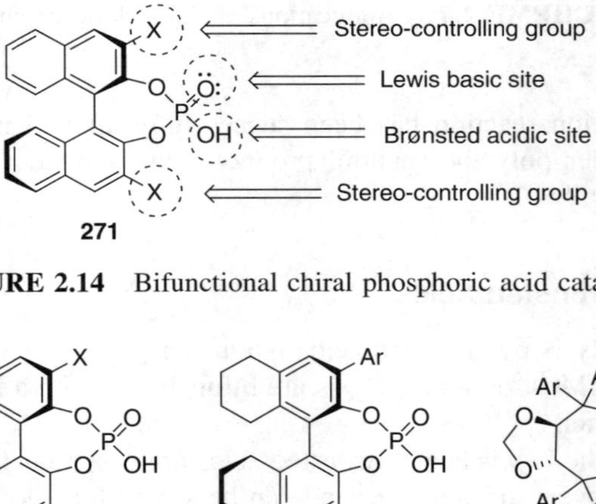

271

FIGURE 2.14 Bifunctional chiral phosphoric acid catalysts.

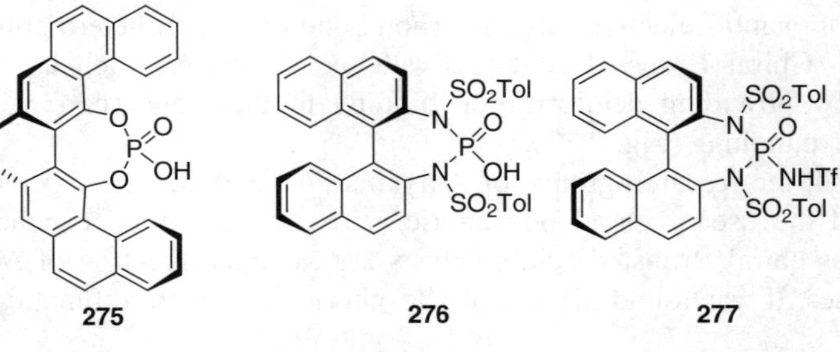

272   **273**   **274**

275   **276**   **277**

FIGURE 2.15 Several representative novel phosphoric acids.

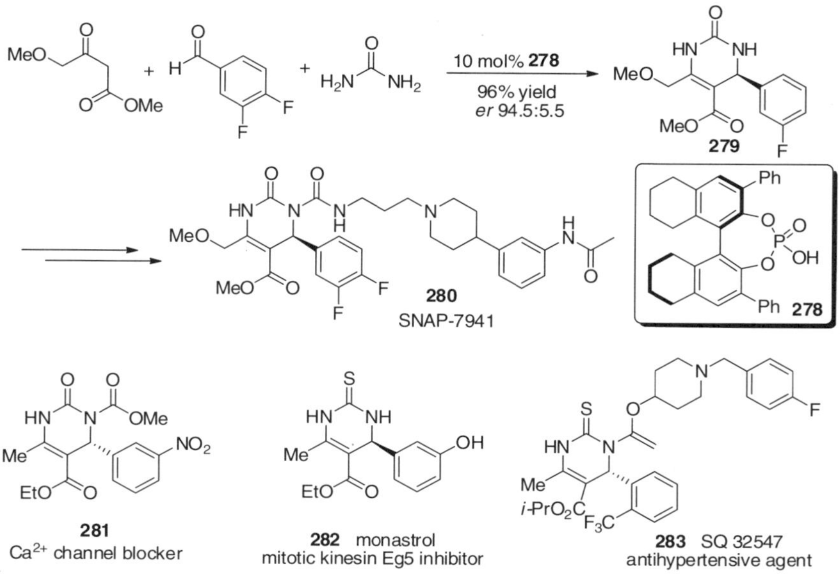

SCHEME 2.45   Asymmetric Biginelli reaction catalyzed by chiral binapthol-derived phosphoric acid.

Rueping discovered that the chiral phosphoric acid catalysts could catalyze the reduction of quinolines, benzoxazines, benzothiazines, and benzoxaziones by transfer hydrogenation [80], with Hantsch reagent. They applied the method to the enantioseletive synthesis of biologically active tetrahydroquinoline alkaloids such as (−)-angustureine (**75**), (+)-galipinine (**77**), and (+)-cuspareine (**78**) as shown in Scheme 2.46.

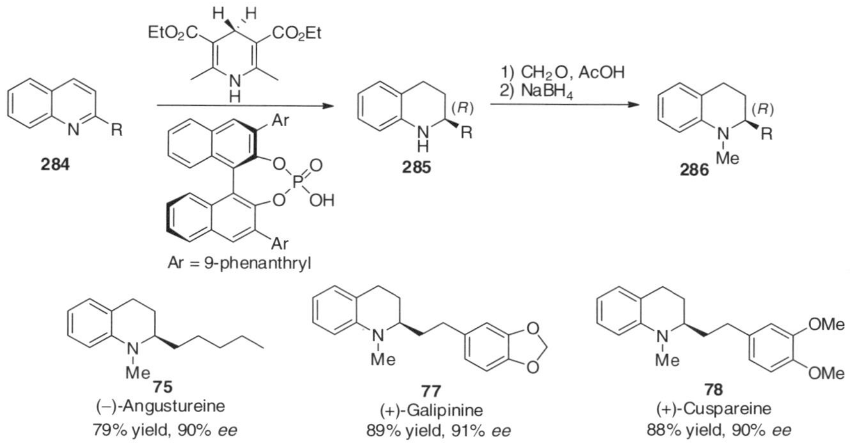

SCHEME 2.46   Synthesis of alkaloids by chiral phosphoric acid-catalyzed transfer hydrogenation.

List reported the direct catalytic asymmetric synthesis of cyclic aminals and hemiaminals via addition of amides and alcohols with imines [81]. This provides an excellent and convenient route to their preparation, which possesses stereogenic cyclic aminals and similar structures in clinically used compounds such as **290–294**. For example, the two benzo(thia)diazine aminals (**290**) and (**291**) are effective diuretic drugs for the treatment of high blood pressure in use since 1957 (Scheme 2.47).

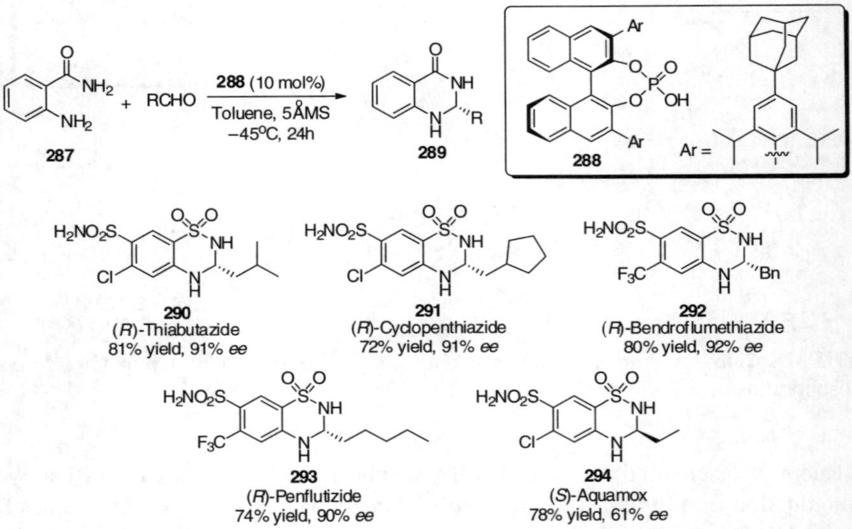
287
RCHO
288 (10 mol%)
Toluene, 5ÅMS
−45°C, 24h
289
288
Ar =
290
(R)-Thiabutazide
81% yield, 91% ee
291
(R)-Cyclopenthiazide
72% yield, 91% ee
292
(R)-Bendroflumethiazide
80% yield, 92% ee
293
(R)-Penflutizide
74% yield, 90% ee
294
(S)-Aquamox
78% yield, 61% ee

SCHEME 2.47 Antihypertensive pharmaceuticals made with direct catalytic asymmetric synthesis of cyclic aminals from aldehydes.

Masson and Zhu have successfully developed the catalytic enantioselective three-component Povarov reaction, leading to the highly efficient asymmetric synthesis of (2, 4-*cis*)-4-amino-2-aryl (alkyl)-tetra-hydro-quinoline analogs such as torcetrapid (**298**) [82] (Scheme 2.48).

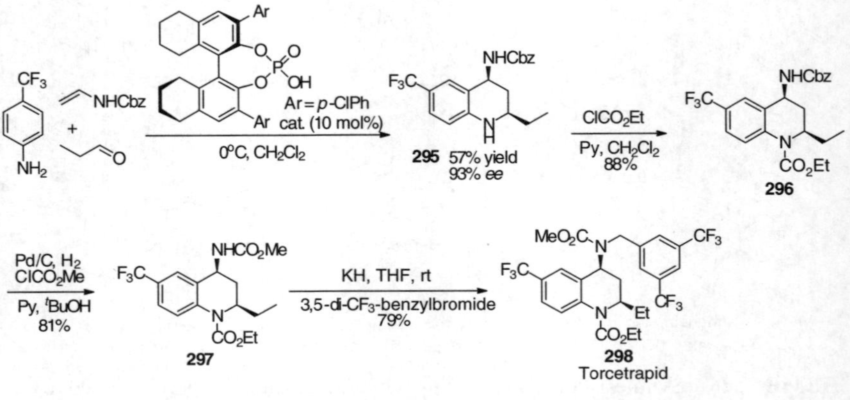
Ar = p-ClPh
cat. (10 mol%)
0°C, CH2Cl2
295 57% yield
93% ee
ClCO2Et
Py, CH2Cl2
88%
296
Pd/C, H2
ClCO2Me
Py, tBuOH
81%
297
KH, THF, rt
3,5-di-CF3-benzylbromide
79%
298
Torcetrapid

SCHEME 2.48 Synthesis of torcetrapib (**298**).

The Friedel–Crafts (F-C) reaction also constitutes one of the most useful carbon-carbon bond formation processes in organic synthesis. Recently, the asymmetric F-C reaction was described with BINOL-derived phosphoric acid as a catalyst. In 2004, Terada and co-workers reported the addition of 2-methoxyfuran (**299**) to different aromatic and heteroaromatic N-Boc aldimines (**300**), affording the Friedel–Crafts products (**301**) in excellent yields and enantioselectivities catalyzed by the optically active BINOL-based phosphoric acid catalyst [83]. The practicality of the process was demonstrated by conducting the reaction on a gram scale in the presence of only 0.5 mol% of the catalyst. Furthermore, in the case of furan imine, the Friedel–Crafts product could be readily converted into the corresponding γ-butenolide (**302**) which is a common building block in the synthesis of various natural products in excellent yields favoring the *syn*-diastereomer (Scheme 2.49).

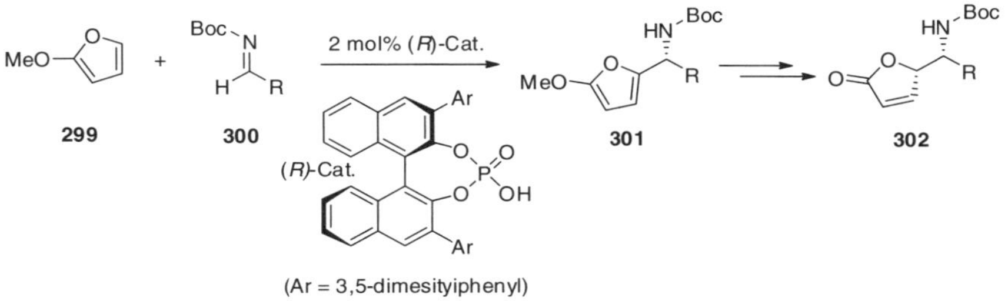

(Ar = 3,5-dimesityiphenyl)

SCHEME 2.49   Friedel–Crafts addition of 2-methoxyfuran and N-Boc aldimines.

The asymmetric Friedel–Crafts reaction of indoles with imines is an important reaction to which chiral phosphoric acid can be applied. The process provides direct, convergent, and versatile access to the synthesis of enantiopure 3-indolyl methanamine structural motif. The latter exists in numerous indole alkaloids and synthetic indole derivatives with significant biological activities. Recently, You and co-workers demonstrated a chiral phosphoric acid catalyst to the enantioselective addition of indoles (**303**) to sulfonyl imines (**304**) [84]. The process features high efficiency, high yields, and excellent enantioselectivities, providing a practical method to synthesize highly enantiopure 3-indolyl methanamine derivatives (Scheme 2.50).

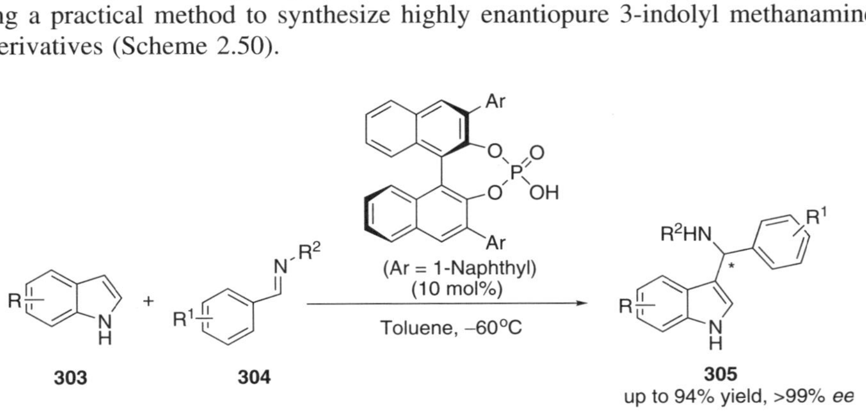

up to 94% yield, >99% *ee*

SCHEME 2.50   Friedel–Crafts addition of indoles and aldimines.

Terada also disclosed that Boc-protected enecarbamates (**304**) could react with indoles to give the Friedel–Crafts products 3-indolyl methanamine derivatives (**306**) in excellent yields and selectivities [85]. (*Z*)- and (*E*)-enecarbamates were shown to afford the products in identical selectivities but with different rates. In the same year, Zhou reported that in chiral phosphoric acid catalyzed Friedel–Crafts reaction employing (*R*)-aryl enamides (**305**) as the imine precursors, the Friedel–Crafts products **307** with one quaternary stereocenter were obtained with excellent selectivities (Scheme 2.51) [86].

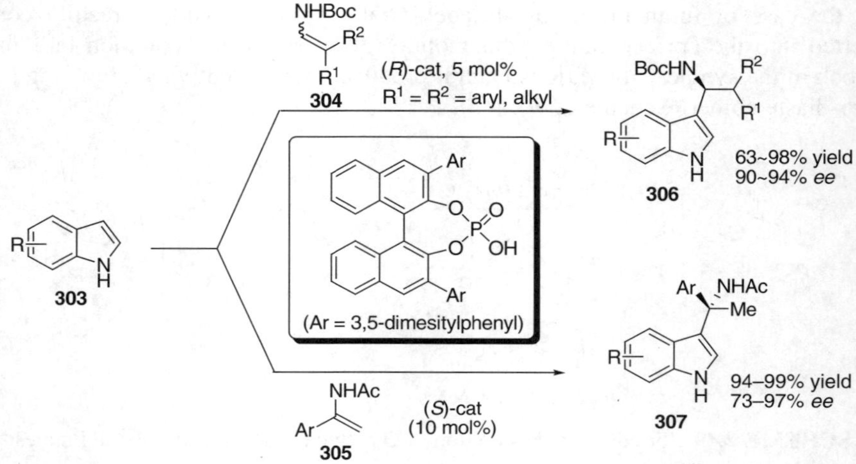

SCHEME 2.51 Friedel–Crafts addition of indoles and enecarbamates.

List recently designed a powerful C_2-symmetric chiral disulfonimide Brønsted acid catalyst, **308**, to catalyze the Mukaiyama aldol reaction with high yield (up to 98%) and excellent *d.e.* (up to 97:3) as shown in Scheme 2.52 [87].

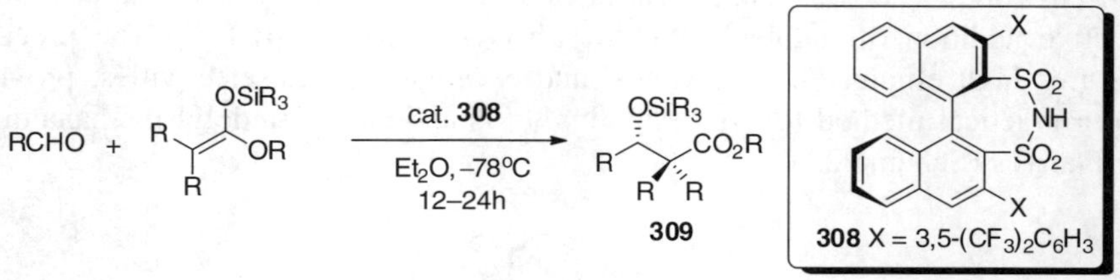

SCHEME 2.52 Mukaiyama aldol reaction catalyzed by chiral disulfonimide (**308**).

2.5.4 Cinchona Alkaloids

Cinchona alkaloids have frequently been used not only in chemical laboratories but also in the pharmaceutical industry because of their availability, stability, and low production cost. Furthermore, they possess a pseudo-enantiomeric counterpart, and their structures can be easily modified to obtain more effective catalysts

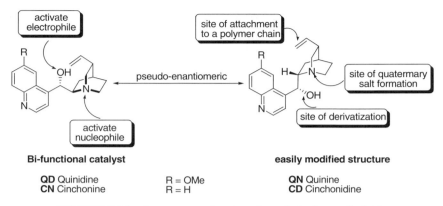

FIGURE 2.16 Structures and acting sites of cinchona alkaloids.

(Fig. 2.16). A particularly striking application of these catalysts has been focused on desymmetrization of meso compounds.

Cinchona alkaloid-catalyzed desymmetrization based on the use of oxygen- and nitrogen-based nucleophiles has been extensively investigated over the past few years. Cinchona alkaloid-catalyzed desymmetrization provides a reliable practical approach for enantioselective synthesis of biologically active molecules, and has also found utilization in the synthesis of chiral aminoalcohols [88], diamines [89], and bisoxazolines [90].

Bolm described a desymmetrization procedure that was applied by scientists at Merck for the ring opening of monocyclic anhydride (**309**) [91]. Accordingly, both methyl hemiester enantiomers (**310** and **311**) could be easily synthesized on a large scale. They are the important intermediate of $\alpha_v\beta_3$ antagonist **312**, a potential candidate for the treatment of osteoporosis (Scheme 2.53).

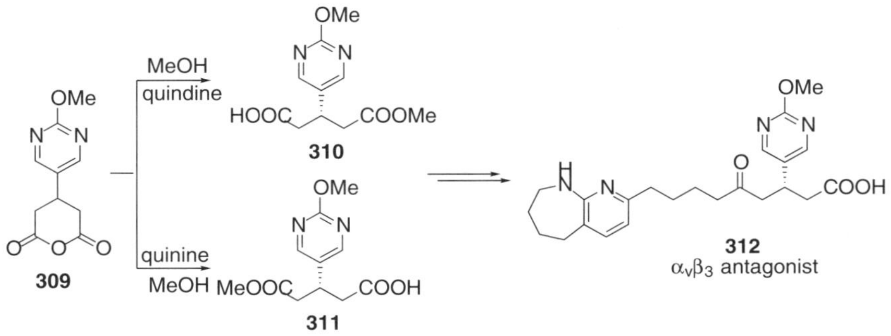

SCHEME 2.53 Desymmetrization of methyl hemiester enantiomers by cinchona alkaloids.

Biotin (**53**) is one of the water-soluble B vitamins (for reviews see [92]). It plays an essential role as a coenzyme in carboxylation reactions related to biochemical processes such as gluconeogenesis and fatty acid biosynthesis. Deficiency of biotin in poultry and swine causes a series of severe symptoms. Deng utilized the diastereoselective anhydride desymmetrization process to synthesize **314**, a key intermediate in the total synthesis of biotin. The best result, 93% *ee*, was achieved when the mono-cinchona alkaloid derivative DHQD-PHN was employed as catalyst in the reaction (Scheme 2.54) [93].

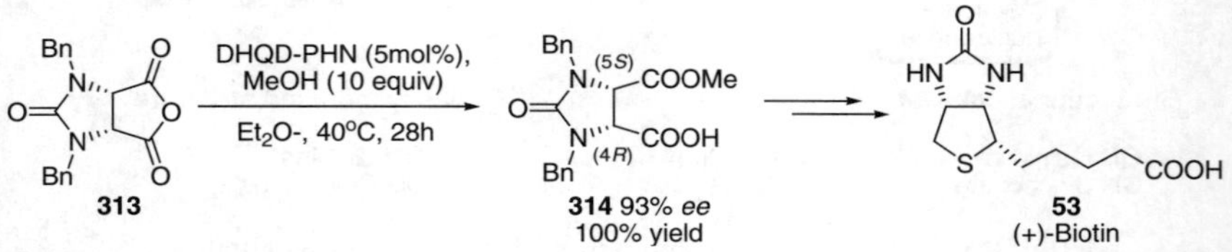

SCHEME 2.54   Asymmetric of (+)-biotin (**53**) by asymmetric desymmetrization.

Deng also reported a general and highly enantioselective catalytic kinetic resolution of racemic monosubstituted succinic anhydrides via (DHQD)$_2$AQN-catalyzed alcoholytic ring opening in 2001 [94]. They demonstrated the usage of this catalytic kinetic resolution in a formal total synthesis of baclofen (**102**), an effective GABA receptor agonist for muscle spasticity (Scheme 2.55). Using the highly efficient catalytic kinetic resolution of racemic **315** as the key step, Deng and Chen prepared **316** in 96% *ee* and 44% overall yield. The conversion of **316** to baclofen **317** was known [95].

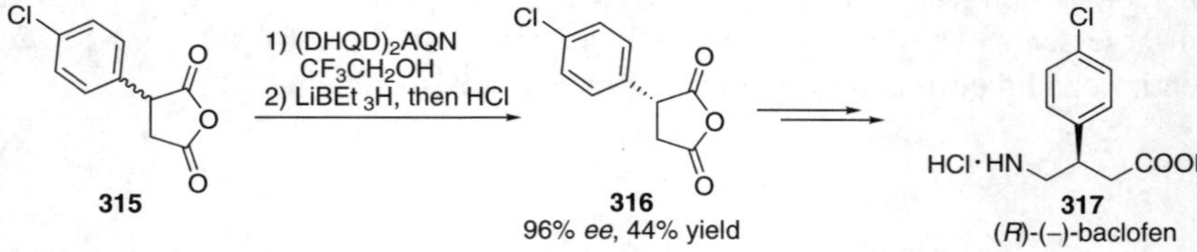

SCHEME 2.55   Synthesis of (−)-baclofen (**317**).

2.5.5   Phase-Transfer Catalysts

Considerable work has been devoted to the development of phase-transfer catalysts (PTC) from readily available resources, in the same way as the preparation of ligands on organometallic system. Asymmetric phase-transfer catalysts have emerged as a powerful and convenient tool for the construction of chiral compounds over the last decade. The majority of the synthetic catalysts such as **318–321** were realized by using chiral quaternary ammonium salts. Very recently Maruoka disclosed the first successful example of chiral quaternary phosphonium salt (**321**) as a phase-transfer catalyst [96] (Fig. 2.17).

The ready accessibility of phase-transfer catalysts and the relatively mild reaction conditions make the asymmetric phase-transfer reactions attractive in

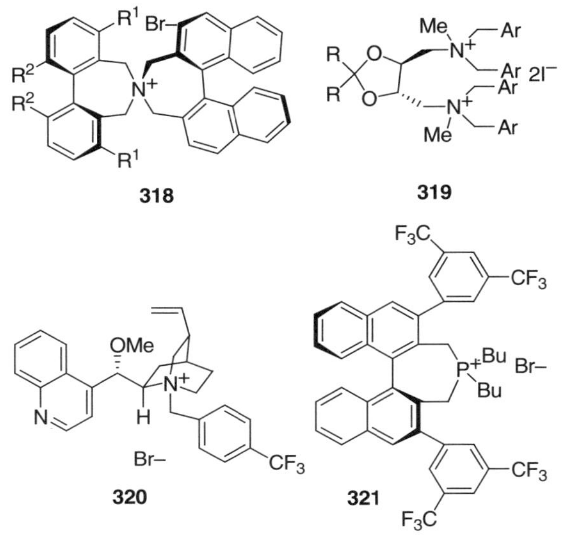

FIGURE 2.17   Examples of PTC catalysts **318–321**.

academic research and for industrial application. These reactions include epoxidation, alkylation, and deasymmetrization. Among PTCs, N-spirobianyl derivatives were reported to have remarkable selectivity and reactivity.

When an atropmeric catalyst such as **325** is utilized for the alkylation of glycine-derived Schiff bases **322** with sterically less demanding electrophiles such as allylic and proparglic bromides (**323**), the corresponding alkylated product **324** can be obtained in high *ee* up to 97% [97] (Scheme 2.56).

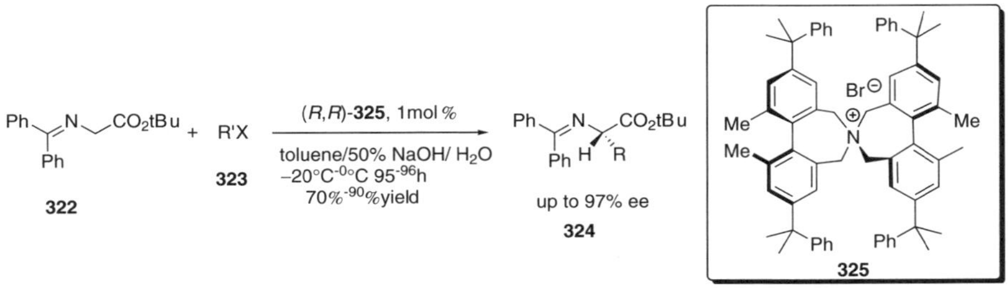

SCHEME 2.56   Asymmetric alkylation reaction.

The phosphonium PTC catalyst **321** could also be used in the asymmetric epoxidation of α, β-unsaturated ketones [98]. Another example is the construction of a quaternary chiral center in 3-arylindoles, which is a featured motif in a number of biological and pharmcentically active natural products and molecules [96] (Scheme 2.57).

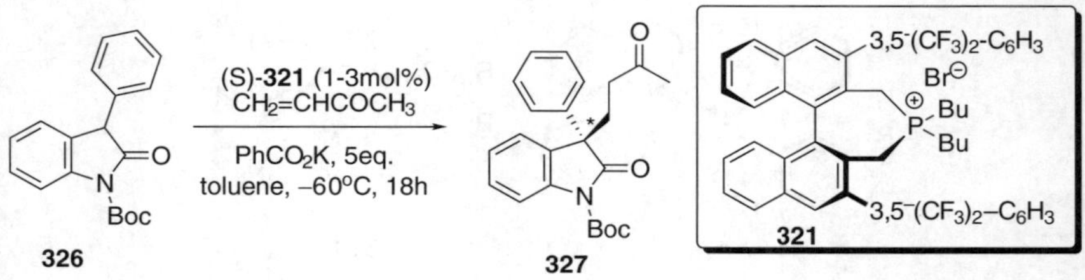

SCHEME 2.57 Example of asymmetric Michael addition catalyzed by PTC.

REFERENCES

1. Yoon, T. P., Jacobsen, E. N. (2003). Privileged chiral catalysts. *Science*, **299**, 1691–1693.

2. Knowles, W. S. (1986). Application of organometallic catalysis to the commercial production of L-DOPA. *J. Chem. Educ*. **63**, 222–225.

3. Akutagawa, S. (1997). Enantioselective isomerization of allylamine to enamine: practical asymmetric synthesis of (−)-menthol by Rh-BINAP catalysts. *Top. Catalysis* **4**, 271–274.

4. Deng, L., Jacobsen, E. N. (1992). A practical, highly enantioselective synthesis of the taxol side chain *via* asymmetric catalysis. *J. Org. Chem*. **57**, 4320–4323.

5. Zhang, W. C., Chi, Y. X., Zhang, X. M. (2007). Developing chiral ligands for asymmetric hydrogenation. *Acc. Chem. Res*. **40**, 1278–1290.

6. Blaser, H. U., Pugin, B., Spindler, F., Thommen, M. (2007). From a chiral switch to a ligand portfolio for asymmetric catalysis. *Acc.Chem. Res*. **40**, 1240–1250.

7. Fukatsu, K., Uchikawa, O., Kawada, M., Yamano, T., Yamashita, M., Kato, K., Hirai, K., Hinuma, S., Miyamoto, M., Ohkawa, S. (2002). Synthesis of a novel series of benzocycloalkene derivatives as melatonin receptor agonists. *J. Med. Chem*. **45**, 4212–4221.

8. Johnson, N. B., Lennon, I. C., Moran, P. H., Ramsden, J. A. (2007). Industrial-scale synthesis and applications of asymmetric hydrogenation catalysts. *Acc. Chem Res*. **40**, 1291–1299.

9. Knowles, W. S., Sabacky, M. J. (1968). Catalytic asymmetric hydrogenation employing a soluble, optically active, rhodium complex. *J. Chem. Soc. Chem. Commun*. **20**, 1445–1446.

10. Horner, L. Siegel, H., Büthe, H. (1968). Asymmetric catalytic hydrogenation with an optically active phosphinerhodium complex in homogeneous solution. *Angew. Chem. Int. Ed*. **7**, 942.

11. Kagan, H. B., Dang, T.P. (1972). Asymmetric catalytic reduction with transition metal complexes I. A catalytic system of rhodium (I) with (−)-2,3-o-isopropylidene-2,3-dihydroxy-1,4-bis(diphenylphosphino) butane, a new chiral diphosphine. *J. Am. Chem. Soc*. **94**, 6429–6433.

12. Imwinkelried, R. (1997). Catalytic asymmetric hydrogenation in the manufacture of d-biotin and dextromethorphan. *Chimia* **51**, 300–302.

13. a) Noyori, R., Ohkuma, T. (2001). Asymmetric catalysis by architectural and functional molecular engineering: practical chemo- and stereoselective hydrogenation of

ketones. *Angew.Chem. Int. Ed*. **40**, 40–73; b) Noyori, R. (2002). Asymmetric catalysis: science and opportunities (Nobel Lecture). *Angew. Chem. Int. Ed*. **41**, 2008–2022.

14. Blaser, H. U. (2002). The chiral switch of (*S*)-metrolachlor: a personal account of an industrial odyssey in asymmetric catalysis. *Adv. Synth. Catal*. **344**, 17–31.

15. Roseblade, S. J., Pfaltz, A. (2007). Iridium-catalyzed asymmetric hydrogenation of olefins. *Acc. Chem. Res*. **40**, 1402–1411.

16. a) Liu, D., Zhang, X. (2005). Practical P-chiral phosphane ligand for Rh-catalyzed asymmetric hydrogenation. *Eur. J. Org. Chem*. 2005, 646–649; b) Liu, D., Gao, W. Z., Wang, C. J., Zhang X. M. (2005). Practical synthesis of enantiopure γ-amino alcohols by rhodium-catalyzed asymmetric hydrogenation of β-secondary-amino ketones. *Angew. Chem., Int. Ed*. **44**, 1687–1689; c) Sun, X. F., Zhou, L., Wang, C. J., Zhang, X. M. (2007). Rh-catalyzed highly enantioselective synthesis of 3-arylbutanoic acids. *Angew. Chem., Int. Ed*. **46**, 2623–2626.

17. Ohkuma, T., Koizumi, M., Doucet, H., Pham, T., Kozawa, M., Murata, K., Katayama, E., Yokozawa, T., Ikariya, T., Noyori, R. (1998). Asymmetric hydrogenation of alkenyl, cyclopropyl, and aryl ketones. $RuCl_2$(xylbinap)(1,2-diamine) as a precatalyst exhibiting a wide scope. *J. Am. Chem. Soc*. **120**, 13529–13530.

18. Matsumoto, T., Murayama, T., Mitsuhashi, S., Miura, T. (1999). Diastereoselective synthesis of a key intermediate for the preparation of tricyclic β-lactam antibiotics. *Tetrahedron Lett*. **40**, 5043–5046.

19. a) Hirao, A., Itsuno, S., Nakahama, S., Yamazaki, N. (1981). Asymmetric reduction of aromatic ketones with chiral alkoxy-amineborane complexes. *J. Chem. Soc. Chem. Commun*. 315–317; b) Itsuno, S., Hirao, A., Nakahama, S., Yamazaki, N. (1983). Asymmetric synthesis using chirally modified borohydrides. Part I. Enantioselective reduction of aromatic ketones with the reagent prepared from borane and (*S*)-valinol. *J. Chem. Soc., Perkin Trans*. 1673–1677; c) Wallbaum, S., Martens, J. (1992). Asymmetric syntheses with chiral oxazaborolidines. *Tetrahedron Asymmetry* **3**, 1475–1504.

20. Quallich, G. J., Woodall, T. M. (1992). Synthesis of 4(*S*)-(3,4-dichlorophenyl)-3,4-dihydro-1(2H)-naphthalenone by S_N2 cuprate displacement of an activated chiral benzylic alcohol. *Tetrahedron* **48**, 10239–10248.

21. a) Wang, W. B., Lu, S. M., Yang, P. Y., Han, X. W., Zhou, Y. G. (2003). Highly enantioselective iridium-catalyzed hydrogenation of heteroaromatic compounds, quinolines. *J. Am. Chem. Soc*. **125**, 10536–10537; b) Lu, S. M., Han, X. W., Zhou, Y. G. (2004). Asymmetric hydrogenation of quinolines catalyzed by iridium with chiral ferrocenyloxazoline derived N, P ligands. *Adv. Synth. Catal*. **346**, 909–912; c) Lu, S. M., Han, X. W., Zhou, Y. G. (2007). An efficient catalytic system for the hydrogenation of quinolines. *J. Organometal. Chem*. **692**, 3065–3069;. d) Yang, P. Y., Zhou, Y. G. (2004). The enantioselective total synthesis of alkaloid (−)-galipeine. *Tetrahedron Asymmetry* **15**, 1145–1149.

22. Lennon, I. C., Ramsden, J. A., Brear, C. J., Broady, S. D., Muir, J. C. (2007). Asymmetric enamide hydrogenation in the synthesis of *N*-acetylcolchinol: a key intermediate for ZD6126. *Tetrahedron Lett*. **48**, 4623–4626.

23. Satoh, K., Inenaga, M., Kanai, K. (1998). Synthesis of a key intermediate of levofloxacin via enantioselective hydrogenation catalyzed by iridium (I) complexes. *Tetrahedron Asymmetry* **9**, 2657–2662.

24. a) Blaser, H. U., Spindler, F. (1997). Enantioselective catalysis for agrochemicals. The case histories of (*S*)-metolachlor (*R*)-metalaxyl and clozylacan. *Top. Catalysis*.

4, 275–282;   b) Stinson, S.C. (1997). Chiral drug market shows signs of maturity. *Chem. Eng. News* **75**, 38–70.

25. a) Cobley, C. J., Henschke, J. P. (2003). Enantioselective hydrogenation of imines using a diverse library of ruthenium dichloride (diphosphine)(diamine) precatalysts. *Adv. Synth. Catal*. **345**, 195–201;   b) Cobley, C. J., Foucher, E., Lecouve, J. P., Lennon, I. C., Ramsden, J. A., Thominot, G. (2003). The synthesis of S18986, a chiral AMPA receptor modulator, via catalytic asymmetric hydrogenation. *Tetrahedron Asymmetry* **14**, 3431–3433.

26. Hentges, S. G., Sharpless, K. B. (1980). Asymmetric induction in the reaction of osmium tetroxide with olefins. *J. Am. Chem. Soc*. **102**, 4263–4265.

27. Lohray, B. B. (1992). Recent advances in the asymmetric dihydroxylation of alkenes. *Tetrahedron Asymmetry* **3**, 1317–1349.

28. Griesbach, R. C., Hamon, D. P. G., Kennedy, R. J. (1997). Asymmetric dihydroxylation in an approach to the enantioselective synthesis of 2-arylpropanoic acid non-steroidal anti-inflammatory drugs. *Tetrahedron Asymmetry* **8**, 507–510.

29. Sharpless, K. B., Amberg, W., Bennani, Y. L., Crispino, G. A., Hartung, J., Jeong, K. S., Kwong, H. L., Morikawa, K., Wang, Z. M. (1992). The osmium-catalyzed asymmetric dihydroxylation: a new ligand class and a process improvement. *J. Org. Chem*. **57**, 2768–2771.

30. Wang, Z. M., Zhang, X. L., Sharpless, K. B. (1993). Asymmetric dihydroxylation of aryl alkyl ethers. *Tetrahedron Lett*. **34**, 2267–2270.

31. Fang, F. G., Xie, S., Lowery, M. W. (1994). Catalytic enantioselective synthesis of 20(*S*)-camptothecin: a practical application of the Sharpless asymmetric dihydroxylation reaction. *J. Org. Chem*. **59**, 6142–6143.

32. Fang, F. G., Bankston, D. D., Huie, E. M., Johnson, M. R., Kang, M. C., LeHoullier, C. S., Lewis, G. C., Lovelace, T. C., Lowery, M. W., McDougald, D. L., Meerholz, C. A., Partridge, J. J., Sharp, M. J., Xie, S. P. (1997). Convergent catalytic asymmetric synthesis of camptothecin analog GI 147211C. *Tetrahedron* **53**, 10953–10970.

33. Weissman, S. A., Rossen, K., Reider, P. J. (2001). Stereoselective synthesis of styrene oxides *via* a mitsunobu cyclodehydration. *Org. Lett*. **3**, 2513–2515.

34. Tan, L., Chen, C.Y., Larsen, R. D., Verhoeven, T. R., Reider, P. J. (1998). An efficient asymmetric synthesis of a potent COX-2 inhibitor L-784,512. *Tetrahedron Lett*. **39**, 3961–3964.

35. a) Sharpless, K. B., Patrick, D. W., Truesdale, L. K., Biller, S. A. (1975). New reaction. Stereospecific vicinal oxyamination of olefins by alkyl imido osmium compounds. *J. Am. Chem. Soc*. **97**, 2305–2307;   b) Herranz, E., Biller, S. A., Sharpless, K. B. (1978). Osmium-catalyzed vicinal oxyamination of olefins by N-chloro-N-argentocarbamates. *J. Am. Chem. Soc*. **100**, 3596–3598;   c) Herranz, E., Sharpless, K. B. (1990). Osmium-catalyzed vicinal oxyamination of olefins by chloramines-T: cis-2-(p-toluenesulfonamido) cyclohexanol and 2-methyl-3-(p- toluenesulfonamido)-2-pentanol. *Org. Syn. Coll*. **7**, 375;   d) Li, G., Chang, H. T., Sharpless, K. B. (1996). Catalytic asymmetric aminohydroxylation (AA) of olefins. *Angew. Chem. Int. Ed*. **35**, 451–454;   e) Rudolph, J., Sennhenn, P. C., Vlaar, C. P., Sharpless, K. B. (1996). Smaller substituents on nitrogen facilitate the osmium-catalyzed asymmetric aminohydroxylation. *Angew. Chem. Int. Ed*. **35**, 2810–2813.

36. Lee, J. C., Kim, G. T., Shim, Y. K., Kang, S. H. (2001). An asymmetric aminohydroxylation approach to the stereoselective synthesis of cis-substituted azetidinone of loracarbef. *Tetrahedron Lett*. **42**, 4519–4521.

37. Zhao, S. H., Samuel, O., Kagan, H. B. (1987). Asymmetric oxidation of sulfides mediated by chiral titanium complexes: mechanistic and synthetic aspects. *Tetrahedron* **43**, 5135–5144.

38. Bowden, S. A., Burke, J. N., Gray, F., McKown, S., Moseley, J. D., Moss, W. O., Murray, P. M., Welham, M. J., Young, M. J. (2004). A new approach to rapid parallel development of four neurokinin antagonists. Part 4. Synthesis of ZD 2249 methoxy sulfoxide. *Org. Process Res. Dev.* **8**, 33–44.

39. Hogan, P. J., Hopes, P. A., Moss, W. O., Robinson, G. E., Patel, I. (2002). Asymmetric sulfoxidation of an aryl ethyl sulfide: modification of Kagan procedure to provide a viable manufacturing process. *Org. Process Res. Dev.* **6**, 225–229.

40. Pitchen, P., France, C. J., McFarlane, I. M., Newton, C.G., Thompson, D. M. (1994). Large scale asymmetric synthesis of a biologically active sulfoxide. *Tetrahedron Lett.* **35**, 485–488.

41. Larsson, E. M., Stenhede, U. J., Sörensen, H., Von Unge, P. O. S., Cotton, H. (1996). Process for synthesis of substituted sulphoxides. World Patent, WO9602535.

42. Onoda, T., Shirai, R., Koiso, Y., Iwasaki, S. (1995). Synthetic study of curacin A: synthesis of partial structure of curacin A and elucidation of the absolute stereochemistry at the methylcyclopropylthiazoline moiety. *Tetrahedron Lett.* **36**, 5765–5768.

43. Seki, M., Kayo, A., Mori, K. (2001). Synthesis of (2*S*,3*R*,11*S*,12*R*,2‴,11‴,12‴*R*)-plakoside A, a prenylated and immunosuppressive marine galactosphingolipid with cyclopropane-containing alkyl chains. *Tetrahedron Lett.* **42**, 2357–2360.

44. Onoda, T., Shirai, R., Koiso, Y., Iwasaki, S. (1996). Synthetic study on curacin A: a novel antimitotic agent from the cyanobacterium *Lyngbya majuscula*. *Tetrahedron* **52**, 14543–14562.

45. Graham, D. W., Ashton, W. T., Barash, L., Brown, J. E., Brown, R. D., Canning, L. F., Chen, A., Springer, J. P., Rogers, E. F. (1987). Inhibition of the mammalian β-lactamase renal dipeptidase (dehydropeptidase-I) by Z-2-(acylamino)-3-substituted-propenoic acids. *J. Med. Chem.* **30**, 1074–1090.

46. Wang, Q., Yang, F., Du, H., Hossain, M. M., Bennett, D., Grubisha, D. S. (1998). The synthesis of *S*-(+)-2,2-dimethylcyploroane carboxylic acid: a precursor for cilastatin. *Tetrahedron: Asymmetry* **9**, 3971–3977.

47. a) Claver, C., Diéguez, M., Pàmies, O., Castillón, S. (2006). Asymmetric hydroformylation. *Top. Organomet. Chem.* **18**, 35–64;   b) Trost, B.M. (1991). The atom economy—a search for synthetic efficiency. *Science* **254**, 1471–1477; c) Trost, B.M. (1995). Atom economy—a challenge for organic synthesis: homogeneous catalysis leads the way. *Angew. Chem. Int. Ed.* **34**, 259–281; d) Breit, B., Wolfgang, S. (2001). Recent advances on chemo-, region- and stereoselective hydroformylation. *Synthesis* 1–36; e) Cuny, G. D., Buchwald, S. L. (1993). Practical, high-yield, regioselective, rhodium-catalyzed hydroformylation of functionalized, α-olefins. *J. Am. Chem. Soc.* **115**, 2066–2068.

48. a) Claver, C., Van Leeuwen, P. W. N. M., Editors. (2000). *Rhodium Catalyzed Hydroformylation*. Dordrecht, The Netherlands: Kluwer Academic Publishers;   b) Agbossou, F., Carpentier, J. F., Mortreux, A. (1995). Asymmetric hydroformylation. *Chem. Rev.* **95**, 2485–2506;   c) Gladiali, S., Bayón, J. C., Claver, C. (1995). Recent advances in enantioselective hydroformylation. *Tetrahedron Asymmetry* **6**, 1453–1474;   d) Diéguez, M., Pàmies, O., Claver, C. (2004). Recent advance in Rh-catalyzed asymmetric hydroformylation using phosphite ligands. *Tetrahedron Asymmetry* **15**, 2113–2122.

49. Cesarotti, E., Rimoldi, I. (2004). Stereoselective synthesis of 1-methylcarbapenem precursors: studies on the diastereoselective hydroformylation of 4-vinyl β-lactam with aminophosphonite-phosphinite and aminophosphine-phosphate rhodium (I) complexes. *Tetrahedron Asymmetry* **15**, 3841–3845.

50. Chen, G., Tokunaga, N., Hayashi, T. (2005). Rhodium-catalyzed asymmetric 1,4-addition of arylboronic acids to coumarins: asymmetric synthesis of (*R*)-tolterodine. *Org. Lett.* **7**, 2285–2288.

51. Mase, T., Houpis, I. N., Akao, A., Dorziotis, I., Emerson, K., Hoang, T., Iida, T., Itoh, T., Kamei, K., Kato, S., Kato, Y., Kawasaki, M., Lang, F., Lee, J., Lynch, J., Maligres, P., Molina, A., Nemoto, T., Okada, S., Reamer, R., Song, J. Z., Tschaen, D., Wada, T., Zewge, D., Volante, R. P., Reider, P. J., Tomimoto, K. (2001). Synthesis of a muscarinic receptor antagonist *via* a diastereoselective Michael reaction, selective deoxyfluorination and aromatic metal-halogen exchange reaction. *J. Org. Chem.* **66**, 6775–6786.

52. Song, Z. J., Zhao, M., Frey, L., Li, J., Tan, L., Chen, C. Y., Tschaen, D. M., Tillyer, R., Grabowski, E. J. J., Volante, R., Reider, P. J., Kato, Y., Okada, S., Nemoto, T., Sato, H., Akao, A., Mase, T. (2001). Practical asymmetric synthesis of a selective endothelin a receptor (ETA) antagonist. *Org. Lett.* **3**, 3357–3360.

53. Barnes, D. M., Ji, J. G., Fickes, M. G., Fitzgerald, M. A., King, S. A., Morton, H. E., Plagge, F. A., Preskill, M., Wagaw, S. H., Wittenberger, S. J., Zhang, J. (2002). Development of a catalytic enantioselective conjugate addition of 1,3-dicarbonyl compounds to nitroalkenes for the synthesis of endothelin-A. Antagonist ABT-546, scope, mechanism and further application to the synthesis of the antidepressant rolipram. *J. Am. Chem. Soc.* **124**, 13097–13105.

54. a) Dalko, P. I., Moisan, L. (2001). Enantioselective organocatalysis. *Angew. Chem. Int. Ed.* **40**, 3726–3748; b) Dalko, P. I., Moisan, L. (2004). In the golden age of organocatalysis. *Angew. Chem. Int. Ed.* **43**, 5138–5175; c) Houk, N. K., List, B. (2004). Asymmetric organocatalysis. *Acc. Chem. Res.* **37**, 487; d) List, B. (2004). Organocatalysis: a complementary catalysis strategy advances organic synthesis. *Adv. Synth. Catal.* **346**, 1021–1022; e) Seayad, J., List, B. (2005). Asymmetric organocatalysis. *Org. Biomol. Chem.* **3**, 719–724; f) Berkessel, A., Gröger, H., Eds. (2005). *Asymmetric Organocatalysis: From Biomimetic Concepts to Applications in Asymmetric Synthesis*. Weinheim: Wiley-VCH; g) List, B. (2006). The ying and yang of asymmetric aminocatalysis. *Chem. Commun.* 819–824; h) List, B., Yang, J. W. (2006). Chemistry: the organic approach to asymmetric catalysis. *Science* **313**, 1584–1586; i) Doyle, A. G., Jacobsen, E. N. (2007). Small-molecule H-bond donors in asymmetric catalysis. *Chem. Rev.* **107**, 5713–5743.

55. a) Hajos, Z. G., Parrish, D. R. (1971). Werkwijze voor de bereiding van 1,3-dioxycycloackanen. German Patent DE 2102623; b) Eder, U., Wiechart, R., Sauer, G. R. (1971). Process for the manufacture of optically active bicycloalkane derivatives. German Patent DE 2014757; c) Hajos, Z. G., Parrish, D. R. (1974). Asymmetric synthesis of bicyclic intermediates of natural product chemistry. *J. Org. Chem.* **39**, 1615–1621.

56. Northrup, A. B., MacMillan, D. W. C. (2002). The first direct and enantioselective cross-aldol reaction of aldehydes. *J. Am. Chem. Soc.* **124**, 6798–6799.

57. Pikho, P. M., Erkkilä, A. (2003). Enantioselective synthesis of prelactone B using a proline-catalyzed crossed- aldol reaction. *Tetrahedron Lett.* **44**, 7607–7609.

58. Brandau, S., Landa, A., Franzén, J., Marigo, M., Jørgensen, K. A. (2006). Organocatalytic conjugate addition of malonates to α, β-unsaturated aldehydes: asymmetric formal synthesis of (−)-paroxetine chiral lactams and lactones. *Angew. Chem. Int. Ed*. 45, 4305–4309.

59. Matsunaga, S., Yoshida, T., Morimoto, H., Kumagai, N., Shibasaki, M. (2004). Direct catalytic asymmetric Mannich-type reaction of hydroxyketone using a Et_2Zn/Linked-BINOL complex: synthesis of either *anti*- or *syn*-B-amino alcohols. *J. Am. Chem. Soc*. **126**, 8777–8785.

60. Paraskar, A. S., Sudalai, A. (2006). Enantioselective synthesis of (−)-cytoxazone and (+)-epi-cytoxazone, novel cytokine modulators via Sharpless asymmetric epoxidation and L-proline catalyzed Mannich reaction. *Tetrahedron* **62**, 5756–5762.

61. Itoh, T., Yokoya, M., Miyauchi, K., Nagata, K., Ohsawa, A. (2003). Proline-catalyzed asymmetric addition reaction of 9-tosyl-3,4-dihydro-B-carboline with ketones. *Org. Lett*. **5**, 4301–4304.

62. a) Wang, W., Wang, J., Li, H. (2005). Direct, highly enantioselective pyrrolidine sulfonamide catalyzed Michael addition of aldehydes to nitrostyrenes. *Angew. Chem., Int. Ed*. **44**, 1369–1371; b) Wang, J., Li, H., Lou, B., Zu, L., Guo, H., Wang, W. (2006). Enantio- and diastereoselective Michael addition reactions of unmodified aldehydes and ketones with nitroolefins catalyzed by a pyrrolidine sulfonamide. *Chem. Eur. J*. **12**, 4321–4332.

63. Aslanian, R., Lee, G., Iyer, R. V., Shih, N. Y., Piwinski, J. J., Draper, R. W., McPhail, A. T. (2000). All asymmetric synthesis of the novel H_3 agonist (+)-(3R,4R)-3-(4-imidazolyl)-4- methylpyrroline dihydrochloride (Sch 50971). *Tetrahedron: Asym*. **11**, 3867–3871.

64. Suri, J. T., Steiner, D. D., Barbas, C. F. III. (2005). Organocatalytic enantioselective synthesis of metabotropic glutamate receptor ligands. *Org. Lett*. **7**, 3885–3888.

65. Chowdari N. S., Barbas III, C. F. (2005). Total synthesis of LFA-1 antagonist BIRT-377 via organocatalytic asymmetric construction of a quaternary stereocenter. *Org. Lett*. **7**, 867–870.

66. a) Curci, R., Fiorentino, M., Serio, M. R. (1984). Asymmetric epoxidation of unfunctionalized alkenes by dioxirane intermediates generated from potassium peroxomonosulphate and chiral ketones. *J. Chem. Soc. Chem. Commun*. 155–156; b) Curci, R., D'Accolti, L., Fiorentino, M., Rosa, A. (1995). Enantioselective epoxidation of unfunctionalized alkenes using dioxiranes generated in situ. *Tetrahedron Lett*. **36**, 5831–5834.

67. a) Adam, W., Curci, R., Edwards. J. O. (1989). Dioxiranes: a new class of powerful oxidants. *Acc. Chem. Res*. **22**, 205–211; b) Murray, R. W. (1989). Chemistry of dioxiranes. 12. Dioxiranes. *Chem. Rev*. **89**, 1187–1201; c) Curci, R., Dinoi, A., Rubino, M. F. (1995). Dioxirane oxidation: taming the reactivity-selectivity principle. *Pure Appl. Chem*. **67**, 811–822; d) Adam, W., Smerz, A. K. (1996). Chemistry of dioxiranes: selective oxidations. *Bull. Soc. Chim. Belg*. **105**, 581–599; e) Adam, W., Saha-Möller, C. R., Zhao, C. G. (2002). Dioxirane epoxidation of alkenes. *Org. React*. **61**, 219–516.

68. a) Denmark, S. E., Wu, Z. (1999). The development of chiral, nonracemic dioxiranes for the catalytic, enantioselective epoxidation of alkenes. *Synlett*. 847–859; b) Frohn, M., Shi, Y. (2000). Chiral ketone-catalyzed asymmetric epoxidation of olefins. *Synthesis*. 1979–2000; c) Shi, Y. (2002). Chiral ketone-catalyzed asymmetric epoxidation of olefins. *J. Synth. Org. Chem. Jpn*. **60**, 342–349; d) Shi, Y. (2004). Organocatalytic

asymmetric epoxidation of olefins by chiral ketones. *Acc. Chem. Res*. **37**, 488–496; d) Yang, D. (2004). Ketone-catalyzed asymmetric epoxidation reaction. *Acc. Chem. Res*. **37**, 497–505.

69. Denmark, S. E., Wu, Z., Grudden, C. M., Matsuhashi H. (1997). Catalytic epoxidation of alkenes with Oxone. 2. Fluoro ketones. *J. Org. Chem*. **62**, 8288–8289.

70. a) Yang, D., Yip, Y. C., Tang, M. W., Wong, M. K., Zheng, J. H., Cheung, K. K. (1996). A C_2 symmetric chiral ketone for catalytic asymmetric epoxidation of unfunctionalized olefins. *J. Am. Chem. Soc*. **118**, 491–492;   b) Yang, D., Wang, X. C., Wong, M. K., Yip, Y. C. Tang, M. W. (1996). Highly enantioselective epoxidation of trans-stilbenes catalyzed by chiral ketones. *J. Am. Chem. Soc*. **118**, 11311–11312; c) Yang, D., Wong, M. K., Yip, Y. C., Wang, X. C., Tang, M. W., Zheng, J. H., Cheung, K. K. (1998). Design and synthesis of chiral ketones for catalytic asymmetric epoxidation of unfunctionalized olefins. *J. Am. Chem. Soc*. **120**, 5943–5952.

71. a) Seki, M., Yamada, S., Kuroda, T., Imashiro, R., Shimizu, T. (2000). A practical synthesis of C_2-symmetric chiral binaphthyl ketone catalyst. *Synthesis*, 1677–1680; b) Seki, M., Furutani, T., Hatsuda, M., Imashiro, R. (2000). Facile synthesis of C_2-symmetric chiral binaphthyl ketone catalysts. *Tetrahedron Lett*. **41**, 2149–2152;   c) Seki, M., Furutani, T., Imashiro, R., Kuroda, T., Yamanaka, T., Harada, N., Arakawa, H., Kusama, M., Hashiyama, T. (2001). A novel synthesis of a key intermediate for diltiazem. *Tetrahedron Lett*. **42**, 8201–8205;   d) Furutani, T., Imashiro, R., Hatsuda, M., Seki, M. (2002). A practical procedure for the large-scale preparation of methyl (2*R*,3*S*)-3-(4-methoxyphenyl)glycidate, a key intermediate for diltiazem. *J. Org. Chem*. **67**, 4599–4601.

72. a) Tu, Y., Wang, Z. X., Shi, Y. (1996). An efficient asymmetric epoxidation method for trans-olefins mediated by a fructose-derived ketone. *J. Am. Chem. Soc*. **118**, 9806–9807;   b) Shu, L. H., Shi, Y. (1999). Asymmetric epoxidation using hydrogen peroxide (H_2O_2) as primary oxidant. *Tetrahedron Lett*. **40**, 8721–8724;   c) Wang, Z. X., Cao, G. A., Shi, Y. (1999). Chiral ketone catalyzed highly chemo- and enantioselective epoxidation of conjugated enynes. *J. Org. Chem*. **64**, 7646–7650;   d) Warren, J. D., Shi, Y. (1999). Chiral ketone-catalyzed asymmetric epoxidation of 2,2-disubstituted vinylsilanes. *J. Org. Chem*. **64**, 7675–7677;   e) Frohn, M., Zhou, X. M., Zhang, J. R., Tang, Y., Shi, Y. (1999). Kinetic resolution of racemic cyclic olefins via chiral dioxirane. *J. Am. Chem. Soc*. **121**, 7718–7719; f) Shu, L. H., Shi, Y. (2001). An efficient ketone-catalyzed asymmetric epoxidation using hydrogen peroxide (H_2O_2) as primary. *Tetrahedron* **57**, 5213–5218;   g) Tian, H. Q., She, X. G., Xu, J. X., Shi, Y. (2001). Enantioselective epoxidation of terminal olefins by chiral dioxirane. *Org. Lett*. **3**, 1929–1931;   h) Zhu, Y. M., Shu, L. H., Tu, Y., Shi, Y. (2001). Enantioselective synthesis and stereoselective rearrangement of enol ester epoxides. *J. Org. Chem*. **66**, 1818–1826;   i) Tian, H. Q., She, X. G., Shi, Y. (2001). Electronic probing of ketone catalysts for asymmetric epoxidation. Search for more robust catalysts. *Org. Lett*. **3**, 715–718;   j) Wang, Z. X., Miller, S. M., Anderson, O. P., Shi, Y. (2001). Asymmetric epoxidation by chiral ketones derived from carbocyclic analogues of fructose. *J. Org. Chem*. **66**, 521–530; k) Zaks, A., Dodds, D. R. (1995). Chloroperoxidase-catalyzed asymmetric oxidation: substrate specificity and mechanistic study. *J. Am. Chem. Soc*. **117**, 10419–10424.

73. Tian, H. Q., She, X. G., Shu, L. H., Yu, H. W., Shi, Y. (2000). Highly enantioselective epoxidation of *cis*-olefins by chiral dioxirane. *J. Am. Chem. Soc*. **122**, 11551–11552.

74. a) Xiong, Z. M., Corey, E. J. (2000). Simple total synthesis of the pentacyclic C_s-symmetric structure attributed to the squalenoid glabrescol and three C_s-symmetric

diastereomers compel structural revision. *J. Am. Chem. Soc*. 122, 4831–4832; b) Simpson, G. L., Heffron, T. P., Merino, E., Jamison, T. F. (2006). Ladder polyether synthesis via epoxide-opening cascades using a disappearing directing group. *J. Am. Chem. Soc*. **128**, 1056–1057.

75. a) Schreiner, P. R. (2003). Metal-free organocatalysis through explicit hydrogen bonding interactions. *Chem. Soc. Rev*. **32**, 289–296;   b) Pihko, P. M. (2004). Activation of carbonyl compounds by double hydrogen bonding an emerging tool in asymmetric catalysis. *Angew. Chem., Int. Ed*. **43**, 2062–2064;   c) Bolm, C., Rantanen, T., Schiffers, I., Zani, L. (2005). Protonated chiral catalysts: versatile tools for asymmetric synthesis. *Angew. Chem. Int. Ed*. **44**, 1758–1763;   d) Pihko, P. M. (2005). Recent breakthroughs in enantioselective Brønsted base catalysis. *Lett. Org. Chem*. **2**, 398–403;   e) Taylor, M. S., Jacobsen, E. N. (2006). Asymmetric catalysis by chiral hydrogen-bond donors. *Angew. Chem. Int. Ed*. **45**, 1520–1543;   f) Akiyama, T., Itoh, J., Fuchibe, K. (2006). Recent progress in chiral Brønsted acid catalysis. *Adv. Synth. Catal*. **348**, 999–1010;   g) Akiyama, T. (2007). Stronger Brønsted acids. *Chem. Rev*. **107**, 5744–5758.

76. Akiyama, T., Itoh, J., Yokota, K., Fuchibe, K. (2004). Enantioselective Mannich-type reaction catalyzed by a chiral Brønsted acid. *Angew. Chem., Int. Ed*. **43**, 1566–1568.

77. Uraguchi, D., Terada, M. (2004). Chiral Brønsted acid-catalyzed direct mannich reactions via electrophilic activation. *J. Am. Chem. Soc*. **126**, 5356–5357.

78. Goss, J. M., Schaus, S. E. (2008). Enantioselective synthesis of SNAP-7941: chiral dihydropyrimidoal inhibitor of MCH1-R. *J. Org. Chem*. **73**, 7651–7656.

79. a) Kappe, C. O. (2000). Recent advances in the Biginelli dihydropyrimidine synthesis. new tricks from an old dog. *Acc. Chem. Res*. **33**, 879–888.   (b) Borowsky, B.; Durkin, M. M., Ogozalek, K., Marzabadi, M. R., DeLeon, J., Heurich, R., Lichtblau, H., Shaposhnik, Z., Daniewska, I., Blackburn, T. P., Branchek, T. A., Gerald, C., Vaysse, P. J., Forray, C. (2002). Antidepressant, anxiolytic and anorectic effects of a Melanin-concentrating hormone-1 receptor antagonist. *Nat. Med*. **8**, 825–830.

80. a) Rueping, M., Antonchick, A. P., Theissmann, T. (2006). A highly enantioselective Brønsted acid catalyzed cascade reaction: organocatalytic transfer hydrogenation of quinolines and their application in the synthesis of alkaloids. *Angew. Chem., Int. Ed*. **45**, 3683–3686;   b) Rueping, M., Antonchick, A. P., Theissmann, T. (2006). Remarkably low catalyst loading in Brønsted acid catalyzed transfer hydrogenation: enantioselective reduction of benzoxazines, benzothiazine and benzoxazinones. *Angew. Chem., Int. Ed*. **45**, 6751–6755.

81. Cheng, X., Vellalath, S., Goddard, R., List, B. (2008). Direct catalytic asymmetric synthesis of cyclic aminals from aldehydes. *J. Am. Chem. Soc*. **130**, 15786–15787.

82. Liu, H., Dagousset, G., Masson, G., Retailleau, P., Zhu, J. P. (2009). Chiral Brønsted acid-catalyzed enantioselective three–component Povarov reaction. *J. Am. Chem. Soc*. **131**, 4598–4599.

83. Uraguchi, D., Sorimachi, K., Terada, M. (2004). Organocatalytic asymmetric aza-Friedel-Crafts alkylation of furan. *J. Am. Chem. Soc*. **126**, 11804–11805.

84. Kang, Q., Zhao, Z. A., You, S. L. (2007). Highly enantioselective Friedel-Crafts reaction of indoles with imines by a chiral phosphoric acid. *J. Am. Chem. Soc*. **129**, 1484–1485.

85. Terada, M., Sorimachi, K. (2007). Enantioselective Friedel-Crafts reaction of electron-rich alkenes catalyzed by chiral Brønsted acid. *J. Am. Chem. Soc*. **129**, 292–293.

86. Jia, Y. X., Zhong, J., Zhu, S. F., Zhang, C. M., Zhou, Q. L. (2007). Chiral Brønsted acid catalyzed enantioselective Friedel-Crafts reaction of indoles and a-aryl enamides: construction of quaternary carbon atoms. *Angew. Chem. Int. Ed*. **46**, 5565–5567.

87. García-García, P., Lay, F., García-García, P., Rabalakos, C., List, B. (2009). A powerful chiral counteranion motif for asymmetric catalysis. *Angew. Chem. Int. Ed*. **48**, 4363–4366.

88. Tanyeli, C., Sünbül, M. (2005). Asymmetric synthesis of 1,4-amino alcohol ligands with a norbornene backbone for use in the asymmetric diethylzine addition to benzaldehyde. *Tetrahedron Asymmetry* **16**, 2039–2043.

89. a) Bolm, C., Schiffers, I., Atodiresei, I., Özcubukcu, S., Raabe, G. (2003). A novel asymmetric synthesis of highly enantiomerically enriched norbornane-type diamine derivatives. *New J. Chem*. **27**, 14;   b) Tanyeli, C., Özcubukcu, S. (2003). The first enantioselective synthesis of chiral norbornane-type 1,4-diamine ligand. *Tetrahedron Asymmetry* **14**, 1167–1170.

90. Atodiresei, I., Schiffers, I., Bolm, C. (2006). Asymmetric synthesis of chiral bisoxazolines and their use as ligands in metal catalysis. *Tetrahedron Asymmetry* **17**, 620–633.

91. a) Keen, S. P., Cowden, C. J., Bishop, B. C., Brands, K. M. J., Davies, A. J., Dolling, U. H., Lieberman, D. R., Stewart, G. W. (2005). Practical asymmetric synthesis of a non-peptidic $\alpha_v\beta_3$ antagonist. *J. Org. Chem*. **70**, 1771–1779;   b) Bishop, B. C., Brands, K. M. J., Cottrell, I. F., Cowden, C. J., Davies, A. J., Keen, S. P., Lieberman, D. R., Stewart. G. W. (2004). Preparation of nonanoic acid derivatives involves reacting a propionaldelyhde compound with a pentanoic acid compound followed by enone reduction and deprotection of the carboxyl and amino groups. *PCT Int. Appl*. WO 2004078109-A2.

92. a) De Clercq, P. J. (1997). Biotin: a timeless challenge for total synthesis. *Chem. Rev*. **97**, 1755–1792;   b) Seki, M. (2006). Biological significance and development of practical synthesis of biotin. *Med. Res. Rev*. **26**, 434–482.

93. Choi, C., Tian, S. K., Deng, L. (2001). A formal catalytic asymmetric synthesis of (+)-biotin with modified cinchona alkaloids. *Synthesis*, 1737–1741.

94. Chen, Y., Deng, L. (2001). Parallel kinetic resolution of monosubstituted succinic anhydrides catalyzed by a modified cinchona alkaloid. *J. Am. Chem. Soc*. **123**, 11302–11303.

95. a) Mazzini, C., Lebreton, J., Alphand, V., Furstoss, R. (1997). A chemoenzymic strategy for the synthesis of enantiopure (R)-(−)-baclofen. *Tetrahedron Lett*. **38**, 1195–1196;   b) Schoenfelder, A., Mann, A., LeCoz, S. (1993). Enantioselective synthesis of (R)-(−)-baclofen. *Synlett*, 63–64.

96. He, R. J., Ding, C. H., Maruoka, K. (2009). Phosphonium salts as chiral phase-transfer catalysts: asymmetric Michael and Mannich reaction of 3-aryloxindoles. *Angew. Chem., Int. Ed*. **48**, 4559–4561.

97. Ooi, T., Kubota, Y., Maruoka, K. (2003). A new N-spiro C_2-symmetric chiral quaternary ammonium bromide consisting of 4,6-disubstituted biphenyl subunit as an efficient chiral phase-transfer catalysts. *Synlett*, 1931–1933.

98. Ooi, T., Ohara, D., Tamura, M., Maruoka, K. (2004). Design of new chiral phase-transfer catalysts with dual functions for highly enantioselective epoxidation of α, β-unsaturated ketones. *J. Am. Chem. Soc*. **126**, 6844–6845.

CHAPTER 3

CHIRAL DRUGS VIA BIOCATALYTICAL APPROACHES

JIANQIANG WANG

ArQule Inc., Woburn, MA

WENYA LU

Department of Chemistry, Iowa State University, Ames, Iowa

Chiral Drugs: Chemistry and Biological Action, First Edition. Edited by Guo-Qiang Lin, Qi-Dong You and Jie-Fei Cheng.
© 2011 John Wiley & Sons, Inc. Published 2011 by John Wiley & Sons, Inc.

3.1   INTRODUCTION

Tremendous progress in the field of asymmetric synthesis, in particular, asymmetric synthesis catalyzed by organometallic catalysts, organocatalysts, or biocatalysts, has taken place in the past two decades [1]. Aside from chiral resolution or synthesis from chiral materials (chiral pool), more and more chiral active pharmaceutical ingredients or intermediates are now being prepared through asymmetric synthesis. Biocatalysts are biomolecules that catalyze or accelerate certain types of chemical reactions in a way similar to other types of small molecule organic catalyst. They are commonly found as enzymes, biocatalytically active cells from organisms such as microorganisms, plants, animals, or humans. The reactions catalyzed by biocatalysts are referred to as biotransformation or biocatalysis or enzymatic reactions.

Biocatalysis has some obvious advantages over chemocatalytic reactions. It generally provides products with superior chemo-, regio-, diastereo-, and enantioselectivities. This enables access to difficult target molecules that might otherwise require multiple protection and deprotection steps with conventional chemical synthetic approaches. Biocatalytical reactions can be carried out under mild conditions, for example, at ambient temperature and atmospheric pressure and at about neutral pH value, which minimizes problems with isomerization, epimerization, racemization, rearrangement, decomposition, and other undesired side reactions that often plague traditional chemical reactions.

Last but not least, the biocatalytical reactions are often referred to as green or environmentally friendly processes. The catalyst itself can be degraded without generating harmful materials. The reaction is mostly run in aqueous media.

Limitations of biocatalysis are also obvious. Instability of the biocatalysts is one of the major problems in current biocatalysis. Enzymes are sensitive to environmental changes in pH, temperature, and physical forces and can undergo degradation gradually even under optimum conditions. Immobilization of enzyme on supports (carriers) or enzyme conformation stabilization through protein and genetic engineering are some of the currently available approaches to overcome the instability issue. Nonconventional media also work sometimes.

Biocatalysts are substrate dependent and therefore are not readily available for application to diverse nonnatural substrates, although about 4,000 different enzymes have been identified and many of these have found their ways into biotechnological and industrial applications. Many available enzymes do not withstand industrial reaction conditions and have to be further optimized as to volumetric productivity or activity to meet the criteria of industrial use. It requires tremendous effort to identify a new enzyme and to optimize the reaction conditions including scope of substrates, which makes biocatalysis highly costly and time consuming. However, these limitations are significantly minimized with the recent technological breakthroughs on large-scale DNA sequencing and microscale high-throughput screening. When these techniques have been combined with recombinant DNA technology and protein engineering-based methodologies (for example, using directed evolution including error-prone PCR and/or DNA shuffling to generate diversified DNA and enzyme libraries) and computer-aided enzyme design, the cycle of enzyme discovery has been dramatically shortened.

Finally, there is some limitation on reaction types as compared to traditional organic reactions, although in theory enzymes could catalyze a whole range of reactions. Use of enzymatic catalysis is therefore a complementary approach to traditional chemistry in the context of chiral drug synthesis.

Biocatalysis can be executed in many ways, in which employing crude or purified enzymes or whole cells is quite common. In recent years there has been growing interest in metabolic engineering of biocatalysis. Metabolic engineering is a term related to the whole cell biotransformation process. The traditional concept of metabolic engineering was defined [1h] as the directed improvement of product formation in cells or cellular properties through the modification of specific biochemical reactions or the introduction of new reactions with the use of recombinant DNA technology. The important novel aspect of metabolic engineering is the emphasis that it places on integrated metabolic pathways as opposed to individual reactions inside living and growing cells [1i]. Thus, metabolic engineering will be able to accomplish de novo metabolic pathways within cells by silencing certain genes, by overexpressing other genes, or by introducing new genes. And such novel pathways enable cells intake substrates and execute multiple step enzymatic reactions inside cells and ultimately output new products of interest. However, these products may not be common in nature. Such a process is also called biosynthesis or bioconversion.

Currently, engineered microorganisms with a diverse set of modified or nonnative enzyme activities are being used both to generate novel products and to provide improved processes for the manufacture of established products, such as in the production of precursors, intermediates, and complete compounds of importance to the pharmaceutical industry, including polyketides, nonribosomal peptides, steroids, vitamins, nonnatural amino acids, and antibiotics, etc. [1i]. For more discussion on metabolic engineering and bioconversion, and their application in the pharmaceutical industry, see Section 3.3.

3.2 CHIRAL DRUGS VIA BIOCATALYTICAL APPROACHES

The application of biocatalysis in the pharmaceutical industry has become a rapidly growing area by taking the advantage of the latest breakthroughs in genomics and proteomics. More and more biocatalytical approaches have been demonstrated to be uniquely efficient in the production of chiral drugs or pharmaceutical intermediates. This chapter intends to provide a snapshot of the current development of biocatalytical approaches and their application in chiral drug discovery, development, and production.

3.2.1 Oxidoreductases

3.2.1.1 Atazanavir (Reyataz, BMS-232632)
Atazanavir **7** is an antiviral drug for the treatment of HIV infection via inhibiting HIV-1 protease. HIV-1 protease is required for the proteolytic cleavage of the viral polyprotein precursors into the individual functional proteins found in infectious HIV-1. Atazanavir binds to the protease active site and inhibits the activity of the enzyme. This inhibition prevents cleavage of the viral polyproteins, resulting in the formation of immature noninfectious viral particles. Atazanavir is the first protease inhibitor approved for once-daily dosing rather than requiring multiple doses per day. It has lesser effects on the patient's lipid profile and may also not be cross-resistant with other protease inhibitors. (2S, 3R)-epoxide **4** is a key intermediate in the chemical synthesis of atazanavir [2] as illustrated in Scheme 3.1. The synthesis of **4** begins with dihydroxylamino derivative **1** and requires several protection and deprotection steps for hydroxyl groups.

An alternative and short synthesis of epoxide **4** was developed by Patel and co-workers [3] vial biocatalytical approach (Scheme 3.2). They used microbial reduction of (S)-[3-chloro-2-oxo-1-(phenylmethyl)propyl] carbamic acid-1,1-dimethylethyl ester **8** followed by ring closure to generate the desired epoxide **4**. Over 120 microorganisms containing alcohol oxidoreductases were screened against the reduction of chloroketone **8** and the results indicated three strains of *Rhodococcus* and *Brevibacterium sp.* SC 16101 (Table 3.1) gave good overall yield and stereoselectivity. *Rhodococcus erythropolis* SC 13845 was then selected and further optimized for the reaction. As a result, the alcohol **9** was obtained in 95% yield with diastereomeric excess of 98.2% and enantiomeric excess of 99.4% (Table 3.2). In contrast, chemical reduction of the chloroketone **8** by NaBH$_4$ primarily resulted in the undesired chlorohydrins.

The (S)-*tert*-leucine moiety is also required for the synthesis of atazanavir (Scheme 3.1). In 1995, scientists at Degussa [4] developed a biocatalytical

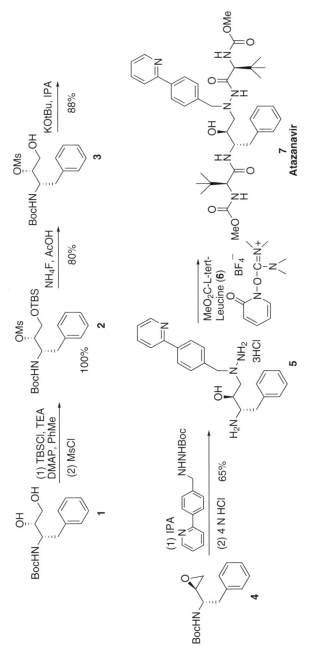

SCHEME 3.1 Chemical synthetic process of atazanavir 7.

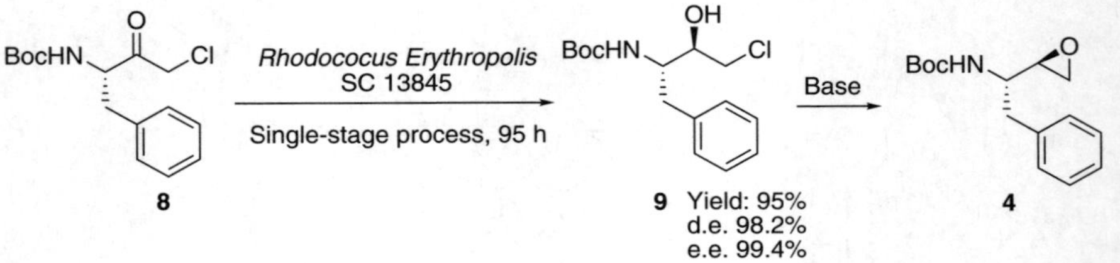

SCHEME 3.2 Biocatalytical approach to epoxide **4**.

TABLE 3.1 Diastereoselective Microbial Reduction of Ketone 8 by Cell Suspensions

Microorganisms	Yield of **9** (%)	de (%)
Brevibacterium sp. SC 16101	99	94
Hansenula anomala SC 13833	33	98
Hansenula saturnus SC 13829	35	98.5
Rhodococcus erythropolis SC 13845	98	98.9
Rhodococcus erythropolis SC 16236	74	98
Rhodococcus sp. SC 13810	91	98
Rhodococcus sp. SC 16002	97	98
Trichoderma viridue SC 13826	12	98

Cells of various microorganisms were suspended separately in 70 mM potassium phosphate buffer (pH 7.0) at 20% (w/v, wet cells) cell concentration and supplemented with 1 mg/ml of **8** and 30 mg/ml of glucose. Reduction was conducted at 28°C and 150 rpm. Periodically, samples of 1 ml were taken and extracted with 5 ml of *tert*-butylmethylether:toluene (60:40). After centrifugation, the organic phase was collected and evaporated with a nitrogen stream. The oily residue obtained was dissolved in 1 ml of ethanol, filtered through a 0.2-mm LID/X filter, and analyzed by HPLC.

TABLE 3.2 Diastereoselective Microbial Reduction of Ketone 8 by a Single-Stage Process

Microorganisms	Time (h)	Yield of **9** (%)	de (%)	ee (%)
Rhodococcus erythropolis SC 13845	21	38		
	48	52		
	72	72		
	93	95	98.2	99.4

synthesis of (S)-*tert*-leucine (**11**) on a ton scale from trimethylpyruvic acid **10** (Scheme 3.3). This process uses a combination of L-leucine dehydrogenase, formate dehydrogenase, and cofactors ammonia and NADH. NAD produced during the reaction is converted back to NADH by a well-established regeneration system, involving formate dehydrogenases from *Candida boidinii*, which oxidizes formate irreversibly to CO_2 and thereby reduces the cofactor NAD^+ to NADH.

3.2.1.2 Indinavir (Crixivan)

Indinavir **17** (Crixivan) is an HIV protease inhibitor developed by Merck for HIV infection. As shown in Scheme 3.4, (1S, 2R)-indene oxide **14** and (−)-*cis*-(1S,

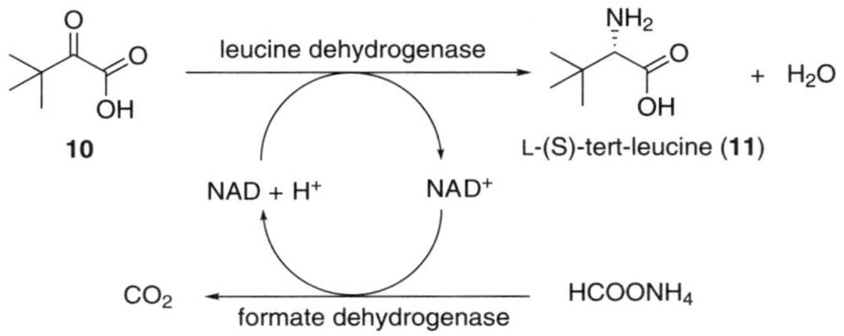

SCHEME 3.3 Industrial process for preparation of L-(*S*)-*tert*-leucine.

2*R*)-1-aminoindan-2-ol **16** are two important chiral intermediates for Crixivan synthesis. Using readily available indene **12** as the starting material, the current chemical process can afford the desired indene oxide **14** in about 87% *ee* (600-kg scale), using 0.7% Jacobsen's [5] (*S,S*)-salen manganese catalyst **13** at 5°C [6]. (1*S*,2*R*)-*cis*-amino indanol **16** was obtained from indene **12** in 50% overall yield and 99% *ee*.

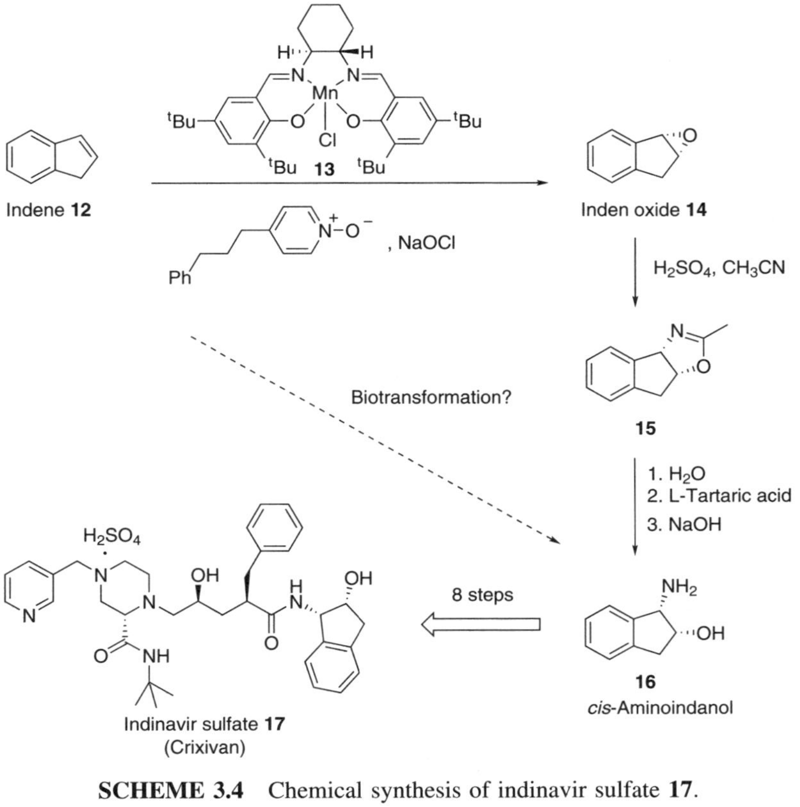

SCHEME 3.4 Chemical synthesis of indinavir sulfate **17**.

Efforts have been being made to apply green process for the synthesis of Crixixan yet haven't got any better results than chemical process does. For example, an epoxide hydrolase from *Diplodia gossypina* could convert racemic epoxide **14** into optically pure epoxide along with the dihydroxyl indandiol through a kinetic resolution (Scheme 3.5). The desired enantiomer (1*S*, 2*R*)-indene oxide **14** was obtained in >99.9% *ee* but with low yield (14%), while indandiol **18** was obtained as a by-product [7].

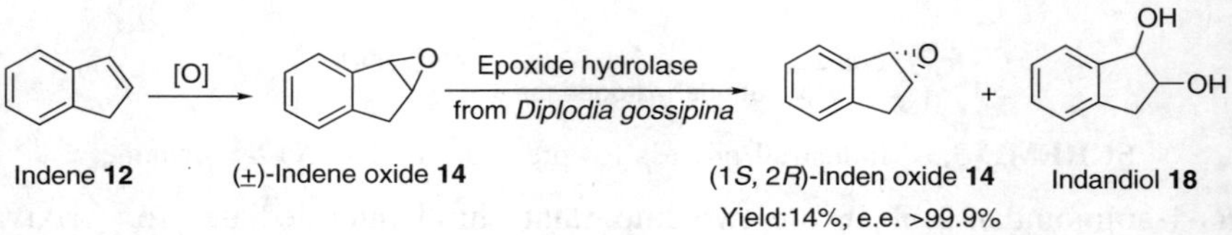

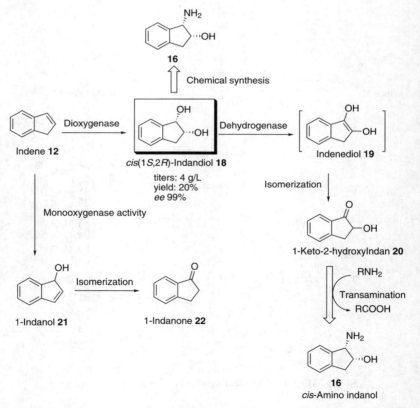

SCHEME 3.5 Enzymatic resolution to prepare (1*S*, 2*R*)-inden oxide **14**.

Scientists at Merck investigated the use of the toluene dioxygenase (TDO) system to synthesize the *cis*-indandiol **18** [8], which can be converted into *cis*-amino indanol **16** via chemical synthesis (Scheme 3.6). *Rhodococcus* was found

SCHEME 3.6 Microbial oxidation to prepare (1*S*, 2*R*)-indandiol **16**.

to be effective in the conversion of indene **12** to **18** and more resistant to the toxic effect of indene **12**. It does not require toluene for induction of the dioxygenase activity. However, both rates of bioconversion and product yields were lower in *Rhodococcus sp*. It was also observed that the dihydroxylation was accompanied by a presumed monohydroxylation reaction, generating 1-indenol **21** and 1-indanone **22** as by-products. The dihydroxylation product, *cis*-indandiol **18**, was further oxidized to the 1-keto-2-hydroxy indan (**20**, 1,2-indenedioltautomer). Fermentations in fed-batch mode at the 23-l scale utilizing *Rhodococcus* sp. B264-1 (MB 5655) yielded titers over 4 g/l of *cis*(1*S*, 2*R*)-indandiol **18** with >99% *ee* at 150 h. Accumulation yields of indandiol were relatively low at approximately 20% of the used indene. MB5655 converted indene to both the *cis*-indandiol **18** and the 1-indenol **21** in a constant ratio of 4:1. The ratio of *cis*-indandiol accumulation to indene consumption was 0.21.

Efforts were also made by use of metabolic engineering approach and a directed evolution technique to avoid the side reactions, to block degradative pathways, and to enhance the key reaction of converting indene to the *cis*-indanediol **18** [9]. A hypothetical engineered microorganism was proposed as having the potential to bioconvert indene to the *cis*-amino indanol via the coordinated activity of three stereospecific enzymes: a dioxygenase, a dehydrogenase, and a transaminase. Metabolic engineering used for constructing de novo metabolic pathways would provide opportunities [10] for the development of biosynthesis of the *cis*-aminoindanol **16** from indene **12**.

3.2.2 Transferases

3.2.2.1 Ribavirin (Rebetol, Copegus, Ribasphere)

Ribavirin **26** has broad-spectrum antiviral activity against influenzas, flaviviruses, HIV, polio, measles, smallpox, HBV, and HCV. Scheme 3.7 [11] demonstrates one-step microbial transformation of orotidine to ribavirin **26** by *E. carotovora* AJ 2992. The mechanism study indicated that *E. carotovora* AJ 2992 contains both orotidine-phosphorolyzing enzyme and purine nucleoside phosphorylase (EC 2.4.2.1, PNPase). These two enzymes worked jointly to generate ribavirin **26** from orotidine **23** and 1,2,4-triazole-3-carboxamide (TCA, **24**), while each enzyme alone could not complete the process. That is, orotidine-phosphorylase catalyzed phosphorolyzing orotidine **23** to orotic acid **25** and ribose-1-phosphate **28**, but could not produce ribavirin **26** in the presence of TCA **24** (Scheme 3.8). PNPase (EC 2.4.2.1) was able to catalyze the formation of ribavirin **26** from ribose-1-phosphate **28** in the presence of TCA **24**, but was unable to split orotidine **23**. Unlike other nucleoside phosphorylases that catalyze the reversible reaction(s), for example, uridine phosphorolysis catalyzed by uridine phosphorylase (EC2.4.2.1) (Scheme 3.8) is reversible, orotidine-phosphorylase from *E. carotovora* AJ 2992 was found to irreversibly catalyze the phosphorolyzation of orotidine **23** to orotic acid **25** and ribose-1-phosphate **28**. This might imply that orotic acid **25** formed by orotidine phosphorolysis might not be able to enter the active center of the enzyme, whereas from uracil uridine phosphorolysis is able to enter.

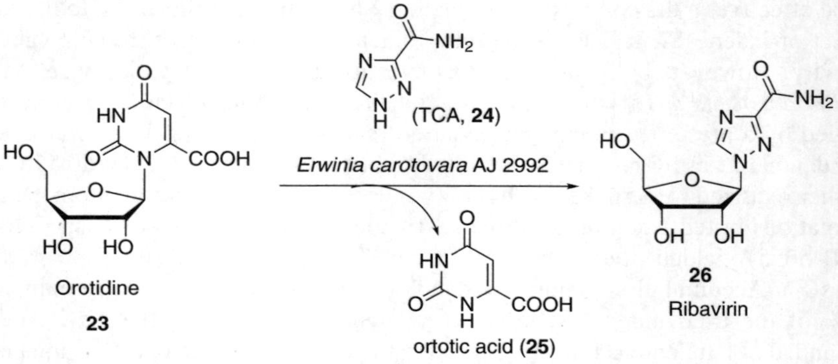

SCHEME 3.7   Microbial production of ribavirin **26** directly from orotidine.

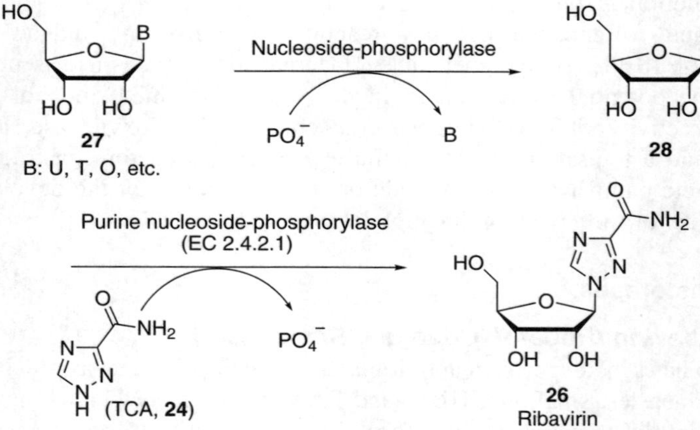

SCHEME 3.8   Microbial transformation of pyrimidine nucleosides to ribavirin **26**.

Ribavirin **26** is a powerful antiviral agent used in combination with α-2β interferon to treat hepatitis C, while it shows dose-dependent hemolysis and anemia. This might be ascribed to a long half-life in the body, caused by trapping of the compound by red blood cells (RBCs) and release of the compound at the end of life of the RBCs. To improve its pharmacokinetics profile and reduce the side effects, two pro-drugs, alanine ester **29** and amidine **30** (Taribavirin), are under development (Fig. 3.1). The ribavirin alanine ester **29** demonstrated improved bioavailability and reduced side effects and is currently in clinical trials. Amidine **30** (Taribavirin), currently in phase III clinical trials, could be a potential pro-drug because it showed less erythrocyte trapping with its basic amidine group, which inhibits drug entry into RBCs.

The biotransformation has been successfully developed in the preparation [12] of the pro-drug intermediate **29** as shown in Scheme 3.9. Chirazyme L-2 (*Candida antarctica* lipase B) acetylates **26** with the oxime ester **32** to form **33** regioselectively. As a comparison, direct chemical acylation of the compound **26** gave a

29

Ribavirin's alanine ester prodrug

30

Taribavirin (ICN 3142) (Phase III)

Ribavirin's 3-carboxamidine prodrug

FIGURE 3.1   Ribavirin pro-drugs under development.

mixture of mono-, bi-, and triacylated products. In fact, although a cost-effective chemical process was developed later, the biocatalytical approach in Scheme 3.9 did benefit the fast delivery of **29** at the beginning of project for initial drug testing, and the know-how obtained through such a biocatalytical process helped a lot on the successful development of chemical process [13].

SCHEME 3.9   Chemo-enzymatic synthesis of the ribavirin pro-drug **29**.

3.2.3 Hydrolases

3.2.3.1 Pregabalin (Lyrica)

Pregabalin **37** is a central nervous system (CNS) drug for neuropathic pain from diabetic neuropathy or postherpetic neuralgia. It is also used for the treatment of fibromyalgia and generalized anxiety disorder. Pregabalin acts via binding to the 2-subunit of the voltage-dependent calcium channel in the CNS, reducing calcium influx into the nerve terminals. It also decreases the release of neurotransmitters such as glutamate and increases neuronal GABA (γ-aminobutyric acid) level by producing a dose-dependent increase of glutamic acid decarboxylase activity.

There are many reports on the synthesis of pregabalin **37**. Main strategies include chemical resolution of racemic pregabalin [14], asymmetric synthesis [15], and biocatalytic resolution of key intermediates (Schemes 3.10 and 3.11). An improved process for manufacture [16] of the pregabalin **37** was based on a dynamic lipase kinetic resolution of the racemic compound **34** as shown in Scheme 3.10. The process is run in aqueous media, and no purification step is required for each intermediate. Furthermore, the undesired resolution by-product (3R)-**34** was recycled through racemization, which made the theoretical resolution yield up to 100% rather than 50% as seen in the conventional resolution process.

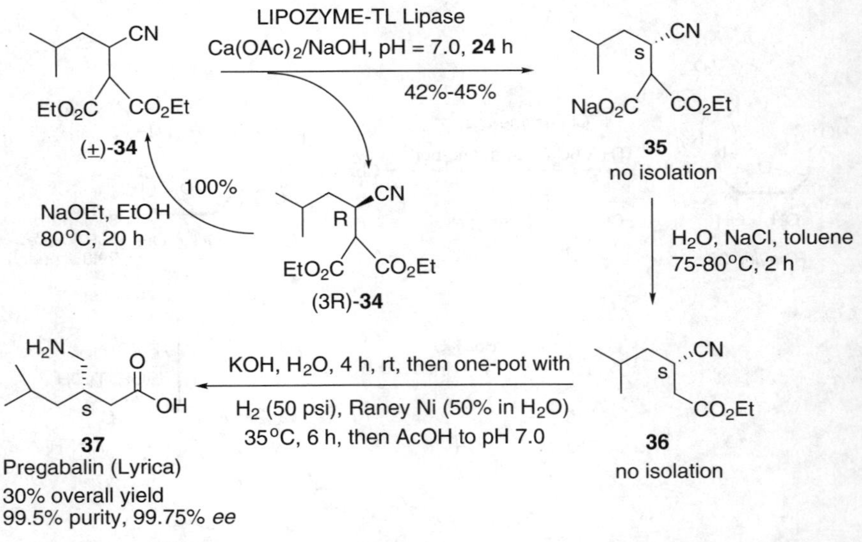

SCHEME 3.10   Chemo-enzymatic synthesis of the pregabalin **37** from racemic **34**.

Another biosynthetic route involved a regio- and stereoselective bioconversion of the aliphatic dinitrile **41** into the corresponding cyanocarboxylic acid **42** (Scheme 3.11). This strategy is similar to the route described above, except that nitrilase was used for the dynamic kinetic resolution [17].

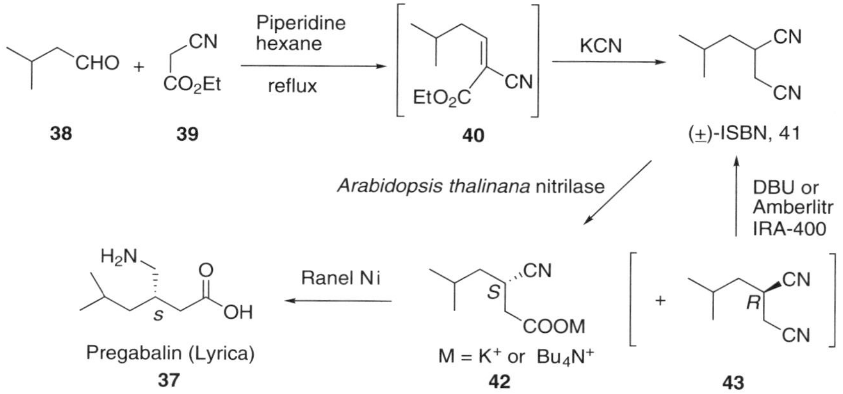

SCHEME 3.11   Chemo-enzymatic synthesis of pregabalin **37** from racemic **41**.

3.2.3.2 Diltiazem (Cardizem, Tiazac, Cartia XT, Dilacor XR)

Diltiazem **53** has antihypertension, anti-angina pectoris, and antiarrhythmic effects by blocking benzothiazepine calcium channel (calcium ion antagonist) to reduce calcium influx. It is a class III agent that has a negligible inotropic effect and causes almost no reflex tachycardia.

There are many approaches [18] for the synthesis of diltiazem hydrochloride. Scheme 3.12 summarizes a chemo-enzymatic synthesis [18d,e] of diltiazem. Thus key intermediate *trans*-(2*R*, 3*S*)-ethyl (*p*-methoxyphenyl) glycidate **47** was obtained in acceptable selectivity and yield via lipase-catalyzed resolution of the racemic *trans*-ethyl (*p*-methoxyphenyl) glycidate **46**. *Mucor meihei* lipase was found to selectively hydrolyze the ester functionality in (2*S*, 3*R*)-**46** into its corresponding acid **48** and leave **47** as the desired product. Reaction proceeded with biphasic media in the presence of toluene and Tris HCl buffer at pH 7.8. However, since acid **48** underwent automatic decarboxylation to yield aldehyde **49** that caused both deactivation of enzyme and problems in purification, Na-bisulphite was therefore added to reaction medium to react with aldehyde **49** and an insoluble hydroxy sulfite salt **50** was formed, which was removed by filtration. Since the regular resolution only gave 50% maximum yield. Chemical process via asymmetric epoxidation of *trans*-*p*-methoxycinnamate **44** catalyzed by organocatalyst was later developed (Scheme 3.12). The desired intermediate **47** was obtained in 89% yield and 78% *ee*. Recrystallization raises its *ee* to >99% in 64% overall yield.

3.2.3.3 Abacavir (Ziagen)

Abacavir **65** is a nucleoside analog reverse transcriptase inhibitor (NRTI) developed by GSK for treatment of HIV infection (AIDS) and hepatitis B (HBV) infection.

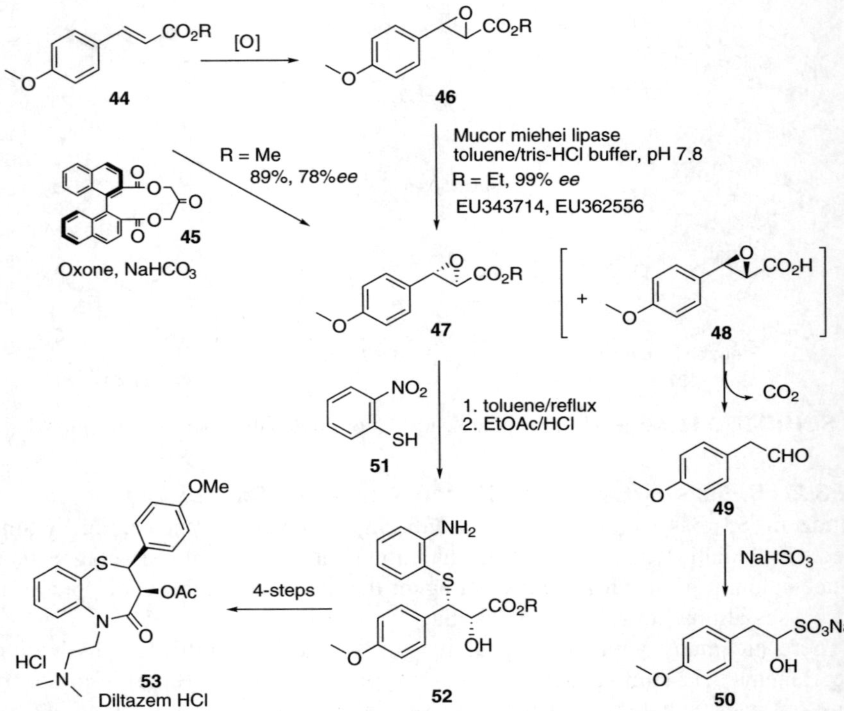

SCHEME 3.12 Chemo-enzymatic synthesis of diltiazem hydrochloride **53** via *trans*-**47**.

Chemo-enzymatic synthesis [19] of abacavir **65** is shown in Scheme 3.13. In 1999, Taylor and co-workers reported a direct resolution of racemic γ-lactam **55** with γ-lactamase to obtain the chiral key intermediate (−)-(1*R*,4*S*)-γ-lactam **55** on an industrial scale. They identified, cloned, and overexpressed a γ-lactamase gene from *Comamonas acidovorans* that was panned out by a novel screening method with classical molecular biology techniques [20]. Overexpression of the γ-lactamase was performed in recombinant *E. coli* in a 500-l fermentor via a 4-day fermentation. Approximately 100 g/l (wet weight) of cell paste (or cell biomass) and 3,000 units/g cell paste were yielded, sufficient to resolve about 5 tons of racemic substrate. The biotransformation with the obtained semipurified γ-lactamase was optimized for high substrate concentration (500 g/l) and high enzyme loading in 100 mM Tris HCl at pH 7.5, 25°C.

Among the indirect approaches by resolution of racemic *N*-substituted γ-lactams, the strategy developed by Mahmoudian and coworkers [21] in 1999 using racemic *N*-Boc-γ-lactam **56** as substrate is noteworthy. A commercially available hydrolytic enzyme, savinase (subtilisin, EC 3.4.21.62, serine-type protease), was used to efficiently resolve racemic **56**. (−)-**56** was obtained with 50% conversion and >99% *ee* when the reaction was run at 100 g/l of racemic **56** in phosphate buffer (pH 8.0) containing up to 50% (v/v) tetrahydrofuran. The

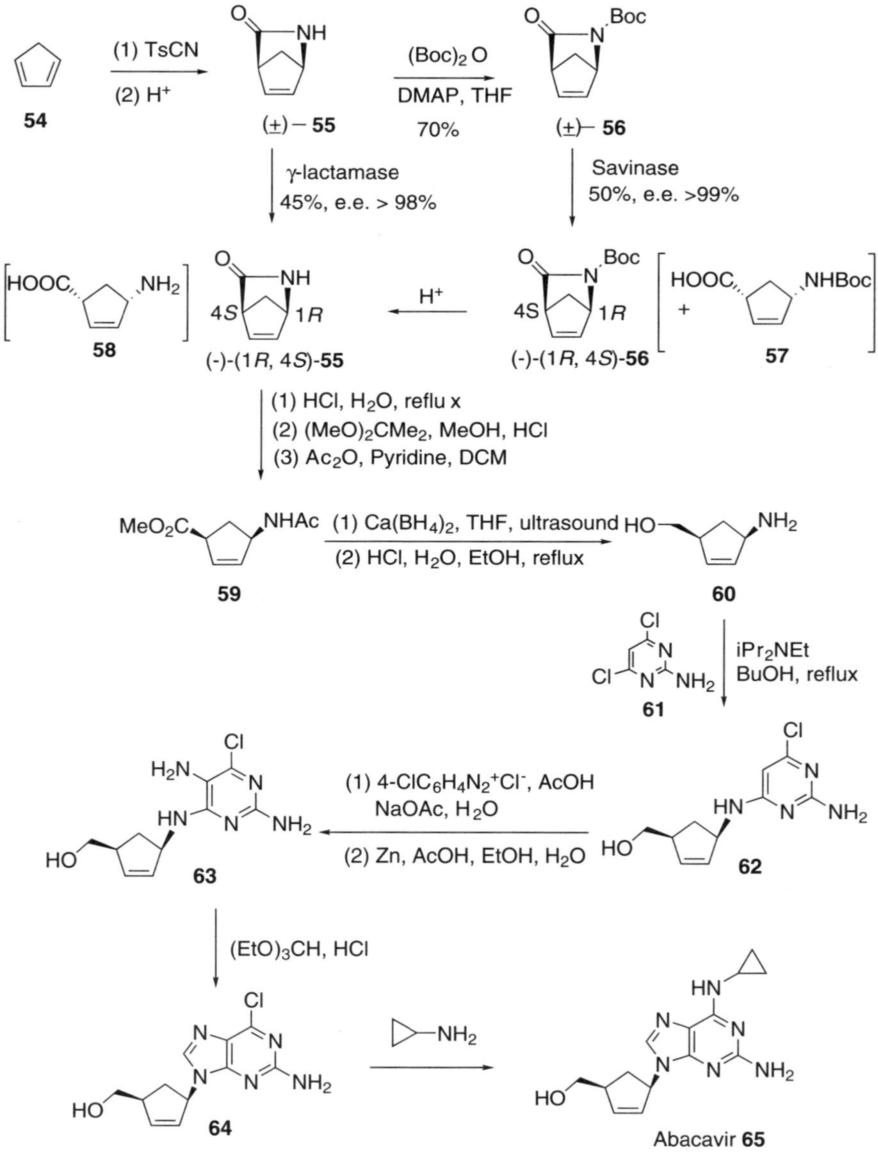

SCHEME 3.13 Chemo-enzymatic synthesis of abacavir **65**.

addition of tetrahydrofuran was aimed at minimizing the chemical hydrolysis of the N-Boc protecting group under aqueous conditions.

3.2.4 Lyases

3.2.4.1 (−)-Ephedrine (Ephedrine Sulfate)

Ephedrine **69** is an active ingredient in some traditional Chinese medicines, such as the herb Ma Huang, which has been used in treatment of asthma and bronchitis for centuries. Ephedrine is a potent sympathomimetic amine that indirectly stimulates the adrenergic receptor system by increasing the activity of noradrenaline at postsynaptic receptors. Increase in blood pressure, stimulation of heart muscle, constriction of arterioles, relaxation of the smooth muscle of bronchi and gastrointestinal tract, and dilation of pupils were observed.

The production of ephedrine represented one of the first chemo-enzymatic processes that also incorporated one of the earliest bioasymmetric C−C formations. As early as 1921 it was found [22] that yeast catalyzed the stereoselective condensation of benzaldehyde (**66**) and pyruvate (**67**) to form L-phenylacetylcarbinol (L-PAC, **68**) (Scheme 3.14). Reductive methylamination of L-PAC **68** generated L-(−)-ephedrine **69**. From (−)-ephedrine **69**, pseudoephedrine **70**, a sympathomimetic drug of phenethylamine and amphetamine chemical classes used as a nasal/sinus decongestant, can be synthesized by a chiral inversion.

The functional enzyme for this biotransformation was pyruvate decarboxylase (PDC, EC 4.1.1.1). PDC uses thiamine pyrophosphate (**71**, TPP, or thiamine diphosphate) as cofactor (Scheme 3.15). It catalyzed the formation of an active acetaldehyde **73** or its resonance-stabilized carbanion **74** via decarboxylation of pyruvic acid in the cytoplasm. The carboligation of carbanion with benzaldehyde produced L-PAC (**68**). PAC production from benzaldehyde and pyruvate has been demonstrated with various degrees of success by growing cells of various yeasts [23] such as viable free and immobilized whole cells of *Saccharomyces cerevisiae* and *Zygosaccharomyces rouxii* [24]. The enzyme-based process using the partially purified PDC allowed an increased supply of benzaldehyde and provided higher production of L-PAC [25].

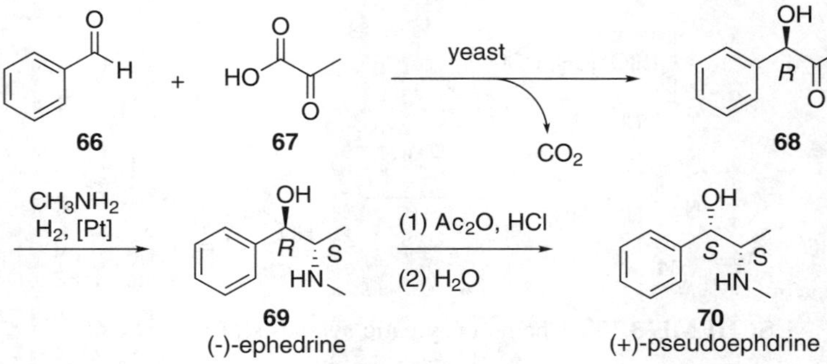

SCHEME 3.14 Chemo-microbial production of (−)-ephedrine **69**.

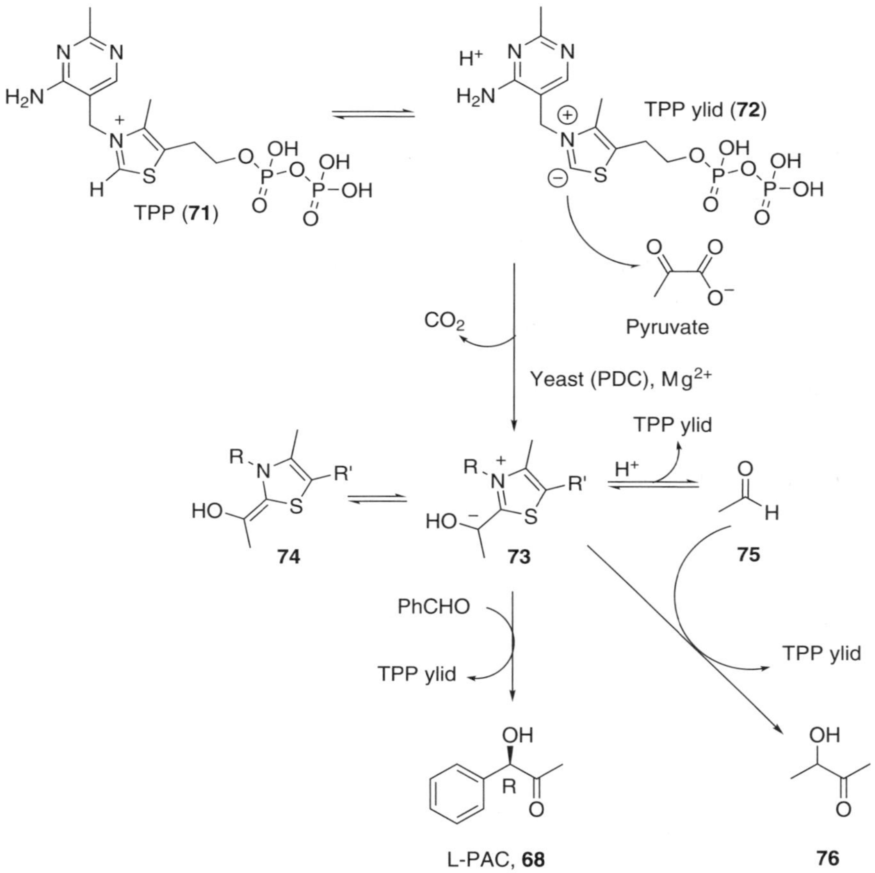

SCHEME 3.15 Mechanism of the microbial production of L-PAC (68).

Although the bioproduction of L-PAC was well established, the process was further improved by switching the enzyme system from yeast-derived PDCs to bacterial PDCs in order to use cheap acetaldehyde rather than expensive pyruvate as the substrate (Scheme 3.16). With acetaldehyde as substrate, the pH value of the medium could be kept constant during biotransformation, whereas the decarboxylation of pyruvate led to acidification of the medium (if not controlled) and gas (CO_2) formation that may cause issues in industrial bioreactor. A concentration of 16–16.6 g/l of L-PAC in the organic phase (1-pentanol, 1-hexanol, or isobutanol) and 1–1.7 g/l of L-PAC in the aqueous phase provided the best productivity when *Zymomonas mobilis* mutant enzyme PDCW392M was used in a biphasic aqueous/organic solvent with acetaldehyde or benzaldehyde as substrates [26]. This still required further improvement in productivity and yield using acetaldehyde as the substrate [27].

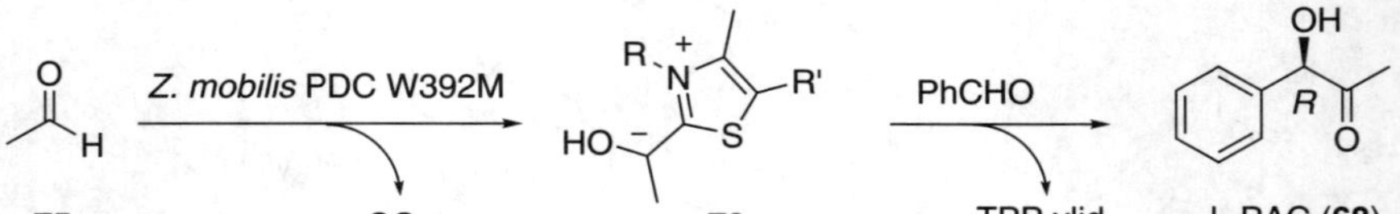

SCHEME 3.16   Enzymatic production of L-PAC by PDC from bacteria.

3.2.4.2   L-Dopa (Levodopa, Parcopa, Syndopa, and More Than 30 Other Brand Names)

L-Dopa **78** is extensively used for the treatment of Parkinson disease and dopa-responsive dystonia. The biocatalytical process for L-dopa production involved one-step biotransformation of the bisphenol **77** and pyruvate in the presence of ammonia by resting cells of *Erwinia herbicola* ATCC21433 (Scheme 3.17). The major functional enzyme for this biotransformation is tyrosine-phenol lyase or β-tyrosinase [28].

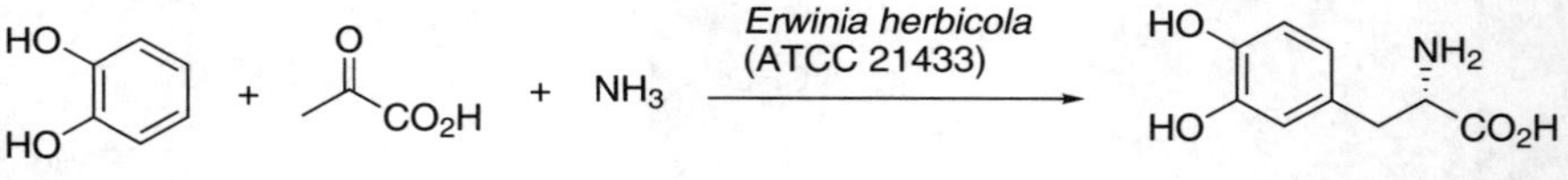

SCHEME 3.17   Microbial production of L-dopa.

3.2.4.3   Zanamivir (Relenza)

Zanamivir **82** (Relenza) is an antiinfluenza drug developed by GSK for treatment and prophylaxis of influenza A and B, the two types most responsible for flu epidemics. By inhibiting viral neuraminidases (sialidases), zanamivir prevents release of progeny influenza virus from infected host cells and spread of the influenza virus from one cell to another within respiratory tract. The virus dies along with the infected host cells. Relenza is administered by inhalation into airways infection sites in the respiratory tract to prevent virus particle aggregation and to destroy the virus directly.

Scheme 3.18 describes the chemo-enzymatic synthesis of Zanamivir. The chemical synthesis of the key intermediate *N*-acetyl-D-neuraminic acid **81** requires tedious protection and deprotection steps, and does not offer much potential for economical large-scale production [29]. Mahmoudian's group [30] found that *N*-acetyl-D-neuraminic acid aldolase enzyme can be used to convert *N*-acetyl-mannosamine **80** to the desired intermediate *N*-acetyl-D-neuraminic acid **81**. However, *N*-acetyl-D-mannosamine **80** is obtained through epimerization of *N*-acetyl-glucosamine **79**; it is always accompanied by the starting material. The ratio of **79** to **80** on equilibration is typically about 1:4 by weight. Since **80** is the substrate of aldolase to *N*-acetyl-D-neuraminic acid **81** while **79** is an inhibitor to the enzyme, removal

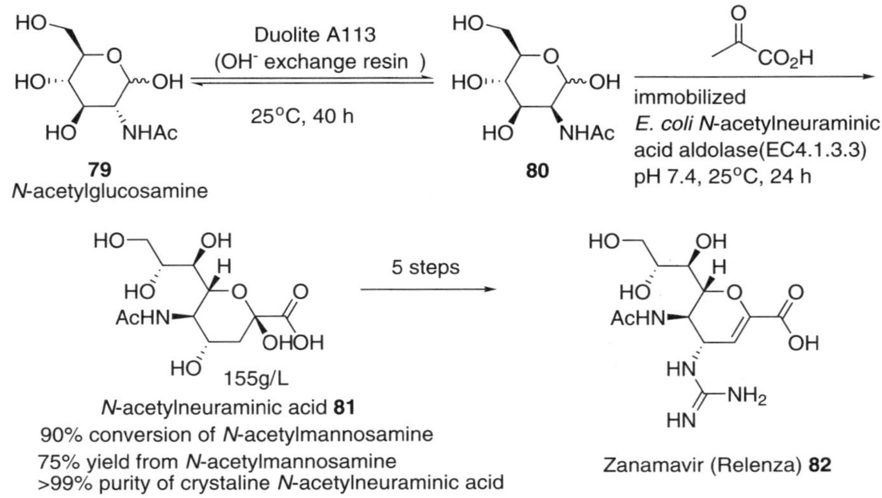

SCHEME 3.18 Process for the preparation of N-acetyl-D-neuraminic acid **81** by aldolase.

of N-acetyl-D-glucosamine **79** and enriching N-acetyl-D-mannosamine (**80**) concentration are essential for a successful enzymatic reaction. In practice, by treatment of N-acetyl-D-glucosamine **79** and N-acetyl-D-mannosamine **80** mixture with isopropanol, the insoluble crystaline N-acetyl-D-glucosamine **79** is filtered and the ratio of N-acetyl-D-mannosamine **80** to N-acetyl-D-glucosamine **79** can be increased to about 1.5:1 to 10:1, typically, on a large scale, 1.5:1 to 4:1. The mixture is then good for enzymatic reactions. N-acetyl-D-glucosamine **79** recovered during the enrichment process may be reused in epimerzation step, optionally in admixture with fresh N-acetyl-D-glucosamine.

The N-acetyl-D-neuraminic acid aldolase was cloned from *E. coli* and over-expressed in an inducible system (tac-promoter) with 10 units/mg protein, representing 30% of cell protein. The cells were simply homogenized and the enzyme directly immobilized from crude extracts onto Eupergit-C beads without any clarification. With the immobilized enzyme, bioconversion of N-acetyl-D-mannosamine **80** to N-acetyl-D-neuraminic acid **81** is carried out in aqueous medium at a pH of 7.4 at 25°C for 24 h. The reaction is terminated (by filtering the immobilized enzyme beads) when a substantial amount of N-acetyl-D-mannosamine **80**, up to about 95%, is converted to N-acetyl-D-neuraminic acid **81**. The concentration of N-acetyl-D-neuraminic acid **81** reaches more than 150 g/l in the reaction mixture, and it can be crystallized from the reaction mixture by the addition of acetic acid. N-acetyl-D-neuraminic acid **81** was obtained in 75% yield from N-acetyl-D-mannosamine **80** with excellent purity (>99%). The production of N-acetyl-D-neuraminic acid **81** is the first ton-scale application of aldolase. Zanamavir **82** can be synthesized after a five-step chemical transformation from N-acetyl-D-neuraminic acid (Scheme 3.18).

3.2.5 Lyases and Oxidoreductases

3.2.5.1 Atorvastatin (Lipitor)

Atorvastatin **86** is a potent inhibitor of cholesterol biosynthesis through inhibition of HMG-CoA reductase, which catalyzes the reduction of 3-hydroxy-3-methylglutaryl-coenzyme A (HMG-CoA) to mevalonate, an early and rate-limiting step in hepatic cholesterol biosynthesis. This inhibition leads to lower blood cholesterol ("bad" cholesterol) by 39–60%, lower triglycerides (a type of fat found in the blood) by 19–37%, and higher HDL ("good" cholesterol) by 5–9%.

Typical synthesis of atorvastatin via a chemical process is shown in Scheme 3.19 [31] by coupling of compound **83** and 3,5-dihydroxylester chain **84**, in which synthesis of chiral chain **84** takes much of the effort. A number of stereoselective syntheses of the side chain **84** have been reported either starting from chiral material or via catalytically asymmetric synthesis. To introduce the chiral centers more efficiently from cheaper starting materials and shorten the synthetic steps, various chemoenzymatic approaches [32] with different biocatalysts have also been developed for stereoselective synthesis of the 3,5-dihydroxyacid side chains. Some of these have been made for application on an industrial scale [32] via optimization of the biocatalytic step.

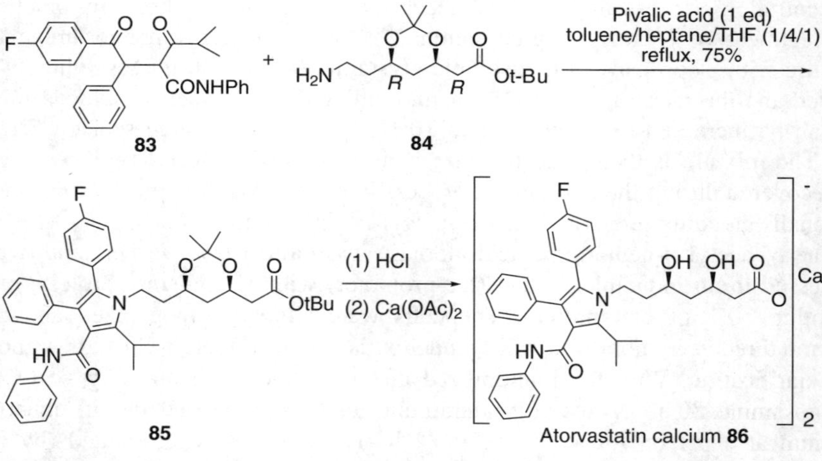

SCHEME 3.19 Production of atorvastatin calcium **86** by convergent chiral synthesis.

Wong and co-workers [33] did pioneer work in 1994 on an elegant asymmetric tandem aldol reaction using substituted acetaldehyde starting materials **87–89** and acetaldehyde **90** to form (3R, 5S)-6-substituted-3,5-dihydroxyhexanal **92** in the presence of 2-deoxyribose-5-phosphate aldolase (DERA, EC 4.1.2.4; Scheme 3.20). The resulting (3R, 5S)-6-substituted-3,5-dihydroxyhexanal **92** exists as a hemiacetal **93**, which can be oxidized to lactone followed by ring opening to give the dihydroxyl product with excellent stereoselectivity.

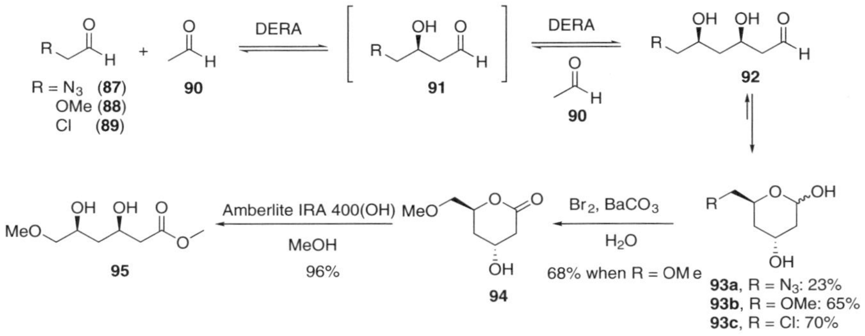

SCHEME 3.20   DERA-catalyzed tandem aldol reaction.

An improved and large-scale synthesis of **93c** with a concentration of 126 g/l at low temperature overnight (up to 240 g/l concentration in a repetitive batch experiment) was developed by DSM [34] as shown in Scheme 3.21.

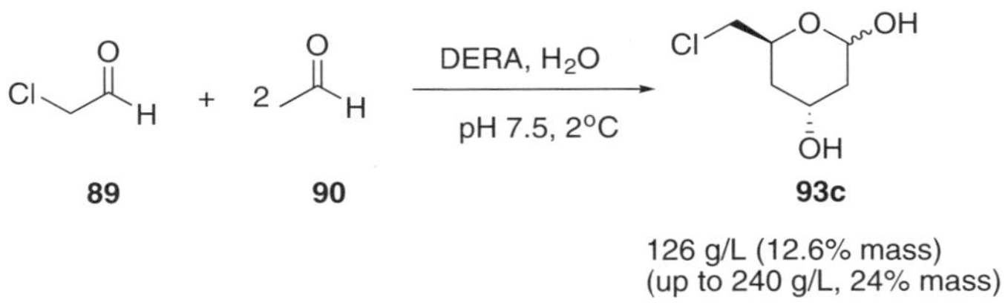

126 g/L (12.6% mass)
(up to 240 g/L, 24% mass)

SCHEME 3.21   Improved process from DSM Research, the Netherlands.

In 2004, Burk and co-workers [35] at Diversa reported a process in which 100 g of **93c** was obtained in a single batch at a rate of 30.6 g/l/h with a final concentration of 93 g/l and enzyme loading of 2 wt% DERA. The improved DERA was identified by screening genomic libraries prepared from environmental DNA. Efficient conversion of **96** to **100**, a versatile intermediate for the synthesis of atorvastatin was also developed (Scheme 3.22).

Wong and co-workers [36] described an improved biotransformation sequence for the synthesis of atrovastatin intermediate **85** using a DERA S238D mutant as catalyst (Scheme 3.23). 3-Azidopropionaldehyde **101** was condensed with acetaldehyde to give the hemiacetal **102**, which was oxidized to the corresponding lactone **103** in 35% overall yield in two steps. Although further optimization is required for the industrial process, this route indeed provides the most efficient way toward the key intermediate **84**.

Scheme 3.24 illustrates Patel's approach [37] for the preparation of Kaneka alcohol **112**, a side chain intermediate for statin synthesis, via the stereoselective microbial reduction of 6-benzyloxy-3,5-dioxohexanote (mixture of **109** and **110**) using a cell suspension of *Acinetobacter calcoaceticus* SC 13876. The resulting diol ethyl ester **111** from microbial reduction was isolated in 85% yield

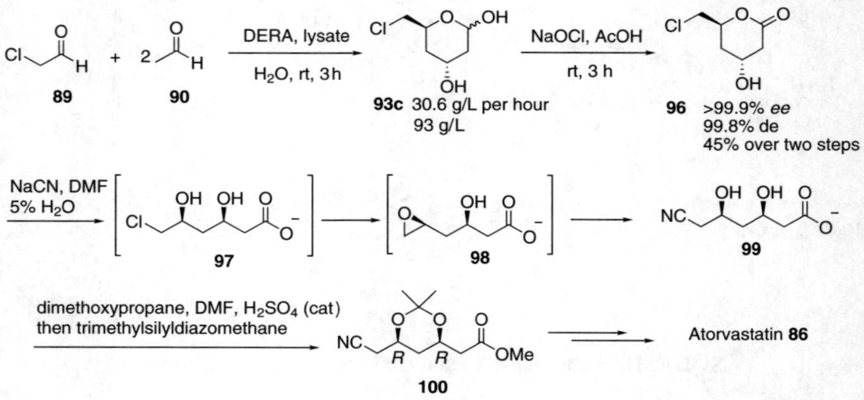

SCHEME 3.22   Process developed by Diversa using a DERA from library screening.

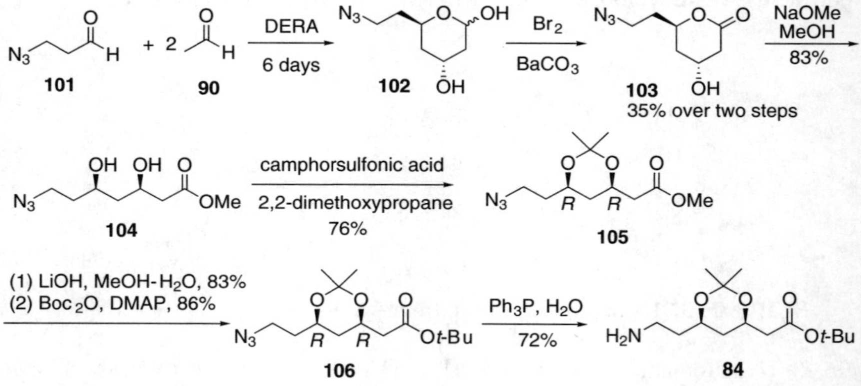

SCHEME 3.23   Improved synthetic route for **84** by Wong.

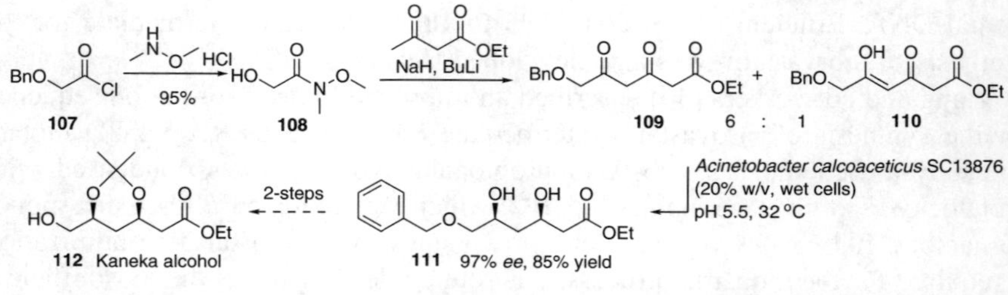

SCHEME 3.24   Patel's ketoreductase approach to Kaneka alcohol **112**.

and 97% *ee*. The same group later on [38] converted *t*-butyl 6-benzyloxy-3,5-dioxohexanote (mixture of **113** and **114**) into diol using *Acinetobacter sp*. SC 13874. The corresponding diol *t*-butyl ester **115** was obtained in 87% *ee* and 51% de (Scheme 3.25). Compound **115** is a precursor to Kaneka alcohol **116**, a more

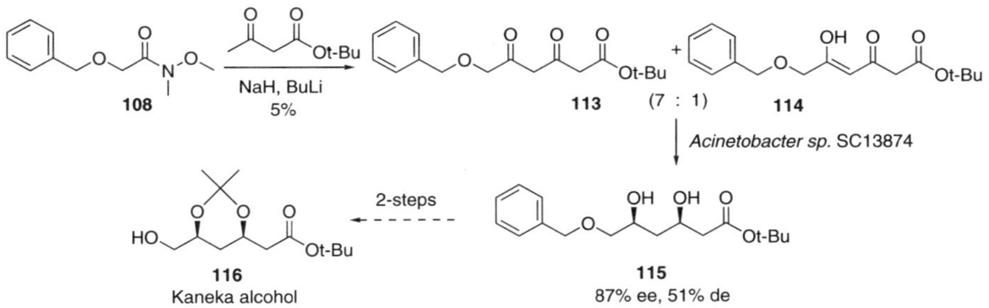

SCHEME 3.25   Patel's ketoreductase approach to Kaneka alcohol **116**.

advanced intermediate than **112** for statin synthesis. The conversion efficiency from **108** to **113** and**114** is still a challenge since only 5% yield for current conversion was obtained.

3.2.6   Oxidoreductases and Hydrolases

3.2.6.1   Dorzolamide (Trusopt, MK-0507)

Dorzolamide **131** is an antiglaucoma agent developed by Merck. It is an inhibitor of human carbonic anhydrase II. Inhibition of carbonic anhydrase results in a reduction in intraocular pressure (IOP) by slowing the formation of bicarbonate ions with subsequent reduction in sodium and fluid transport and decrease in aqueous humor secretion. The aim of controlling the IOP is to erase the main risk factor in pathogenesis of optic nerve damage and glaucomatous visual field loss.

Scheme 3.26 outlines a practical asymmetric synthesis [39] of MK-0507 with >32% overall yield. One of the key steps of this chemically synthetic route involves introduction of the first chiral center from **118** to **120** by maximizing S_N2 displacement over elimination via solvent effect. DMF is proved to be the best solvent for the preferred S_N2 displacement with >97% *ee* of **120**. The use of acid-catalyzed hydrolysis of methyl ester **120** to **121** (100% yield) over basic hydrolysis is also highlighted because saponification of **120** is completely ineffective caused by the basic β-elimination of **120** to 2-mercaptothiophene and subsequent reversible Michael addition instead of the ester hydrolysis. Another key step is the construction of the second chiral center by a Ritter reaction that exhibits an unexpected tendency to proceed with retention of chirality from **125** to **126**. The *de* value was improved from 50% of precursor **125** to 78% of Ritter reaction product **126**.

The remaining problem in this synthetic route is the reduction of keto group in **122** by LiAlH$_4$. The reduction leads to the predominant (95%) formation of the unwanted *cis*-isomer, and also the subsequent epimerization with sulfuric acid at 0°C was incomplete. Therefore, a biocatalytic solution was sought as shown in Scheme 3.27.

Instead of using **122** as the substrate, whole cell bioreduction was developed by use of more water-soluble (6*S*)-methylketosulfone **133** (Scheme 3.27) [40].

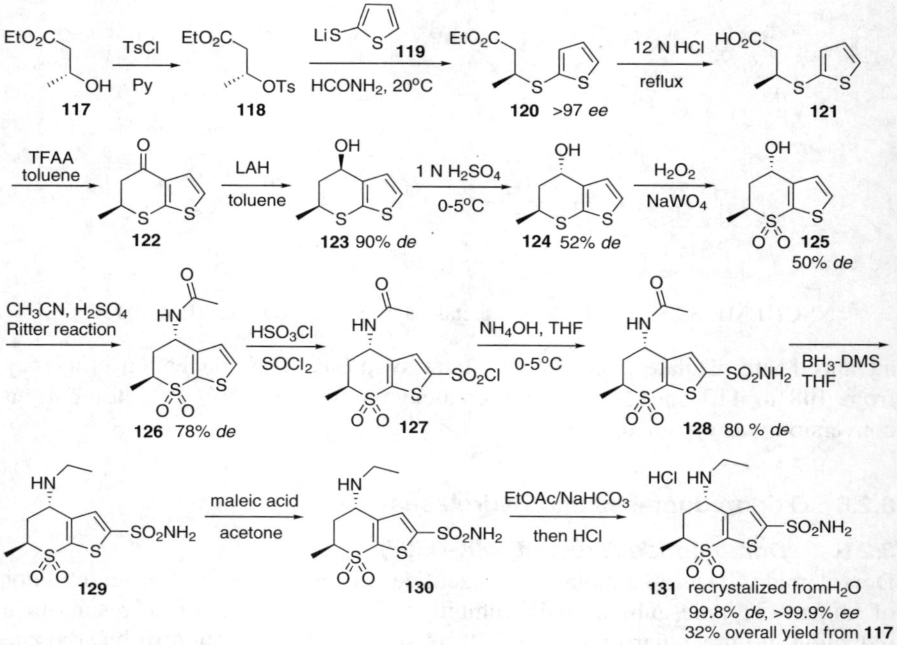

SCHEME 3.26 Chemical synthesis of dorzolamide hydrochloride (MK-0507, **131**).

Fungus *Neurospora crassa* IMI 19419 was found to be capable of reducing keto-sulfone **133** to desired (4*S*, 6*S*)-hydroxysulfone **125**. Control of pH at 4.0 and slow addition of the substrate to maintain its concentration as low as 0.2 g/l are critical to the high yield and high selectivity of (4*S*, 6*S*)-hydroxysulfone **125**, as ketosulfone **133** tends to epimerize to the (6*R*)-methylketosulfone **135** via a ring-opening reaction at pH >5. Under these conditions, a high *trans*-to-*cis* ratio (99.8% *de*) and almost complete conversion (isolated yield is 80%) were achieved. And industrial-scale production of the intermediate **125** was then developed [40]. It is worth noting that the starting material poly-(3*R*)-hydroxybutyrate (PHB, **132**) was also obtained by biocatalytic fermentation.

In addition, intermediate **121** could also be prepared from (*S*)-3-(thiophen-2-ylthio) butanenitrile (**136**) by whole cell biohydrolysis with nitrilase [41], although conventional hydrolysis of cyanide failed to give the desired product. Enzymatically, two methods for hydrolysis of the nitrile into corresponding carboxylic acids have been used (Scheme 3.28). One method directly uses nitrilase, and the other is to combine nitrile hydratase (NHase) with amidases. Nitrilase (E.C. 3.5.5.1) is a thiol type of enzyme, directly converting a nitrile into corresponding carboxylic acid in an aqueous solution without formation of an amide. Nitrile hydratase (E.C. 4.2.1.84) has either a nonheme iron atom or a noncorrinoid cobalt atom and catalyzes the hydration of nitriles to amides [42].

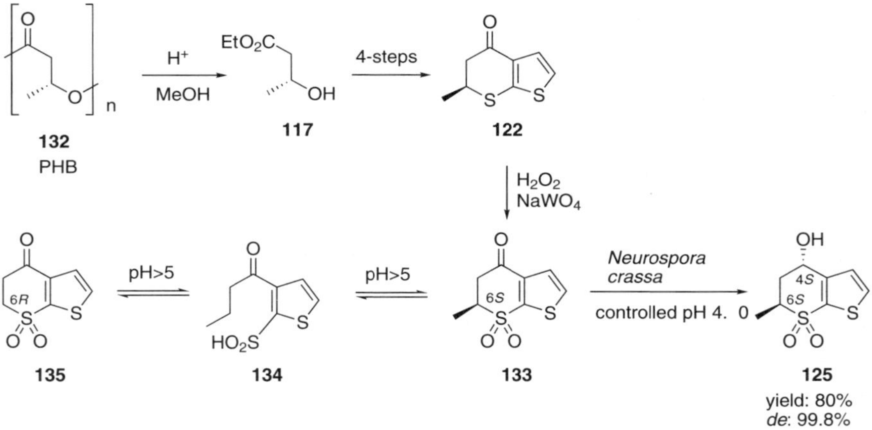

SCHEME 3.27   Biosynthetic approach to compound **125** by microbial reduction.

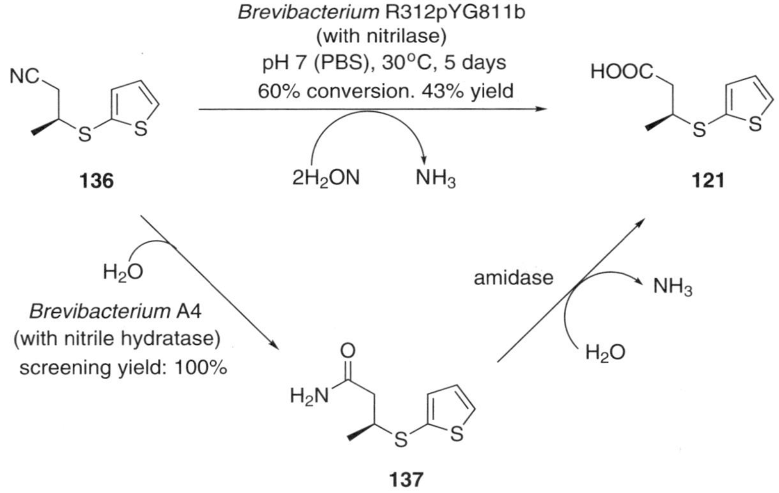

SCHEME 3.28   Microbial synthesis of compound **121** from the nitrile precursor **136**

Among 53 strains known for their ability to grow and assimilate nitriles and amides as a carbon source, 12 strains displayed a high activity of converting nitrile **136** to its acid **121** with >75% conversion. Gram-scaled preparation of the acid **121** (1.2 g, 60% conversion, 43% yield) from **136** was achieved by *Brevibacterium* R312pYG811b at pH 7.0, 30°C at a substrate loading of 5 g/l (well tolerated up to 20 g/l) in the presence of 3% acetonitrile (acetonitrile is not substrate here) as cosolvent after 5-day incubation. Results also indicated that *Brevibacterium* R312pYG811b grows faster without inducer. The substrate **136** is a more efficient inducer of nitrilase production than isobutyronitrile, while **136** significantly inhibits the cell growth.

3.2.6.2 Captopril (Capoten)

Captopril **141**, originally developed by Squibb (now Bristol-Myers Squibb), is the first oral ACE inhibitor drug used to treat high blood pressure (hypertension), congestive heart failure, and kidney problems caused by diabetes and to improve survival after a heart attack. There are two chiral centers in its structure, and its potency relies on the configuration of the mercaptoalkanoyl moiety. The compound with the (S, S)-configuration is about 100 times more active than its corresponding (R, S)-isomer in inhibiting ACE. Traditional chemical synthesis of captopril used resolution of a diastereisomer mixture (SS+RS) by cyclohexyl diamine salt. Yet, chemo-enzymatic synthesis included (1) microbial production of the key intermediate, L-(+)-β-hydroxyisobutyric acid **139**, via microbial oxidation of isobutyric acid **138** (Scheme 3.29) [43]; (2) lipase-catalyzed enantioselective preparation of D-(−)-3-chloro-2-methylpropionic ester **144** from its racemic precursor **143** (Scheme 3.30) [44], resulting in 36% yield and 98% ee with lipase from *Candida cylindracea*; (3) enzymatic production of the key intermediate (S)-3-acetylthio-2-methyl propanoic acid **149** via lipase-catalyzed enantioselective hydrolysis of the thioester of racemic **148** (Scheme 3.31) [45] (in which lipase from *Rhizopus oryzae* ATCC 24563 and lipase PS-30 from *Pseudomonas cepacia* are the top 2 lipases among various lipases screened, with >24% yield and >95% ee; and (4) resolving the racemic 3-acylthio-2-methylpropionic acid ester by enantiospecific ester hydrolysis [46] utilizing extracellular lipases of microbial origin.

3.2.6.3 Baclofen (Kemstro, Lioresal)

Baclofen acts as an agonist of γ-aminobutyric acid (GABA) receptor to relax muscles and to reduce pain and stiffness. Baclofen is a racemic β-parachlorophenyl GABA, a lipophilic analog to the neurotransmitter GABA. Unlike GABA, it can cross the blood-brain barrier (BBB) as an oral CNS drug. Racemic baclofen is widely used as an antispastic agent. In fact, (R)-(−)-baclofen **156** is more active (also more toxic) than (S)-(+)-baclofen **157**. (S)-baclofen **157** antagonizes the

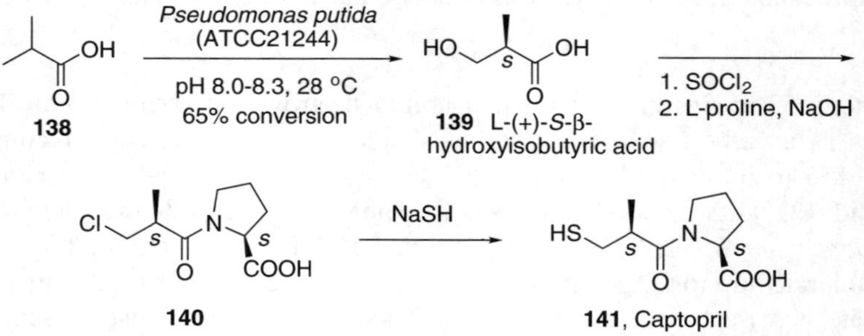

SCHEME 3.29 Chemo-enzymatic synthesis of captopril **141** from isobutyric acid **138**.

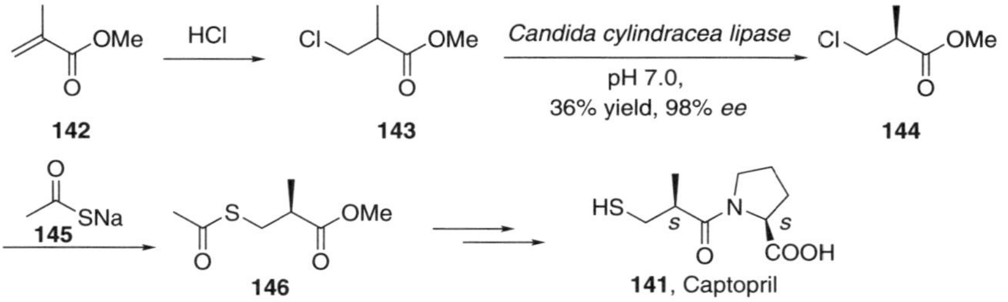

SCHEME 3.30 Chemo-enzymatic synthesis of captopril **141** from **142**.

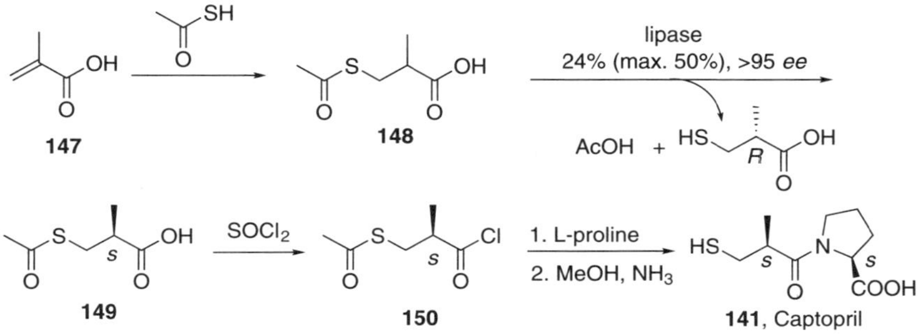

SCHEME 3.31 Chemo-enzymatic synthesis of captopril **141** from **147**.

activity of the (R)-form, and its presence makes it necessary to administer a higher dose of the drug. Patients would greatly benefit from the administration of enantiomerically pure (R)-baclofen **156**, even though the (R)-isomer is also responsible for the toxic and other side effects.

There are a lot of approaches for the production of (R)-(−)-baclofen **156**, for example, (1) isolation from a racemic mixture of (R)- and (S)-baclofen by *Streptomyces halstedii* [47], (2) asymmetric synthesis [48], and (3) chemo-enzymatic synthesis.

In 1991, Chenevert and Desjardins reported a five-step synthesis of both enantiomers from 4-chlorocinnamic acid **151**, which is outlined in Scheme 3.32. The key step was the highly stereoselective desymmetrization of dimethyl 3-(4-chlorophenyl) glutarate **154** with chymotrypsin, leading to the (R)-acid **155** in 85% yield with >98% *ee*. This is one of the most efficient synthetic routes reported so far due to the fact that symmetric diester **154** is commercially available now and that both (R)-(−)-baclofen **156** and (S)-(+)-baclofen **157** can be synthesized [49] from the intermediate **155**.

The chemo-enzymatic synthesis shown in Scheme 3.33 [50] utilized enantiopure lactone **160** as a key building block. Compound **160** was prepared by whole cell enzymatic Baeyer–Villiger oxidation and the enzyme was derived from a culture of fungus *Cunninghamella echinulata* NRLL 3655. Oxidation of compound **159** allowed stereoselective incorporation of an oxygen into one

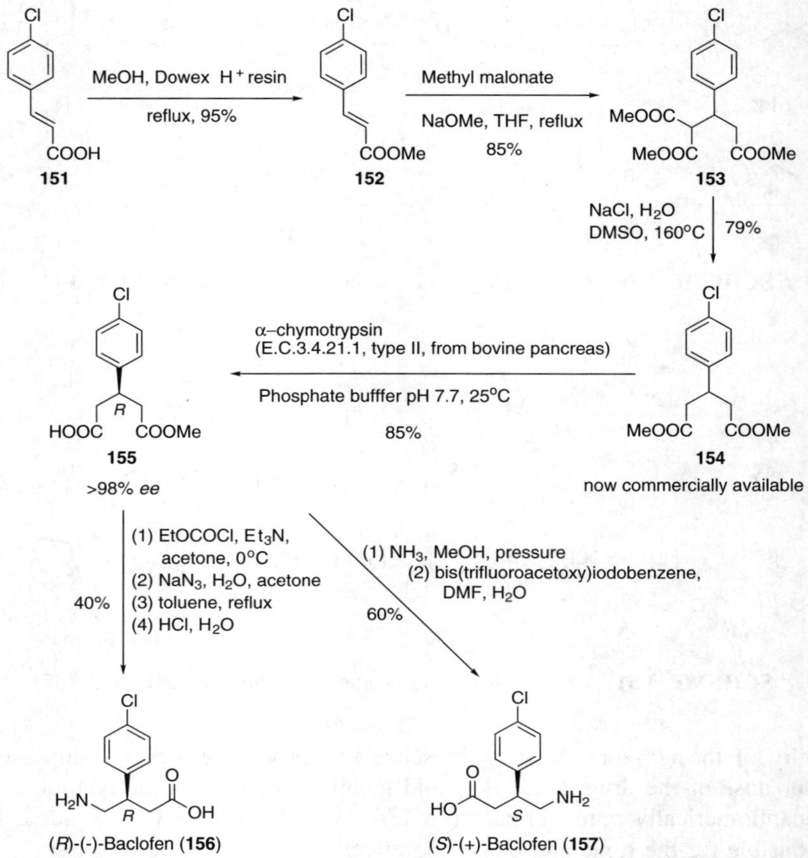

SCHEME 3.32   Production of baclofen via resolution of the symmetric diester **154**.

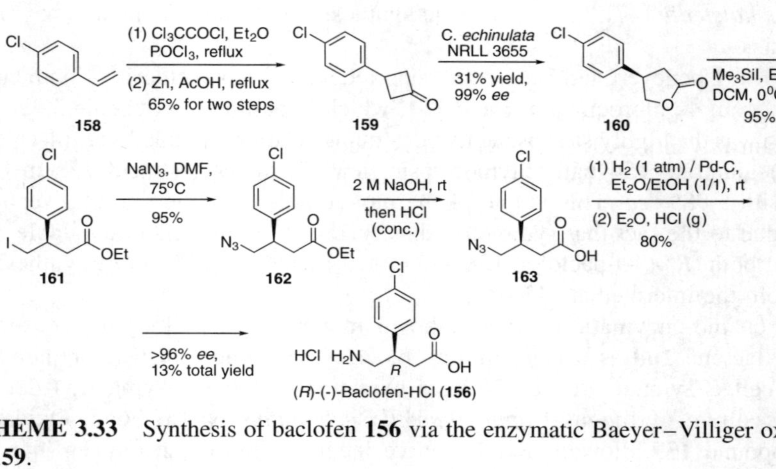

SCHEME 3.33   Synthesis of baclofen **156** via the enzymatic Baeyer–Villiger oxidation of **159**.

single enantiotopic C–C bond, yielding the (*R*)-(−)-chlorophenyl lactone **160** in high enantiomeric purity (*ee* >99%, yield 30%). The obtained lactone **160** was transformed to baclofen in five steps.

3.3 BIOSYNTHESIS

3.3.1 Introduction

3.3.1.1 *Biosynthesis and Biocatalysis*

Natural products represent an extremely rich source of biologically active compounds that find wide-ranging applications in nearly all human therapeutics, including infectious, neurological, cardiovascular, metabolic, and ontological diseases [51]. In the past two decades, more than one-third of all small-molecule drugs have been natural products or their derivatives, while more than 75% of the approved antibacterial drugs are natural products or natural product derivatives [52]. Pharmaceutical natural products originate from a variety of organisms, particularly plant, fungi, and bacteria. Owing to their structural complexity, the majority of clinically used natural product therapeutics are produced by biosynthesis rather than by total chemical synthesis. Biosynthesis (sometimes called biogenesis) is an enzyme-catalyzed process in cells of living organisms by which substrates are converted to other and even more complex products. The biosynthesis process often consists of several enzymatic steps in which the product of one step is used as the substrate of the following step. Examples of such multistep biosynthetic pathways are those for the production of natural products, peptides, fatty acids, and enzymes. In contrast to biocatalysis (biotransformation), which is usually complementary to total chemical synthesis of pharmaceuticals, biosynthesis can directly produce complex nature products via fermentation. A generalized comparison of biocatalysis and biosynthesis is outlined in Figure 3.2.

Biosynthesis can be realized either by the native host (bacteria, fungi, plants, and animals) or by fermentation of cells from different organisms. However, the former approach always requires a large quantity of natural resources and needs long time period. Therefore, sustainable fermentation processes that facilitate the production of a wide variety of natural products by scalable cultivation of microorganisms, plant cells, and mammalian cells are much more preferable. A significant advantage of this approach is the ability to increase the time-space yield beyond natural levels via strain selection and improvement.

3.3.1.2 *Strain Improvement and Metabolic Engineering*

Classical strain improvement was mainly based on random strain mutagenesis and selection. The mutant strains improved production rates of the specific product compared with the wild-type parents. Random mutagenesis of the strain is accomplished by exposure to radiation or chemical agents or extreme conditions, followed by screening for colonies with the desired novel phenotype. These strategies have been successfully used in the biotechnological production of amino acids such as glutamic acid and lysine and antibiotics such as penicillins. For

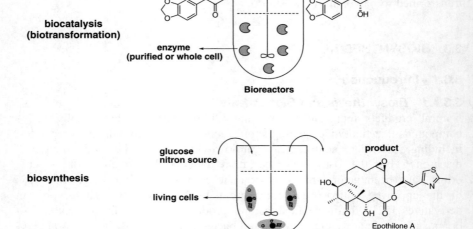

FIGURE 3.2 Comparison of biosynthesis and biocatalysis.

example, the harvest titer of penicillin was increased from 0.5–1 g/l in 1950 to 40–50 g/l in 2000. However, the resulting mutations were largely underdefined and the selection procedures were labor intensive.

In the past decades, the development of recombinant DNA technology and the subsequent emergence of the field of metabolic engineering have brought about a viable alternative to improve the yield of biosynthesis. Metabolic engineering is defined as the rational modification of specific existing metabolic pathways or the introduction of novel pathways using recombinant DNA technology for the overproduction of specific compounds [53].

Strain improvement via metabolic engineering often involves two basic procedures in an iterative fashion: metabolic network analysis and rational genetic design [54]. Knowledge of the metabolic network provides the basis for the subsequent rational target design, while the rational design in turn is tested by the next round of metabolic network analysis, which may identify a new potential target for further genetic modification.

Compared to the use of the traditional mutagenesis procedures, metabolic engineering utilizes a strategy of rational strain improvement using recombinant DNA technology rather than random mutagenesis. The introduction of target genes into the host and the subsequent gene expression are implemented with vectors (usually plasmids), which can be classified as low copy, moderate copy, or high copy [55]. For many applications, high-copy-number plasmids are often favored, because normally a high gene dosage corresponds to the high expression levels and the elevated titers of the metabolites. The plasmid copy number is determined by the origin of replication (ori); thus mutations in this region can lead to an increased number of plasmid copies. For example, pBBR1 is a broad-host-range plasmid, isolated from *Bordetella pertussis* [56], which has been widely used. To increase the copy number of pBBR1, its replication control region was subjected to mutagenesis, resulting in three- to sevenfold higher copy numbers than the original plasmid [57]. The strength of the promoter plays a pivotal role in regulating the expression of the target gene in engineered host cells, thereby affecting the overall flux through the pathway of interest. Thus it is essential to place the target gene under a promoter that allows tight control over the gene's expression [58]. The promoters used for metabolic engineering application include constitutive and inducible promoters. Their utilities and effects in metabolic engineering are dependent on the engineering systems in which they work.

Metabolic network analysis was realized by a series of tools such as metabolic flux analysis (MFA), DNA microarray, and metabolite profiling et al. MFA is an analytical technique that is used to quantify the metabolic fluxes through all intracellular reactions, thereby dissecting the functional aspects of a metabolic network into greater detail and elucidating the metabolic state of the cells in vivo [59]. The application of microarrays in metabolic engineering enables the identification of target genes (such as genes encoding repressors, enhancers, and other regulatory factors) for rational genetic alteration, in addition to the metabolic enzyme genes [54].

3.3.1.3 Modular Enzymes

The biosynthesis of natural products is catalyzed by a cascade of enzymes. In the past decades, elucidation of natural product biosynthesis via new tools of modern genetics enabled the discovery of some large modular megasynthases such as polyketide synthases (PKSs) and nonribosomal peptide synthases (NRPSs) [60]. It was found that these synthases consist of modules; each module loads and synthesizes the growing oligomeric metabolite by one starter unit. Modules consist of domains; each domain performs a single catalytic role in the assembly-line organization. Repeated catalytic activity was observed from module to module, but differentiation existed as the growing natural products moved along these assembly lines, depending on the presence or absence of selective domains [61]. For example, the biosynthesis (Fig. 3.3 Wild) of deoxyerythromycin B, the macrocyclic core of erythromycin, was achieved by three synthases, seven modules, and 28 domains. These synthases are called "modular" not only because of their structural organization but also because those domains could be replaced by other domains. As shown by C. Khosla [62], the loading domain of deoxyerythromycin B synthase was replaced by another loading domain from NRPS and an adenylation (A) and a thiolation (T) domain, yielding an engineered "hybrid" synthase. In the presence of exogenous benzoate, the engineering strain of *E. coli* produced the expected 6dEB analog (Fig. 3.3 Hybrid). It is believed that the discovery

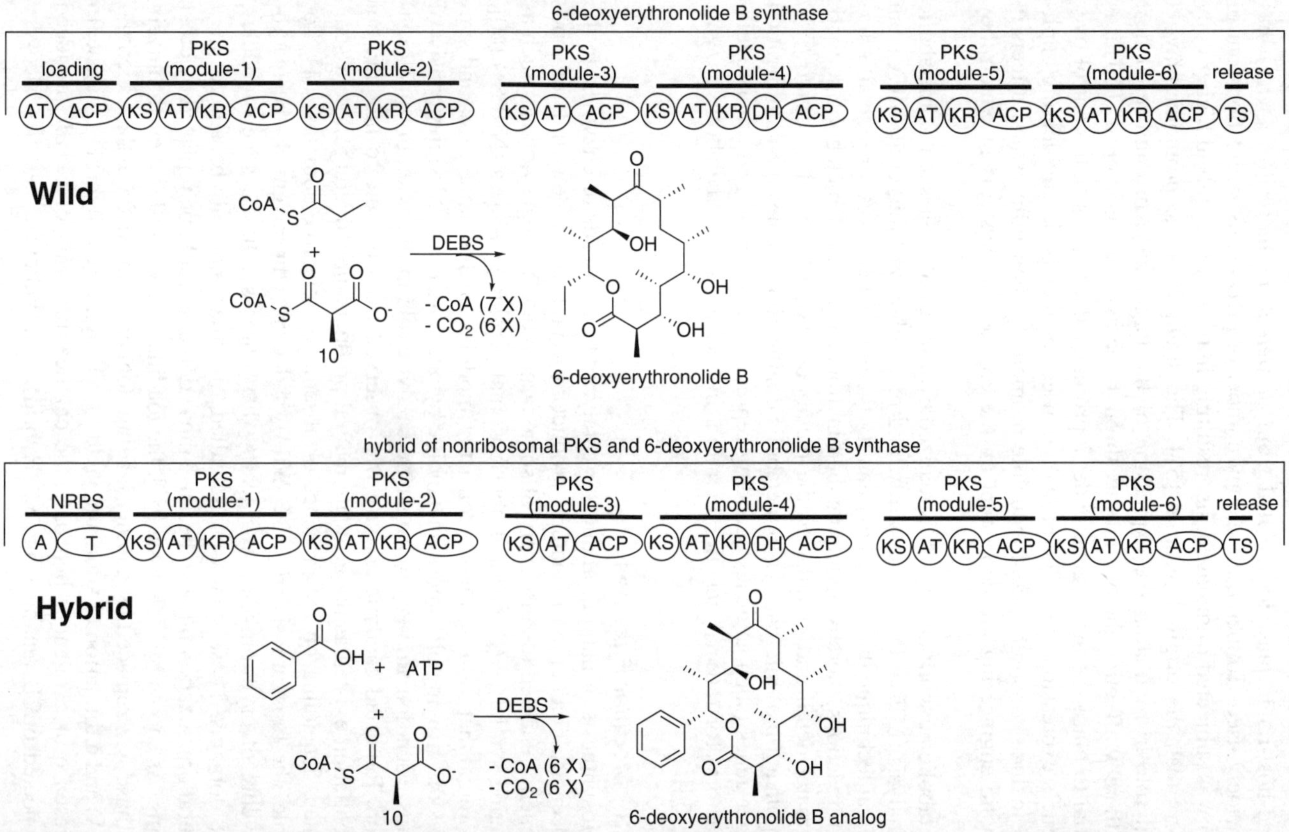

FIGURE 3.3 Natural 6-deoxyerythronolide B synthase. Catalytic domains: KS, ketosynthase; AT, acyl transferase; ACP, acyl carrier protein; KR, ketoreductase; ER, enoylreductase; DH, dehydratase; TE, thioesterase. H, A hybrid of NRPS and 6-deoxyerythronolide B synthase; an adenylation (A) and a thiolation (T) domain.

TABLE 3.3   Selected Marketed Penicillins and Cephalosporins

Penicillins	R	Cephalosporins	R_1	R_2
Benzylpenicillin (penicillin G)		Cefapirin		
Phenoxy-methylpeni-cillin (penicillin V)		Cefalexin		Methyl
Cloxacillin		Cephradine		Methyl
Ticarcillin		Cefdinir		Methylene
Ampicillin		Ceftibuten		H
Amoxicillin		Cefepime		
Nafcillin		Cefapirome		
Piperacillin		Cefatobiprole		

of modular enzymes would have a significant impact on the role of enzymes in synthetic and process chemistry.

3.3.2 Development and Production of β-Lactams

3.3.2.1 Penicillins and Cephalosporins

The β-lactams are the most widely used group of antibiotics applied for treatment of infections, with about 120 compounds approved for human use [63]. β-Lactams, structurally, can be divided into penicllins, cephalosporins, monobactams, carbapenems, and clavams.

Penicillins and cephalosporins are the best-characterized group of β-lactams so far, and these groups are also considered to be the main β-lactam families. There are now at least 30 marketed penicillins and more than 50 marketed cephalosporins (Table 3.3).

The biosynthetic pathways for the penicillins, cephalosporins, and cephamycins are well characterized as shown in Scheme 3.34. The first step in the biosynthesis of β-lactams involves the condensation of the three amino acids L-α-aminoadipate, L-cysteine, and L-valine to form the tripeptide δ-L-α-aminoadipyl-L-cysteinyl-D-valine (ACV), catalyzed by the enzyme ACV synthetase, which also performed an epimerization of L-valine to D-valine [64]. The second step in the biosynthesis of β-lactams is the ring closure of ACV to form isopenicillin N (IPN), catalyzed by isopenicillin N synthase, and in this reaction the characteristic penam ring structure is formed [65]. In the third step of the penicillin pathway, the hydrophilic L-α-aminoadipate side chain of IPN is exchanged with the CoA-thioester, an activated form of a hydrophobic acyl group such as phenylacetic acid or phenoxyacetic acid, leading to penicillin G or penicillin V, respectively. This side chain replacement is carried out by the enzyme acyl-CoA:IPN acyltranferase.

In the biosyntheses of cephalosporin C and cephamycin C, IPN is epimerized to penicillin N by isopenicillin N epimerase (IPNE). Deacetoxycephalosporin C synthase (expandase) converts the penicillin N to deacetoxycephalosporin C (DAOC) via an oxidative ring expansion that involves oxygen and 2-oxoglutarate. DAOC is then hydroxylated to deacetylcephalosporin C (DAC) by another 2-oxoglutarate-linked dioxygenase, deacetylcephalosporin C synthase (DACS). In *S. clavuligerus*, the expandase and hydroxylase activities are performed by a distinct enzyme, whereas one bifunctional enzyme is responsible for both activities in *A. chrysogenum*. In *S. clavuligerus*, DACS constitutes the second branch point in the biosynthesis of β-lactams, dividing the pathways for cephalosporin C and cephamycin C biosynthesis. Acetyl-CoA:DAC acetyltransferase (DAT) is responsible for the last step in the formation of cephalosporin C, where an acetyl group is added to the hydroxyl group of DAC. In the biosynthesis of cephamycin C, the C-3 acetoxy group of DAC is replaced by a carbamyl group by DAC-carbaoyl transferase (DACCT), followed by hydroxylation and methylation of the C-7

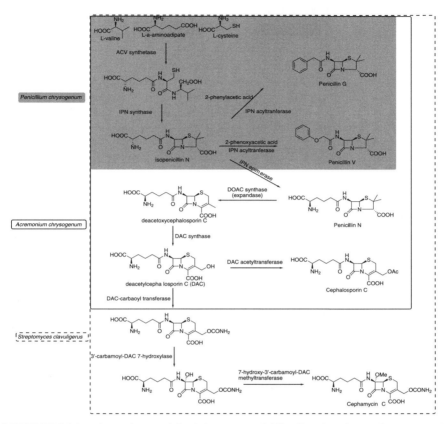

SCHEME 3.34   Biosyntheses of the β-lactams penicillin G and V (in *P. chrysogenum*), cephalosporin C (in *A. chrysogenum*), and cephamycin C (in *S. clavuligerus*).

catalyzed by 3′-carbamoyl-DAC 7-hydroxylase and 7-hydroxy-3′-carbamoyl-DAC methyltransferase [66].

   Penicillin G and V are produced by fermentation of *P. chrysogenum* using a fed-batch process carried out aseptically in stainless steel tank reactors of 30,000–100,000 gallon capacity. The yield and productivity of this classic fermentation process has been improved through many rounds of mutations and selection of better production strains. A number of strain development programs have been carried out, and most of the large producers of penicillin have developed very efficient strains that give very high yields and final titers. The fermentation usually involves two to three initial seed growth phases followed by a fermentation production phase having a time cycle ranging from 120 to 200 h. High dissolved oxygen levels are critical, especially during peak growth periods that often occur at the 40–50 h time period of the cycle. The fermentation mode is fed batch, and crude sugar and precursor are fed throughout the cycle.

Current penicillin fermentations are highly computerized and automated. Temperature, pH, dissolved oxygen, carbon dioxide, sugar, precursor, and ammonia are closely monitored and controlled for optimal antibiotic production [67].

Various carbon sources, now mainly glucose and sucrose, have been adopted for the fermentation. About 65% of the carbon is metabolized for cellular maintenance, 20–25% for growth, and 10–12% for penicillin production [68]. Sugar and precursor are fed continuously, and the sugar is also used to help regulate the pH of the fermentation to between 6.4 and 6.8 during the active penicillin production phase. Corn steep liquor and cottonseed or soybean meal, ammonia, and ammonium sulfate represent major nitrogen sources. The essential precursor substances are phenylacetic acid (for penicillin G) or phenoxyacetic acid (for penicillin V) that are either fed or batched. Mini-harvest protocols, involving removal of 20–40% of the fermentor contents with replacement with fresh sterile medium, are often employed in penicillin fermentations [69].

Skatrud et al. described the first use of recombinant DNA for improvement of an antibiotic-producing strain in 1989 [70]. Since then, recombinant DNA technology has been used extensively for construction of the improved antibiotic producers. The most natural choice for genetic engineering is to overexpress the single β-lactam genes or entire clusters, for example, the penicillin biosynthetic cluster. This approach has indeed led to substantial increases in the production of β-lactams. Some β-lactams are produced by organisms that have low productivities, and other types of β-lactams are produced by chemical modification of fermentation end products. In both cases it is often beneficial to extend the pathway in well-characterized production organisms, thereby increasing the yields of the product or eliminating environmentally damaging chemical steps. However, modification of these strains genetically often results in new problems, for example, precursor requirements and strain instability.

Most penicillins are produced enzymatically and/or chemically from penicillin G or V. As shown in Scheme 3.35 [71], penicillin G is hydrolyzed by penicillin G acylase to afford 6-aminopenicillanic acid (6-APA). 6-APA is then acylated by reaction with Hane anhydride or acyl chloride to afford various penicillins [72]. In 1998, an enzymatic route was developed with D-p-hydroxyphenylglycine 2-hydroxyethyl ester as the donor [73].

Most cephalosporins are enzymatically or chemically synthesized from cephalosporin C. Cephalosporin C is produced by high-yielding strains of *A. chrysogenum* in large-scale, fed-batch fermentations with a harvest titers in the range of 20–25 g/l. Production-scale fermentations are fed-batch with carbon supplied as simple or complex carbohydrate feeds during the growth phase of the fermentation. As the fermentation progresses, sugar feeds are reduced and are usually replaced by higher-energy oils such as soybean oil or peanut oil. Energy conservation from oil as a substrate is considerably less efficient and leads to slower growth, with the vegetative mycelium becoming largely transformed into multicellular arthrospores. The arthrospore stage leads to greater oxygen availability to the organism and results in rapid cephalosporin production [69]. Cephalosporin C is hydrolyzed to 7-aminocephalosporanic

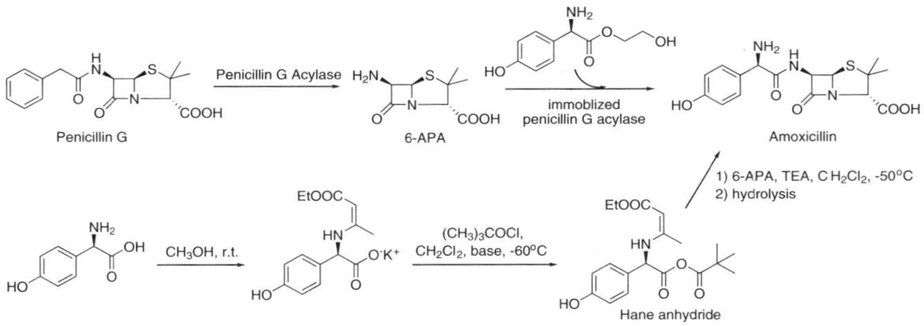

SCHEME 3.35   Comparisons of enzymatic and chemical production routes to amoxicillin.

acid (7-ACA) by the D-amino acid oxidase and glutaryl acylase. Catalase is added to remove hydrogen peroxide (H_2O_2), which quickly inactivates the enzymes; this step allows these three enzyme reactions to be performed in one pot (Scheme 3.36) [74].

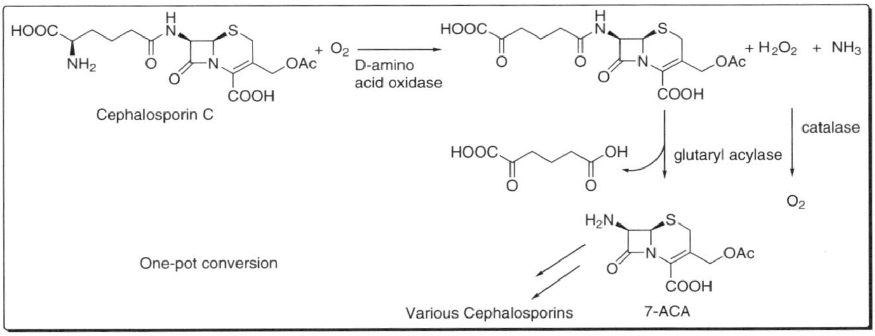

SCHEME 3.36   Production of 7-aminocephalosporanic acid.

About one-third of commercial cephalosporins are derived from 7-aminodeacetoxycephalosporanic acid (7-ADCA). Because of the lower cost of penicillin, 7-ADCA is usually produced from penicillin G by ring expansion via sequential sulfoxidation, esterification, and dehydration/expansion of the penicillin nucleus [75]. The ring expansion is also realized through recombinant *Penicillium* encoded with expandase (Scheme 3.37) [76].

3.3.2.2 Monobactams

Aztreonam (trade name Azactam) is a synthetic monocyclic β-lactam antibiotic (a monobactam), with the nucleus based on a simpler monobactam isolated from *Chromobacterium violaceum* displaying resistance to some β-lactamases [77]. It was approved by the U.S. FDA in 1986. In 2002, an in vitro synergistic additive effect of aztreonam with other antibiotics, for example, ciprofloxacin and levofloxacin, against *Pseudomonas aeruginosa* was observed [78]. Scheme 3.38

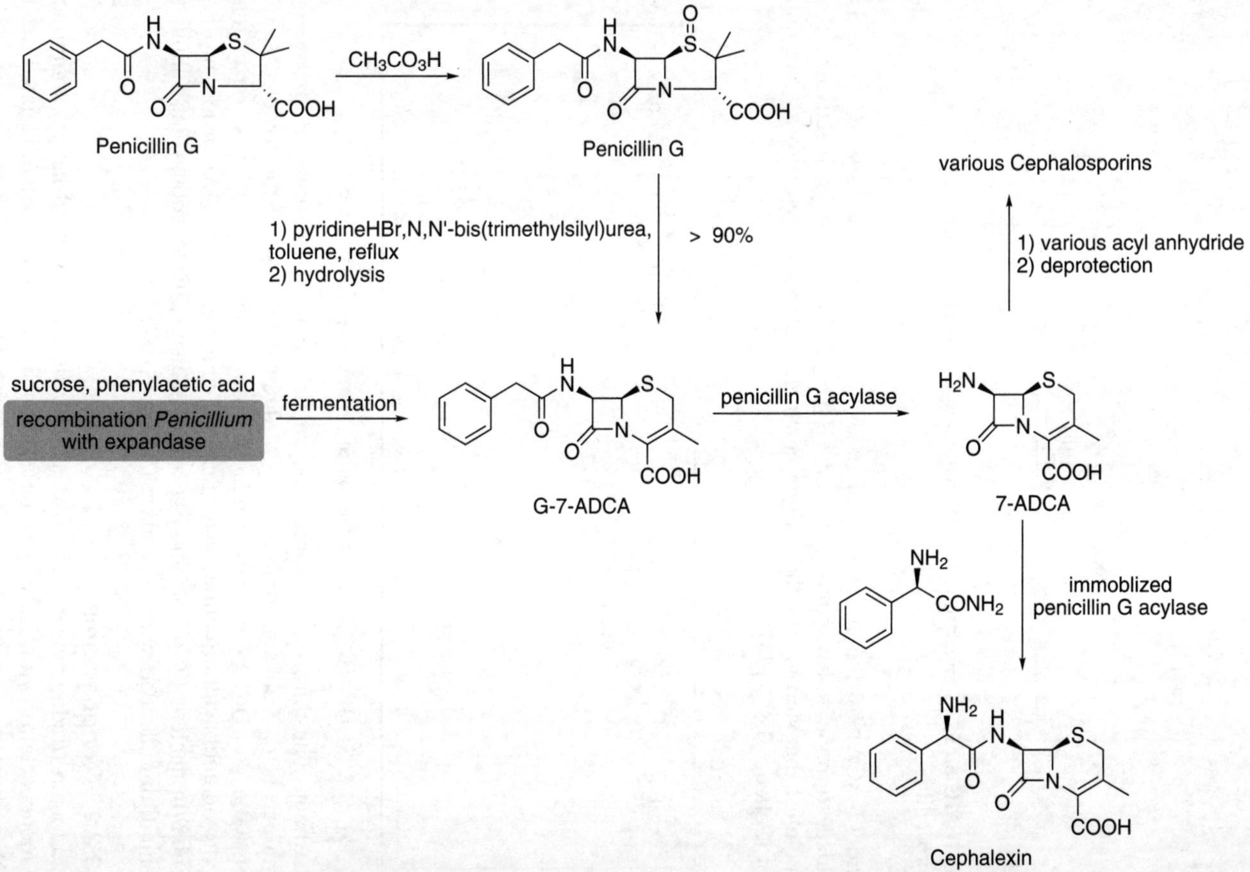

SCHEME 3.37 An enzymatic production route to Cephalexin.

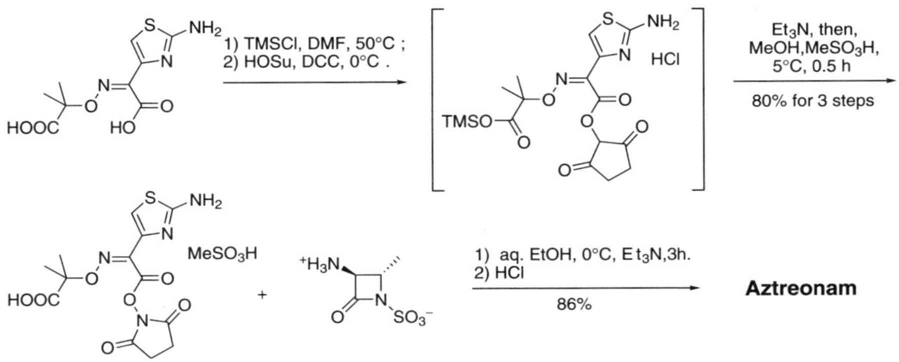

SCHEME 3.38 Chemical synthesis of aztreonam.

monobactams nucleus   Tigemonam   Aztreonam

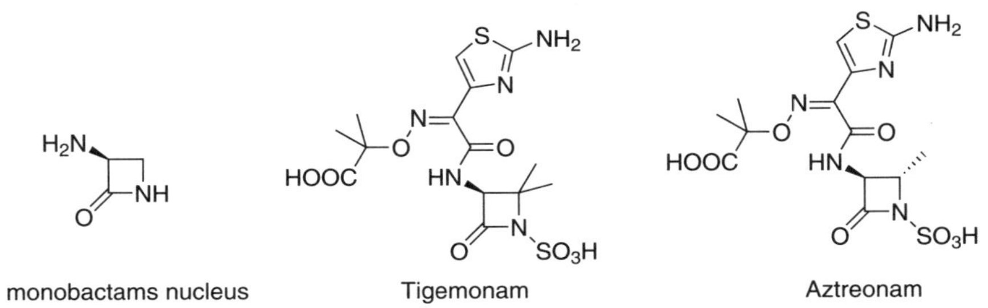

FIGURE 3.4 Two marketed monobactam antibiotics and their synthesis.

shows the chemical synthesis of aztreonam. Tigemonem is synthesized through a similar procedure (Fig. 3.4).

3.3.2.3 Carbapenems

The carbapenems have a β-lactam ring fused to a five-membered ring that does not contain sulfur. Sulfur is present in the molecule outside the ring in all carbapenems produced by *Streptomycetes*. A large number of carbapenems have been discovered (Fig. 3.5). Thienamycin was discovered at Merck by Kahan and co-workers by a highly sensitive screening protocol based on the inhibition of peptidoglycan synthesis in 1979 [79]. It is one of the most potent broad-spectrum, nontoxic antibacterial compounds.

Townsend has made great efforts to elucidate the carbapenem biosynthesis [80]. A chemical route to prepare thienamycin and its derivatives was also reported by his group [81] (Scheme 3.39). However, carbapenem fermentation is still low titer and difficult to be recovered and purified. Most of the marketed carbapenem antibiotics are chemically synthesized.

3.3.2.4 Clavams and β-Lactamase Inhibitors

A variety of bacteria have been found to be resistant to the penicillins and cephalosporins, which is attributed to their ability to produce β-lactamase. Clavulanic acid possesses only weak antibiotic activity, but it is an excellent inhibitor of

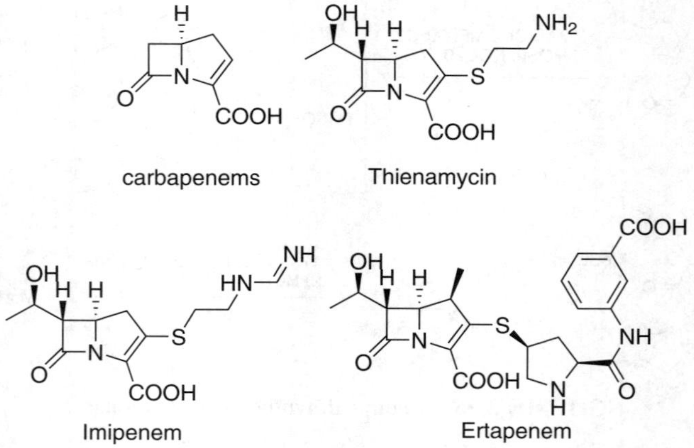

FIGURE 3.5   Some examples of marketed carbapenem antibiotics.

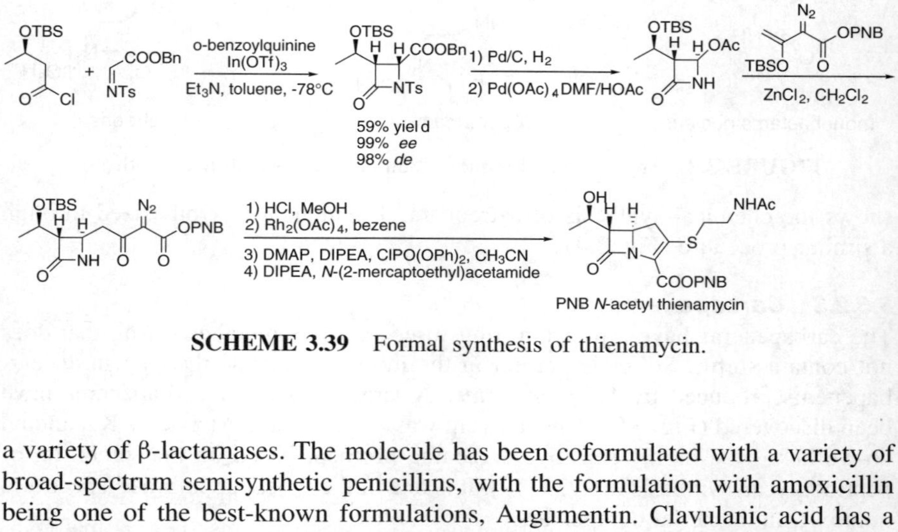

SCHEME 3.39   Formal synthesis of thienamycin.

a variety of β-lactamases. The molecule has been coformulated with a variety of broad-spectrum semisynthetic penicillins, with the formulation with amoxicillin being one of the best-known formulations, Augumentin. Clavulanic acid has a world sales value of ~2 billion per year [82]. Some examples of β-lactamase inhibitors are shown in Figure 3.6.

Clavulanic acid is produced by submerged cultivation of *Streptomyces clavuligerus* [83]. The purification of clavulanic acid can be realized by adsorption of the antibiotic on active carbon from the culture filtrates followed by elution with aqueous acetone or solvent extraction at pH 2.0 using butanol with back-extraction into water at pH 7.0, leading to the crude clavulanic acid. In 2009, a phosphate- and polyethylene glycol-based aqueous two-phase system was employed to purify clavulanic acid from culture media. A 100% yield and a 1.5-fold purification factor were obtained when the polyethylene glycol had clavams   Clavulanic Acid

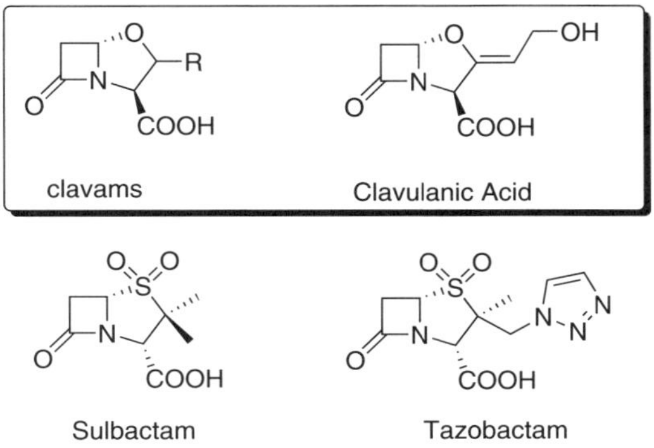

Sulbactam   Tazobactam

FIGURE 3.6   Some examples of marketed β-lactamase inhibitors.

a low molecular mass, pH close to the isoelectric point, and lower top phase volume [84].

Sulbactam and tazobactam (Scheme 3.40) are irreversible inhibitors of β-lactamase. They both are chemically synthesized from penicillin [85]. A formulation of sulbactam/ampicillin was developed in 1987 and marketed in the United States under the tradename Unasyn. Tazobactam is combined with piperacillin to form Tazocin or Zosyn, which was marketed by Wyeth.

3.3.3   Development and Production of Polyketide Pharmaceuticals

Polyketides are structurally a very diverse family of natural products with diverse biological activities and pharmacological properties built from malonyl-CoA monomers. They are the building blocks for a broad range of natural products or are further derivatized. There are about 10,000 polyketides that have thus far been identified from bacteria, fungi, and plants. Many polyketides display a variety of pharmacological bioactivities, which include antibacterial activity (tetracycline, erythromycin, rifamycin), antifungal activity (amphotericin), anticancer activity (doxorubicin), and immunosuppressant activity (rapamycin) [86].

The biosynthesis of polyketides is catalyzed by large multienzyme complexes called polyketide synthases (PKSs). The polyketides are synthesized from starter units such as acetyl-CoA and propionyl-CoA. Extender units such as malonyl-CoA or/and methylmalonyl-CoA are repetitively added via a decarboxylative process to a growing chain. Ultimately, the polyketide chain is released from the PKS by cleavage of thioester (TE), usually accompanied by chain cyclization [87]. There are at least three types of bacterial PKSs. Type I PKSs are large, highly modular proteins. Type I PKSs are multifunctional enzymes that are organized into modules, each of which harbors a set of distinct, noniteratively acting activities responsible for the catalysis of one cycle of polyketide chain elongation [88]. Type II PKSs are aggregates of monofunctional proteins. Type III PKSs do not use acyl carrier protein (ACP) domains.

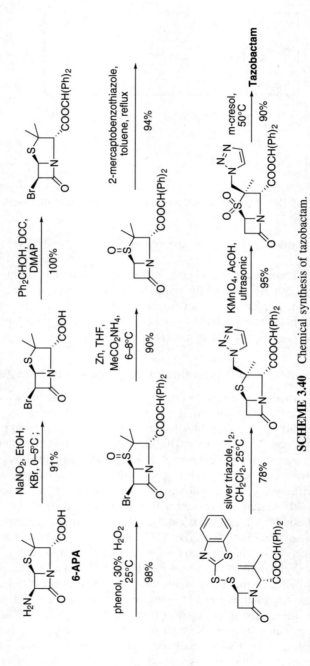

SCHEME 3.40 Chemical synthesis of tazobactam.

3.3.3.1 Erythromycins

Erythromycin is a macrocyclic compound containing a 14-membered lactone ring with 10 asymmetric centers and two sugars that is produced from a strain of the actinomycete *Saccharopolyspora erythraea*. Erythromycin has an antimicrobial spectrum similar to or slightly wider than that of penicillin, and is often used for people who are allergic to penicillins. It was first marketed by Eli Lilly, and it is today commonly known as erythromycin ethylsuccinate (EES). EES, along with two semisynthetic erythromycins, Clarithromycin and Azithromycin (Fig. 3.7), are the world's best-selling antibiotics.

Erythromycin is produced by the fermentation of *Saccharopolyspora erythraea* [89]. A group in Egypt optimized the cultivation medium through the change of carbon and nitrogen sources to cheaper sources in order to reduce the cost of medium and to utilize sugar cane molasses, a major sugar industry by-product. It was found that the addition of sugar cane molasses, a sole carbon source at a concentration of 60 g/l, accompanied by corn steep liquor in combination with ammonium sulfate gave the maximal erythromycin production. The antibiotic production in this medium reached about 600 mg/l, which was about 33% higher than the value obtained in glucose-based medium. On the other hand, the addition of *n*-propanol in a concentration of 1% (v/v) increased the antibiotic production (up to about 720 mg/l after 144 h) [90]. Zhang et al. investigated the effects of feeding different available nitrogen sources from 80 h in erythromycin biosynthesis phase on the erythromycin A (Er-A) production in

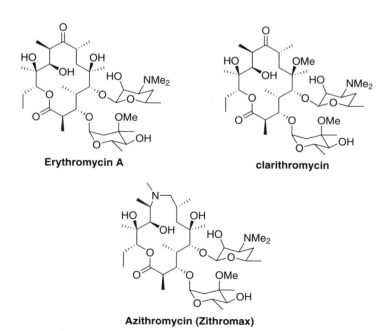

Erythromycin A

clarithromycin

Azithromycin (Zithromax)

FIGURE 3.7 Some examples of the marketed erythromycins.

a 50-l fermentor. When ammonium sulfate was fed at a high feeding rate, the maximal Er-A production and ratio of Er-A to Er-C were 7,953 U/ml and 98.18:1 at 184 h, respectively, which were higher than those of the control (6742 U/ml and 5.47:1). The ammonium sulfate feeding process was successfully scaled up from a 50-l to a 25,000-l fermentor [91].

The macrolide aglycon of erythromycin, 6-deoxyerythronolideB (6-dEB), is produced by 6-deoxyerythronolide B synthase (DEBS) from *Saccharopolyspora erythraea*, which is the best-known member of the modular PKS family [92] (Scheme 3.41).

The successful total biosynthesis of 6-dEB in a genetically engineered *E. coli* strain represented a milestone for engineered biosynthesis of macrolides. Subsequently, heterologous biosynthesis of the bioactive erythromycin C was achieved by coexpressing the deoxysugar biosynthetic genes from the megalomicin (which is structurally related to erythromycin) gene cluster in a 6-dEB-producing *E. coli* host [93]. Khosla and Lee then used random mutagenesis and high-throughput screening of the *E. coli*. producer to yield mutant strains that have improved erythromycin C titers [94].

3.3.3.2 *Doxorubicin and Tetracenomycin C*

Doxorubicin (trade name Adriamycin) is a chemotherapy drug used to treat bladder, breast, head, and neck cancer and leukemia. Daunorubicin is more abundantly found as a natural product because it is produced by a number of different wild-type strains of streptomyces [95]. In 1969, the first doxorubicin-producing species, *Streptomyces peucetius* subspecies *caesius* ATCC 27952 was discovered by Arcamone et al. by mutating a strain producing daunorubicin [96] (Fig. 3.8). In 1996, Strohl's group discovered and characterized DoxA, the gene encoding a cytochrome P450 hydroxylase [97]. Cytochrome P450 hydroxylase catalyzed not only the oxidation of doxorubicin to daunorubicin but also multiple steps in doxorubicin biosynthesis. This result may indicate that all daunorubicin-producing strains have the necessary genes to produce doxorubicin [98]. Hutchinson's group found that several strains of *Streptomyces* could produce doxorubicin under special environmental conditions [99]. His group constructed a doxorubicin-overproducing strain, *Streptomyces peucetius* ATCC 29050, by introducing DoxA-encoding plasmids and mutations to deactivate enzymes that shunt doxorubicin precursors to other products, such as baumycin-like glycosides [100].

A semisynthetic production of doxorubicin from daunorubicin is also reported by Solvias AG [101]. Valrubicin is being developed by Endo Pharmaceuticals for the treatment of a distinct form of bladder cancer as of March 2009. For the structures of daunorubicin and its analog, see Figure 3.8.

The biosynthesis of doxorubicin [102] (Scheme 3.42) is initiated by the loading of the propionyl starter unit from propionyl-CoA to CLF. Propionyl is then added to a two-carbon ketide unit, decarboxylated from malonyl-ACP, to produce the five-carbon β-ketovaleryl ACP. The five-carbon diketide is delivered by the ACP to the cysteine sulfhydryl group at the KS active site by thioester exchange, and the ACP is released from the chain. The free ACP picks up another malonate group from malonyl-CoA, also by thioester

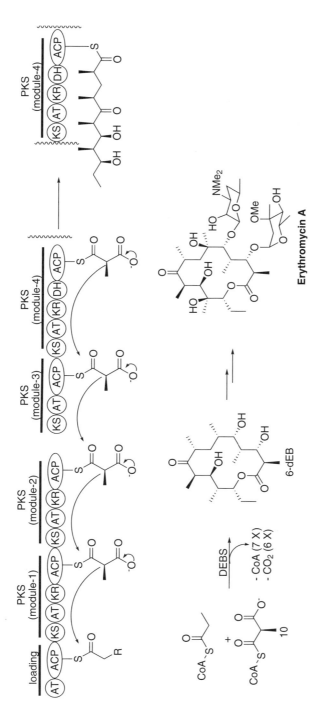

SCHEME 3.41 Synthesis pathway of erythromycin by the 6-deoxyerythronolide B synthase.

121

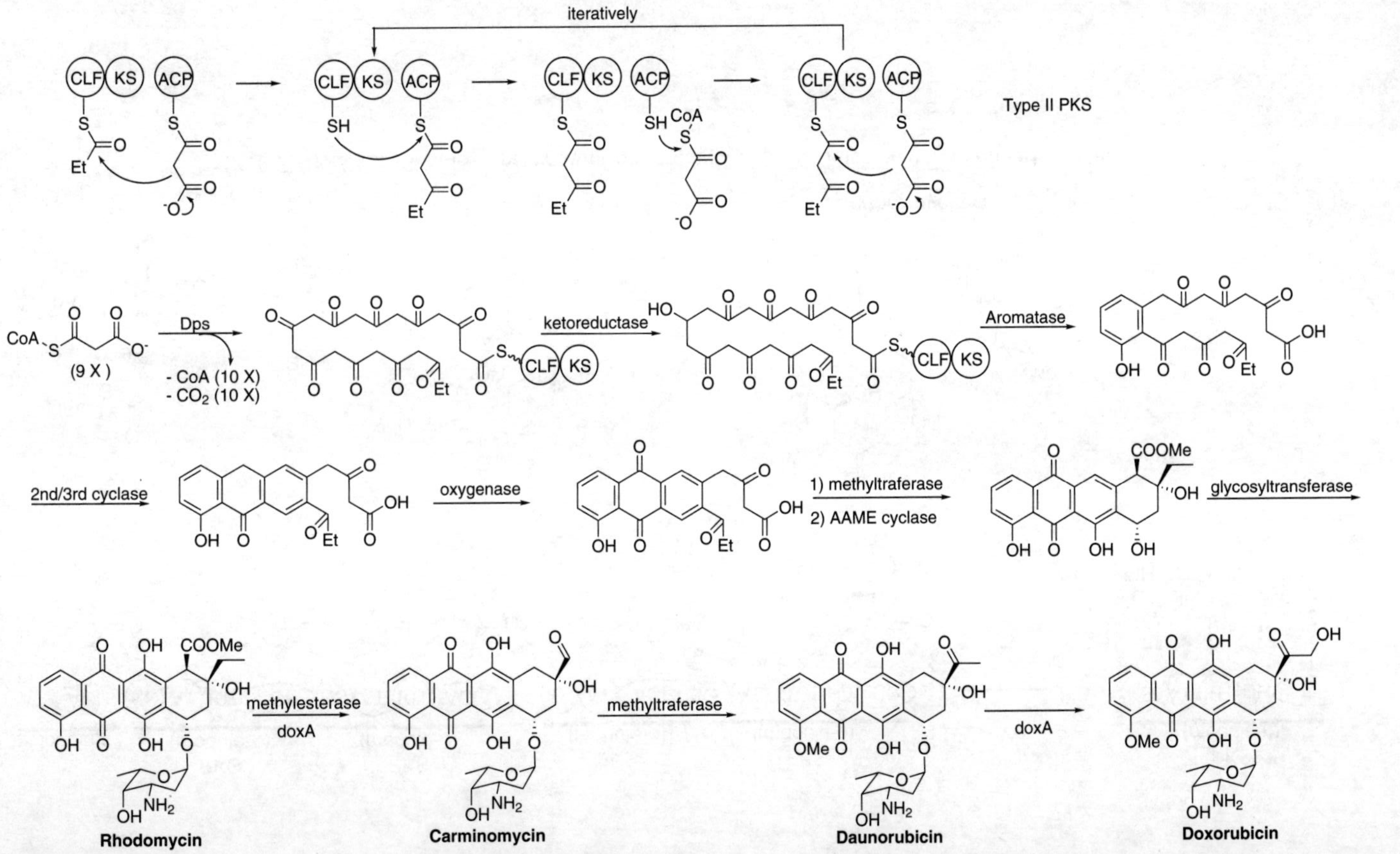

SCHEME 3.42 The biosynthesis of doxorubicin by type II PKS. Catalytic domains: CLF, chain length factor; KS, ketosynthase; ACP, acyl carrier protein.

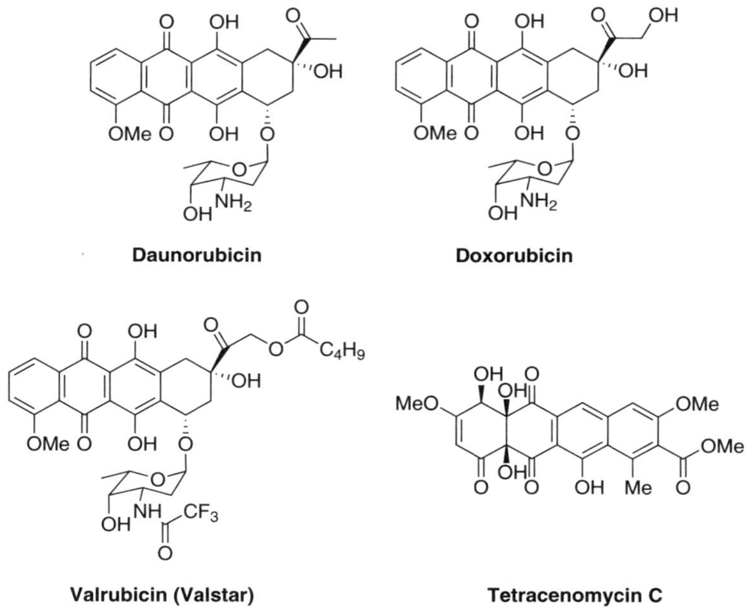

FIGURE 3.8   Structures of daunorubicin, doxorubicin, tetracenomycin C, and valrubicin.

exchange, with release of the CoA. The ACP brings the new malonate to the active site of the KS where is it decarboxylated, possibly with the help of the CLF subunit, and joined to produce a seven-carbon triketide, now anchored to the ACP. Again the ACP hands the chain off to the KS subunit, and the process is repeated iteratively until the decaketide is completed.

Tetracenomycin C is a cytotoxic antibiotic produced by *Streptomyces glaucescens* that is notable for its broad activity against actinomycetes [103]. The biosynthesis of the backbone of tetracenomycin C is accomplished by a typical type II polyketide synthase, a multienzyme complex that carries a single set of iteratively acting activities (Scheme 3.43) [104].

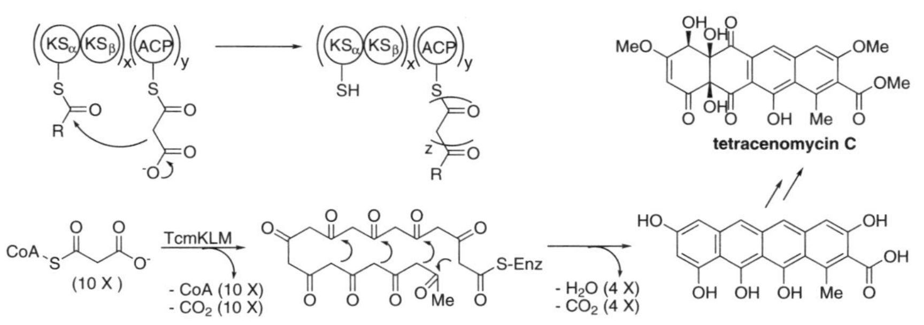

SCHEME 3.43   Biosynthesis of the backbone of tetracenomycin C by type II PKS.

3.3.4 Development and Production of Paclitaxel and Docetaxel

Paclitaxel (Taxol as trademarked by Bristol-Myers Squibb) (Fig. 3.9), gained first marketing approval from the U.S. Food and Drug Administration (FDA) for the treatment of refractory ovarian cancer in 1992 and metastatic breast cancer in 1994 [105]. Paclitaxel was first isolated from the bark of the Pacific yew, *Taxus brevifolia*, in 1971 [106]. However, this method could not meet the increasing demand for Taxol on the market because these yews grow very slowly and are a rare and endangered species. Currently, paclitaxel is produced by two methods. One method is semisynthesis from paclitaxel precursors, baccatin III or 10-deacteylbaccatin III, that are more readily available from the needles of various yew species as a more sustainable resource (Scheme 3.44) [107].

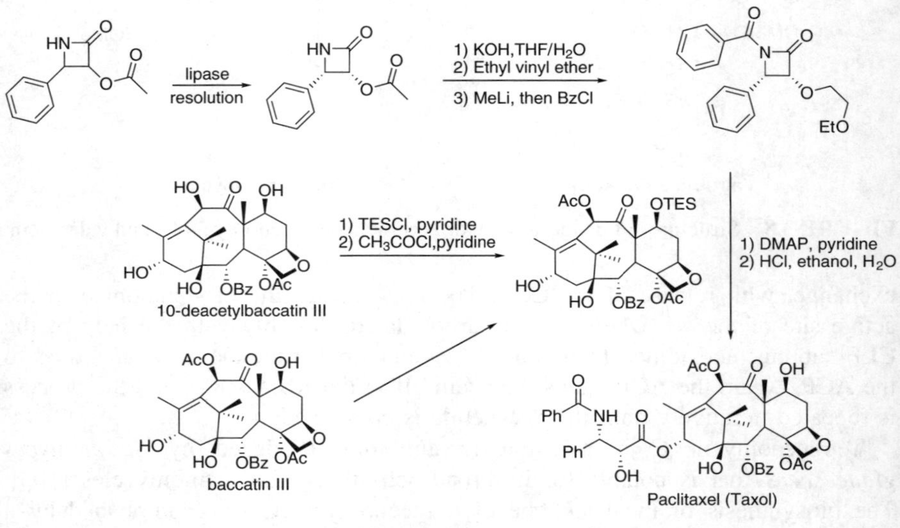

SCHEME 3.44   Semisynthesis of paclitaxel.

The other method is plant cell culture, which now is efficient enough for scale-up [108]. For example, all paclitaxel production for Bristol-Myers Squibb uses plant cell fermentation technology developed by the biotechnology company Phyton Biotech, Inc and carried out at their plant in Germany [109].

All *Taxus* species produce paclitaxel, although the mixture of taxoids accumulated can vary widely between species and between tissues of the same species [110]. Thus far, a number of *Taxus* cell cultures have been developed from different *Taxus* species. To enhance time-space yield of paclitaxel in *Taxus* cell suspension culture, different strategies were applied, including, for example, the use of elicitors, novel bioreactors [111], immobilized yew cells, perfusion, and continuous cultivation [112]. The synthesis of second metabolites (such as paclitaxel) in plants is believed to be part of the plant defense responses to environmental stresses and pathogen attacks. Most strategies for improving the yield of paclitaxel in plant cell suspension culture are based on this mechanism [113]. Many elicitors and signal molecules, for example, methyl jasmonate (MJ), arachidonic acid, silver ion, chitosan, fungal mycelia extracts, have been used to improve taxoid production by plant cell suspension culture based on this principle [114]. MJ, which specifically induces the upregulation of secondary metabolic genes involved in stress, wounding, and pathogen ingress, might be the most effective of this group [115].

In 2005, Bentebibel et al. reported a paclitaxel productivity of 2.71 mg/l per day using immobilized cells elicited with methyl jasmonate in a 5-l turbine-stirred bioreactor; this was fivefold higher than that of the control cells [116].

In 1993, Stierle and Strobel discovered that a newly described fungus, *Taxomyces andreanae*, living in the yew tree can produce paclitaxel [117]. Since then, various endophytic fungi associated with *Taxus* were found to be able to produce paclitaxel, including the following species: *Taxomyces sp.*, *Trichoderma*, *Tubercularia sp.*, *Monochaetia sp.*, *Fusarium lateritium*, *Pestalotiopsis microspora*, *Pestalotiopsis guepinii*, *Pithomyces sp.*, *Pestalotia bicilia*, *Papulaspora sp.1*, *Pseudomonas aureofaciens*, *Pleurocytospora taxi*, *Cephalosporium spp.*, *Chaetomium*, *Martensiomyces spp.*, *Mycelia sterilia*, *Nodulisporium sylviforme*, *Rhizoctonia* sp., *Penicillium*, *Alternaria sp.*, *Alternaria taxi*, *Ectostroma sp.1*, *Botrytis sp.1*, *Alternaria alternate*, *Botrytis taxi*, and recently *Aspergillus niger*. Although the production of Taxol via plant cell culture has been established, this process still has a variety of disadvantages, such as long period and variability of the product accumulation among cell lines, within cell aggregates, and over the course of the cell culture [118]. Those investigations might offer important information and a new resource for the production of Taxol by endo-fungus fermentation [119]. Strain improvement and optimization of the media for some strains could produce Taxol at an amount of 200–300 µg/l [120]. However, the amounts are still small compared to those produced by the various plant species.

In 2009, Li and Tao established a coculture of the suspension cells of *Taxus chinensis* var. mairei and its endophytic fungus, *Fusarium mairei*, in a 20-l cobioreactor for paclitaxel production. The cobioreactor consists of two-unit tanks (10 l each) with a repairable separate membrane in the center, culturing *Taxus* suspension cells in one tank and growing fungi in another. The endophytic fungus *F. mairei* could produce ~200 µg/l of paclitaxel in its paclitaxel-producing medium. Moreover, it strongly stimulated the *Taxus* cell to produce paclitaxel in the cobioreactor, which can possibly be attributed to the induction of the oligosaccharide, since *F. mairei* may produce an oligosaccharide with molecular masses ~2 kDa. By optimization of the coculture conditions, the *Taxus* cell cultures in the cobioreactor produced 25.63 mg/l of paclitaxel within 15 days; this was equivalent to a productivity of 1.71 mg/l per day and 38-fold higher than that by uncoupled culture (0.68 mg/l within 15 days) [121].

Docetaxel (Taxotere as trademarked by Sanofi-Aventis) (Fig. 3.9), a semisynthetic analog of paclitaxel, is a clinically well-established antimitotic chemotherapy medication used mainly for the treatment of breast, ovarian, and non-small-cell lung cancer [122]. Docetaxel has an approved claim for treatment of patients who have locally advanced or metastatic breast or non-small-cell lung cancer who have undergone anthracycline-based chemotherapy that failed to stop cancer progression or have relapsed. Administered as a 1-hour infusion every 3 weeks generally over a 10-cycle course, docetaxel is considered as or more effective than doxorubicin, paclitaxel, and fluorouracil as a cytotoxic antimicrotubule agent. The semisynthesis of docetaxel is accomplished by condensation of protected 10-deacetylbaccatin and its side chain (Scheme 3.45) [123]. In 2006, Bioxel Pharma, Inc. (Canada) filed two patents covering a unique synthetic route to docetaxel and other taxane drugs. The synthesis

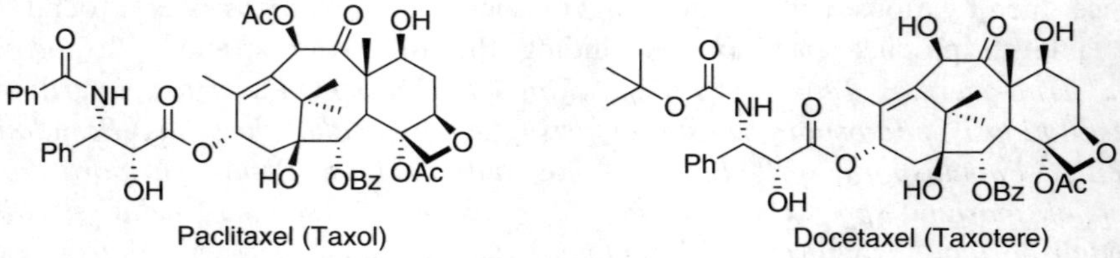

Paclitaxel (Taxol)   Docetaxel (Taxotere)

FIGURE 3.9   Structure of paclitaxel and docetaxel.

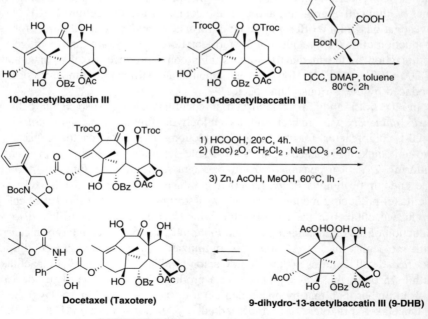

10-deacetylbaccatin III   Ditroc-10-deacetylbaccatin III

DCC, DMAP, toluene
80°C, 2h

1) HCOOH, 20°C, 4h.
2) (Boc)$_2$O, CH$_2$Cl$_2$, NaHCO$_3$, 20°C.

3) Zn, AcOH, MeOH, 60°C, Ih.

Docetaxel (Taxotere)   **9-dihydro-13-acetylbaccatin III (9-DHB)**

SCHEME 3.45   Semisynthesis of docetaxel.

process started with 9-dihydro-13-acetylbaccatin III (9-DHB), a taxane precursor extracted from the needles of Canada yew, *Taxus canadensis* [124].

3.3.5 Summary

Biosynthesis not only produces pharmaceuticals with much lower cost but also affords new chemical entities that allow the development of more novel pharmaceuticals. Moreover, in the past decade, combinatorial biosynthesis, defined as the application of various biotechnological manipulations to engineer biosynthetic pathways of natural products aiming to expand their structural diversities in natural biosynthetic machineries, offers a promising alternative way to drug discovery [125]. The emerging insights into many natural product biosyntheses will provide blueprints to engineer these pathways and biocatalysts. Therefore, biotechnology will play a more and more important role in pharmaceuticals production and drug discovery through successful integration of bioinformatics, biochemical, and genetic tools.

REFERENCES

1. a) Bommarius, A. S., Riebel, B. R. (2004). *Biocatalysis: Fundamentals and Applications*. Wiley-VCH.   b) Aehle, W. Ed. (2007). *Enzymes in Industry: Production and Applications*. 3 rd ed., Wiley-VCH.   c) Tao, J. H., Lin, G. Q., Liese, A. (eds). (2008). *Biocatalysis for the Pharmaceutical Industry: Discovery, Development, and Manufacturing*. John Wiley & Sons, Inc.;   d) Patel, R. N. (ed). (2006). *Biocatalysis in the Pharmaceutical and Biotechnology Industries*. CRC;   e) Liese, A., Seelbach, K., Wandrey, C. (eds). (2006). *Industrial Biotransformations*. Wiley-VCH; f) Hou, C. T. (eds). (2005). *Handbook of Industrial Biocatalysis*. CRC;   g) Illanes, A. (ed). (2008). *Enzyme Biocatalysis: Principles and Applications*. Springer.   h) Stephanopoulos, G. (1999). Metabolic fluxes and metabolic engineering. *Metab. Eng*. **1**, 1–11;   i) Buckland, B. C., Robinson, D. K., Chartrain, M. (2000). Biocatalysis for pharmaceuticals-status and prospects for a key technology. *Metab. Eng*. **2**, 42–48.

2. Xu, Z., Singh, J., Schwinden, M. D., Zheng, B., Kissick, T. P., Patel, B., Humora, M. J., Quiroz, F., Dong, L., Hsieh, D. M., Heikes, J. E., Pudipeddi, M., Lindrud, M. D., Srivastava, S. K., Kronenthal, D. R., Mueller, R. H. (2002). Process research and development for an efficient synthesis of the HIV protease inhibitor BMS-232632. *Org. Proc. Res. Dev*. **6**, 323–328.

3. Patel, R. N., Chu, L., Muller, R. (2003). Diastereoselective microbial reduction of (*S*)-[3-chloro-2-1-(phenylmethyl)propyl]carbamic acid, 1,1- dimethylethyl ester. *Tetrahedron Asymmetry* **14**, 3105–9109.

4. a) Bommarius, A. S., Schwarm, M., Stingl, M., Kotterhahn, K., Huthmacher, K., Drauz, K. Synthesis and use of enantiomerically pure *tert*-leucine. (1995). *Tetrahedron Asymmetry* **6**, 2851–2888;   b) Wichmann, R., Wandrey, C., Bückmann, A. F., Kula, M. R. (1981). Continuous enzymatic transformation in an enzyme membrane reactor with simultaneous NAD(H) regeneration. Reprinted from *Biotechnol. Bioeng*.

vol. XXIII, no. 12, p. 2789–2802; c) Wichmann, R. Wandrey, C. (2000). Continuous enzymatic transformation in an enzyme membrane reactor with simultaneous NAD(H) regeneration. *Biotechnol. Bioeng*. **67**, 791–804.

5. a) Jacobsen, E. N., Zhang, W., Muci, A. R., Ecker, J. R., Deng, L. (1991). Highly enantioselective epoxidation catalysts derived from 1,2-diaminocyclohexane. *J. Am. Chem. Soc*. **113**, 7063–7064; b) Larrow, J. F., Jacobsen, E. N. (1994). Kinetic resolution of 1,2-dihydronaphthalene oxide and related epoxides via asymmetric C–H hydroxylation. *J. Am. Chem. Soc*. **116**, 12129–12130.

6. a) Senanayake, C. H., Roberts, F. E., DiMichele, L. M., Ryan, K. M., Liu, J., Fredenburgh, L. E., Foster, B. S., Douglas, A. W., Larsen, R. D., Verhoeven, T. R., Reider, P. J. (1995). The behavior of indene oxide in the Ritter reaction: a simple route to *cis*-aminoindanol. *Tetrahedron Lett*. **36**, 3993–3996; b) Senanayake, C. H., Smith, G. B., Ryan, K. M., Fredenburgh, L. E., Liu, J., Roberts, F. E., Hughes, D. L., Larsen, R. D., Verhoeven, T. R., Reider, P. J. (1996). The role of 4-(3-phenylpropyl)pyridine *N*-oxide (P3NO) in the manganese-salen-catalyzed asymmetric epoxidation of indene. *Tetrahedron Lett*. **37**, 3271–3274; c) Senanayake, C. H., DiMichele, L. M., Liu, J., Fredenburgh, L. E., Ryan, K. M., Roberts, F. E., Larsen, R. D., Verhoeven, T. R., Reider, P. J. (1995). Regio- and stereocontrolled syntheses of cyclic chiral *cis*-amino alcohols from 1,2-diols or epoxides. *Tetrahedron Lett*. **36**, 7615–7618.

7. Zhang, J., Reddy, J., Roberge, C., Senanayake, C., Greasham, R., Chartrain, M. (1995). Chiral bio-resolution of racemic indene oxide by fungal epoxide hydrolases. *J. Ferment. Bioeng*. **80**, 244–246.

8. Buckland, B. C., Drew, S. W., Connors, N. C., Chartrain, M. M., Lee, C., Salmon, P. M., Gbewonyo, K., Zhou, W., Gailliot, P., Singhvi, R., Olewinski, R. C., Sun, W. J., Reddy, J., Zhang, J., Jackey, B. A., Taylor, C., Goklen, K. E., Junker, B., Greasham, R. L. (1999). Microbial conversion of indene to indandiol: a key intermediate in the synthesis of Crixivan. *Meta. Eng*. **1**, 63–74.

9. Zhang, N., Stewart, B. G., Moore, J. C., Greasham, R. L., Robinson, D. K., Buckland, B. C., Lee, C. (2000). Directed evolution of toluene dioxygenase from *Pseudomonas putida* for improved selectivity toward *cis*-indandiol during indene bioconversion. *Metab. Eng*. **2**, 339–348.

10. Buckland, B. C., Robinson, D. K., Chartrain, M. (2000). Biocatalysis for pharmaceuticals-status and prospects for a key technology. *Metab. Eng*. **2**, 42–48.

11. a) Shirae, H., Yokozeki, K. (1991). Purifications and properties of orotidine-phosphorolyzing enzyme and purine nucleoside phosphorylase from *Erwinia carotovora* AJ 2992. *Agric. Biol. Chem*. **55**, 1849–1857; b) Shirae, H., Yokozeki, K., Kubota, K. (1988). Enzymatic production of ribavirin from pyrimidine nucleosides by *Enterobacter aerogenes* AJ 11125. *Agric. Biol. Chem*. **52**, 1233–1237; c) Shirae, H., Yokozeki, K., Kubota, K. (1988). Enzymatic production of ribavirin from orotidine by *Erwinia carotovora* AJ 2992. *Agric. Biol. Chem*. **52**, 1499–1504; d) Shirae, H., Yokozeki, K., Kubota, K. (1988). Enzymatic production of ribavirin from purine nucleosides by *Brevibacterium acetylicum* AJ 1442. *Agric. Biol. Chem*. **52**, 1777–1783; d) Shirae, H., Yokozeki, K., Kubota, K. (1988). Enzymatic production of ribavirin. *Agric. Biol. Chem*. **52**, 295–296.

12. Tamarez, M., Morgan, B., Wong, G. S. K., Tong, W., Bennett, F., Lovey, R., McCormick, J. L., Zaks, A. (2003). Pilot-scale lipase-catalyzed regioselective acylation of ribavirin in anhydrous media in the synthesis of a novel prodrug intermediate. *Org. Process Res. Dev*. **7**, 951–953.

13. Patel, R. N. (2006). *Biocatalysis in the Pharmaceutical and Biotechnology Industries*. CRC Press, p. 648–651.

14. Hoekstra, M. S., Sobieray, D. M., Schwindt, M. A., Mulhern, T. A., Grote, T. M., Huckabee, B. K., Hendrickson, V. S., Franklin, L. C., Granger, E. J., Karrick, G. L. (1997). Chemical development of CI-1008, an enantiomerically pure anticonvulsant. *Org. Proc. Res. Dev*. **1**, 26–38.

15. Burk, M., de Koning, P. D., Grote, T. M., Hoekstra, M. S., Hoge, G., Jennings, R. A., Kissel, W. S., Le, T. V., Lennon, I. C., Mulhern, T. A., Ramsden, J. A., Wade, R. A. (2003). An enantioselective synthesis of (S)-(+)-3-aminomethyl-5-methylhexanoic acid via asymmetric hydrogenation. *J. Org. Chem*. **68**, 5731–5734.

16. Hu, S., Martinez, C. A., Tao, J., Tully, W. E., Kelleher, P., Yves, D. (2005). Preparation of pregabalin and related compounds. US20050283023.

17. a) Xie, Z., Feng, J., Garcia, E., Bernett, M., Yazbeck, D., Tao, J. (2006). Cloning and optimization of a nitrilase for the synthesis of (3S)-3-cyano-5-methyl hexanoic acid. *J. Mol. Catal. B Enzym*. **41**, 75–80;   b) Burns, M. P., Weaver, J. K., Wong, J. W. (2005). Stereoselective bioconversion of aliphatic dinitriles into cyano carboxylic acids. WO20050100580.

18. a) Jacobsen, E. N., Deng, L., Furukawa, Y., Martinez, L. (1994). Enantioselective catalytic epoxidation of cinnamate esters. *Tetrahedron* **50**, 4323–4334;   b) Hulshof, L. A., Roskan, J. H. (1994). Phenylglycidate stereoisomers, conversion products thereof with e.g. 2-nitrothiophenol and preparation of diltiazem. EP0343714/1994;   c) Shibatani, T., Nakamichi, K., Matsumae, H. (1990). Method for preparing optically active 3-phenylglycidic acid esters. EP0362556/1990;   d) Seki, M., Furutani, T., Imashiro, R., Kuroda, T., Yamanaka, T., Harada, N., Arakawa, H., Kusama, M., Hashiyama, T. (2001). A novel synthesis of a key intermediate for diltiazem. *Tetrahedron Lett*. **42**, 8201–8205;   e) Schwatz, A., Madan, P. B., Mohacsi, E., O'Brien, J. P., Todaro, L. J., Coffen, D. L. (1992). Enantioselective synthesis of calcium channel blockers of the diltiazem group. *J. Org. Chem*. **57**, 851–856.

19. Evans, C. T., Roberts, S. M., Shoberu, K. A., Sutherland, A. G. (1992). Potential use of carbocyclic nucleosides for the treatment of AIDS: chemo-enzymatic synthesis of the enantiomers of carbovir. *J. Chem. Soc. Perkin Trans*. **1**, 589–592.

20. a) Taylor, S. J. C., Sutherland, A. G., Lee, C., Wisdom, R., Thomas, S., Roberts, S. M., Evans, C. T. (1990). Chemoenzymatic synthesis of (−)-carbovir utilizing a whole cell catalysed resolution of 2-azabicyclo[2.2.1]hept-5-en-3-one. *J. Chem. Soc. Chem. Commun*. 1121;   b) Evans, C. T., Roberts, S. M. (1991). Chiral azabicyloheptanone and a process for their preparation. EP0424064/1991;   c) Taylor, S. J. C., Brown, R. C., Keene, P. A., Taylor, I. N. (1999). Novel screening methods-the key to cloning commercially successful biocatalysts. *Bioorg. Med. Chem*. **7**, 2163–2168.

21. Mahmoudian, M., Lowdon, A., Jones, M., Dawson, M., Wallis, C. (1999). A practical enzymatic procedure for the resolution of *N*-substituted 2-azabicyclo[2.2.1]hept-5-en- 3-one. *Tetrahedron Asymmetry* **10**, 1201–1206.

22. a) Neuberg, C., Hirsch, J. (1921). Über ein kohlenstoffketten knÜpfendes ferment (carboligase). *Biochem. Z*. **115**, 282–310;   b) Neuberg, C., Ohle, H. (1922). Zur kenntnis der carboligase. *Biochem. Z*. **127**, 327–337.

23. Netrval, J., Vojtisek, V. (1982). Production of phenylacetyl carbinol in various yeast species. *Eur. J. Appl. Microbiol. Biotechnol*. **16**, 35–38.

24. a) Mandwal, A. K., Tripathi, C. K. M., Trivedi, P. D., Joshi, A. K., Agarwal, S. C., Bihari, V. (2004). Production of L-phenylacetyl carbinol by immobilized cells of *Saccharomuces cerevisiae*. *Biotechnol. Lett*. **26**, 217–221; b) Mahmoud, W. M., El-Sayed, A. H. M. M., Coughlin, R. W. (1990). Production of L-phenylacetylcarbinol by immobilized yeast cells, I & II. *Biotechnol. Bioeng*. **36**, 47–54 & 55–63.

25. Rosche, B., Sandford, V., Breuer, M., Hauer, B., Rogers, P. L. (2002). Enhanced production of *R*-phenylacetylcarbinol (*R*-PAC) through enzymatic biotransformation. *J. Mol. Catal. B Enzym*. **19-20**, 109–115.

26. Rosche, B., Breuer, M., Hauer, B., Rogers, P. L. (2004). Biphasic aqueous/organic biotransformation of acetaldehyde and benzaldehyde by *Zymomonas mobilis* pyruvate decarboxylase. *Biotechnol. Bioeng*. **86**, 788–794.

27. a) Breuer, M., Ditrich, K., Habicher, T., Hauer, B., Keßeler, M., Stürmer, R., Zelinski, T. (2004). Industrial Methods for the production of optically active intermediates, *Angew. Chem. Int. Ed*, **43**, 788–824; b) Panke, S., Wubbolts, M. (2005). Advances in biocatalytic synthesis of pharmaceutical intermediates. *Curr. Opin. Chem. Biol*. **9**, 188–194.

28. Yamamoto, A., Yokozeki, K., Kubota, K. (1987). Enzymic manufacture of aromatic amino acids. JP 01010995.

29. a) De Ninno, M. P. (1991). The synthesis and glycosidation of *N*-acetyl-D-neuraminic acid. *Synthesis* **8**, 583–593; b) Mahmoudian, M. (2006). A decade of biocatalysis at Glaxo Wellcome. In *Biocatalysis in the Pharmaceutical and Biotechnology Industries*, Ed R. N. Patel, Wiley, chapter 3.

30. a) Dawson, M. J., Noble, D., Mahmoudian, M. (1994). Process for the preparation of *N*-acetyl-D-neuraminic acid. WO1994/029476; b) Mahmoudian, M., Noble, D., Drake, C. S., Middleton, R. F., Montgomery, D. S., Piercey, J. E., Ramlakhan, D., Todd, M., Dawson, M. J. (1997). An efficient process for production of *N*-acetylneuraminic acid using *N*-acetylneuraminic acid aldolase. *Enzyme Microbial Technol*. **20**, 393–400.

31. a) Roth, B. D. (2002). The discovery and development of atorvastatin, a potent novel hypolipidemic agent. *Prog. Medicinal Chem*. **40**, 1–22; b) Baumann, K. L., Bulter, D. E., Deering, C. F., Mennen, K. E., Millar, A., Nanninga, T. N., Palmer, C. W., Roth, B. D. (1992). The convergent synthesis of CI-981, an optically active, highly potent, tissue selective inhibitor of HMG-CoA reductase. *Tetrahedron Lett*. **33**, 2283–2284.

32. a) Müller, M. (2005). Chemoenzymatic synthesis of building blocks for statin side chains. *Angem. Chem. Int. Ed*. **44**, 362–365; b) Panke, S., Wubbolts, M. (2005). Advances in biocatalytic synthesis of pharmaceutical intermediates. *Curr. Opin. Chem. Biol*. **9**, 188–194.

33. Gijsen, H. J. M., Wong, C. H. (1994). Unprecedented asymmetric aldol reaction with three aldehyde substrate catalyzed by 2-deoxyribose-5-phosphate aldolase. *J. Am. Chem. Soc*. **116**, 8422–8423.

34. Kierkels, J. G. T., Mink, D., Panke, S., Lommen, F. A. M., Heemskerk, D. (2003). Process for the preparation of 2,4-dideoxyhexoses 2,4,6,-trideoxyhexoses. WO 2003/006656.

35. Greenberg, W. A., Varvak, A., Hanson, S. R., Wong, K., Huang, H., Burk, M. J. (2004). Development of an efficient, scalable, aldolase-catalyzed process for enantioselective synthesis of statin intermediates. *Proc. Natl. Acad. Sci. U. S. A*. **101**, 5788–5793.

36. Liu, J., Hsu, C. C., Wong, C. H. (2004). Sequential aldol condensation catalyzed by DERA mutant Ser238Asp and a formal total synthesis of atovastatin. *Tetrahedron Lett.* **45**, 2439–2441.

37. Patel, R. N., Banerjee, A., McNamee, C. G., Brzozowski, D., Hanson, R. L., Szarka, L. J. (1993). Enantioselective microbial reduction of 3,5-dioxo-6-(benzyloxy)hexanoic acid ethyl ester. *Enzyme Microb. Technol.* **15**, 1014–1021.

38. Guo, Z., Chen, Y., Goswami, A., Hanson, R. L., Patel, R. N. (2006). Synthesis of ethyl and t-butyl (3R,5S)-dihydroxy-6-benzyloxyhexanoates *via* diastereo- and enantioselective microbial reduction. *Tetrahedron Asymmetry* **17**, 1589–1602.

39. Blacklock, T. J., Sohar, P., Butcher, J. W., Lamamnec, T., Grabowski, E. J. J. (1993). An enantioselective synthesis of the topically-active carbonic anhydrase inhibitor MK-0507: 5,6-dihydro-(*S*)-(ethylamino)-(*S*)-6-methyl-4*H*-thieno[2,3-*b*]thiopyran-2-sulfonamide 7,7-dioxide hydrochloride. *J. Org. Chem.* **58**, 1672–1679.

40. Holt, R. A., Richard, S. R. (1994). Enzymatic asymmetric reduction process to produce 4H-thieno(2,3-6)thio pyrane derivatives. WO1994/005802.

41. Gelo-Pujic, M., Marion, C., Mauger, C., Michalon, M., Schlama, T., Turconi, J. (2006). Biohydrolysis of (*S*)-3-(thiophen-2-ylthio)butanenitrile. *Tetrahedron Lett.* **47**, 8119–8123.

42. Kobayashi, M., Shimizu, S. (2000). Nitrile hydrolases. *Curr. Opin. Chem.* **4**, 95–102.

43. a) Goodhue, C. T., Schaeffer, J. R. (1971). Preparation of L-(+)-hydroxyisobutyric acid by bacterial oxidation of isobutyric acid. *Biotechnol. Bioeng.* **13**, 203–214; b) Shimazaki, M., Hasegawa, J., Kan, K., Nomura, K., Nose, Y., Kondo, H., Ohashi, T., Watanabe, K. M. (1982). Synthesis of captopril starting from an optically active beta-hydroxy acid. *Chem. Pharm. Bull.* **30**, 3139–3146.

44. Elferink, V. H. M., Kierkels, J. G. T., Kloosterman, M., Roskam, J. H. (1990). Process for the enantioselective preparation of D-(−)-3-hal-2-methylpropionic acid or derivatives thereof and the preparation of captopril there from. EP0369553/1990.

45. Patel, R. N., Howell, J. M., McNamee, C. G., Fortney, K. F., Szarka, L. J. (1992). Stereoselective enzymatic hydrolysis of α-[(acetylthio)methyl]benzenepropanoic acid and 3-acetylthio-2-methylpropanoic acid. *Biotechnol. Appl. Biochem.* **16**, 34–47.

46. a) Gu, Q. M., Reddy, D. R., Sih C. J. (1986). Bifunctional chiral synthons via biochemical methods. VIII. Optically-active 3-aroylthio-2-methylpropionic acids. *Tetrahedron Lett.* **27**, 5203–5206; b) Chen, C. S., Fujimoto, Y., Girdaukas, G., Sih, C. J. (1982). Quantitative analyses of biochemical kinetic resolutions of enantiomers. *J. Am. Chem. Soc.* **104**(25), 7294–7299; c) Sih, C. J. (1987). Process for preparing optically active 3-acylthio- 2-methylpropionic acid derivatives. WO1987/005328.

47. Levadoux, W., Groleau, D., Trani, M., Lortie, R. (1998). Streptomyces microorganism useful for the preparation of (R)-baclofen from the racemic mixture. US1998/5843765.

48. a) Chênevert, R., Desjardins, M. (1994). Chemoenzymatic enantioselective synthesis of baclofen. *Can. J Chem.* **72**, 2312–2317;   b) Schoenfelder, A., Mann, A., Le Coz, S. (1993). Enantioselective synthesis of (*R*)-(−)-baclofen. *Synlett* **7**, 63–64;   c) Herdeis, C., Hubmann, H. P. (1992). Synthesis of homochiral R-baclofen from S-glutamic acid. *Tetrahedron Asymm.* **3**, 1213–1221;   d) Hubmann,

H. P., Herdeis, C. (1994). Process for preparing enantiomer-pure (beta)-substituted (gamma)-aminobutyric acid derivatives, new enantiomer-pure intermediate stages of said process and their ues. WO 1994/02443; e) Vaccher, C., Berthelot, P., Flouquet, N., Debaert, M. (1991). Preparative liquid chromatographic separation of isomers of 4-amino-3-(4-chlorophenyl)butyric acid. *J. Chromatogr. A* **542**, 502–507; f) Resende, P., Almeida, W. P., Coelho, F. (1999). An efficient synthesis of (*R*)-(−)-baclofen. *Tetrahedron Asymm*. **10**, 2113–2118.

49. Chenevert, R., Desjardins, M. (1991). Chemoenzymatic synthesis of both enantiomers of baclofen. *Tetrahedron Lett*. **32**, 4249–4250.

50. Maxxini, C., Lebreton, J., Alphand, V. (1997). A chemoenzymatic strategy for the synthesis of enantiopure (*R*)-(−)-baclofen. *Tetrahedron Lett*. **38**, 1195.

51. a) Li, J. W. H., Vederas, J. C. (2009). Drug discovery and natural products: end of an era or an endless frontier? *Science* **325**, 161–165; b) Butler, M. S. (2008). Natural products to drugs: natural product-derived compounds in clinical trials. *Natural Product Rep. Art*. **25**, 475–516.

52. Newman, D. J., Cragg, G. M. (2007). Natural products as sources of new drugs over the last 25 years. *J Natural Products* **70**, 461–477.

53. Stephanopoulos G. (1999). Metabolic fluxes and metabolic engineering. *Metabol. Eng*. **1**, 1–11.

54. Raab, R. M., Tyo, K., Stephanopoulos, G. (2005). Metabolic engineering. *Adv. Biochem. Eng. Biotechnol*. **100**, 1–17.

55. Friehs, K. (2004). Plasmid copy number and plasmid stability. *Adv. Biochem. Eng. Biotechnol*. **86**, 47–82.

56. Antoine, R., Locht, C. (1992). Isolation and molecular characterization of a novel broad-host-range plasmid from *Bordetella bronchiseptica* with sequence similarities to plasmids from Gram-positive organisms. *Mol. Microbiol*. **6**, 1785–1799.

57. Tao, L., Jackson, R. E., Cheng, Q. (2005). Directed evolution of copy number of a broad host range plasmid for metabolic engineering. *Evol. Eng*. **7**, 10–17.

58. Keasling, J. D. (1999). Gene-expression tools for the metabolic engineering of bacteria. *Trends Biotechnol*. **17**, 452–460.

59. Iwatani, S., Yamada, Y., Usuda, Y. (2008). Metabolic flux analysis in biotechnology processes. *Biotechnol. Lett*. **30**, 791–799.

60. Cane, D. E., Walsh, C. T., Khosla, C. (1998). Harnessing the biosynthetic code: combinations, permutations, and mutations. *Science* **282**, 63–68.

61. Khosla, C., Harbury, P. B. (2001). Modular enzymes. *Nature* **409**, 247–252.

62. Pfeifer, B. A., Admiraal, S. J., Gramajo, H., Cane, D. E., Khosla, C. (2001). Biosynthesis of complex polyketides in a metabolically engineered strain of *E. coli*. *Science* **291**, 1790–1792.

63. a) Walsh, C. T. (2003). *Antibiotics: Actions, Origins, Resistance*. Washington, DC: ASM Press; b) Neu, H. C. (1992). The crisis in antibiotic resistance. *Science* **257**, 1064–1073.

64. Kennedy, J., Turner, G. (1996). δ-L-α-aminoadipyl-L-cysteinyl-D-valine synthetase is a rate limiting enzyme for penicillin production in *Aspergillus nidulans*. *Mol. Gen. Genet*. **253**, 189–197.

65. Kreisberg-Zakarin, R., Borovok, I., Yanko, M., Aharonowitz, Y., Cohen, G. (1999). Recent advances in the structure and function of isopenicillin N synthase. *Antonie van Leeuwenhoek* **75**, 33–39.

66. Martin, J. F. (1998). New aspects of genes and enzymes for β-lactam antibiotic biosynthesis. *Appl. Microbiol. Biotechnol.* **50**, 1–15.

67. Waites, M. J., Morgan, N. L., Rockey, J. S., Higton, G. (2001). *Industrial microbiology: An Introduction.* Oxford, UK: Blackwell.

68. Nistelrooij, V., Krijgsman, J., DeVroom, E., Oldenhof, C. (1998). Penicillin update. In: Mateles RI (ed). *Penicillin: A Paradigm for Biotechnology.* Chicago, IL: Candida, p. 85–91.

69. Elander, R. P. (2003). Industrial production of b-lactam antibiotics. *Appl Microbiol Biotechnol.* *61*, 385–392.

70. Skatrud, P. L., Tietz, A. J., Ingolia, T. D., Cantwell, C. A., Fisher, D. L., Chapman, J. L., Queener, S. W. (1989). Use of recombinant DNA to improve production of cephalosporin C by *Cephalosporium acremonium.* *Nat. Biotechnol.* **7**, 477–485.

71. a) Clausen, K., Dekkers, R. M. (1996). β-Lactam antibiotic production by enzymatic acylation (Bist-brocades B.V.), WO 9602663;   b) Diender, M. B., Straathof, A. J. J., van der Does, T., Zomerdijk, M., Heijnen, J. J. (2000). Course of pH during the formation of amoxicillin by a suspension-to-suspension reaction. *Enzyme Microbial Technol.* **27**, 576–582.

72. van der Drift, J. K., Henniger, P. W., van Veen, G. J. (1978). 6-(D-α-Amino-p-hydroxyphenylacetamido)-penicillanic acid preparation. US 4128547.

73. Usher, J. J., Romancik, G. (1998). Synthesis of β-lactam antibacterials using soluble side chain esters and enzymatic acylase. (Bristol-Meyers Squibb Co.), WO 9804732.

74. Lopez-Gallego, F., Batencor, L., Hidalgo, A., Mateo, C., Fernandez-Lafuente, R., Guisan, J. M. (2005). One-pot conversion of cephalosporin C to 7-aminocephalosporanic acid in the absence of hydrogen peroxide. *Adv. Synth. Catal.* **347**, 1804–1810.

75. Tan, H. S., Verweij, J. (1975). One-step, high yield conversion of penicillin sulfoxides to deacetoxycephalosporins. *J. Org. Chem.* **40**, 1346–1347; b) Verweij, J., Tan, H. S., Kooreman, H. J. (1977). Preparation of 7-substituted amino-desacetoxycephalosporanic acid compounds (Gist-brocades N.V.), US Patent 4003894.

76. a) Faarup, P. (1977). Method of preparing a sparingly soluble complex of cephalexin (Novo Industri A/S), US Patent 4003896,;   b) Kemperman, G. J., de Gelder, R., Dommerholt, F. J., Raemakers-Franken, P. C., Klunder, A. J. H., Zwanenburg, B. (1999). Clathrate-type complexation of cephalosporins with β-naphthol. *Chem. Eur. J.* **5**, 2163–2168;   c) Wegman, M. A., van Langen, L. M., van Rantwijk, F., Sheldon, R. A. (2002). A two-step, one-pot enzymatic synthesis of cephalexin from D-phenylglycine nitrile. *Biotechnol. Bioeng.* **79**, 356–361.

77. Singh, J., Denzel, T. W., Fox, R., Kissick, T. P., Herter, R., Wurdinger, J., Schierling, P., Papaioannou, C. G., Moniot, J. L., Mueller, R. H., Cimarusti, C. M. (2002). Regioselective activation of aminothiazole (iminoxyacetic acid) acetic acid: an efficient synthesis of the monobactam aztreonam. *Org. Process Res. Dev.* **6**, 863–868.

78. Pendland, S. L., Messick, C. R., Jung, R. (2002). In vitro synergy testing of levofloxacin, ofloxacin, and ciprofloxacin in combination with aztreonam, ceftazidime, or piperacillin against *Pseudomonas aeruginosa.* *Diagnostic Microbiol. Infect. Dis.* **42**, 75–78.

79. Kahan, J. S., Kahan, F. M., Goegelman, R., Currie, S. A., Jackson, M., Stapley, E. O., Miller, T. W., Hendlin, D., Mochales, S., Hernandez, S., Woodrull, H. B., Birnbaum, J. (1979). Thienamycin, a new betalactam antibiotic. 1. Discovery, taxonomy, isolation and physical properties. *J Antibiot.* **32**, 1–12.

80. Stapon, A., Li, R., Townsend, C. A. (2003). Carbapenem biosynthesis: confirmation of stereochemical assignments and the role of CarC in the ring stereoinversion process from L-proline. *J Am Chem Soc.* **125**, 8486–8493.

81. Bodner, M. J., Phelan, R. M., Townsend, C. A. (2009). A catalytic asymmetric route to carbapenems. *Org. Lett.* **11**, 3606–3609.

82. Demain, A. L., Vaishnav, P. (2006). Involvement of nitrogen-containing compounds in β-lactam biosynthesis and its control. *Crit. Rev. Biotechnol.* **26**, 67–82.

83. a) Lawrence, G. C., Lilly, G. (1980). Clavulanic acid production. UK Patent GB1571888; b) Xiang, S. H., Li, J., Yin, H., Zheng, J. T., Yang, X., Wang, H. B., Luo, J. L., Bai, H., Yang, K. Q. (2009). Application of a double-reporter-guided mutant selection method to improve clavulanic acid production in *Streptomyces clavuligerus. Metab. Eng.* **11**, 310–318.

84. Silva, C. S., Bovarotti, E., Rodrigues, M. I., Hokka, C. O., Barboza, M. (2009). Evaluation of the effects of the parameters involved in the purification of clavulanic acid from fermentation broth by aqueous two-phase systems. *Bioprocess Biosyst. Eng.* **32**, 625–632.

85. a) Taniguchi, M., Sasaoka, M., Matsumara, K., Kanahara, I., Kaze, K., Suzuki, D., Shimabayashi, A. A. (1989). Preparation of β-lactamase inhibiting triazole-containing β-lactam. EP 0331146; b) Xu, W. L., Li, Y. Z., Zhang, Q. S., Zhu, H. S. (2005). A new approach to the synthesis of tazobactam using an organosilver compound synthesis of tazobactam. *Synthesis* **3**, 442–446.

86. McDaniel, R., Licari, P., Khosla, C. (2001). Process development and metabolic engineering for the overproduction of natural and unnatural polyketides. *Metab. Eng.* **73**, 31–52.

87. Shen, B. (2003). Polyketide biosynthesis beyond the type I, II and III polyketide synthase paradigms. *Curr. Opin. Chem. Biol.* **7**, 285–295.

88. Staunton, J., Weissman, K. J. (2001). Polyketide biosynthesis: a millennium review. *Natural Product Rep.* **18**, 380–416.

89. Smith, R. L., Bungay, H. R., Pittenger, R. C. (1962). Growth-biosynthesis relationships in erythromycin fermentation. *Appl. Microbiol.* **10**, 293–296.

90. El-Enshasy, H. A., Mohamed, N. A., Farid, M. A., El-Diwany, A. I. (2008). Improvement of erythromycin production by *Saccharopolyspora erythraea* in molasses based medium through cultivation medium optimization. *Bioresource Technol.* **99**, 4263–4268.

91. Zou, X., Hang, H. F., Chu, J., Zhuang, Y. P., Zhang, S. L. (2009). Enhancement of erythromycin a production with feeding available nitrogen sources in erythromycin biosynthesis phase. *Bioresource Technol.* **100**, 3358–3365.

92. Cortes, J., Haydock, S. F., Roberts, G. A., Bevitt, D. J., Leadlay, P. F. (1990). An unusually large multifunctional polypeptide in the erythromycin-producing polyketide synthase of *Saccharopolyspora erythraea. Nature* **348**, 176–178.

93. Peiru, S., Menzella, H. G., Rodriguez, E., Carney, J., Gramajo, H. (2005). Production of the potent antibacterial polyketide erythromycin C in *Escherichia coli. Appl. Environ. Microbiol.* **71**, 2539–2547.

94. Lee, H. Y., Khosla, C. (2007). Bioassay-guided evolution of glycosylated macrolide antibiotics in *Escherichia coli*. *PLoS Biol*. **5**, e45.

95. Lomovskaya, N., Otten, S. L., Doi-Katayama, Y., Fonstein, L., Liu, X. C., Takatsu, T., Inventi-Solari, A., Filippini, S., Torti, F., Colombo, A. L., Hutchinson, C. R. (1999). Doxorubicin overproduction in *Streptomyces peucetius*: cloning and characterization of the dnrU ketoreductase and dnrV genes and the doxA cytochrome P-450 hydroxylase gene. *J. Bacteriol*. **181**, 305–318.

96. Arcamone, F., Cassinelli, G., Fantini, G., Grein, A., Orezzi, P., Pol, C., Spalla, C. (2000). Adriamycin, 14-hydroxydaunomycin, a new antitumor antibiotic from *S. peucetius* var. *caesius*. *Biotechnol. Bioeng*. **67**, 704–713.

97. Dickens, M. L., Strohl, W. R. (1996). Isolation and characterization of a gene from *Streptomyces sp*. strain C5 that confers the ability to convert daunomycin to doxorubicin on *Streptomyces lividans* TK24. *J. Bacteriol*. **178**, 3389–3395.

98. Walczak, R. J., Dickens, M. L., Priestley, N. D., Strohl, W. R. (1999). Purification, properties, and characterization of recombinant Streptomyces sp. strain C5 DoxA, a cytochrome P-450 catalyzing multiple steps in doxorubicin biosynthesis. *J. Bacteriol*. **181**, 298–304.

99. Grimm, A., Madduri, K., Ali, A., Hutchinson, C. R. (1994). Characterization of the *Streptomyces peucetius* ATCC 29050 genes encoding doxorubicin polyketide synthase. *Gene* **151**, 1–10.

100. Lomovskaya, N., Otten, S. L., Doi-Katayama, Y., Fonstein, L., Liu, X. C., Takatsu, T., Inventi-Solari, A., Filippini, S., Torti, F., Colombo, A. L., Hutchinson, C. R. (1999). Doxorubicin overproduction in *Streptomyces peucetius*: cloning and characterization of the dnrU ketoreductase and dnrV genes and the doxA cytochrome P-450 hydroxylase gene. *J. Bacteriol*. **181**, 305–318.

101. http://www.solvias.com.

102. Hutchinson, C. R. (1997). Biosynthetic studies of daunorubicin and tetracenomycin C. *Chem. Rev*. **97**, 2525–2535.

103. Weber, W., Zähner, H., Siebers, J., Schröder, K., Zeeck, A. (1979). Metabolic products of microorganisms. 175. Tetracenomycin-C. *Arch. Microbiol*. **121**, 111–116.

104. Shen, B. (2000). Biosynthesis of aromatic polyketides. *Top. Curr. Chem*. **209**, 1–51.

105. Suffness, M., Wall, M. E. (1995). Discovery and development of taxol. In: Suffness M (ed) *Taxol–Science and Applications*. Boca Raton, FL: CRC Press, p. 3–25.

106. Wani, M. C., Taylor, H. L., Wall, M. E., Coggon, P., McPhail, A. T. (1971). Plant antitumor agents VI: The isolation and structure of taxol, a novel antileukemic and antitumor agent from *Taxus brevifolia*. *J. Am. Chem. Soc*. **93**, 2325–2327.

107. a) Wuts, P. G. M. (1998). Semisynthesis of taxol. *Curr. Opin. Drug Discov. Dev*. **1**, 329–337; b) Holton, R. A. (1990). Method for preparation of taxol. EP 0400971 A2.

108. Tabata, H. (2004). Paclitaxel production by plant-cell-culture technology. *Adv. Biochem. Eng. Biotechnol*. **7**, 1–23.

109. http://www.phytonbioth.com.

110. Itokawa, H. (2003). Taxoids occurring in the genus taxus. In: Itokawa H & Lee K-H (eds). *Taxus—The Genus Taxus*. London, UK: Taylor & Francis, p. 35–78.

111. Son, S. H., Choi, S. M., Lee, Y. H., Choi, K. B., Yun, S. R., Kim, J. K., Park, H. J., Kwon, O. W., Noh, E. W., Seon, J. H., Park, Y. G. (2001). Large-scale growth and taxane production in cell cultures of *Taxus cuspidate* (Japanese yew) using a novel bioreactor. *Plant Cell Rep*. **19**, 628–633.

112. Seki, M., Ohzora, C., Takeda, M., Furusaki, S. (1997). Taxol (paclitaxel) production using free and immobilized cells of *Taxus cuspidata*. *Biotechnol. Bioeng*. **53**, 214–219.

113. Zhao, J., Davis, L. C., Verpoorte, C. (2005). Elicitor signal transduction leading to production of plant secondary metabolites. *Biotechnol. Adv*. **23**, 283–333.

114. A) Yukimune, Y., Tabata, H., Higashi, Y., Hara, Y. (1996). Methyl jasmonate induced over-production of paclitaxel and baccatin III in Taxus cell suspension cultures. *Nat. Biotechnol*. **14**, 1129–1132; b) Zhang, C. H., Wu, J. Y. (2003). Ethylene inhibitors enhance elicitor-induced paclitaxel production in suspension cultures of *Taxus spp*. cells. *Enzyme Microbial Technol*. **32**, 71–77.

115. Reymond, P., Bodenhausen, N., van Poecke, R. M. P., Krishnamurthy, V., Dicke, M., Farmer, E. (2004). A conserved transcript pattern in response to a specialist and a generalist herbivore. *Plant Cell* **16**, 3132–3147.

116. Bentebibel, S., Moyano, T., Palazon, J., Cusidó, R. M., Bonfill, M., Eibl, R., Piñol, T. (2005). Effects of immobilization by entrapment in alginate and scale-up on paclitaxel and baccatin III production in cell suspension cultures of *Taxus baccata*. *Biotechnol. Bioeng*. **89**, 647–655.

117. Stierle, A., Strobel, G. (1993). Taxol and taxane production by *Taxomyces andreanae*, an endophytic fungus of Pacific yew. *Science* **260**, 214–216.

118. Roberts, S. C. (2007). Production and engineering of terpenoids in plant cell culture. *Nat. Chem. Biol*. **3**, 387–395.

119. Zhao, K., Ping, W., Li, Q., Hao, S., Zhao, L., Gao, T., Zhou, D. (2009). *Aspergillus niger var*. taxi, a new species variant of taxol-producing fungus isolated from *Taxus cuspidata* in China. *J. Appl. Microbiol*. **107**, 1202–1207.

120. Xu, F., Tao, W. Y., Cheng, L., Guo, L. J. (2006). Strain improvement and optimization of the media of taxol-producing fungus, *Fusarium mairei*. *Biochem. Eng. J*. **31**, 67–73.

121. Li, Y. C., Tao, W. Y., Cheng, L. (2009). Paclitaxel production using co-culture of Taxus suspension cells and paclitaxel-producing endophytic fungi in a co-bioreactor. *Appl. Microbiol. Biotechnol*. **83**, 233–239.

122. Lyseng-Williamson, K. A., Fenton, C. (2005). Docetaxel: a review of its use in metastatic breast cancer. *Drugs* **65**, 2513–2531.

123. Commerçon, A., Bernard, F., Bézard, D., Bourzat, J. D. (1992). Improved protection and esterification of a precursor of the Taxotere® and taxol side chains. *Tetrahedron Lett*. **33**, 5185–5188.

124. Gaetan, C., Mettilda, L. (2009). Preparation of taxanes from 9-dihydro-13-acetylbaccatin III, US 0062376 A1.

125. a) Walsh, C. T. (2002). Combinatorial biosynthesis of antibiotics: challenges and opportunities. *Chembiochem* **3**, 125–134; b) Liu, W., Yu, Y. (2008). Combinatorial biosynthesis of pharmaceutical natural product. In *Biocatalysis for the Pharmaceutical Industry–Discovery, Development, and Manufacturing*. Ed. by Tao, J. H., Lin, G. Q., Liese, A. John Wiley & Sons, chapter 11, p. 229–246.

CHAPTER 4

RESOLUTION OF CHIRAL DRUGS

QI-DONG YOU

China Pharmaceutical University, Nanjing, China

Chiral Drugs: Chemistry and Biological Action, First Edition. Edited by Guo-Qiang Lin, Qi-Dong You and Jie-Fei Cheng.
© 2011 John Wiley & Sons, Inc. Published 2011 by John Wiley & Sons, Inc.

Chiral molecules can be obtained through asymmetric synthesis, manipulation of chiral materials, or chiral resolution. Chiral resolution is one of the important manufacturing approaches to preparation of optically pure chiral drugs. Although asymmetric synthesis has become more and more efficient and accessible, the practicability and reproducibility of chiral resolution make it an indispensable method in the pharmaceutical industry.

The chiral resolution process creates an asymmetric environment in which the two enantiomers can be physically separated. Based on the separation procedures used, chiral resolution is divided into several categories such as crystallization resolution (physical resolution and chemical resolution), kinetic resolution, biological resolution (a substantial part of biologically catalytic kinetic resolution), and chromatography resolution.

4.1   THE CHARACTERISTICS OF RACEMATES

Theoretically, a racemate that contains equal amounts of two enantiomers should possess the same physical properties as those of its enantiomers except optical rotation. However, this is not always true in reality because of some interactions between the enantiomers. These interactions could be negligible when a racemate exists in dilute solution and vapor phase. The interactions, however, become remarkable if racemates are in solid state, pure solution, or concentrated solution. Because of the packing forces among crystallized racemates, the formed crystals have three different types:

1. *Racemic mixture:* This is also known as a racemic conglomerate. The conglomerate is an equimolar mechanical mixture of crystals, each of which contains only one of the two enantiomers in a racemate. It is estimated that about 5–10% of racemates exist in racemic conglomerates.

   The driving force to form a racemic conglomerate results from the different affinities among the enantiomers. When the affinity between two enantiomers is lower than that between the same configurations, the racemic conglomerate forms. During the crystallization of a racemate, each enantiomer of the racemate accumulates on its own and precipitates from the solution spontaneously in the form of a pure crystal. Once one enantiomer precipitates to form the primary crystal, other molecules of the same configuration will accumulate on the surface of the primary crystal, which results in crystal mixtures of two enantiomers crystallized in separate phases, respectively. This conglomerate also has an asymmetric character, and the respective crystals appear as mirror images of each other. The characteristics of the racemic mixture are similar to those of the general mixture, with a melting point lower than the single pure enantiomer and solubility greater than the single pure enantiomer (Fig. 4.1).

2. *Racemic compound:* When two enantiomers are present in equal quantities in a well-defined arrangement within one crystal lattice, a homogeneous crystalline addition compound is formed that is called a racemic compound. This is mainly due to the greater affinity of one enantiomer for the other enantiomer than for the same enantiomer. During crystallization, enantiomers precipitate equally and coexist in the same crystal lattice. Because the interaction between the enantiomers is stronger, the melting point of the racemic compound is often higher than that of the pure single enantiomer (Fig. 4.2). Other physical properties such as solubilities and infrared spectra are also different from those of the pure enantiomers.

3. *Pseudoracemate:* Because of the small difference in the affinity between enantiomers of like or opposite configuration, when two enantiomers coexist in an unordered manner and are nonequivalent in the crystal, a racemic solid solution is formed called a pseudoracemate, which occurs in rare cases. However, once this happens, the enantiomers of different configurations precipitate with each other in any proportion and form a mixture. The melting point curve of the pseudoracemate can be either convex or concave but ideally is a straight line (Fig. 4.3). The physical properties of the pseudoracemate are basically the same as those of pure enantiomers.

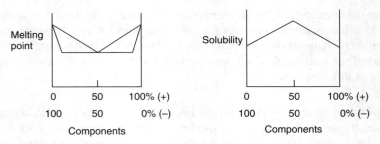

FIGURE 4.1 Melting point and solubility of racemic mixture.

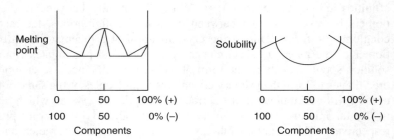

FIGURE 4.2 Melting point and solubility of racemic compound.

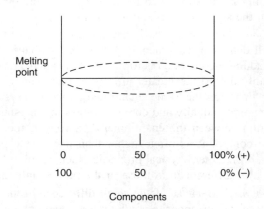

FIGURE 4.3 Melting point of pseudoracemate.

There are several ways to determine racemic compounds, racemic mixtures, and pseudoracemates: 1) infrared spectroscopy (IR), 2) the powder X-ray diffraction method (XRD), and 3) differential thermal analysis (DSC). As racemic compounds are pairs of enantiomers equivalently coexisting in crystal lattice, so their IR spectra, XDR spectra, and DSC spectra have obvious differences from pure enantiomers. The latent heat of fusion for the racemic compounds is approximately onefold greater than that of single enantiomers in the DSC spectrum. While the crystal lattice of racemic mixtures contains only one configuration molecule, its IR spectrum, XDR spectra, and DSC spectra have no significant difference from pure enantiomers.

4.2   CHEMICAL RESOLUTION BY CRYSTALLIZATION

Resolution by crystallization is the most useful and important method for preparation of chiral drugs. These are two ways to perform crystallization-based resolution: direct crystallization and indirect crystallization. The direct crystallization method makes the chiral compound crystallize out directly from a racemate solution by utilizing its characteristic aggregate-forming properties. The indirect crystallization method converts the enantiomers of chiral compounds into diastereomers and then crystallizes one of the diastereoisomers from the resolution by using the solubility differences of two diastereomers. The chromatographic method can also be used easily for resolution of the formed diastereomers.

4.2.1   Resolution by Direct Crystallization

4.2.1.1   Resolution by Spontaneous Crystallization
In 1882, Louis Pasteur conducted a chiral resolution by seeding a supersaturated solution of sodium ammonium tartrate with a d-crystal on one side of the vessel and an l-crystal on the opposite side: Crystals of opposite handedness form on the opposite sides of the vessel. This type of resolution is called spontaneous resolution and is demonstrated by racemic methadone [1].

Spontaneous resolution is a process of forming a racemic conglomerate on crystallization of a racemate, where two pairs of enantiomers in a racemate could precipitate to form different conglomerate crystals. The formed conglomerate crystals are enantiomorphous, and each crystal could be separated manually.

The prerequisite of spontaneous resolution is that racemates are able to form a conglomerate, and therefore it is possible to separate them on the basis of their enantiomorphous crystals. Typically 5–10% of all racemates are known to be crystallized as mixtures of conglomerates. The compounds forming non-conglomerates can be derivatived into new compounds to form solid of manmade conglomerates (usually its salt). A racemic liquid sample could also be converted to conglomeritic solids with this method. For example, the liquid α-phenylethylamine (**1**) was derivatized into its sulfate (**3**) or cinnamate (**4**), and α-phenylethanol (**2**) was esterified with 3,5-dinitrobenzoic acid to form an ester (**5**), which were able to form conglomeritic crystals (Scheme 4.1).

Leucovorin calcium (**6**) can be easily converted into a folinic acid derivative. Although it does not increase anticancer activity, leucovorin calcium can rescue bone marrow and gastrointestinal mucosa cells when used in combination with methotrexate as compared to methotrexate alone. There are two chiral centers in the structure of leucovorin: One is located in the moiety of L-glutamic acid,

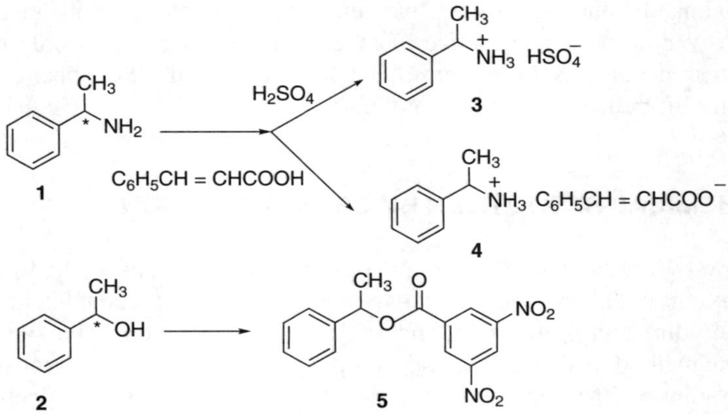

SCHEME 4.1 Converting nonconglomerate compounds (**1**) and (**2**) to conglomerates.

with S-configuration, and another on the C6 of the tetrahydropteridine ring. It is reported that the compound with S-configuration at C6, known as L-leucovorin calcium, showed good efficacy [2]. Cosulich et al. reported that 6R- and 6S-leucovorin calcium were separated according to the different water solubilities of the two isomers [3], but this method is not reproducible. Lin et al. developed a new method to prepare highly optically pure L-leucovorin (*d.e.* > 98%) by crystallization of the solution of a diastereometric mixture of leucovorin calcium with a certain proportion of anhydrous calcium chloride [4] (Scheme 4.2).

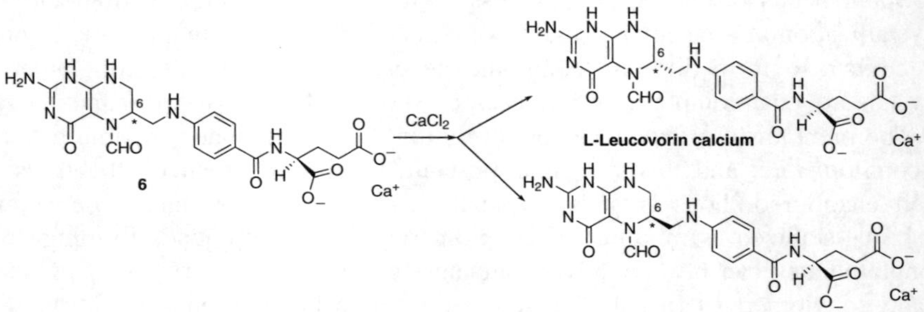

SCHEME 4.2 Spontaneous resolution of leucovorin calcium with anhydrous calcium chloride.

4.2.1.2 *Resolution by Preferential Crystallization (Resolution by Entrainment)*

Spontaneous resolution requires that crystals should have certain shapes for separation. This method clearly has limitations. However, if a pure enantiomer seed is added to such a supersaturated solution to break the solution equilibrium, more enantiomers of this type will preferentially precipitate. This is the process called preferential crystallization.

Preferential crystallization is one of the most straightforward and efficient separation and purification processes, given its advantages of low cost and formation of solid products. In this method, one of the enantiomers is seeded in a saturated or supersaturated racemate solution, resulting in a slight excess of the same isomer of the seeds to make an asymmetric environment, so that the same isomers of the seeds are crystallized out from the solution. For example, addition of a trace amount of $(-)$-hydrobenzoin to an ethanol solution of $(\pm)$-hydrobenzoin resulted in formation of $(-)$-enantiomer crystals, and after 15 cycles 97% optical purity was obtained. During the 1960s–1970s, preferential crystallization was employed for preparation of L-glutamic acid, with annual production of 13,000 tons. Not only is preferential crystallization a production process of choice in industry, but it can also be used in the laboratory to separate optically active compounds in tens of grams.

Although preferential crystallization is limited to separation of racemates that are able to form conglomerate crystals [5], it is an efficient, simple, and fast resolution method. Addition of crystal seeds can result in two pairs of enantiomers with different crystallization rates to control the dynamic process. The yield of product can be increased by extending the crystallization time at the expense of optical purity. Further improvement of the optical purity of the crystals from the preferential crystallization could be achieved through repeated recrystallization.

Preferential crystallization can be performed in a cyclic operation mode in practical industrial process. First, racemates are made into a supersaturated solution, and then either pure enantiomer crystal (such as the dextroisomer) is added as a seed. At this time, the concentration of the dextroisomer in solution is higher than that of the levoisomer; therefore, the dextroisomer could precipitate preferentially to form crystals when it cools down. The amounts of crystal are far greater than the added dextroisomer. The dextroisomer crystals formed are quickly separated to give an optically pure mass. Since a large amount of dextroisomer is separated, the amounts of levoisomer in the solution are now more than those of the dextroisomer, leaving the solution supersaturated in levoisomer. The operation is repeated as mentioned above to generate a large amount of pure levoisomer. Such cyclical separation, known as cross-induced crystallization resolution, can be performed many times.

This preferential crystallization approach was used in the resolution of D-threo-*1*-p-nitrophenyl-2-amino-1,3-propanediol (**8**), an intermediate of the antibiotic chloramphenicol (**7**) [6], and the manufacture of the antihypertension drug L-methyldopa (**9**) [7] (Fig. 4.4).

Synoradzki et al. used cyclic preferential crystallization combined with the racemization method for industrial preparation of calcium pantothenate (PTTCa) (**10**) [8]. (*R*)-PTTCa is denoted as vitamin B5. It formed a solvate in a 4-to-1 ratio of methanol and water during the process of crystallization from the racemic solution when seeded with optically pure crystals. Therefore, under optimal conditions (8°C and 3 μM optically pure seeding 0.1% calculated on PTTCa), PTTCa

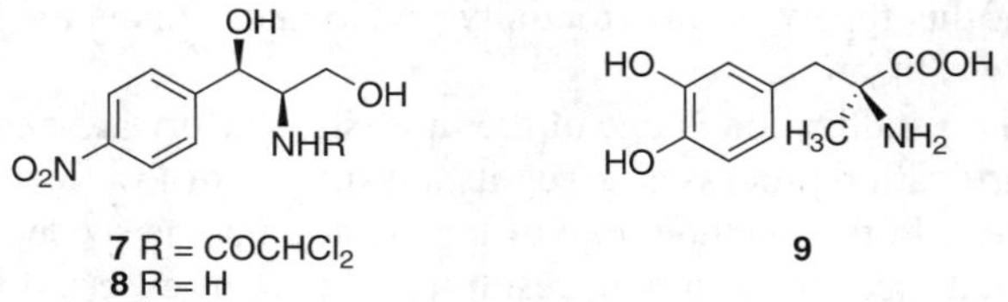

7 R = COCHCl$_2$
8 R = H

9

FIGURE 4.4 Structures of chloramphenicol (**7**) and its intermediate (**8**) and L-methyldopa (**9**).

and water contents in solution were 28–31% and 23–26%, respectively. After 13 runs, the obtained *ee* value was around 97–100%, with average yield per batch of 26–33% on the scale of hundreds of kilograms. It was obtained from 1.2 kg (*R,S*)-PTTCa for 1 kg (*R*)-PTTCa in 83% overall yield. (*S*)-PTTCa is biologically inactive and racemizes in the presence of sodium methoxide under reflux in methanol. The racemization is due to the difference in solubility between respective pantothenates but not the different basic properties of calcium and sodium methoxides. In contrast to sodium pantothenate, calcium pantothenate is only slightly soluble in methanol, which leads to negligible racemization. The use of sodium methoxide was also tried but unfortunately was rather inconvenient for technological purposes, since sodium must be removed before (*R,S*)-PTTCa crystallization. After optimization, calcium methoxide was used as a base to racemize (*S*)-PTTC quantitatively. Typically, the suspension was heated at 90°C for 30–60 min followed by 110°C for another 1.5 h when racemization occurred. The racemized PTTCa with 84–90% yield would enter the next cycle of resolution [8] (Fig. 4.5).

The efficiency of the preferential crystallization can be expressed by a resolution index (RI) defined by the ratio of the weight of pure enantiomer obtained ($W_{product}$) to the initial excess of this enantiomer (W_{excess}).

$$RI = [W_{product} \times ep - W_{seed}]/E_{excess} \qquad (4.1)$$

Here, ep is defined as enantiomeric purity and is equal to *ee*. RI = 1 indicates that only the added enantiomer is crystallized, and hence the entrainment of the seeded enantiomer did not occur. RI should be greater than 2 for an efficient preferential crystallization [9].

4.2.1.3 Resolution by Reverse Crystallization

In preferential crystallization, the crystal cones are formed by addition of insoluble additives and/or seeds into the solution of the racemic mixture to accelerate or facilitate the growth of the same crystal of uniconfiguration. Reverse crystallization is the addition of one soluble stereoisomer into a saturated solution of racemates, this soluble stereoisomer will be absorbed to the surface of isomer with the same configuration in the solution of racemates, inhibiting the growth of this isomer crystal. Meanwhile, the crystallization rate of the opposite configuration will accelerate to create more crystals from the racemic solution [10].

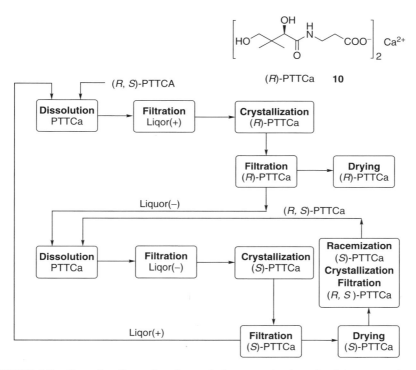

FIGURE 4.5   Operation flow of cycle resolution-racemization of calcium pantothenate.

This reverse method was used to resolve (R, R)-$(+)$-sodium ammonium tartrate from aqueous solution of racemic sodium ammonium tartrate by adding a small amount of (S)-$(-)$-sodium ammonium malate or (S)-$(-)$-asparagines [11].

The structure and configuration of the added soluble stereoisomer in reverse crystallization must be related to the compounds to be resolved. Thus these added substances can embed into the lattice of growing crystal to replace its normal component and prevent the crystal growth. Reverse crystallization is a kinetic process of crystal growth in that addition of one soluble stereoisomer induces a difference in crystallization rate between two stereoisomers. Extension of the crystallization time will ultimately give rise to the racemic crystal. Some examples of the use of reverse crystallization are listed in Table 4.1 [12].

4.2.1.4 *Resolution by Crystallization in Optically Active Solvent*
Resolution by crystallization in optically active solvent refers to crystallization of enantiomeric substances in optically active solvent or in achiral solvents containing variable amounts of optically active cosolute.

In the late nineteenth century, Van't Hoff et al. noticed that there was a difference in the solubility of enantiomers in an optically active solvent [13]. Over the years, people have been trying to take advantage of this solubility difference

TABLE 4.1   Examples of Reverse Crystallization

Racemate	Chiral Additives	First Precipitate Enantiomer
Threonine	(S)-glutamic acid, (S)-glutamine, (S)-asparagine, (R)-cysteine, (S)-phenylalanine, (S)-histidine, (S)-lysine, (S)-aspartic acid	(R)-threonine
Glutamic acid hydrochloride	(S)-lysine, (S)-ornithine, (S)-histidine, (S)-serine, (S)-threonine, (S)-cysteine, (S)-tyrosine, (S)-leucine	(R)-glutamine
Asparagine monohydrate	(S)-glutamic acid, (S)-aspartic acid, (S)-serine, (S)-lysine, (S)-glutamine, (S)-ornithine, (S)-histidine	(R)-aspartic acid
(p-Hydroxyphenyl) glycine-p-toluenesulfonate	(S)-phenylglycine, (S)-tyrosine, (S)-methoxy-phenyl glycine, (S)-tryptophan, (S)-phenylalanine, (S)-lysine, (S)-DOPA, (S)-methyl-dopa	(R)-p-hydroxyph enylglycine
Histidine hydrochloride	(S)-tryptophan, (S)-phenylalanine	(R)-histidine
Phenyl hydroxy propionic acid	(S)-phenyl lactic acid	(R)-phenyl-hydroxy acid

for separation of enantiomers. However, in practice, only crystal mixtures with some extent of excess of enantiomers were obtained by crystallization in optically active solvent.

The solubility of enantiomer in optically active solvents can be divided into two categories [13]:

1. General chiral organic compounds. Despite some solubility differences of these compounds in an inert optically active solvent, this was not significant enough to form a remarkable asymmetric environment.
2. Chiral organometallic compounds. The solubility of this kind of compound is of great difference in an optically active solvent containing hydroxyl group (or achiral solvents containing optically active ionic compounds). The solubility difference arises possibly because (+) or (−) ions of chiral organometallic molecules are chelated to optically active solvents to form tight-binding diastereoisomeric complexes.

4.2.2 Crystallization Resolution via Diastereomer Formation

4.2.2.1 Diastereomer Formation and Resolution Principles

In this process, two enantiomers of racemates [(±)A in Scheme 4.3] could form diastereomer compounds [(+)A·(+)B and(−) A·(+)B in Scheme 4.3] with an optically active reagent (resolving agent) [the (+)B in Scheme 4.3] and therefore can be separated by resolution or chromatography. Unlike enantiomers, diastereoisomer pairs have significantly different physical properties, especially differences of solubility. So far, this method has been widely used to separate many optically active drugs. The enantiomers able to be separated by this method range from acids, alkalines, alcohols, phenols, aldehydes, and ketones to amides and amino acids.

$$(\pm)A + (+)B \rightarrow (+)A·(+)B + (−)A·(+)B$$

SCHEME 4.3   Resolution of enantiomers, (±)A, with resolving agent, (+)B.

According to the Scheme 4.3, the (±)A and (+)B form diastereoisomers (+)A·(+)B and (−)A·(+)B. The diastereomeric salt resulting from constituents with like sign of optical rotation [(+)A, (+)B] is called a p salt, while that from constituents with unlike sign of optical rotation [(−)A, (−)B] is called an n salt [14].

There are two prerequisites for the chiral resolutions based on different solubilities of the diastereomers [15]: 1) at least one of the formed diastereomers should be crystallized and 2) the solubility difference of two diastereomer salts must be significant. It was reported that the use of 2-methoxyl strychnine as the resolving agent to separate racemic 2,2′-dimethyl-6,6′-dicarboxyl biphenyl in a nonspecific solvent is unsuccessful [16]. Interestingly, switching to a more polar solvent such as methanol-acetone (7:3) resulted in almost absolute optically pure (S)-(+)-isomer. With a large solubility difference, enantiomer salts can be separated by simply washing with warm solvent or grinding without recrystallization [17].

The resolution process might not be satisfactory under certain situations: 1) Two diastereomers interact with each other to generate salts or addition compounds, which causes the two diastereomers to be crystallized at the same time and form racemic crystals. 2) In solid state, two diastereomers partially mix to form a solid solution, making it impossible to separate.

In Scheme 4.3, the optically active resolving agent B was used to separate the racemate A. Likewise, the optically active A can also be used as a resolving agent to separate the racemate B; this process is called reciprocal resolution. For example, chiral (−)-ephedrine [(−)-Eph] was used for resolution of racemic N-Cbz-alanine [(+)-Z-Ala], and vice versa [18].

An improved classical approach is called mutual resolution [19], in which the racemate [(±)-B] is used as a resolving agent instead of an intrinsic optically active isomer [(−)- or (+)-B]. Therefore, both the resolving agent and the compounds needed to be resolved are racemates and four diastereomers, (+)A · (+)B, (+)A · (−)B, (−)A · (+)B, and (−)A · (−)B, could be formed. If (+)A · (+)B/(−) A · (−)B has lower solubility, then (+)A · (+)B [or (−) A · (−)B] seed can be

TABLE 4.2 Resolution Types of Formation of Diastereomer Salt

Resolution Types	Mixtures to Be Separated	Resolving Agents	Diastereoisomer Production	
			Salts of lower solubility	Salts of higher solubility
Conventional resolution method	(±)-Z-Ala	(−)-Eph	(−)-Z-Ala·(−)-Eph	(+)-Z-Ala·(−)-Eph
	(±)-Z-Ala	(+)-Eph	(+)-Z-Ala·(+)-Eph	(−)-Z-Ala·(+)-Eph
Conventional reciprocal resolution method	(±)-Z-Ala	(−)-Eph	(−)-Z-Ala·(−)-Eph	(+)-Z-Ala·(−)-Eph
	(±)-Eph	(+)-Z-Ala	(+)-Z-Ala·(+)-Eph	(+)-Z-Ala·(−)-Eph
Mutual resolution method [19][a]	(+)-Z-Ala	(±)-Eph	(+)-Z-Ala·(+)-Eph	(+)-Z-Ala·(−)-Eph
Mutual resolution method	(±)-Z-Ala	(±)-Eph	(+)-Z-Ala·(+)-Eph[b]	(+)-Z-Ala·(-)-Eph[d]

[a]This method is used for the resolution of alanine containing rich (+)-Z-Ala;
[b]addition of (+,+) salt as the seed;
[c]addition of (−,−) salt as the seed;
[d]the diastereoisomer salts of (+)-Z-Ala·(−)-Eph and (−)-Z-Ala·(+)-Eph have not been obtained.

Z-Ala =

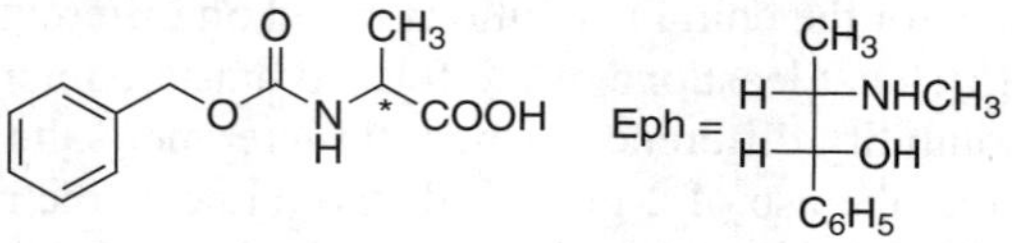

Eph =

added to precipitate the crystal salt of the same configuration. At this point, the liquid portion still contains one lower-solubility salt and two higher-solubility enantiomer salts. These three optically active compounds can be separated by the same crystallization method. This mutual resolution approach does not need optically pure resolving agent and eliminates tedious operations, making the resolution more economical and practical. It is worth noting that the prerequisite of this mutual resolution is that reverse resolution using resolving agents must be feasible. Table 4.2 summarizes the resolution of ephedrine and N-benzyl N-Boc-alanine racemates by conventional resolution, reciprocal resolution, and mutual resolution [19].

In industry, reciprocal resolution has been used to resolve racemic fosfomycin by using (±)-α-phenylethylamine as the resolving agent. The resulting (−)-fosfomycin·(+)-α-phenylethylamine crystal was treated with sodium hydroxide to afford (−)-fosfomycin (**11**) [20] (Fig. 4.6).

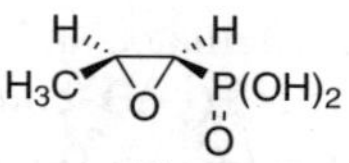

11

FIGURE 4.6 Structure of (−)-fosfomycin (**11**).

In the above-discussed resolution methods, the amounts of resolving agents are stoichiometric. To improve the resolution efficiency, a half-quantity resolution method has been reported [21].

Basically, a half-equivalent of resolving agent relative to the racemic compound is used in the half-quantity resolution method, with another half-equivalent of nonoptically active acid and alkali. The advantages of this method are the use of a smaller amount of resolving agent and the potential increase of solubility differences in some cases. Because four salts in the solution are balanced, this method is known as the "balance method."

This method also can directly use hydrochloride salt of racemic alkali (or other acid salt) and a half-equivalent of optically active ammonium salt (or other inorganic salts). It opposite case will be true when the separated compounds are racemic acids.

This technology has been successfully used for the resolution of the non-steroidal anti-inflammatory drug naproxen. In the resolution process, using the racemic naproxen and a half-equivalent of the chiral resolving agent N-alkyl-glucosamine, as well as the other half-amount of achiral amines, four different salts are formed, that is, (S)-naproxen and chiral amine salt, (S)-naproxen and achiral amine salt, (R)-naproxen and chiral amine salt, and (R)-naproxen and achiral amine salt. Among these, only (S)-naproxen and chiral amine salt crystallize in the form of solid salt, which is then hydrolyzed by acid to give (S)-naproxen (**12**) [22] (Fig. 4.7).

The half-quantity resolution method is also applied for the preparation of the intermediate of the anti-TB drug 2-amino-butyl alcohol (**13**) [23]. These racemates were separated with a half-equivalent of 2,3-dibenzoyl-tartaric acid and a half-equivalent of hydrochloric acid in 95% ethanol. The optimized process was reported to use sulfuric acid to replace hydrochloric acid [24]. A better result was reported when 2-amino-butanol was treated with d-tartaric acid with 4:1 molar ratio in toluene-methanol mixed solvent, affording a neutral salt composed of two parts of $(+)$-2-amino-butanol and one part of tartaric acid [25].

4.2.2.2 Selection of Resolving Agent

In addition to physical and chemical properties, the structural characteristics of a resolving agent have a great impact on the resolution result. For example, a resolving agent containing several functional groups is superior to those with a single functional group, and the effects of an aromatic resolving agent are better than aliphatic agents. The main reason for this distinction is that the resolution agents of multifunctional groups have different impacts on the interaction of opposite ion pairs of diastereoisomer salts, while the aromatic resolution agents

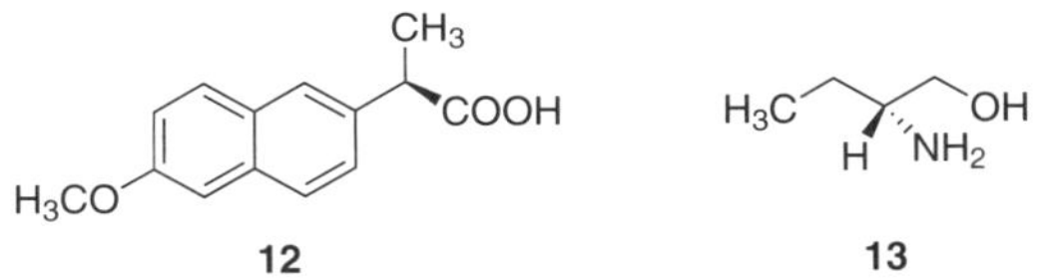

FIGURE 4.7 Structure of (S)-naproxen (**12**) and 2-amino-butyl alcohol (**13**).

produce additional van der Waals forces between the aromatic rings, which results in different solubility of diastereomers.

To conduct the resolution more favorably, the resolving agent structure should meet the following criteria.

1. Alkaline or acid groups should be close to the chiral center [26]. A long distance would decrease the optical purities. For example, the resolving agent (**14**) (Fig. 4.8) needs 28 cycles of crystallization to provide the product with reasonable optical purity.
2. Introduction of one or more polar groups close to the key functional group will lead to more contact points when forming the diasteromeric salts, which will further lead to enhanced resolution capacity [27].
3. Increasing acid and alkaline strength of a resolving agent will help formation of the diastereomer salts, especially when the separated racemate is a weak acid or weak base [28]. For example, the pK_b value of α-phenylethylamine in 95% ethanol is 4.5, while that of dimethyl-strychnine is 5.9. The salts of a weak basic or weak acidic resolving agent with a racemate dissociate extremely easily and are thus not very stable.

Resolution of racemates with a resolving agent is widely used in the pharmaceutical industry, some examples are shown in Table 4.3 [29].

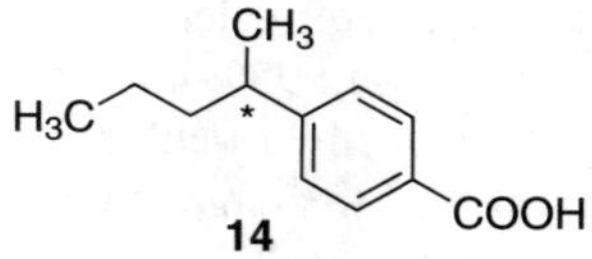

FIGURE 4.8   Structure of the resolving agent (**14**).

TABLE 4.3   Examples of Preparation of Common Drugs by Resolution

Chiral Drugs	Resolution Agent
Ampicillin	D-camphor sulfonic acid
Ethambutol	L-(+)-tartaric acid
Chloramphenicol	D-camphor sulfonic acid
Propoxyphene	D-camphor sulfonic acid
Dexbrompheniramine	D-phenyl succinic acid
Fosfomycin	α-Phenylethylamine
Thiamphenicol	L-(+)-tartaric acid
Naproxen	Cinchonine
Diltiazem	α-Phenylethylamine

4.2.2.2.1 Resolving Agents for Acid or Lactones. The resolving agents for acid or lactone racemates are usually basic, most of which are natural products such as alkaloids (Table 4.4) or terpenoids containing an amino group, such as (+)-3-aminomethyl pinane (**24**), rosin enamine (**25**), (1*R*)-3-endo-amino-borneol (**26**), and endo-borneol amine (**27**) (Fig. 4.9).

Many synthetic amine compounds are also being used for resolution of acids or lactones, such as *N*-methyl-glucosamine (**28**), *N*-octyl-glucosamine (**29**), (1*R*, 2*S*)-(−)-ephedrine (**30**), (1*S*, 2*R*)-2-amino-1,2-diphenyl ethanol (**31**), (1*S*, 2*S*)-(+)-2-amino-1-phenyl-1,3-propanediol (**32**), (1*S*, 2*S*)-2-amino-1-(4-nitro)phenyl-1,3-propanediol (**33**), (*S*)-(−)-α-phenylethylamine (**34**), (*S*)-*N*-benzyl-α-phenylethylamine (**35**), (*S*)-(4-isopropyl)-α-phenylethylamine (**36**), (*S*)-(4-nitro)-α-phenylethylamine (**37**), (*S*)-(−)-α-naphthylethylamine (**38**), and *cis*-*N*-benzyl-2-(hydroxymethyl)-cyclohexylamine (**39**) (Fig. 4.10).

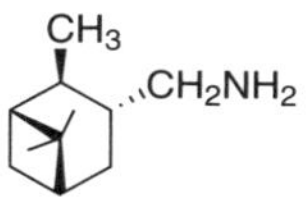

24

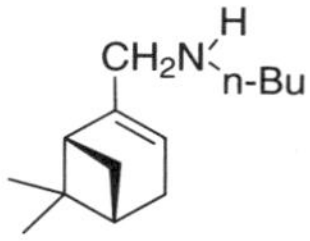

25

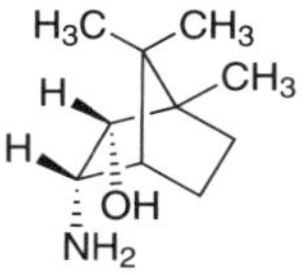

26

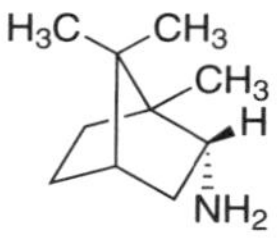

27

FIGURE 4.9 Terpenoids used in resolution of acids and lactones.

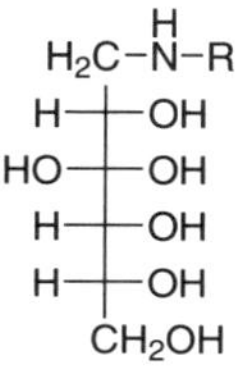

28 R = CH_3
29 R = C_8H_{17}

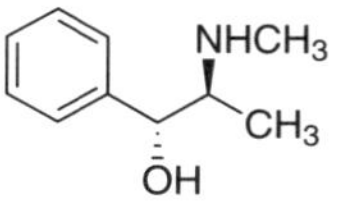

30

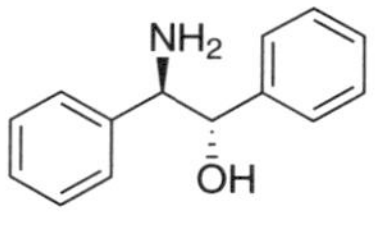

31

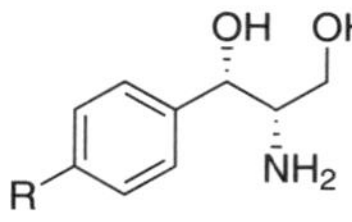

32 R = H
33 R = NO_2

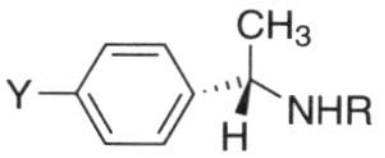

34 R = H, Y = H
35 R = $C_6H_5CH_2^-$, Y = H
36 R = H, Y = $(CH_3)_2CH^-$
37 R = H, Y = NO_2

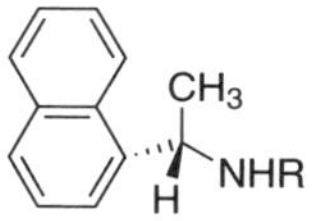

38

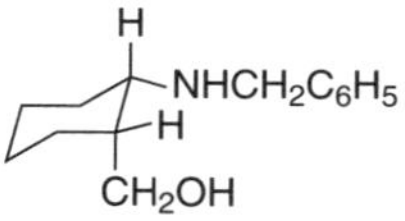

39

FIGURE 4.10 Chiral amines used for resolving acids and esters.

Amino acids and their alkaline derivatives can also be used as resolving agents for the resolution of chiral carboxylic acids, such as (*S*)-(+)-arginine (**40**), (*S*)-(+)-phenylglycine (**41**), (*S*)-p-hydroxyphenylglycine (**42**), (*S*)-(−)-prolyl acid (**43**), L-phenyl amide (**44**), (*R*)-(−)-2-phenylglycinol (**45**), and (*S*)-(−)-3-phenylalaninol (**46**) (Fig. 4.11).

TABLE 4.4 Alkaloids Used as Resolving Agents

Names		Physical Constants	Solubility	
			Chloroform	Ether
Brucine, **15**		m.p: 178°C, $[\alpha]_D$ −127° (CHCl₃), −85°(abs C₂H₅OH), pK$_{a1}$: 6.04, pK$_{a2}$: 11.7.	1 g/5 ml	1 g/187 ml
Strychnine, **16**		m.p.: 275–285°C, $[\alpha]_D^{18}$ −104.3° (c 0.254, EtOH), $[\alpha]_D^{25}$ −139 ° (c 0.4, CHCl₃), pKa (25°C): 8.26	1 g/6.5 ml	Very slightly soluble
Quinidine, **17**		m.p.: 171°C, $[\alpha]_{546}^{20}$ +301.1° (CHCl₃: EtOH, 97.5:2.5)		

Compound	Structure	Properties	Solubility	
quinine, **18**		m.p.: 177°C(partly dec). $[\alpha]_D^{15}$ −169° (c2, 97% EtOH), $[\alpha]_D^{17}$ −117° (c 1.5, CHCl$_3$), $[\alpha]_D^{15}$ −285° (c 0.4 M, 0.1 N H$_2$SO$_4$).pK_{a1}: (18°C) 5.07; pK_{a2}: 9.7.	1 g/1.2 ml	1 g/250 ml
Cinchonine, **19**		m.p.: about 265°C, $[\alpha]_D$ +229° (EtOH)	1 g/110 ml	1 g/500 ml
Cinchonidine, **20**		m.p.: 210°C, $[\alpha]_D^{20}$ 109.2° (EtOH), pK_{a1}: 5.80, pK_{a2}: 10.03	Soluble	Sparingly soluble

TABLE 4.4 (*Continued*)

Names	Physical Constants	Solubility	
		Chloroform	Ether

Quinotoxine, **21**

Cinchonicine, **22**, cinchotoxine

m.p.: 58–60°C, $[\alpha]_D^{15}$ +47° (EtOH)

Dehydro-abietylamine, **23**

m.p.: 41°C, 42.5 - 45°C, b.p. 192–193°C/1 mmHg, η_D^{40} 1.5462

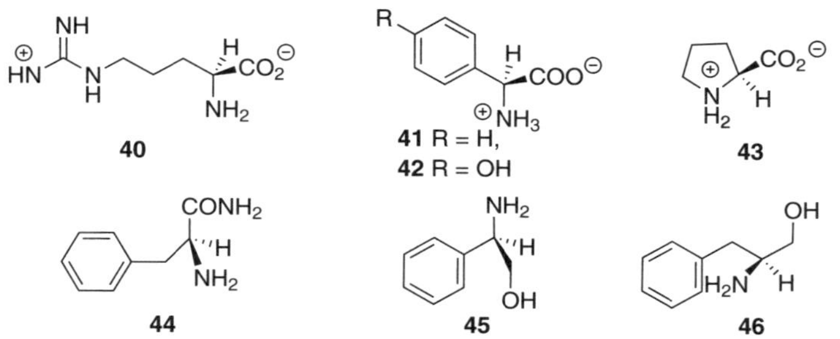

FIGURE 4.11   Amino acids and their alkaline derivatives used for resolving acids and esters.

These alkaline resolving agent can react with racemic carboxylic acids or esters to form the covalent diastereomers such as amides and esters. After resolution and subsequent hydrolysis, pure enantiomers can be obtained. However, when racemic compounds are converted into amide diastereomers, harsh conditions are required for the hydrolysis. The racemization of chiral carbon next to the carbonyl group might occur; therefore, this method is not generally used. If the structure of the chiral amine resolving agent contains a α-amino β-hydroxyl moiety (such as **46**), the amide hydrolysis can be promoted through the neighboring group effect.

4.2.2.2.2 Resolving Agents for Resolution of Alkalines. The resolving agents used for the racemic amines are normally chiral carboxylic acids, and the commonly used acidic resolving agents are listed in Figure 4.12 (**47**–**68**).

Tartaric acid (**47**) and its acyl derivatives (**48, 49**) are common reagents used for the resolution of basic compounds. The acidities of compounds (**48**) and (**49**) are stronger than that of their parent compound. In addition, introduction of aroyl group provides an additional functional group to achieve a better recognition of the diastereomers. Tartaric acid and its derivatives are multivalent organic acids that are used stoichiometrically with separated materials in the resolution process, and the final product is their diastereomeric salt [30]. Lin et al. have successfully separated the racemic antihistamine drug fexofenadine and its hydrochloride with tartaric acid (**47**) and its acyl derivatives (**48, 49**) [31].

Mandelic acid (**52**), its derivatives (**53, 54**), and Mosher acid (**55**) have been used to resolve racemic primary and secondary amines. However, in the case of tertiary amines which are stronger bases, a relatively stronger acid, such as (S)-(+)-1,1'-binaphthyl phosphoric acid (**58**, $pK_a = 2.5$), should be used as the resolution agent [32]. Deoxycholic acid (**59**) is usually used to resolve more water-soluble amines, with good results [33].

Menthyl chloroformate (**68**) can convert amines into carbamate diastereomers before the resolution process. Chiral isocyanate (**67**) was used to separate racemic amines by forming diastereomeric ureas [34,35]. The resolving agents (**54**), (**64**), and (**65**) were employed for acylation of chiral amines, to generate the diastereomer amides for the separation.

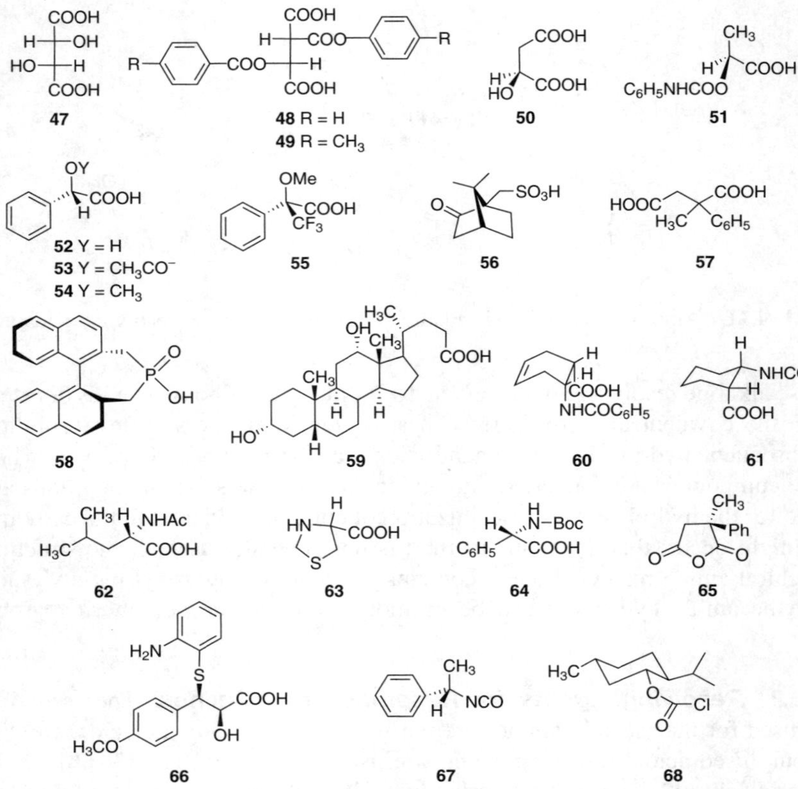

FIGURE 4.12 Resolving agent used for resolution of racemic alkali.

As reported, the racemic antiarrhythmic drug mexiletine was separated with tetrahydropyran-protected mandelic acid (THPMA), furnishing (*R*)- and (*S*)-enantiomers with optical purity up to 99% *ee* [35].

4.2.2.2.3 *Resolving Reagents for Resolution of Amino Acids.*

Because free amino groups in amino acids can interact with acidic functionality in the resolution process, acylation derivatives, such as *N*-acetyl, *N*-formyl, *N*-benzoyl, *N*-tosyl, *N*-acyl-phthalimide, *N*-Cbz, *N*-(*p*-nitrophenyl) sulfoximine acyl amides, are normally used to derivatize the amino acids. The resulting amino-protected racemic amino acids contain free carboxylic acid moiety and can be resolved with alkaline agents such as strychnine (**16**), quinine (**18**), and ephedrine (**30**), through the formation of diastereomeric salts [33].

Occasionally, chiral acid resolving agents can be used directly for the resolution of unprotected amino acids. For example, mandelic acid has been used to separate racemic amino acid [36]. 10-Camphor sulfonic acid (**56**) is another example used for the separation of racemic 2-*tert*-butylglycine [37].

4.2.2.2.4 *Resolving Reagents for Resolution of Hydroxyl-Containing Compounds.* Hydroxyl-containing compounds include alcohols, diols, thiols, dithiols, and phenols. Alcohols are usually derivatized into esters before resolution. Previously, it was reported that treatment of phthalic anhydride and alcohol afforded phthalic monoester, where the resulting free carboxyl acid can form diastereomeric salt with chiral alkaline resolving agents [38]. However, this approach is not used as extensively as before. As an alternative, alcohols react directly with chiral carboxylic acids to form diastereomer esters that can be separated either by crystallization or chromatography. Mandelic acid (**52**) and its derivatives (**53, 54**), Mosher acid (**55**), ω-camphene acid (**69**) [39], naproxen (**12**), and *trans*-1,2-cyclohexane anhydride (**70**) are the chiral carboxylic acids commonly used as resolving agents. In addition, (*S*)-2-(*N*-tosylamino)-3-phenylpropanoic acid [40] (**72**) and 10-camphor sulfonic acid can also be converted into acid chloride for resolution. In the case of diols, only one hydroxyl group is needed to be esterified. ω-Camphene acid (**69**) can also be used for separation of chiral compounds containing phenolic hydroxyl moiety.

Chiral isocyanates, such as (*R*)-naphthalene ethyl isocyanate (**72**), could react with amine compounds to form a diastereomic carbamate, which is used for the purpose of racemate resolution. These approaches were applied to alcohols, especially tertiary alcohols, with good results [41] (Fig. 4.13).

4.2.2.2.5 *Resolving Reagents for Resolution of Aldehyde and Ketone.*
The carbonyl groups in racemic aldehydes and ketones can form covalently linked diastereomers with amine or hydroxyl groups, which is the key principle in the resolution of this type of racemates(Fig. 4.14).

Two neighboring chiral hydroxyl groups in tartaric acid and its derivatives (**73, 74**) or (*R,R*)-2,3-butane-diol can react with the carbonyl groups of aldehydes and ketones to generate diastereomeric acetal and ketal compounds [42]. Chiral amines (**75**) and amino acid esters and hydrazines (**76, 77, 78**) can form diastereomeric Schiff bases [43] and hydrazones with carbonyl groups of aldehydes or ketones, respectively. A classic example is the resolution of the antifertility drug gossypol (**79**), which is separated by using aldehydes to form diastereomeric Schiff bases with a chiral amine.

As examples, aldehydes and ketones are separated by forming diastereoisomeric oxazolidines [44] or imidazolines [45] with α-hydroxy amines such as ephedrine or 1,2-diamine (**80**).

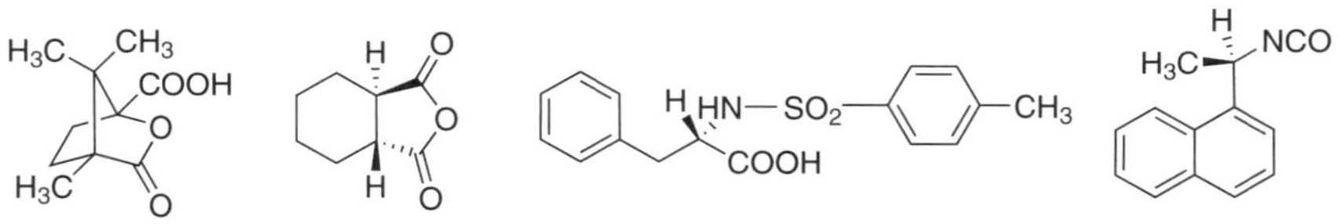

| 69 | 70 | 71 | 72 |

FIGURE 4.13 Resolution reagent for racemic hydroxyl-containing compounds.

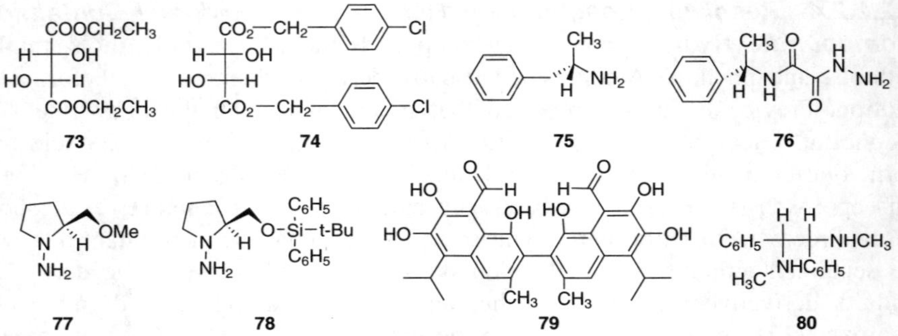

FIGURE 4.14 Resolution reagent for separation of racemic aldehyde and ketone racemates.

4.2.3 Combinatorial Resolution

In the past century, chemical resolving agents have been selected by random screening. It is desirable to develop a simple and quick way to carry out the resolution of racemic compounds.

The technology of combinatorial resolution has been reported in recent years [46]. Its principle is to utilize a family of resolving agents comprising a set of similar structure types of chiral derivatives to replace a single chiral resolving agent for the resolution of racemic compounds. Experiments showed that addition of a group of such resolving agents into the racemic compound in solution led to quick precipitation and furnished crystals of diastereoisomeric salts. The optical purity (*ee*) of the resolved compounds can reach up to 90%, with almost quantitative yield.

These resolving agent families are either structurally modified derivatives of commonly used chiral resolving agents,or those compounds containing substitutions. For instance, the α-phenylethylamine resolving agent family of PE-I, PE-II, and PE-III and α-amino alcohols PG are widely used in acidic compound resolution; tartaric acid derivative resolving agent family T and TA, *p*-substituted mandelic acid M, *N*-substituted phenylglycine PGA, and *o*-substituted phenylpropanediolglycol phosphate P are usually used for the resolution of alkaline compounds. Additionally, P, PGA, and M can also be used in the resolution of amino alcohols (Fig. 4.15).

Compared with classical methods, combinatorial resolution technology has a faster crystallization rate and higher yield and purity. For example, 1-(*p*-tolyl)ethylamine (**81**) was resolved with 4-methylmandelic acid (Fig. 4.15, M, X = Me) to give a chiral product with only 57% *ee*.

During the resolution process, substrates and the resolving agent family are stoichiometrically dissolved in the same solvent. No matter how the resolving agent family is composed, the proportion of each component in the family is fixed. It is noteworthy that, theoretically, the amount of each component of the resolving agent family contained in the solution should be equal in the resultant

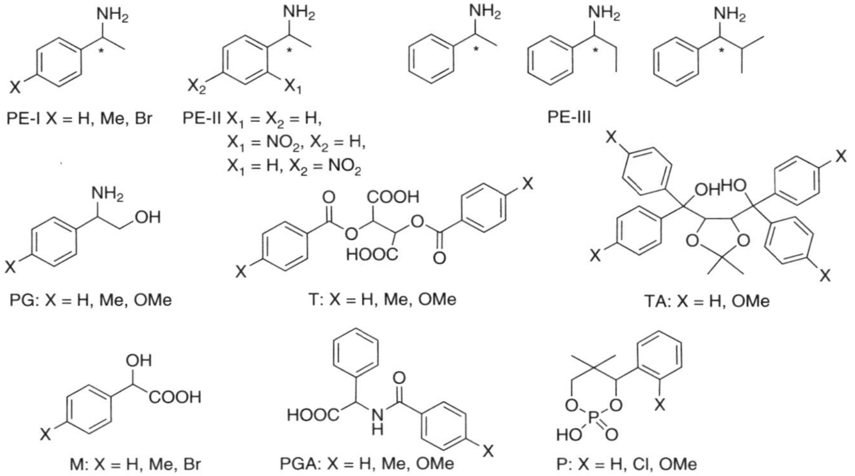

FIGURE 4.15 Resolving agent families for combinatorial resolution.

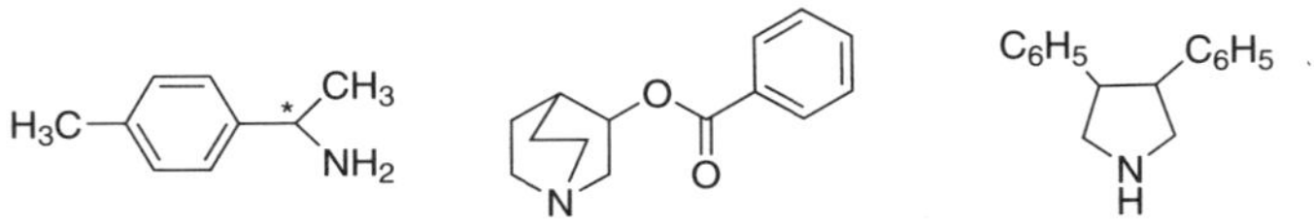

FIGURE 4.16 Structures of compounds **81–83**.

diastereoisomer precipitation. However, this is not always the case. For example, in the treatment of compounds (**82**) and (**83**), the resolving agent T is composed of three tartaric acid benzoic esters with different substitutions (Fig. 4.16). The ratio of the three components of T is 1:10:4 and 1:3:3, respectively, in the formed crystals.

In combinatorial resolution, the resolving agent family is neither a random combination of different resolving agents nor a mixture of resolving agents with similar functional groups (for example, chiral carboxylic acids such as tartaric acid, lactic acid, and malic acid). Each resolving agent in its family has a high degree of structural similarity and stereochemical homogeneity (mainly referring to the same chirality and enantiomeric purity of the compounds).

Deng et al, separated a set of racemic adrenergic β-agonists by using combinatorial resolution. Thus tartaric acid family T (d-tartaric acid: dibenzoyl-d-tartaric acid: 2-p-methylbenzoyl-d-tartaric acid = 1:1~2:1~2 mol) and adrenergic β-agonists (adrenaline, isoprenaline, salbutamol, terbutaline, salmeterol, etc.) were refluxed for 5–6 h in alcohols, ketones, their mixture, or ethyl acetate. After cooling down to room temperature, the diastereomer salts of one adrenergic β-agonist stereomer with T precipitated [47].

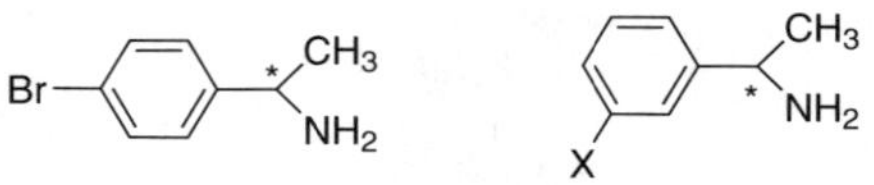

FIGURE 4.17   Structures of compounds **84** and **85**.

The resolving agent family could also be applied to a racemic mixture. For instance, in the resolution of 4-bromo-α-methylbenzylamine (**84**), (*S*)-mandelic acid and racemic *p*-methyl mandelic acid were mixed together, yielding (**84**) with *ee* 90% as well as *p*-methyl mandelic acid with *ee* 95%. This result is reminiscent of reverse resolution, in which a single resolving agent is used to separate a racemic mixture. For example, optically pure *p*-methyl mandelic acid is used for resolution of a racemic mixture of 3-methoxy-, 3-chloro-, or 3-bromo-α-methylbenzylamine (**85**) (1:1:1) to afford an optically active mixture with 98% *ee* and 2:4:4 ratio [46] (Fig. 4.17).

4.3   COMPOSITE RESOLUTION AND INCLUSION RESOLUTION

In crystallization resolution, direct resolution requires that the substrates can form an aggregate, which can be separated by physical means through the formation of crystals. In general, diastereomeric resolution takes advantage of functional groups such as carboxyl, amino, hydroxyl, ester, aldehyde, or ketone to form diastereomers containing ionic or covalent bonds with chiral resolving reagents such as chiral carboxylic acids or chiral amines. The diastereomers are then separated by crystallization. However, the crystallization resolution method has some limitations for those without obvious and applicable functional groups. The newly developed composite resolution and inclusion resolution might be a good choice. Composite resolution and inclusion resolution make use of differences of interactions of hydrogen bonding or van der Waals interaction to achieve the resolution. This is a crystallization method in principle but is also one of the chemical resolution methods.

4.3.1   Composite Resolution

Composite resolution is suitable for resolution of racemic olefins and aromatic compounds with π electrons, as well as elemento-organic compounds with lone electron pair atoms such as the organosulfur, organoarsenical, and organophosphorus compounds. In the resolution process, olefinic or aromatic compounds containing delocalized π electrons can form an electron transfer complex with π electron-rich chiral reagents or form coordinates with chiral organometallic compounds. These electron transfer complexes and metallic coordinates have diastereomeric characteristics and are easily separated. The empty orbitals or

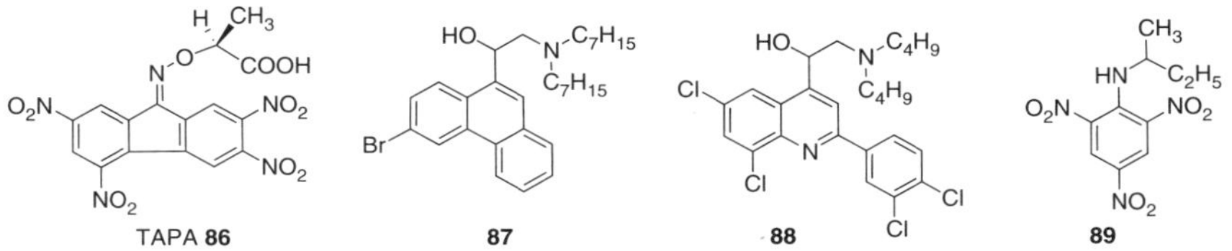

FIGURE 4.18   Structures of compounds **86–89**.

lone pairs in organosulfur, organoarsenical, and organophosphorus compounds can form complexes with Lewis acid or Lewis basic chiral reagents.

The resolution method using the formation of π electron complex or π electron transfer complex is mostly applied to the resolution of aromatic compounds. These agents are π electron-rich chiral acids, such as α-(2,4,5,7-tetranitro-9-fluorenone sub-ammonia-oxy) propionic acid (also known as TAPA, **86**). TAPA is a multi-nitro aromatic compound with a coplanar structure, which can form the π − π electron transfer complex or π − π electron complex with aromatics or polycyclic aromatic compounds. These compounds are darker in color and have better crystal shapes and melting points, which can be used to separate aromatic ethers, aromatic esters, and phosphate compounds, especially for those hydrocarbons lacking functional groups, such as pentacene, hexahelicene, α-bromohexahelicene, N-containing spirocompound, and *meta, para*-cyclic alkane-substituted benzene. Because of steric hindrance around the chiral centers or rigidity in the structure, these are usually difficult to separate by conventional resolution methods [48].

TAPA is also used for the resolution of chiral side chains containing aromatic amines, heteroaromatic compounds, and aromatic hydrocarbons. These aromatic amines or nitrogen-containing heteroaromatics weakly alkaline and therefore are difficult to separate by forming diastereomer salts with chiral acid resolving agents. For example, compounds (**87**) and (**88**) were successfully separated with TAPA (Fig. 4.18).

Resolution with TAPA is carried out by π electron-containing acids as the resolving agent. On the other hand, π electron-containing bases such as 2-naphthyl camphene amine were used to separate weak acid compounds such as *N*-sec-butylpicrcyl amine (**89**), which could not be separated by π electron-containing acids.

4.3.2 Inclusion Resolution

Inclusion resolution is a relatively new method for racemic compound resolution in which racemates are wrapped by chiral resolving agents through noncovalent interactions and two enantiomers are then separated by crystallization.

The principle of inclusion resolution relies on the cavities existing in the resolving agent molecule (host molecule). Such cavities accommodate certain shapes and sizes of guest molecules and form diastereomeric clathrates to achieve separation. These host molecules with cavities generate chirality match and chirality selectivity at the molecular level. Several factors may affect the chirality match and chirality selectivity between the host and guest molecules:

1. Host and guest molecules have a tight binding capacity, which is a prerequisite for enantioselectivity.
2. Host and guest molecules have good compatibility, meaning that the chiral cavities of host molecules and guest molecules produce a lock-key relationship.
3. Guest molecules need to have certain flexibility to make them better fit the cavities of host molecules.

There are two types of wrapped complexes, the cavitates and the clathrates. In cavitates the chiral substrates (guest molecules) are partially or entirely wrapped in the chiral cavity of host molecules, whereas in clathrates guest molecules are wrapped by a few host molecules to form cagelike or tunnel shapes. The basic principle of inclusion resolution is that different strengths of hydrogen bonding and van der Waals interactions formed by guest compounds of different configurations with cavities of host compounds produce favorable inclusion with one of the enantiomers.

4.3.2.1 Cavity Inclusion Resolution

The resolving agents used in cavity inclusion resolution are chiral cyclic polyether (crown ether) and cyclodextrin, which form cavities inside their molecules.

Compound (**90**) is a crown ether formed by two binaphthols. Because of the chirality in the structure, binaphthol has three isomers, the chiral (R, R)-isomer and (S, S)-isomer as well as the achiral (R, S)-isomer. The crown ether compound (**90**) is used for the separation of amino acids. For instance, the symmetric crown ether (**90a**) (R, R' = H) can separate phenyl glycine methyl ester [49] and the asymmetric ether (**90b**) (R = CH$_3$, R' = H) can be used for the separation of phenylglycine perchlorate [50]. (R, R)-crown ether (**90**) and d-phenylglycine perchlorate form thermal stable crystals during the process of resolution; while the (S, S)-enantiomer of the crown ether cannot form crystals with phenylglycine perchlorate. Therefore, this process provides a good example of selectivity in resolution. Conversely, d-phenylglycine perchlorate is also used to resolute racemic crown ether (**90**) [51]. As for the (R, S)-isomer of meso compounds, a narrow cavity is formed because of binaphthol configuration difference, which does not favor the formation of inclusion complexes [52] (Fig. 4.19).

Another example is macrocyclic ether compound (**91**), specifically designed for the resolution of naproxen. Both naproxen and naproxen methyl ester can form an inclusion complex with (**91**). The inclusion complex of (**91**) with naproxen methyl ester is more stable than that with naproxen because the ionic pair formed by the COO$^-$ of naproxen with N$^+$ of (**91**) is less strong than the $\pi - \pi$ complex. Meanwhile, the difference of the diastereoisomer in inclusion complex is narrower than that of the $\pi - \pi$ complex [53].

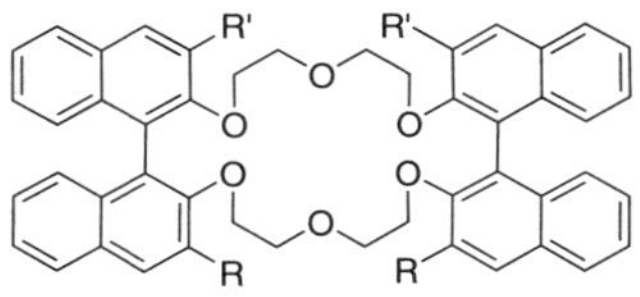

90a R, R' = H
90b R = CH₃, R' = H

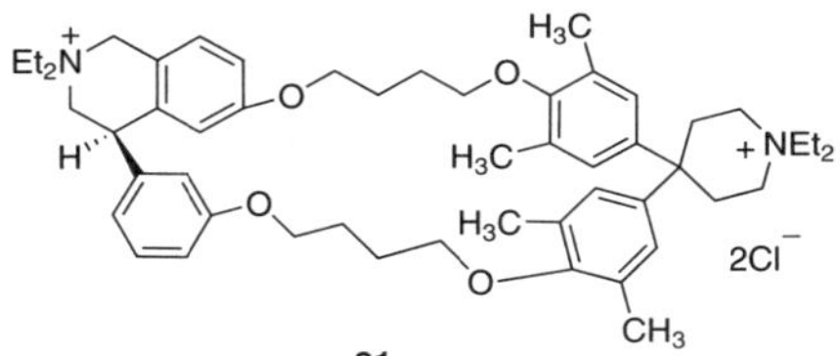

91

FIGURE 4.19 Structures of compounds **90** and **91**.

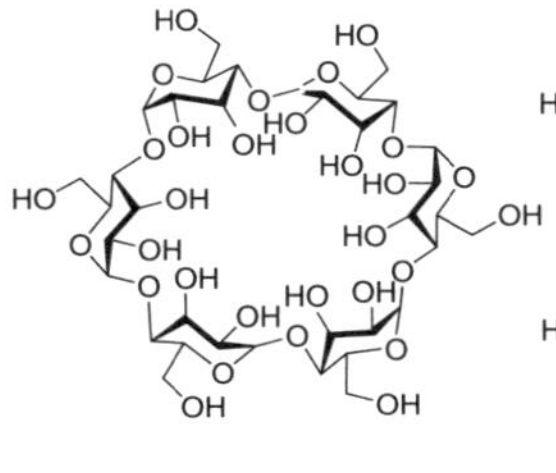

92a α-cyclodextrin 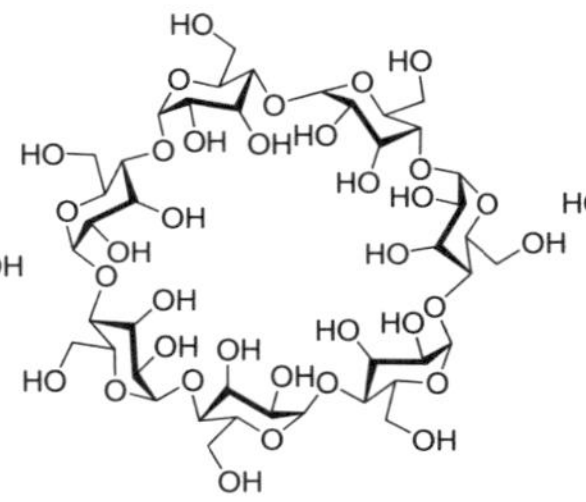   **92b** β-cyclodextrin 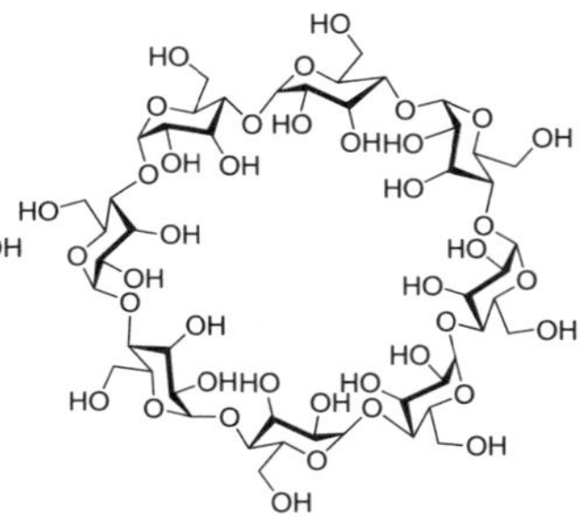   **92c** γ-cyclodextrin

FIGURE 4.20 Structures of α-, β-, and γ-cyclodextrin.

Cyclodextrin (**92**) is the most widely used reagent for inclusion resolution. It is a water-soluble macrocyclic oligo-glucose containing six, seven, and eight D-(+)-glucopyranosyl structural units formed by α-(1,4)-glycosidic bond connected head to tail, generally known as α-, β- and γ-cyclodextrin (**92a–92c**) (Fig. 4.20).

Cyclodextrin takes a cylindrical shape, with both end openings like a big and a small circle, respectively, and with a cavity in between. The glucose units in the cylinder maintain the original stable chair conformation of the glucopyranosyl ring system, and its rigid secondary C2-OH and C3-OH extend into the big circle. In addition, C2-OH of glucose unit forms a hydrogen bonding with a proximate C3′-OH of another glucose unit. The primary hydroxyl of C6-OH hooks up to the small circle through a C6-C5 bond. The rotation of the C6-C5 bond then leads to C6-OH introversion, which partially covers the mouth of the small circle.

All the hydroxyl groups of glucose units locate at the outside of the cylinder, so cyclodextrin shows strong hydrophilicity and dissolves well in water. The cylinder wall of cyclodextrin contains the acetal-type C-O-C bonds and C-H bonds and therefore is hydrophobic. As the glycosidic bonds are not stable under acid conditions, cyclodextrin shows better stability in basic medium than in acidic medium. Cyclodextrin is composed of varying numbers of d-glucopyranosyls to form a cyclic structure with a strong chirality. Its aqueous solution is dextrorotatory. Moreover, different cyclodextrin molecules form different sizes of cavities. As such, some of the hydrophobic groups of different chiral molecules would

TABLE 4.5   Nature and Characteristics of Three Kinds of Cyclodextrins

Cyclodextrin	Numbers of Glucose	Molecular Weight	Solubility (g/100 ml)	$[\alpha]_D^{25}$	Peripheral Diameter (nm)	Cavity Diameter (nm)	Depth of Cavity (nm)
α	6	972	14.5	+150.5	1.46	≈0.49	≈0.79
β	7	1135	1.85	+162.5	1.54	≈0.62	≈0.79
γ	8	1297	23.2	+177.4	1.75	≈0.79	≈0.79

interact with hydrophobic groups inside the cylinder wall and be trapped in the cavity to form diastereomers composed of 1:1 ratio of host-guest molecules. Since cyclodextrin is a chiral molecule, there are differences of diastereomers by chirality match and chirality recognition when cyclodextrin is mixed with racemates (Table 4.5).

There are many hydroxyl groups in the structure of β-cyclodextrin. It will increase the selectivity of the host molecule if hydroxyl groups form stable hydrogen bonding interactions with the substrate. One typical example is the resolution of sulfoxide compounds. Figure 4.21 depicts the interactions of β-cyclodextrin with (R)-(−)-methyl benzyl sulfoxide as well as with (R)-(+)-tertbutyl benzyl sulfoxide. The diameter size and hydrophobic characteristics of the cyclodextrin cavity are in accordance with the size and characteristics of the phenyl ring. Therefore, when compounds with phenyl group interact with cyclodextrins, the phenyl group inserts properly into the cavity of cyclodextrins. Meanwhile, the S = O group of sulfoxide forms a hydrogen bonding with hydroxyl groups of cyclodextrin, thereby increasing the stereoselectivity [54].

β-Cyclodextrin was used for the resolution of the nonsteroidal antiinflammatory drug fenoprofen (**93**). As expected, the phenyl ring inserts into the β-cyclodextrin cavity, but the (R)- and (S)-isomers showed different crystal morphology when they formed crystals with β-cyclodextrin. In the crystallization

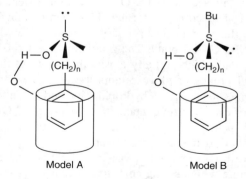

Model A   Model B

FIGURE 4.21   The role of β-cyclodextrin with (R)-(−) -methyl benzyl sulfoxide and (R)-(+)-*tert*-butyl benzyl sulfoxide.

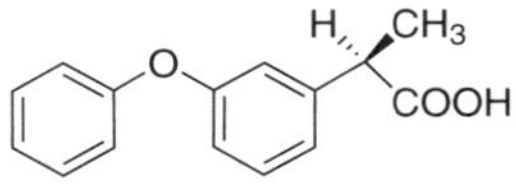

FIGURE 4.22   Structures of antiinflammatory drug fenoprofen (**93**).

process, the inclusion complex of (*R*)-isomers and β-cyclodextrin was a linear dimer with head-to-head linking, while the inclusion complex of (*S*)-isomers and β-cyclodextrin was also a linear dimer but linked head to tail [55] (Fig. 4.22).

In addition, the derivatized cyclodextrin was also used as the stationary phase of gas chromatography for separation of chiral compounds [56].

4.3.2.2 Clathrate Compound Resolution

In general, clathrate compounds are formed when a guest molecule is surrounded by multiple host molecules. Host molecules constitute a tunnel in the lattice when the crystal forms, which traps the guest molecule. Host molecules of clathrates include urea (**94**) and tri-*o*-thymotide (TOT, **95**), as exemplified in Figure 4.23.

Urea is an achiral compound. Under normal conditions, the formed urea crystal assumes a compact pyramidal structure due to the intermolecular hydrogen bonds. However, this becomes a cubic structure in the presence of guest molecules. The cube contains a long spiral tunnel structure that holds the guest molecule to form the tunnel clathrate. Because of the spiral characteristics, the tunnel can be left-handed or right-handed. Although urea itself is an achiral compound, it can be used for the resolution of chiral compounds because of the spiral tunnel in crystals.

Clathrate compounds formed by urea have been used mainly in the separation of chiral linear alkane molecules, where the chiral center has small steric substituents such as CH_3, Cl, Br, NH_2, SH, OH, and epoxy. Ureas and alkanes form stable molecular structures through van der Waals interactions. This method has been successfully applied to the resolution of 2-chlorooctane, 2-methyldecane, and 3-methyloctane.

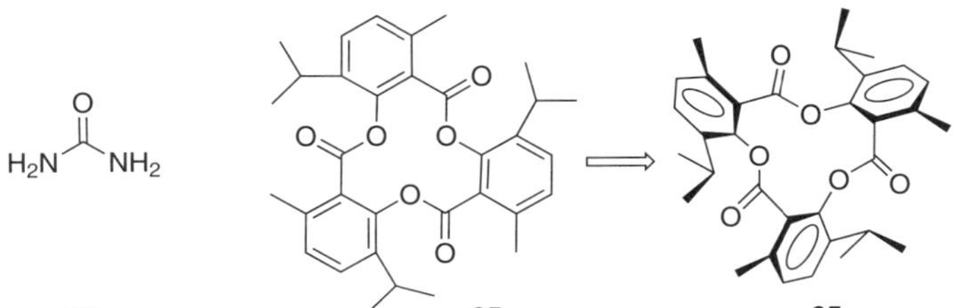

FIGURE 4.23   Structures of **94** and **95**.

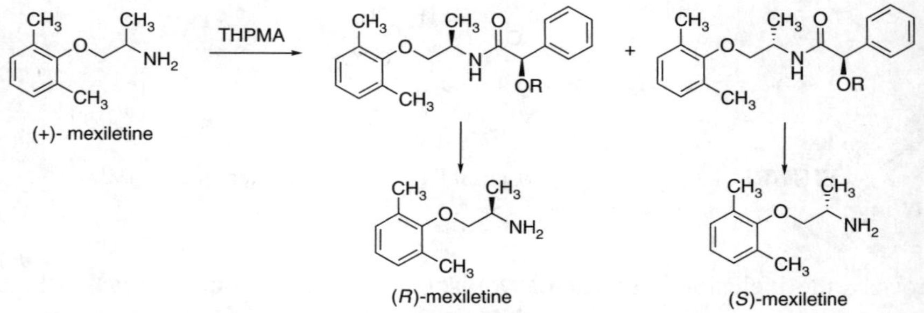

SCHEME 4.4 Resolution of mexiletine with THPMA.

TOT is a 12-membered ring lactone compound prepared by condensation of three molecules of *o*-thymol acid. The molecule has a three-bladed propeller shape that displays C_2 symmetry, so there is a pair of enantiomers. However, TOT undergoes racemization in solution through rapid conversion from one enantiomer to the other. During the process of crystallization, chiral TOT crystal can be obtained by addition of a single chiral crystal to induce the formation of crystals under certain conditions. However, racemic mixtures are the only existing form in solution. Although TOT is racemic in solution, TOT will form an optically pure crystal with guest molecules.

There are two conditions for TOT to form crystals: 1) When the guest molecule contains less than six nonhydrogen atoms, the length is less than 9 AA, and the ratio between guest molecule and host molecule in crystal is 2:1, the optical purity of the resulted chiral compounds ranges from 1% to 83% *ee*. 2) The host-guest molecule ratio tends to be different when the shape of the host molecule is longer than 9 AA. Therefore TOT can only be applied to the resolution of linear alkanes with low molecular weight, such as 2- or 3-halogenated octane.

Other guest molecules such as brucine (**15**) can also form an inclusion complex. Initial studies indicated that alkynyl-containing tertiary alcohol (**96**) formed a stable complex with brucine in a 1:1 ratio [57] (Scheme 4.5).

Further studies showed that there was a strong hydrogen bond interaction between the alcoholic hydroxyl group (**96**) and brucine. The phenomena were also observed in diols, propenols, and cyanohydrins [58].

Sparteine (**97**) can also be applied to the resolution of alkynols such as **98** and vice versa [59].

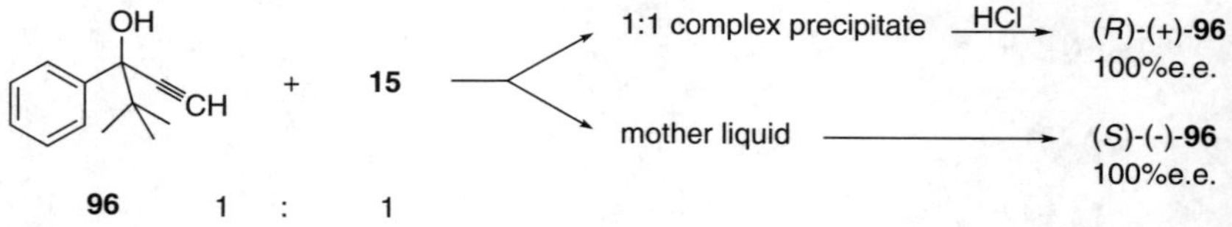

SCHEME 4.5 Resolution of **96** utilizing complexation with brucine (**15**).

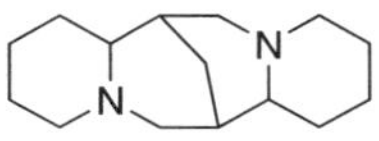

97 sparteine

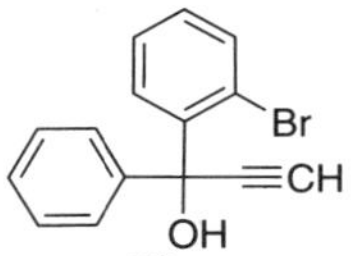

98

$CF_3CHBrCl$

99

CHBrClF

100

FIGURE 4.24 Structure of compounds **97–100**.

As reported, brucine was used to separate chiral halogenated hydrocarbons such as the anesthesia agent halothane (**99**) and chiral halomethane (**100**) by forming an inclusion complex. CHs of chiral halogenated hydrocarbons formed stable hydrogen bonding with nitrogen atom of alkaloids [60] (Fig. 4.24).

4.3.3 Inclusion Complex Resolution

Inclusion complex resolution is similar to the clathrate compound resolution discussed in Section 4.3.2. It was reported for the first time by Toda in 1984 [57]. The basic principle is that chiral host molecules selectively form supramolecular complexes with one enantiomer of racemic guest molecule through weak intermolecular forces such as hydrogen bonding or intermolecular $\pi - \pi$ force. This supramolecular complex, also known as an inclusion complex, precipitates from the solution and separates from the other enantiomer.

The reason that inclusion complexes are formed is that host compounds have efficient and strong molecule recognition toward guest compounds. The host compounds are chiral, and the recognized guest molecules are racemic mixtures. Therefore, the interactions of host and guest molecules are in fact a chiral recognition.

Host compounds interact with guest compounds through hydrogen bondings or van der Waals forces during inclusion resolution. There are three structural types of host resolving agents: 1) dihydroxy compounds such as BINOL (**101**), 2,2′-dihydroxyl-9,9′-dianthranide(**102**), and (4*R*,5*R*) and (4*S*,5*S*) 4,5-di(diphenylmethyl)-2,2-dimethyl-1,3-dioxolane (TADDOL, **103**), etc.; 2) hydrocarbon compounds such as (**104**); and 3) amide compounds such as (**105**) (Fig. 4.25).

Diols are the most commonly used chiral compounds in inclusion complex resolution. Because diols are generally bulky and hydrogen bonds can be formed

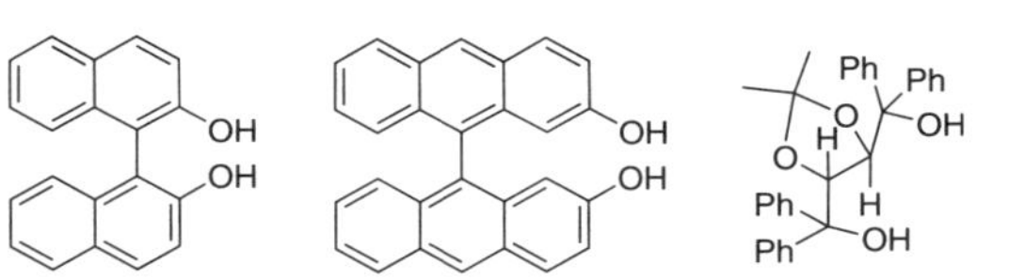

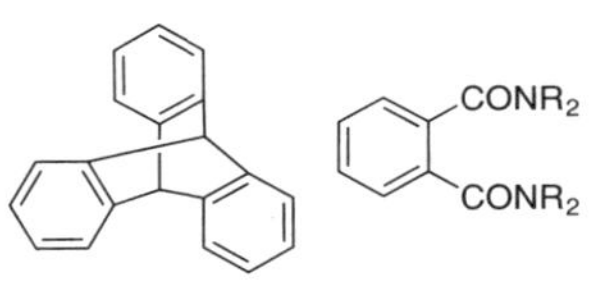

101 **102** **103** (4*R*,5*R*) or (4*S*,5*S*) **104** **105**

FIGURE 4.25 Host resolving agents **101–105**.

between them, guest molecules will be accommodated in between diols. Thus network crystals form in a "host compound—OH•••guest compound•••HO—host compound" pattern.

Diols such as compounds (**103**) and (**106**) were used to separate phosphonate ester (**107**), amine oxides (**108**), alcohols (**109**), β-hydroxyl carboxylic esters (**110**), α,β-epoxy carboxylic esters (**111**), and α,β-epoxy ketones (**112**) [58–61] (Fig. 4.26).

Deng and co-workers have successfully separated sulfoxide compound (**113**) by using a diol (**103**) as the host molecule [62]. However, the use of such diols was not successful in resolving an antiulcerative drug (omeprazole, **114**) that contains the same chiral sulfoxide group. Later, the use of chiral compounds such as BINOL (**101**) instead of diols successfully provided (*S*)-(−)- and (*R*)-(+)- omeprazole enantiomers, respectively[63]. X-ray crystal crystallography studies suggested that the inclusion complex crystal was formed through hydrogen bonding between the oxygen of the sulfoxide (**113**) and hydroxyl group in resolving agent (**103**) with a 1:1 ratio. In general, the oxygen atom on the sulfoxide of omeprazole forms intermolecular hydrogen bonding with N–H of benzimidazole in another omeprazole and exists as a dimer. It is speculated that BINOL (**101**) forms a supramolecular chirality with the dimer by hydrogen bondings [63]. BINOL was also applied to the resolution of racemic lansoprazole (**115**) with *ee* value greater than 98% and in yield of over 60% [64] (Fig. 4.27).

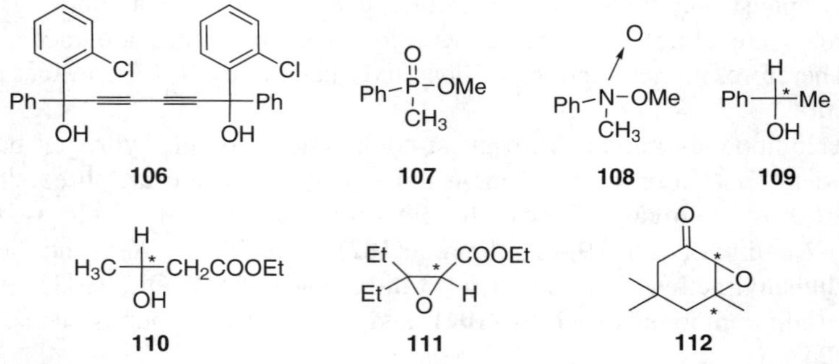

FIGURE 4.26   Structures of compounds **106–112**.

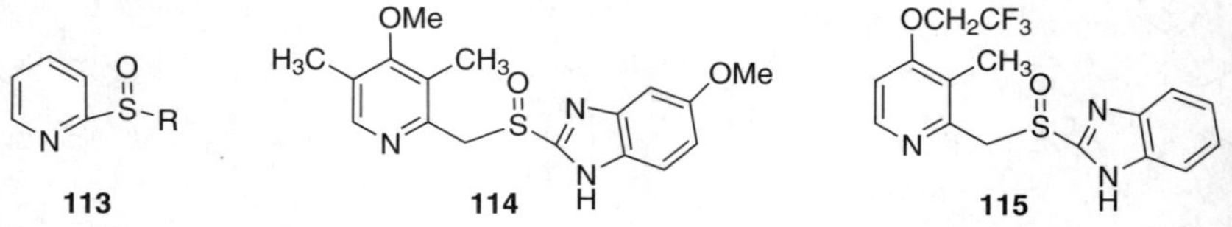

FIGURE 4.27   Structures of sulfoxide compound (**113**), omeprazole (**114**), and lansoprazole (**115**).

The nucleoside antivirus drug (lamivudine, **116**) has two chiral centers. Both *cis*-(−) and *cis*-(+) optical isomers inhibit human immunodeficiency virus (HIV)-1 and -2 in vitro. However, the *cis*-(−)-enantiomer is considerably less cytotoxic than the its optical isomers and is applied clinically against HIV and hepatitis B. It was shown that (*S*)-BINOL (**101**) forms the cocrystal inclusion complex with optical *cis*(−)- and *trans*(−)-isomers (**116**), whereas *cis*-(+)- and *trans*-(+)-isomers which were separated from their corresponding isomers were left in the solution. The optical purity of lamivudine could reach as high as 99.9% [65] (Scheme 4.6).

In general, host-guest compounds in inclusion complex resolution can be purified and separated through recrystallization or column chromatography. Toda's group also reported that the separation was accomplished by distillation or fractionation. Solid (*R,R*)-TADDOL (**103**) was mixed with liquid *p*-methylphenylethylamine (**117**) in a 2:1 ratio before distillation under reduced pressure. TADDOL and (−)-(**117**) isomer form an inclusion crystal with a 1:1 ratio under a heating condition. The remaining (+)-(**117**) isomer is obtained by distillation with up to 98% optical purity and 51% yield. With further

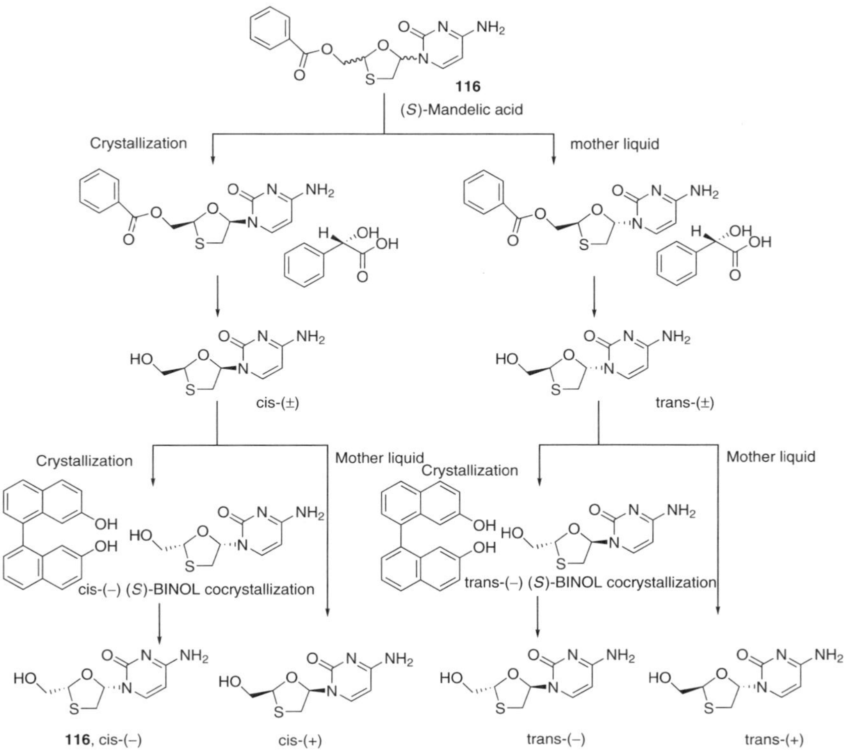

SCHEME 4.6   Inclusion complex resolution of lamivudine (**116**).

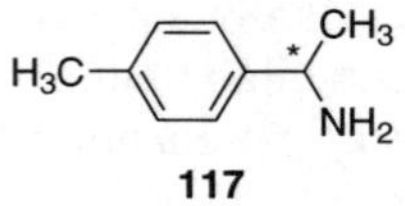

FIGURE 4.28 Structure of p-methylphenylethylamine (**117**).

increase of temperature, the inclusion crystal is decomposed and $(-)$-(**117**) is then produced with 100% optical purity and 49% yield. In this manner, chiral TADDOL can be recycled and used repeatedly. This method is very promising for industry application [66] and application to many kinds of inclusion systems [61] (Fig. 4.28).

4.4 KINETIC RESOLUTION

Kinetic resolution is a technology for separation based on the difference of reaction rates between the two enantiomers in a racemic mixture (S_R, S_S) with chiral catalysts. The enantiomer that reacts faster will be converted into product (S_R) while the relatively slower enantiomer (S_S) remains unchanged in the reaction system. As a consequence, resolution can be accomplished [67,68] (Scheme 4.7).

The efficiency of kinetic resolution is represented by the ratio of optical isomers (E). This ratio refers to two reaction constant rates, the faster reactive isomer K_A versus the slower reactive isomer K_B. Calculation of enantiomeric excess, ee, is expressed in Scheme 4.8, with regard to product (S_R) and unreacted enantiomer after purification (S_S) as ee_P and ee_Q, respectively.

From the stoichiometry point of view, the efficiency of kinetic resolution is evaluated by conversion ratio and productive ratio, as shown in the following formulas:

In the productive ratio formula "mols of obtained desired enantiomers" is the product (S_R), as well as unreacted enantiomer (S_S).

$$S_R \xrightarrow[K_A]{\text{Fast}} P_R \qquad S_R, S_S: \text{Substrate enantiomer}$$

$$S_S \xrightarrow[K_B]{\text{Slow}} P_S \qquad P_R, P_S: \text{Product enantiomer}$$

SCHEME 4.7 Kinetic resolution.

$$ee_Q = \left(\frac{S_S - S_R}{S_S + S_R} \right) \times 100\% \quad S_S > S_R \qquad ee_P = \left(\frac{P_R - P_S}{P_R + P_S} \right) \times 100\% \quad P_R > P_S$$

SCHEME 4.8 Calculation of ee_P and ee_Q.

Kinetic resolution can be achieved by chemical or enzymatic methods (see Chapter 3). In the chemical methods, the reaction can either be catalytic or stoichiometric. Catalytic reactions are more attractive, and, in some cases, enzymatic methods can obtain higher E values than chemical methods.

4.4.1   Chemical Methods in Kinetic Resolution

There are generally two types of chemical methods in kinetic resolution: 1) asymmetric reaction using chiral auxiliary or substrates and 2) asymmetric reaction catalyzed by chiral catalysts.

4.4.1.1   *Asymmetric Reaction Using Chiral Auxiliary or Substrates*
Addition of equivalent chiral auxiliary to react selectively with one optical isomer of racemic compounds will accomplish kinetic resolution. For example, in case of naproxen (**13**), racemic naproxen was first derivatized into anhydride (**118**), followed by reaction with optically pure (*R*)-1-(4-pyridinyl) ethanol (**119**) to afford (*R,R*)- naproxen (*R*)-1-(4- pyridinyl) glycollate (**120**) [69], as well as one equivalent of (*S*)-naproxen. This kinetic resolution method of reacting chiral alcohols with acids was also applied to the resolution of pyrethroid acid [70] and aryloxy propionic acid [71] (Scheme 4.9).

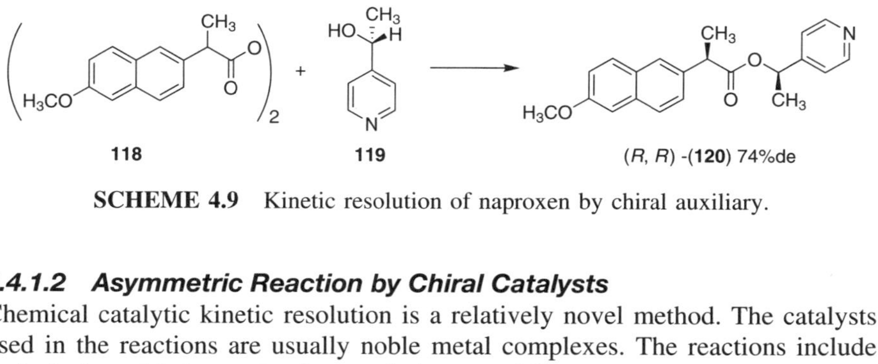

SCHEME 4.9   Kinetic resolution of naproxen by chiral auxiliary.

4.4.1.2   *Asymmetric Reaction by Chiral Catalysts*
Chemical catalytic kinetic resolution is a relatively novel method. The catalysts used in the reactions are usually noble metal complexes. The reactions include hydrolysis, oxidation, reduction, acylation. etc. [68].

One of the most typical methods is Sharpless asymmetric oxidation. When a pro-chiral allylic alcohol compound is treated with stoichiometric peroxide in the presence of a catalytic amount of tartrate and tetraisopropoxy titanium, the oxidation occurs enantiomerically in one face of the oxidized epoxide to afford the epoxide with over 90% *ee* value and 70–90% yield [72,73] (Scheme 4.10).

Terminal racemic epoxides can undergo hydrolytic kinetic resolution (HKR) to provide dihydroxy compounds and chiral unreacted starting epoxide in the presence of Jacobsen's chiral Co(III) or Cr(III) catalysts (also called salen catalysts) [74]. For instance, epoxide compound (**121**) was treated with chiral catalysts (**124**) to afford (*R*)-**121** and diol compound (*S*)-**122** with 45% yield. (*S*)-**122** was subsequently cyclized into (*S*)-**123** (94% *ee* and 50% yield). (*R*)-**121** also

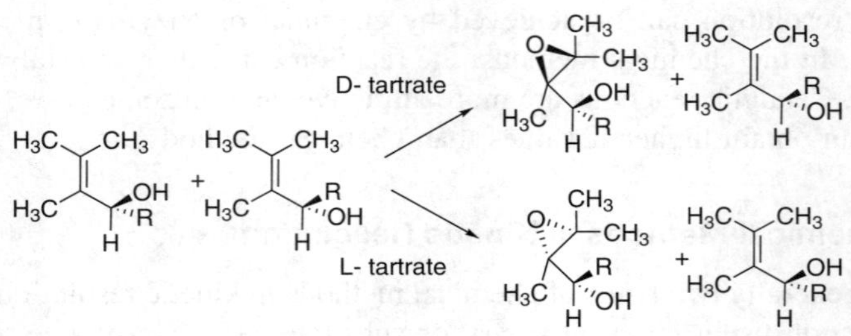

SCHEME 4.10   Sharpless asymmetric oxidation.

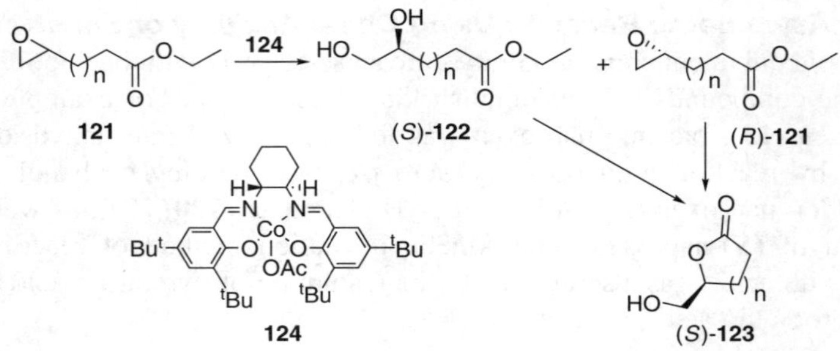

SCHEME 4.11   Hydrolytic kinetic resolution of epoxides by Salen catalysts.

underwent configuration-reversing ring opening and subsequent ring closure in the presence of trifluoroacetic acid solution to afford (*S*)-**123** (96% *ee* and 44% yield) [75] (Scheme 4.11).

In hydrolytic kinetic resolution, not only can chiral (Salen)Co(I) catalysts catalyze the resolution of racemic terminal epoxide compounds to generate chiral 1,2-diol, but the terminal chiral epoxides can also be treated with various substituted amines or alkoxides to yield optically pure β-aminol and β-alkoxy alcohol compounds. β-Aminols and β-alkoxy alcohols are two important units existing in bioactive molecules. The method was applied to the resolution of the intermediate (**125**) of anti-Alzheimer disease drug T588 and the key intermediate (**126**) of levalbuterol (**127**) [76] (Scheme 4.12).

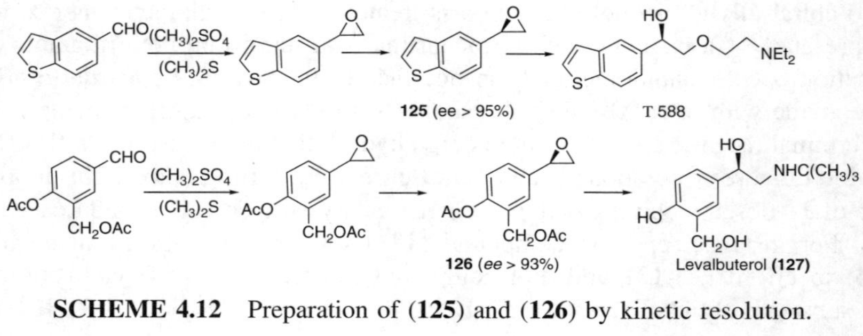

SCHEME 4.12   Preparation of (**125**) and (**126**) by kinetic resolution.

4.4.2 Enzymatic Catalytic Kinetic Resolution

Enzymes can recognize a form of chirality of the large molecule and therefore can be used to resolve racemic compounds effectively through enzymatic catalysis. Enzymatic transformation features high stereoselectivities and mild reaction conditions and is environmentally benign. However, the disadvantage is the low concentration of substrate and the corresponding product because of their low solubilities in water, leading to some difficulties in separation. In addition, the specificity of the enzyme in aqueous solution is difficult to adjust, limiting applications in the aqueous media. Since the 1980s, non-aqueous-phase enzymatic catalysis has been considerably developed. The advantages of non-aqueous-phase enzymatic catalysis are obvious. Lipase, esterase, protease, and other hydrolytic enzymes are usually used in enzymatic kinetic resolution as biocatalysts, for the preparation of optically active alcohols, acids, and esters.

4.4.2.1 Acidic Drugs

The acidic chiral drugs are prepared by forming the corresponding esters, followed by selective enzymatic hydrolysis of one enantiomer of the esters. (*S*)-enantiomers possess higher biological activity. α-Aryl propionic acids are a class of widely used nonsteroidal anti-inflammatory drugs. Lipase in *Candida cylindracea* can selectively hydrolyze the (*S*)-naproxen ester, such as its methyl ester (**128**) [77] or ethoxy ethyl ester (**129**) [78], to afford highly optical purity of (*S*)-naproxen (**13**). Other examples are racemic naproxen methyl ester (**128**) [77] resolved by carboxylesterase NP, to obtain (*S*)-ibuprofen (**130**) up 98%*ee* [79] and by horse liver esterase to afford the (*S*)-product up 96% *ee* [80] (Scheme 4.13).

(*R*)-Chiral carboxylic acid, serving as the synthon of novel antihypertension drug 16-fluoro prostaglandin, angiotensin-converting enzyme inhibitor cilazapril, and leukotriene antagonist Ro 233544 [81], are obtained by enzymatic hydrolysis of chiral carboxylic esters catalyzed by lipase PS. *Aspergillus* lipase, *Streptomyces griseus* lipase, and subtilisin are used for the preparation of chiral carboxylic

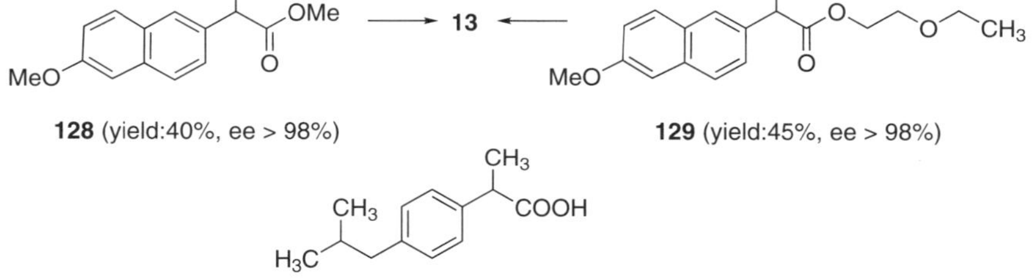

128 (yield:40%, ee > 98%)   **129** (yield:45%, ee > 98%)

130 (ee > 98%)

SCHEME 4.13   Resolution of α-aryl propionic acid, the NSAIDs.

acid used as synthon for the preparation of, for example, vitamin E [82,83] and antibiotics florfenicol [84] and chloramphenicol [85].

It was reported that chiral esters can react with ammonia to form the corresponding amide in the presence of catalytic lipase [86]. This method particularly applies to the synthesis of unsaturated fatty acid amides, which are difficult to obtain by chemical methods. Since many amides have lower solubility in organic solvents, they are easy to precipitate during the ammonolysis reaction. Not only does this help to achieve separation of the product in situ, but reaction equilibrium will also favor production formation. Ibuprofen chloracetate (**131**) can form (*S*)-ibuprofen chloracetate (**132**) and (*R*)-buprofen amide (**133**) through lipase SP-435 ammonolysis reaction [87] (Scheme 4.14).

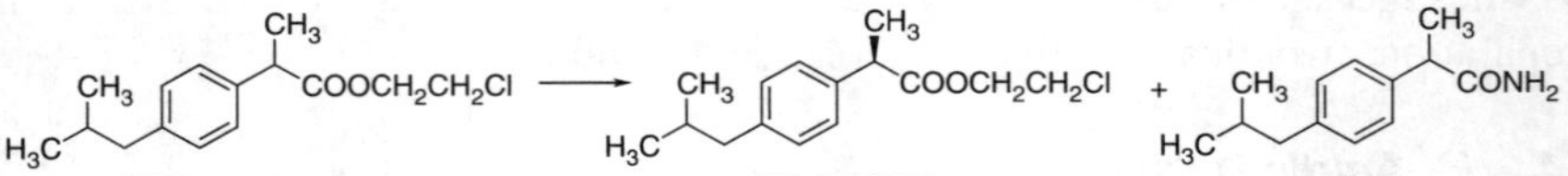

SCHEME 4.14   Ammonolysis of ibuprofen chloracetate (**131**) by lipase SP-435.

4.4.2.2 Hydroxyl-Containing Drugs

The hydroxyl-containing chiral drugs can be obtained by conversion into the corresponding esters, followed by selective enzymatic hydrolysis of one of the two enantiomer esters while one of the enantiomer esters remains untouched.

This method is widely used in preparation of aryloxy phenylpropanolamine β-receptor blockers such as propranolol and betaxolol by resolution. (*S*)-propranolol (**135a**) is active in blocking β-receptors, clinically used for the treatment of cardiovascular diseases, while the (*R*)-propranolol (**135b**) has a contraceptive activity. It was reported that (*S*) (**135a**) and (*R*)-propranolol (**135b**) can be separated by lipase PS-catalyzed kinetic resolution of (**134**) with >95% *ee* [88] (Scheme 4.15).

This resolution approach has been successfully applied in the preparation of the side chain of anticancer drug Taxol [89] and 5-hydroxytryptamine antagonists [90].

4.4.2.3 Amide Drugs

Penicillinacylase (PA) isolated from *Escherichia coli* is widely used in the synthesis of 6-aminopenicillanic acid and semisynthetic β-lactam antibiotics in industry. Chiral synthon of loracarbef can be resolved by immobilized PA [91]. The substituted γ-aminobutyric acid derivative (*S*)-enantiomers are active as antiepileptic and anticonvulsant agents, which are obtained from racemic *N*-phenylacetyl γ-aminobutyric acid derivatives (**136**) through PA-mediated resolution. Performance of the resolution process at room temperature leads to (*R*)-enantiomer (**137**), while that at 45°C results in (*S*)-enantiomer (**138**) [92] (Scheme 4.16).

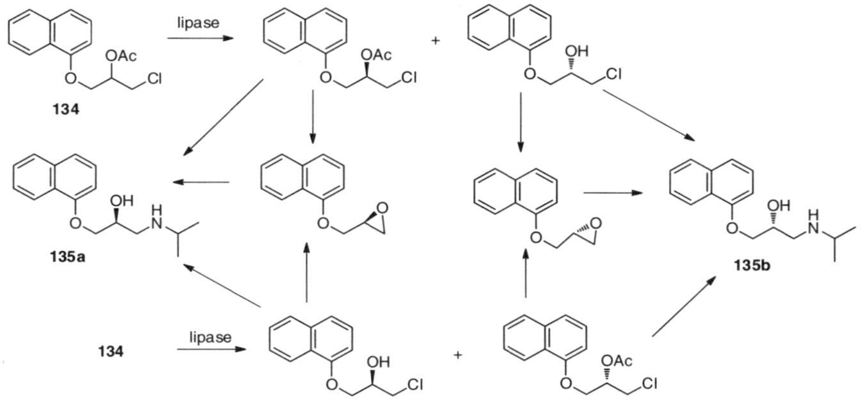

SCHEME 4.15   Enzyme-catalyzed kinetic resolution of (**134**).

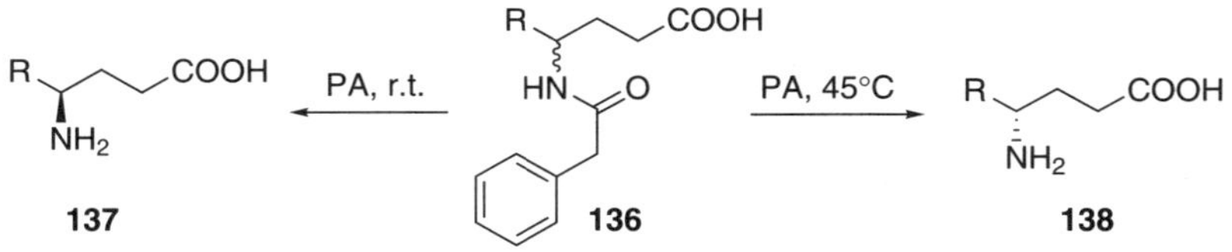

SCHEME 4.16   Preparation of γ-aminobutyric acid derivatives by enzyme-catalyzed kinetic resolution.

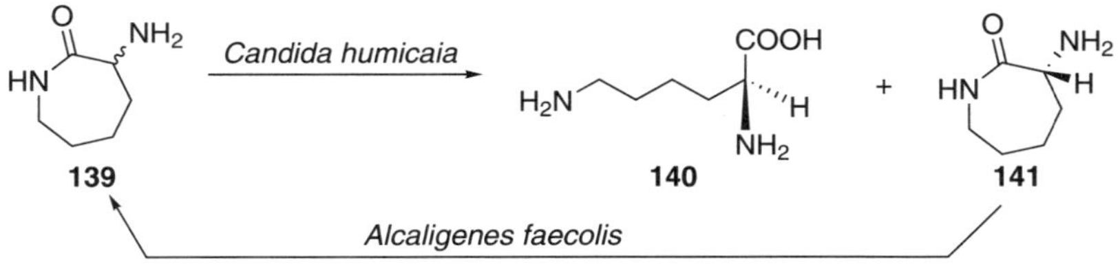

SCHEME 4.17   Preparation of L-lysine (**140**) by amide hydrolase.

L-Lysine (**140**) was produced on an industrial scale by amide hydrolase (*Candida humicaia*) catalyzed hydrolysis of DL-α-amino-ε-caprolactam (**139**). This process generated a large amount of by-product D-α-amino-ε-caprolactam (**141**), which was then racemized by another enzyme (*Alcaligenes faecolis*), returning back to DL-α-amino-ε-caprolactam (**139**) for the next run [93] (Scheme 4.17).

4.4.2.4 *Miscellaneous Drugs*

The nucleoside antiviral drug (−)-carbovir (**143**) is more potent than its enantiomer. It was obtained through selective hydrolysis of diaminocarbovir (**142**) by adenosinedeaminase (ADA) [94]. Another potent antiviral drug, lamivudine (3TC, **144**), is the *l*-isomer, which has antiviral activity similar to its enantiomer, but with a lower cytotoxicity. Human cytidine deaminase (CDA) immobilized

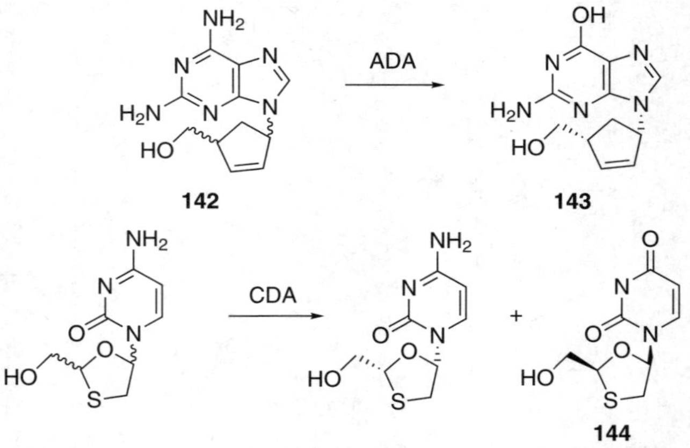

SCHEME 4.18 Preparation of (−) -carbovir (**143**) and lamivudine (**144**) by enzyme-catalyzed kinetic resolution.

on macroporous particles, Eupergit-C, could be used for selective removal of 4-amino group in the racemic mixture; therefore the desired enantiomer lamivudine remained. This method was used successfully on a kilogram scale [95] (Scheme 4.18).

4.5 DYNAMIC RESOLUTION

Most of the above-discussed crystallization resolution, complex resolution, and inclusion resolution are physical means to separate racemates by the formation of crystals, which are based on the physicochemical property (mainly solubility) differences among the diastereomers. The theoretical maximum yield of the optically pure product is 50%.

Kinetic resolution is based on the assumption that one enantiomer is converted into a product faster than the antipode in a racemic mixture. Compared to crystallization resolution, the kinetic resolution method has many advantages. However, the traditional method of kinetic resolution achieves a maximum yield of optically pure product of only 50%. In addition, the purity of recycled substrate and product enantiomers is influenced by the extent of reaction conversion. The larger numbers of the substrates are converted into products with the lower *ee*. The final resolution selectivity is related to reaction time and other conditions.

To overcome these shortcomings, dynamic resolution techniques have been developed. Dynamic resolution is based on the racemization of chiral substrates or dynamic equilibrium of chiral intermediates. One of the chiral substrates or chiral intermediates is converted into its corresponding enantiomers. By this means, maximum resolution of a single chiral compound can be achieved.

Dynamic resolution includes dynamic kinetic resolution (DKR) and dynamic thermodynamics resolution (DTR).

4.5.1   Dynamic Kinetic Resolution

(DKR is a combination of classical kinetic resolution and chiral substrate racemization. Because certain substrates are not stable and racemization tends to occur in the presence of enzymes or under the chemical conditions, these substrates can be racemized at the time of kinetic resolution. The process of racemization is a dynamic equilibrium, in which slower reactive substrate is converted into faster reactive substrate through the balance. As a consequence, the required compound can be continuously provided (Scheme 4.19) [96–99].

Theoretically, it is possible that dynamic kinetic resolution can lead to 100% of the product. To achieve the best result and higher optical purity of the product, the K_A/K_B ratio should be larger than 20 and the speed constant k_{inv} should be larger than K_A [98].

4.5.1.1   Chemical Methods in Dynamic Kinetic Resolution

Chemical methods of dynamic kinetic resolution can be classified into three categories: 1) asymmetric dynamic kinetic resolution; 2) chiral auxiliary-promoted dynamic kinetic resolution, which is achieved by chiral reagents or steric factors existing in substrate molecules through asymmetric chemical reactions to undergo resolution; and 3) chiral catalysis of dynamic kinetic resolution, which is achieved by chiral metal catalyst-induced asymmetric reactions.

4.5.1.1.1   Asymmetric Transformational Dynamic Kinetic Resolution via Asymmetric Transfer.   There are two levels of racemic mixture asymmetric transformation. 1) In first-level asymmetric transformation, in the presence of external chiral reagents, the enantiomers' equilibrium shifts in solution, resulting in a nonequal relationship and hence leading to asymmetric transformation and resolution of racemic mixture by crystallization. This transformation usually occurs in the diastereomers. 2) In second-level asymmetric transformation, in the equilibrium of the mixture, one of the enantiomers undergoes spontaneous

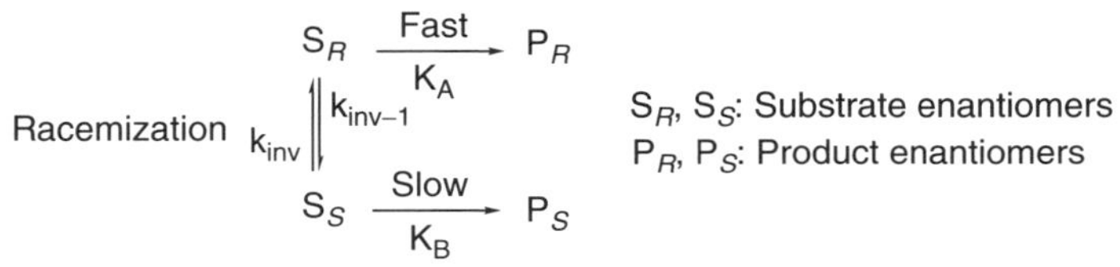

SCHEME 4.19   Principles of dynamic kinetic resolution.

but slow crystallization. On the other hand, the solution equilibrium will be continuously broken when pure enantiomer of seed crystal is added, because its crystallization rate is slower than the balance shifts, resulting in asymmetric transformation and crystallization resolution of racemic mixture. This process is also known as "crystallization-induced asymmetric transformation," which transforms the racemates into a single pure enantiomer [100] (Scheme 4.20).

The combination of asymmetric transformation and crystallization resolution of racemates is mostly suitable for the carbonyl-containing compounds with α chiral methine carbon atom. Under basic conditions, the α chiral carbon of the carbonyl group racemizes through enolization. For example, racemic *p*-methoxyphenyl benzyl ketone (**145**) under basic conditions underwent asymmetric transformation [101]. Asymmetric transformation and crystallization of ketone (**146**), a precursor of plant growth regulator paclobutrazol, was also accomplished under basic conditions [102] (Scheme 4.21).

Zhou successfully utilized this method to produce (*S*)-(−)-2-methylamino-1-phenyl-1-acetone from (±)-2-methylamino-1- phenyl-1-acetone (**147**), which was further converted into ephedrine hydrochloride (**150a**) and pseudoephedrine hydrochloride (**150b**). Thus (±)-2-methylamino-1-phenyl-1-acetone (**147**) was mixed with (2*R*,3*R*)- tartaric acid derivatives (DBTA) in the alcohol-ester, or water alcohol-ester solvents. The salt (**148a**) formed by (*S*)-(−)-2-methylamino-1-phenyl-1-acetone and (2*R*, 3*R*)-DBTA precipitated from the solution, thus breaking the equilibrium of the system. Therefore, there was no need for the other enantiomer (*R*)-(+)-2-methylamino-1-phenyl-1-acetone to form the salt (**148b**) with (2*R*, 3*R*)-DBTA, which is racemized through enolization back to (**147**), which subsequently is resolved to generate the compound (**148a**) of the

$$A_R \rightleftharpoons A_S \text{ (Solution)}$$

$$(A_R) \qquad (A_S) \text{ (crystal)}$$

SCHEME 4.20 Crystallization-induced asymmetric transformation of racemates.

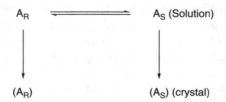

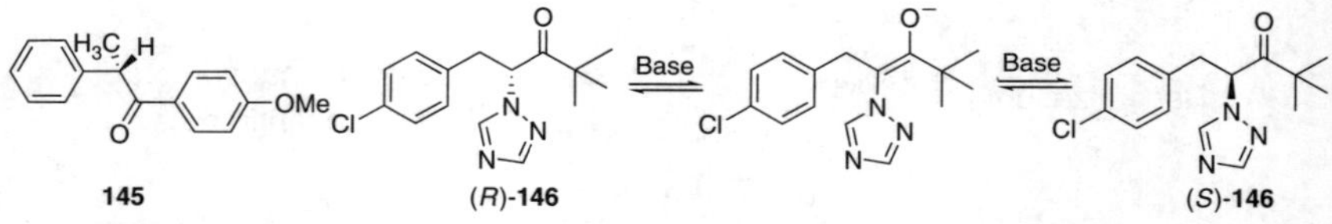

SCHEME 4.21 Asymmetric transformation and crystallization of ketone (**145**) and (**146**).

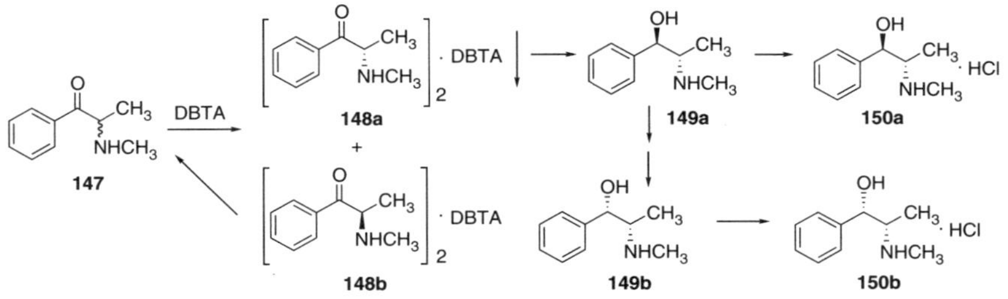

SCHEME 4.22 Dynamic resolution of (**147**).

required configuration. The diastereoisomer purity of (**148a**) is 97.2% (*d.e.*) [101] (Scheme 4.22).

Asymmetric transformation and resolution of naproxen is carried out under melting conditions. For instance, the naproxen methyl ester (**128**) and sodium methoxide are melted at 70°C and then rapidly cooled to 67°C, while crystallization occurs after crystal seed of (+)-naproxen methyl ester was added in (+)-naproxen methyl ester was attained in yield of 87% [104]. Crystallization of one of the enantiomers will be accompanied by the racemization of the other enantiomer. This was the case in the use of naproxen ethylammonium salt solution; the resolution yield reached 90% [105].

4.5.1.1.2 Dynamic Kinetic Resolution via Chiral Auxiliary. α-Halogen atoms of a carboxylic acid, such as bromine or iodine, easily undergo SN$_2$ nucleophilic substitution reaction because of its active chemical nature. At the same time, by addition of polar solvents, alkalis, or other halide salts, halogenated carbon atoms will be racemized to achieve dynamic kinetic resolution purposes. This concept was used in the synthesis of phenoxybutanoic acid derivatives. Chiral lactic acid amide as the chiral auxiliary agent formed the ester (**151**) with α-bromobutanoic acid and then sodium phenolate as the nucleophile to form α-phenoxyester intermediate, which by saponification affords (*R*)-2-(4-chlorophenoxy) butyric acid (**152**) with up to 96–98% [106,107] (Scheme 4.23).

Chiral lactamide has been used for resolution of arylpropionic acid-based drugs, such as ibuprofen, flurbiprofen, and fenoprofen. The arylpropionic acid drug forms onium salt with 4-dimethylaminopyridine, which is then quickly

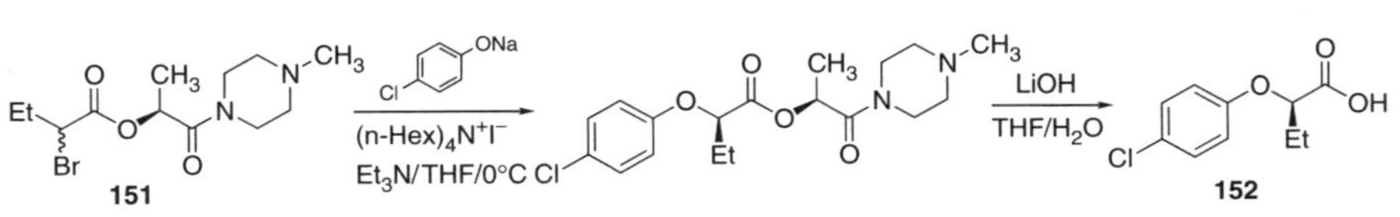

SCHEME 4.23 Preparation of (**152**) by chiral auxiliary dynamic kinetic resolution.

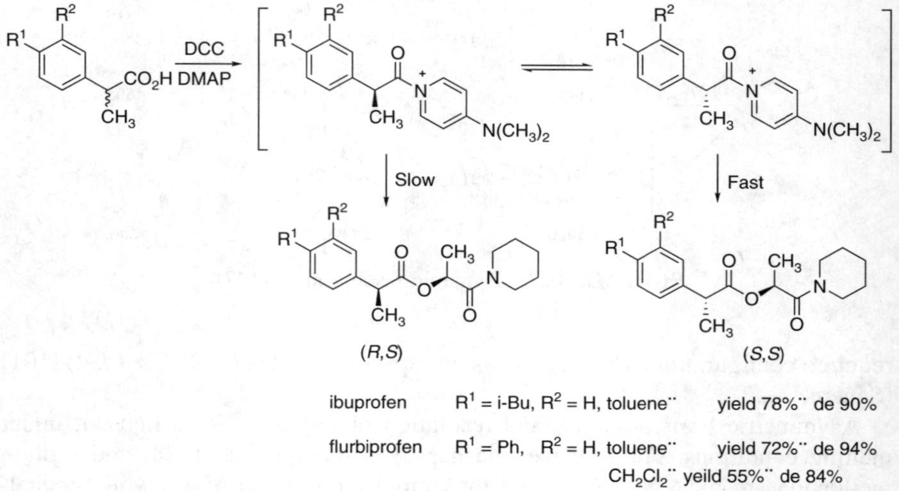

SCHEME 4.24 Asymmetric synthesis of arylpropionic acids and aryloxy acids.

treated with chiral lactamide to provide an ester. This rapid esterification reaction promotes the dynamic equilibrium of onium salt. It is reported that performance of such a dynamic resolution in toluene gave a better result compared with dichloromethane [107] (Scheme 4.24).

This method is also applicable to the synthesis of substituted α-amino acids and peptides. As inexpensive chiral auxiliary agents, sugars are esterified with α-halo acids, which are then reacted with amines to produce substituted α-amino acids with a yield of up to 99% and *de* values of up to 94% [108,109] (Scheme 4.25).

Proline was used as a chiral auxiliary in forming amide (**153**) with α-halogenated acid. (**153**) was then reacted with the corresponding amines to afford proline-containing dipeptide (**154**) with 93% yield and 98% *de*. A further example of this method is the synthesis of antihypertensive angiotensin-converting enzyme inhibitor (**155**). Reduction of (**154**) resulted in chiral ligand (**156**) [110] (Scheme 4.26).

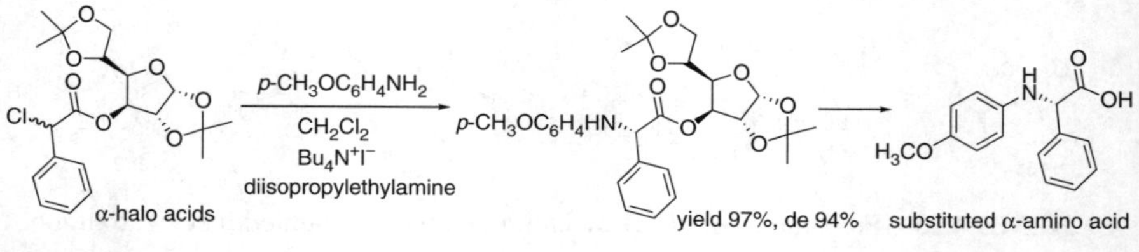

SCHEME 4.25 Using sugars as chiral auxiliary agents to prepare substituted α-amino acids.

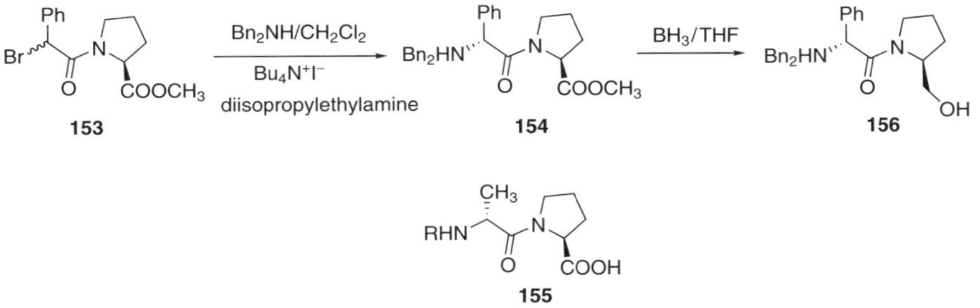

SCHEME 4.26   Using proline as a chiral auxiliary to form chiral amides.

4.5.1.1.3 *Dynamic Kinetic Resolution via Chiral Catalysis.* Two types
of catalysts have been used in chiral catalytic dynamic kinetic resolution: chiral
metal catalysts and chiral organic catalysts.

The metals used in catalysis for dynamic kinetic resolution include ruthenium,
platinum, palladium, iridium, cobalt, copper, nickel, molybdenum, titanium, and
lanthanum [99]. Among these, the most widely used and effective catalyst for
asymmetric catalytic hydrogenation is chiral ruthenium (Ru) complex. In 1989,
Noyori and co-workers successfully obtained chiral β-hydroxy ester by catalytic
hydrogenation of β-keto ester compounds via kinetic resolution. This method was
rather useful for preparation of β-lactam antibiotics [111] (Scheme 4.27).

Genet and co-workers also investigated the chiral ruthenium-catalyzed reaction. They found that the selectivity of the asymmetric reaction is mainly related
to the nature of the chiral ruthenium catalyst, reaction conditions, and substrates
[112]. In 2003, Genet and co-workers applied this approach to the synthesis
of diltiazem, a calcium channel blocker. Chiral ruthenium catalyst was used to
asymmetrically reduce α-chloro-β-keto esters to afford chlorohydroxy ester (**157**).
(**157**) was then converted into glycidic ester (**158**) through a DBU-catalyzed
intramolecular SN$_2$ reaction, thus achieving the total synthesis of diltiazem [113]
(Scheme 4.28).

Chiral organic catalysts have been extensively investigated in recent years.
Small-molecule organic compounds are attractive because of their lower cost,
easy recovery, stability, and lower toxicity. These organic catalysts including
quinoline, quinine, and proline and their derivatives, and many other artificial
compounds.

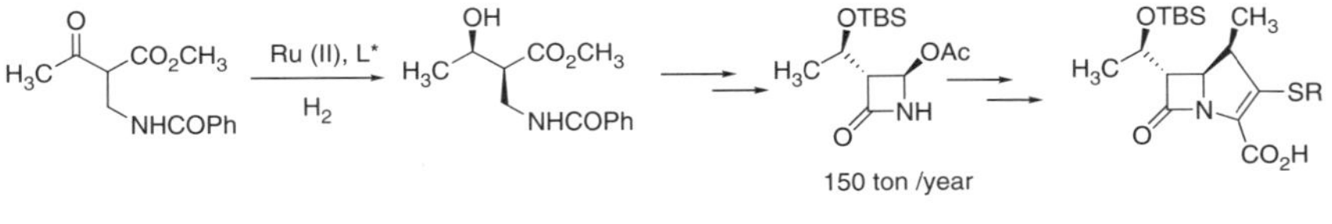

SCHEME 4.27   Preparation of chiral β-hydroxy ester by catalytic hydrogenation with
chiral metal catalysts.

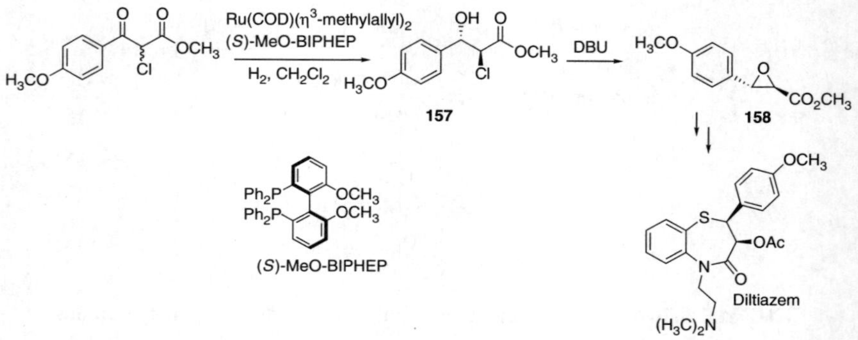

SCHEME 4.28   Asymmetrically reducing α-chloro-β-keto esters to chlorohydroxy ester.

Berkessel and co-workers used a dual-chiral functional groups-substituted thiourea compound (**160**) as the catalyst to perform alcoholysis of oxazolidone compounds (**159**). Natural and nonnatural amino acids (**161**) can be obtained with high optical purities by using this methodology [114,115] (Scheme 4.29).

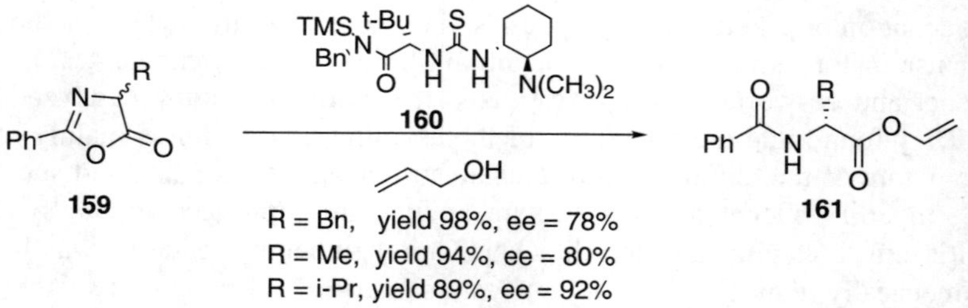

R = Bn, yield 98%, ee = 78%
R = Me, yield 94%, ee = 80%
R = i-Pr, yield 89%, ee = 92%

SCHEME 4.29   Alcoholysis of oxazolidone catalyzed by chiral organic catalyst.

4.5.2   Enzyme-Catalyzed Dynamic Kinetic Resolution

There are a number of examples of chiral compounds undergoing enzyme-catalyzed dynamic kinetic resolution but, however, few applications in chiral drug synthesis. Most examples are related to the resolution of the arylpropionic acid drugs. (S)-(−)-ketorolac (**163**) has a stronger anti-inflammatory activity than (R)-(+)-ketorolac. Racemic ketorolac ethylester (**162**) was treated, in carbonate (pH = 9.7) buffer solution, with protease that was extracted from *Streptomyces griseus*. After kinetic resolution, (S)-(−)-ketorolac could be obtained with 92% yield and 85% *ee* [116] (Scheme 4.30).

Hydrolysis of racemic naproxen ester with lipase afforded (S)-naproxen (**12**). This kinetic resolution is a very practical method. To improve the efficiency of kinetic resolution, *Candida rugosa* lipase was immobilized onto polyethylene particles. Polystyrene and 2% divinylbenzene cross-linked resin was used as a

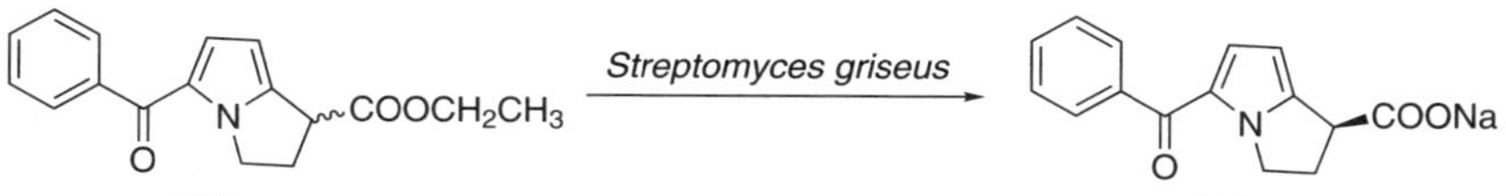

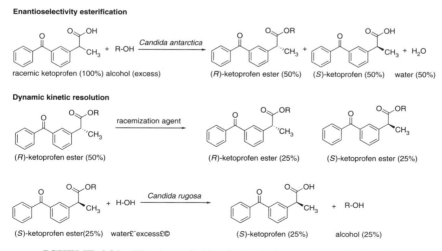

SCHEME 4.30 Preparation of (S)-(−)-ketorolac (**163**) by enzyme-catalyzed dynamic kinetic resolution.

racemic reagents immobilized to organic base 1,5,7-azabicyclo[4,4,0]decane-5-ene, which racemized nonhydrolyzed (R)-Naproxen 2,2,2-trifluoro-ethyl ester. Providing (S)-naproxen (**12**) in 96% yield and 58% *ee* [117].

To further improve the enzyme resolution efficiency, optical yield and practicality, a combination of lipase with enzyme-membrane reactor was developed as a new resolution technique. (Fig. 4.14). (R, S)-ketoprofen was chosen as the substrate in the biphasic (water phase and organic phase) membrane reactor; lipase was fixed on the membrane, which selectively esterified (R)-ketoprofen. Therefore, lipophilic (R)-ketoprofen was able to enter into the organic phase, while (S)-ketoprofen was not esterified and stayed in the water phase. (S)-ketoprofen was isolated through the process of kinetic resolution, while (R)-ketoprofen ester in organic phase was racemized by racemic reagent preadded to racemic ketoprofen ester. The resultant (R, S)-ketoprofen ester entered into another membrane reactor, where the (S)-ketoprofen ester was selectively hydrolyzed by hydrolase and entered into the water phase. This resolution process combined "esterification-racemization-hydrolysis" multistep reactions with enzyme-membrane reactor technology to achieve dynamic kinetic resolution (Fig. 4.29). In theory, the optical resolution yield can be 100%, making this process more economical and feasible [118] (Scheme 4.31).

Enantioselectivity esterification

racemic ketoprofen (100%)   alcohol (excess)   (R)-ketoprofen ester (50%)   (S)-ketoprofen (50%)   water (50%)

Dynamic kinetic resolution

(R)-ketoprofen ester (50%)   (R)-ketoprofen ester (25%)   (S)-ketoprofen ester (25%)

(S)-ketoprofen ester(25%)   water£¨excess£©   (S)-ketoprofen (25%)   alcohol (25%)

SCHEME 4.31 The dynamic kinetic resolution process of ketoprofen.

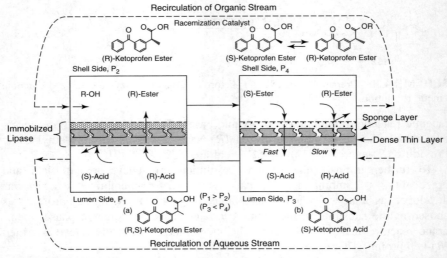

FIGURE 4.29   Enzyme-membrane reactor technology applied to ketoprofen dynamic kinetic resolution.

4.5.3   Dynamic Thermodynamic Resolution

Dynamic thermodynamic resolution is a technology that combines induced diastereoisomer equilibrium and kinetic resolution to achieve maximum stereoselectivity in the product-formation stage. Dynamic thermodynamic resolution unifies multiple controllable reaction steps in a one-pot operation system and achieves stereoselectivity [119]. It can be illustrated by reaction as shown in Scheme 4.32.

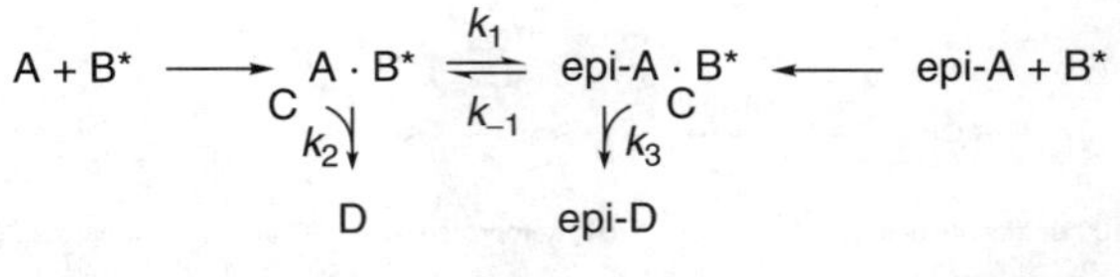

$$A + B^* \longrightarrow A \cdot B^* \underset{k_{-1}}{\overset{k_1}{\rightleftharpoons}} \text{epi-}A \cdot B^* \longleftarrow \text{epi-}A + B^*$$
$$\Big\downarrow k_2 \qquad \qquad \Big\downarrow k_3$$
$$D \qquad \qquad \text{epi-}D$$

SCHEME 4.32   Dynamic thermodynamic resolution chart.

When the enantiomer **A, epi-A**, and chiral compound **B*** (**B*** can be chiral ligands, chiral auxiliary agents, or chiral solvent) interact with each other through chemical reactions or physical interactions, diastereomeric complexes **A · B*** and **epi-A · B*** are formed and are converted into each other by an equilibrium (equilibrium constant of k_1, k_{-1}). However, it is also possible that they cannot transform into each other, and therefore the equilibrium cannot be achieved. If reactant **C** is added into the system, diastereomeric complexes **A · B*** and **epi-A · B*** will react with **C** to form **D** and **epi-D** (reaction rate constants k_2, k_3), respectively, and **B*** is dissociated at the same time.

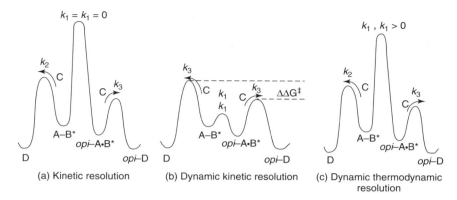

FIGURE 4.30   Resolution energy diagram of enantiomers **A** and **epi-A**.

If **A · B*** and **epi-A · B *** cannot convert into each other in the system, and one of the complexes reacts with reactant **C** faster than other complexes (i.e., $k_1 = k_{-1} = 0$, $k_2 \neq k_3$), this process is called kinetic resolution (resolution energy chart shown in Fig. 4.30a).

However, if the equilibrium rate between **A · B*** and **epi-A · B*** is much larger than that for their reaction with **C** (i.e., $k_1, k_{-1} \gg k_2, k_3$) in the system, the ratio of products depends on energy differences of transition state between these two diastereomeric complexes and the products ($\Delta\Delta G^{\ddagger}$), this process is called dynamic kinetic resolution (resolution energy chart shown in Fig. 4.30b).

When **C** is added to a reaction mixture where **A · B*** and **epi-A · B*** are interchangeable, the original equilibrium cannot be achieved (i.e., $k_1, k_{-1} \neq 0$, $k_2, k_3 \gg k_1, k_{-1}$) (resolution energy chart shown in Fig. 4.30c).

The ratio of **A · B*** to **epi-A · B*** depends on their corresponding thermodynamic stabilities, which also determine the ratio of the final product **D/epi-D**. The reaction rates of **A · B *** and **epi-A · B *** with **C**, k_2 and k_3, will impact on the ratio of the products, but the impact is relatively insignificant.

Dynamic thermodynamic resolution is in fact a multistep resolution process. The initial formation of diastereomeric complexes **A · B*** and **epi-A · B*** is controlled by the thermodynamic kinetic equilibrium, while the subsequent resolution process is a dynamic stereoselective reaction.

In dynamic thermodynamic resolution, conversion and thermodynamic equilibrium between **A · B*** and **epi-A · B*** can be fine-tuned by adjusting the reaction temperature, thereby controlling the ratio of the final product **D/epi-D**. For example, N-pivaloyl-o-ethyl aniline (**164**) was generated with the corresponding dilithium salts (**165**) by treatment of lithium reagents. It was then cooled to—78°C, and chiral ligand (−)-cytisine was added. The intermediate species (**166**) and (**epi-166**) were formed while the whole reaction was maintained at—78°C. Finally, addition of a twofold amount of electrophilic reagent trimethylchlorosilane (TMSCl) led to the products (R)-**167** and (S)-**167** with enantiomeric ratio 56:44. If dilithium salts (**165**) and (−)-cytisine were stirred at

$-25°C$ for 45 min at the beginning of the reaction, and then cooled to $-78°C$ before addition of a two-fold amount of trimethylchlorosilane (TMSCl), products (R)-**167** and (S)-**167** were obtained with enantiomeric ratio of 92:8 in 30 min [120,121] (Scheme 4.33).

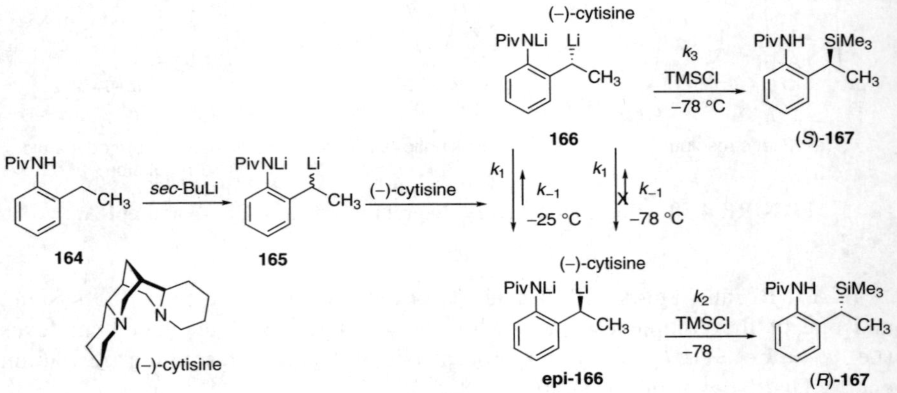

SCHEME 4.33 Results for reaction between diastereomeric organolithium species (**166**) and (epi-**166**) and TMSCl under isothermal and variable temperature protocols.

When the reaction was conducted at $-78°C$, the chelating process of (**165**) and $(-)$-cytisine was controlled by a dynamic process. The amounts of generated intermediates (**166**) and (**epi-166**) were almost the same. However, under this condition, (**166**) and (**epi-166**) are not interchangeable. When they reacted with excessive TMSCl, the same amount of products (R)-**167** and (S)-**167** could be obtained. But when (**165**) was chelated with $(-)$-cytisine at $-25°C$, the formation process of intermediates (**166**) and (**epi-166**) was under thermodynamic control. Thermodynamically stable (**epi-166**) possessed a large proportion in the equilibrium. The proportion of obtained products (R)-**167** and (S)-**167** would have a larger difference when the reaction was cooled to $-78°C$ followed by addition of excessive TMSCl [122] (energy process diagram shown in Fig. 4.31) [119].

Under the above-mentioned reaction conditions, the amount of electrophilic reagent TMSCl was excessive. This could sufficiently reflect a quantitative relationship between diastereomeric complexes (**166**) and (**epi-166**) under different conditions when TMSCl reacted with (**166**) and (**epi-166**). However, reactivity between (**166**) and (**epi-166**) would be greatly affected if the amount of TMSCl was reduced. For example, when the amount of electrophilic reagent TMSCl was reduced to one-tenth of the reactants, the product ratio of (R)-**167** and (S)-**167** would be 91:9 under the condition of $-78°C$. However, when dilithium salts (**165**) and $(-)$-cytisine were stirred for 45 min at $-25°C$ followed by cooling to $-78°C$, the product ratio of (R)-**167** and (S)-**167** would be 99:1. As shown from these data, when (**166**) and (**epi-166**) reacted with TMSCl, the activation

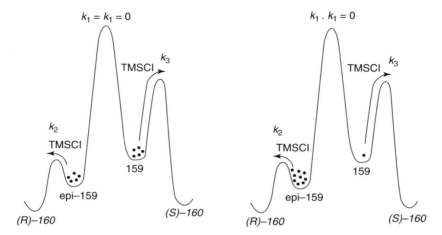

$k_1 = k_1 = 0$ $k_1 . k_1 = 0$

TMSCI k_3 TMSCI k_3

k_2 k_2
TMSCI   TMSCI
                159   159
epi–159   epi–159
(R)–160   (R)–160
                (S)–160   (S)–160

(a) Reaction under the conditions of –78°C   (b) Controlled by dynamic thermodynamics
                                                   reach equilibrium under –25°C

FIGURE 4.31   Energy chart of (**166**) and (**epi-166**) reaction with TMSCl.

free energy barrier of (**epi-166**) was relatively low, which was easily attacked by nucleophilic reagents [121].

REFERENCES

1. Zaugg, H. E. (1995). A mechanical resolution of dl-methadone base. *J. Am. Chem. Soc*. **77**, 2910.

2. Sirotnak, F. M., Chello, P. L., Moccio, D. M., Kisliuk, R. L., Combepine, G., Gaumont, Y., Montgomery, J. A. (1979). Stereospecificity at carbon 6 of formyltetrahydrofolate as a competitive inhibitor of transport and cytotoxicity of methotrexate in vitro. *Biochem. Pharmacol*. **28**, 2993–2997.

3. Cosulich, D. B., Smith, J. M. Jr., Broquist, H. P. (1952). Diastereoisomers of leucovorin. *J. Am. Chem. Soc*. **74**, 4215–4216.

4. Lin, G. Q., Sun, J. Q., Xia, K. M. (2003). High specific rotation *L*-Leucovorin calcium and its resolution. CN 1401647A.

5. Inagaki, M. (1977). Some aspects on the optical resolution by preferential crystallization based on phase equilibrium. *Chem. Pharm. Bull*. **25**, 2497–2503.

6. Velluz, L., Amiard, G., Joly, R. (1953). Resolution of DL-Threo-1-p-nitrophenyl-2-amino propane-1,3-diol. *Bull. Soc. Chim. Fr*. **20**, 342–344.

7. Anonymous. (1965). *Chem. Eng. (N.Y.)* **72**, 247.

8. Synoradzki, L., Hajmowicz, H., Wisialski, J., Mizerski, A., Rowicki, T. (2008). Calcium pantothenate. Part 3. (1) Process for the biologically active enantiomer of the same via selective crystallization and racemization. *Org. Process Res. Dev*. **12**, 1238–1244.

9. Coquerel, G., Bouaziz, R., Brienne, M. J. (1988). Optical resolution of (±)-fenfluramine and (±)-norfenfluramine by preferential crystallization. *Chem. Lett.* **7**, 1081–1084.

10. Addadi, L., Weinstein, S., Gati, E., Weissbuch, I., Lahav, M. (1982). Resolution of conglomerates with the assistance of tailor-made impurities. Generality and mechanistic aspects of the &rule of reversal". A new method for assignment of absolute configuration. *J. Am. Chem. Soc.* **104**, 4610–4617.

11. Purvis, J. L. (1957). Resolution of racemic amino acids. US Patent 2 790 001 [*Chem. Abstr.* (1957). 51, 13910i].

12. Su, J. Y., Zeng, L. M. (1999). *Organic Stereochemistry*. Guangzhou: Sun Yat-Sen University Press, p. 82.

13. Jacques, J., Collet, A., Wilen, S. H. (1981). Resolution by Direct Crystallization. *Enantiomers, Racemates, and Resolutions*. New York: Wiley-Interscience, p. 245–246.

14. Jacques, J., Collet, A., Wilen, S. H. (1981). Formation and Separation of Diastereomers. *Enantiomers, Racemates, and Resolutions*. New York: Wiley-Interscience, p. 251.

15. Leclercq, M., Jacques, J. (1975). Study of optical antipode mixtures. X. Separation of complex diastereoisomeric salts by isomorphism. *Bull. Soc. Chim. Fr.* **9–10**, 2052–2056.

16. Kanoh, S., Muramoto, H., Kobayashi, N., Motoi, M., Suda, H. (1987). Practical method for the synthesis and optical resolution of axially dissymmetric 6,6'-dimethylbiphenyl-2,2'-dicarboxylic acid. *Bull. Chem. Soc. Jpn.* **60**, 3659–3662.

17. Fizet, C. (1986). Optical resolution of (RS)-pantolactone through amide formation. *Helv. Chim. Acta.* **69**, 404–409.

18. Overby, L. R., Ingersoll, A. W. (1960). Resolution of N-carbobenzoxy amino acids. Alanine, phenylalanine and tryptophan. *J. Am. Chem. Soc.* **82**, 2067–2069.

19. Ingersoll, A. W. (1925). A method for the complete, mutual resolution of inactive acids and bases. *J. Am. Chem. Soc.* **47**, 1168–1173.

20. Merck & Co. Inc. (1978). (−)-cis-1,2-Epoxypropylphosphonic acid. Tokkyo Koho 24 407 [*Chem. Abstr.* (1979), **90**, 6534g].

21. Jacques, J., Collet, A., Wilen, S. H. (1981). Formation and Separation of Diastereomers. *Enantiomers, Racemates and Resolutions*. New York: Wiley-Interscience, p. 307–309.

22. Harrington, P. J., Lodewijk, E. (1997). Twenty years of naproxen technology. *Org. Process Res. Dev.* **1**, 72.

23. Qi, J. Z., Wang, Q. Z. (1986). Study on anti-TB drug ethambutol ø: the new resolution agents for resolution of aminobutanol. *Pharm. Industry.* **17**, 241–244.

24. Qi, J. Z., Wang, Q. Z. (1986). Study on anti-TB drug ethambutol ε: the mutual resolution of aminobutanol and tartaric acid. *Pharm. Industry* **17**, 145–147.

25. Laszlo, M., Lajos, K. (1981) Optically active 2-aminobutanol and its salts with acids. *Hung. Teljes* **19**, 369 (CA 95: 97030g).

26. Woodward, R. B., Cava, M. P., Hunger, A., Ollis, W. D., Daeniker, H. U., Schenker, K. (1963). The total synthesis of strychnine. *Tetrahedron* **19**, 247–288.

27. Wilen, S. H. (1971). Resolving agents and resolution in organic chemistry. *Top. Stereochem.* **6**, 107–176.

28. Leclercq, M., Jacques, J. (1979). Titre de l'article. *Nouv. J. Chim*. **3**, 629.

29. Bayley, C. R., Vaidya, N. A. (1992). *Chirality in Industry*. Collins, A. N, Sheldrake G N, Crosby J, editors. New York: Wiley-Interscience, p. 71.

30. Jacques, J., Collet, A., Wilen, S. H. (1981). Experimental Aspects and Art of Resolution. *Enantiomers, Racemates, and Resolutions*., New York: Wiley-Interscience, p. 387.

31. Lin, G. Q., Peng, J. S. (2003). Preparation of optical pure fexofenadine and its hydrochloride by chemical resolution method. CN 1442407A.

32. Imhof, R., Kyburz, E., Daly, J. J. (1984). Design, synthesis, and x-ray data of novel potential antipsychotic agents. Substituted 7-phenylquinolizidines: stereospecific, neuroleptic, and antinociceptive properties. *J. Med. Chem*. **27**, 165–175.

33. Kelly, R. C., Schletter, I., Stein, S. J., Wierenga, W. (1979). Total synthesis of alpha.-amino-3-chloro-4,5-dihydro-5-isoxazoleacetic acid (AT-125), an antitumor antibiotic. *J. Am. Chem. Soc*. **101**, 1054–1056.

34. Rozwadowska, M. D., Brossi, A. (1989). Optically active tetrahydro-alpha-phenyl-6,7-dimethoxyisoquinoline-1-methanols from (1-phenylethyl) ureas. Absolute configuration of (-)- and (+)-isomers of the erythro series. *J. Org. Chem*. **54**, 3202.

35. Aav, R., Parve, O., Pehk, T., Claesson, A., Martin, I. (1999). Preparation of highly enantiopure stereoisomers of 1-(2,6-dimethyl- phenoxy)-2-aminopropane (mexiletine). *Tetrahedron Asymmetry* **10**, 3033–3038.

36. Tashiro, Y., Aoki, S. (1985). Optically active phenylalanine. Eur Patent Appl EP 133 053 [*Chem Abstr*. (1985). **103**, 37734p] .

37. Viret, J., Patzelt, H., Collet, A. (1986). Simple optical resolution of terleucine. *Tetrahedron Lett*. **27**, 5865–5868.

38. Klyashchitskii, B. A., Shevets, V. I. (1972). Splitting of racemic alcohols into optical antipodes. *Russ. Chem. Rev*. **41**, 1315–1334.

39. Billington, D. C., Baker, R., Kulagowski, J. J., Mawer, I. M. (1987). Synthesis of myo-inositol 1-phosphate and 4-phosphate, and of their individual enantiomers. *J. Chem. Soc. Chem. Commun*. **4**, 314–316.

40. Hashimoto, S., Kase, S., Suzuki, A., Yanagiya, Y., Ikegami, S. (1991). A practical access to optically pure (*S*)-1-octyn-3-ol. *Synth. Commun*. **21**, 833–837.

41. Corey, E. J., Danheiser, R. L., Chandrasekaran, S., Keck, G. E., Gopalan, B., Larsen, S. D., Siret, P., Gras, J. L. (1978). Stereospecific total synthesis of gibberellic acid. *J. Am. Chem. Soc*. **100**, 8034–8036

42. Fessner, W. D., Prinzbach, H. (1986). D3-trishomocubanetrione synthesis and optical resolution. *Tetrahedron* **42**, 1797–1803.

43. Arcamone, F., Bernardi, L., Patelli, B. (1976). Optically active anthracyclinones. Ger Offen. 2 601 785 [*Chem. Abstr*. (1976). **85**, 142918j] .

44. Just, G., Luthe, C., Potvin, P. (1982). A method for the systematic resolution of unbranched α-acetoxyalkyl- and aralkylaldehydes: synthesis of 11R and 11S -hete. *Tetrahedron Lett*. **23**, 2285–2288.

45. Mangeney, P., Alexakis, A., Normant, J. F. (1988). Resolution and determination of enantiomeric excesses of chiral aldehydes via chiral imidazolidines. *Tetrahedron Lett*. **29**, 2677–2680.

46. Vries, T., Wynberg, H., van Echten, E., Koek, J., ten Hoeve, W., Kellogg, R. M., Broxterman, Q. B., Minnaard, A., Kaptein, B., van der Sluis, S., Hulshof, L., Kooistra, J. (1998). The family approach to the resolution of racemates. *Angew. Chem. Int. Ed*. **37**, 2349–2354.

47. Deng, J. G., Peng, X. H., Hua, Z. M. (2004). Preparation of optical pure β-adrenaline agonist by combinatorial resolution method. CN 1173929C.

48. Jaeques, J., Collet, A., Wilen, S. H. (1981). Formation and Separation of Diastereomers. In: *Enantiomers, Racemates, and Resolutions*. New York: Wiley-Interscience, p. 273.

49. Goldberg, I. (1977). Structure and binding in molecular complexes of cyclic polyethers. 4. Crystallographic study of a chiral system: an inclusion complex of a macrocyclic ligand with phenylglycine methyl ester. *J. Am. Chem. Soc*. **99**, 6049.

50. Knobler, C. B., Gaeta, F. C. A., Cram, D. J. (1988). Source of chiral recognition in coraplexes with phenylglycine as guest. *J. Chem. Soc. Chem. Commun*. 330–333.

51. Peacock, S. S., Walba, D. M., Gaeta, F. C. A., Helgeson, R. C., Cram, D. J. (1980). Host-guest complexation. 22. Reciprocal chiral recognition between amino acids and dilocular systems. *J. Am. Chem. Soc*. **102**, 2043–2052.

52. Brienne, M. J., Jacques, J. (1975). Remarques sur la chiralite de certains clathrates du type cage. *Tetrahedron Lett*. **16**, 2349–2352.

53. Dharanipragada, R., Ferguson, S. B., DIederich, F. (1988). A novel optically active host: design, computer graphics, synthesis, and diastereomeric complex formation in aqueous solution. *J. Am. Chem. Soc*. **110**, 1679–1690.

54. Mikolajczyk, M., Drabowicz, J. (1978). Optical resolution of chiral sulfinyl compounds via cyclodextrin inclusion complexes. *J. Am. Chem. Soc*. **100**, 2510.

55. Hamilton, J. A., Chen, L. (1988). Crystal structure of an inclusion complex of beta-cyclodextrin with racemic fenoprofen: direct evidence for chiral recognition. *J. Am. Chem. Soc*. **110**, 5833–5841.

56. Schuring, V. (1994). Enantiomer separation by gas chromatography on chiral stationary phases. *J. Chromatogr. A* **666**, 111–129.

57. Toda, F., Tanaka, K., Ueda, H. (1984). A new optical resolution method of tertiary acetylenic alcohol utilizing complexation with brucine. *Tetrahedron Lett*. **22**, 4669–4672.

58. Toda, F. (1987). Isolation and optical resolution of materials utilizing inclusion crystallization. *Top. Curr. Chem*. **140**, 43.

59. Toda, F., Tanaka, K., Ueda, H., Oshima, T. (1983). Chiral recognition in complexes of tertiary acetylenic alcohols and sparteine; mutual optical resolution by complex formation. *J. Chem. Soc. Chem. Commun*. **13**, 743–744.

60. Hassel, O. (1970). Structural aspects of interatomic change-transfer bonding. *Science* **170**, 497–502.

61. Kaupp, G. (1994). Resolution of racemates by distillation with inclusion compounds. *Angew. Chem. Int. Ed. Engl*. **33**, 728–729.

62. Zhu, J., Qin, Y., He, Z., Fu, F. M., Zhou, Z. Y., Deng, J. G., Jiang, Y. Z., Chau, T. Y. (1997). Resolution of alkyl pyridyl sulfoxides by complexation with a chiral host compound derived from tartaric acid. *Tetrahedron Asym*. **8**, 2505–2508.

63. Deng, J., Chi, Y., Fu, F., Cui, X., Yu, K., Zhu, J., Jiang, Y. Z. (2000). Resolution of omeprazole by inclusion complexation with a chiral host BINOL. *Tetrahedron Asym*. **11**, 1729–1732.

64. Deng, J. G., Chi, Y. X., Zhu, J. (2002). Preparation of optical pure benzimidazole compounds as anti-ulcerative drugs by inclusion resolution method. CN 1087739C.

65. Roy, B. N., Singh, G. P., Srivastava, D., Jadhav, H. S., Saini, M. B., Aher, U. P. (2009). A novel method for large-scale synthesis of Lamivudine through cocrystal formation of racemic lamivudine with (S)-(-)-1,'-bi(2-naphthol) [(S)-(BINOL)]. *Org. Process Res. Dev*. **13**, 450–455.

66. Toda, F., Tohi, Y. (1993). Novel optical resolution methods by inclusion crystallization in suspension media and by fractional distillation. *J. Chem. Soc. Chem. Commun*. **15**, 1238–1240.

67. Keith, J. M., Larrow, J. F., Jacobsen, E. N. (2001). Practical considerations in kinetic resolution reactions. *Adv. Synth. Catal*. **343**, 5–26.

68. Robinson, D. E. J. E., Bull, S. D. (2003). Kinetic resolution strategies using nonenzymatic catalysts. *Tetrahedron Asym*. **14**, 1407–1446.

69. Franck, A., Ruchardt, C. (1984). Optically active naproxen by kinetic resolution. *Chem. Lett*. **8**, 1431–1434.

70. Ruchardt, C., Gärtner, H., Saltz, U. (1984). Optically active α-arylcarboxylic acids by kinetic resolution: pyrethroid acids. *Angew. Chem. Int. Ed. Engl*. **23**, 162–164.

71. Saltz, U., Ruchardt, C. (1984). Preparation of optical active 2-(aryloxy) propionic acids by kinetic resolution. *Chem. Ber*. **117**, 3457–3462.

72. Martin, V. S., Woodard, S. S., Katsuki, T., Yamada, Y., Ikeda, M., Sharpless, K. B. (1981). Kinetic resolution of racemic allylic alcohols by enantioselective epoxidation. A route to substances of absolute enantiomeric purity? *J. Am. Chem. Soc*. **103**, 6237–6240.

73. Katuki, T., Sharpless, K. B. (1980). The first practical method for asymmetric epoxidation. *J. Am. Chem. Soc*. **102**, 5974–5976.

74. Tokunaga, M., Larrow, J. F., Kakiuchi, F., Jacobsen, E. N. (1997). Asymmetric catalysis with water: efficient kinetic resolution of terminal epoxides by means of catalytic hydrolysis. *Science* **277**, 936–938.

75. Liu, Z. Y., Ji, J. X., Li, B. G. (2000). The conversion of racemic terminal epoxides into either (+)- or (−)-diol γ- and δ-lactones. *J. Chem. Soc. Perkin Trans*. **1**, 3519.

76. Shen, K. S., Xiong, F., Hu, J., Yang, L. P. (2003). Application of chiral (salen) cocatalyzed hydrolytic kinetic resolution of racemic terminal epoxides to the syntheses of chiral drugs. *Chin. J. Org. Chem*. **23**, 542–545.

77. Gu, Q. M., Chen, C. S., Sih, C. J. (1986). A facile enzymatic resolution process for the preparation of (+)-S-2-(6-hethoxy-2-naphthyl)propionic acid (naproxen). *Tetrahedron Lett*. **27**, 1763–1766.

78. Battistel, E., Bianchi, D., Cesti, P., Pina, C. (1991). Enzymatic resolution of (S)-(+)-naproxen in a continuous reactor. *Biotechnol. Bioeng*. **38**, 659–664.

79. Mutsaers, J. H. G. M., Kooremain, H. J. (1991). Preparation of optically pure 2-aryl- and 2-aryloxypropionates by selective enzymic hydrolysis. *Rec. Trav. Chim. Pays-Bas* **110**, 185–188.

80. Ahmar M, Girard C, Bloch R. (1989). Enzymatic resolution of methyl 2-alkyl-2-arylacetates. *Tetrahedron Lett*. **30**, 7053–7056.

81. Kalaritis, P., Regenye, R. W., Partridge, J. J., Coffen, D. L. (1990). Kinetic resolution of 2-substituted esters catalyzed by a lipase ex. *Pseudomonas fluorescens*. *J. Org. Chem*. **55**, 812–815.

82. Sugai, T., Kakeya, H., Ohta, H. (1990). Preparation of enantiomerically enriched compound using enzymes. 3. Enzymic preparation of enantiomerically enriched tertiary alpha-benzyloxy acid esters. Application to the synthesis of (S)-(−)-frontalin. *J. Org. Chem*. **55**, 4643–4647.

83. Sugai, T., Watanabe, N., Ohta, H. (1991). A synthesis of natural α-tocopherol intermediate. *Tetrahedron Asymmetry* **2**, 371–376.

84. Clark, J. E., Fischer, P. A., Schumacher, D. P. (1991). An enzymatic route to florfenicol. *Synthesis* **10**, 891–894.

85. Chenevert, R., Thiboutot, S. (1989). Synthesis of chloramphenicol via an enzymatic enantioselective hydrolysis. *Synthesis* **6**, 444–447.

86. Zoete, M. C., Rantwijk, F., Sheldon, R. A. (1994). Lipase catalyzed transformations with unnatural acyl acceptors. *Catalysis Today* **22**, 563–590.

87. Zoete, M. C., Danlen, A. C. K., Sheldon, R. A. (1994). A new enzymatic reaction: enzyme catalyzed ammonolysis of carbolic esters. *Biocatalysis* **10**, 307–316.

88. Bevinakatti, H. S., Banerji, A. A. (1991). Practical chemoenzymatic synthesis of both enantiomers of propranolol. *J. Org. Chem*. **56**, 5372–5375.

89. Lee. D., Kim, M. J. (1998). Lipase-catalyzed transesterification as a practical route to homochiral syn-1,2-diols. The synthesis of the Taxol side chain. *Tetrahedron Lett*. **39**, 2163–2166.

90. Nieduzak, T. R., Carr, A. A. (1990). Enzymatic resolution of arylpiperidine derivative utilizing lipase-catalyzed hydrolysis. *Tetrahedron Asymmetry* **1**, 535–536.

91. Zmijewski, M. J., Briggs, B. S., Thompson, A. R., Wright, I. G. (1991). Enantioselective acylation of a beta-lactam intermediate in the synthesis of loracarbef using penicillin G amidase. *Tetrahedron Lett*. **32**, 1621–1622.

92. Margolin, A. L. (1993). Enzymes in the synthesis of chiral drugs. *Enzyme Microb. Technol*. **15**, 266–280.

93. Crosby, J. (1991). Synthesis of optically active compounds: a large scale perspective. *Tetrahedron* **47**, 4789–4846.

94. Vince, R., Brownell, J. (1990). Resolution of racemic carbovir and selective inhibition of human immunodeficiency virus by the (−) enantiomer. *Biochem. Biophys. Res. Commun*. **168**, 912–916.

95. Mahmoudian, M., Baines, B. S., Drake, C. S., Hale, R. S., Jones, P., Piercey, J. E., Montgomery, D. S., Purvis, I. J., Storer, R., Dawson, M. J., Lawrence, G. C. (1993). Enzymatic production of optically pure (2′R-cis)-2′-deoxy-3′-thiacytidine (3TC, lamivudine): a potent anti-HIV agent. *Enzyme Microb. Technol*. **15**, 749–755.

96. Gihani, M. T. E. I., Williams, J. M. J. (1999). Dynamic kinetic resolution. *Curr. Opin. Chem. Biol*. **3**, 11–15.

97. Turner, N. J. (2004). Enzyme catalysed deracemisation and dynamic kinetic resolution reactions. *Curr. Opin. Chem. Biol*. **8**, 114–119.

98. Pellissier, H. (2003). Dynamic kinetic resolution. *Tetrahedron* **59**, 8291–8327.

99. Pellissier, H. (2008). Recent developments in dynamic kinetic resolution. *Tetrahedron* **64**, 1563–1601.

100. Brands, K. M. J., Davies, A. J. (2006). Crystallization-induced diastereomer transformations. *Chem. Rev*. **106**, 2711–2733.

101. Chandrasekhar, S., Ravindranath, M. (1989). "Preferential spontaneous resolution" of p-anisyl α-methylbenzyl ketone. *Tetrahedron Lett*. **30**, 6207.

102. Black, S. N., Williams, L. J., Davey, R. J., Moffatt, F., Jones, R. V. H, Mcewan, D. M, Sadler, D. E. (1989). The preparation of enantiomers of paclobutrazol: a crystal chemistry approach. *Tetrahedron* **45**, 2677.

103. Huang, C.J., Zhou, H. Y. (1999). Preparation of [(S)-(−)-α-(methylamino) propiophenone]₂. (2R, 3R)-tartaric acid derivatives. CN1293039C.

104. Arai, K. (1986). Isomerization-crystallization method in optical resolution. *Yuki Gosei Kagaku Kyokaishi* **44**, 486.

105. Piselli, F. L. (1989). A process for the optical resolution of 2-(6-methoxy-2-naphthyl) propionic acid (naproxen). Eur Patent Appl, EP 298 395 [*Chem. Abstr.* (1989). **111**, 7085a].

106. Ammazzalorso, A., Amoroso, R., Bettoni, G., De Fillppis, B., Glampietro, L., Maccallini, C., Tricca, M. L. (2004). Dynamic kinetic resolution of α-bromoesters containing lactamides as chiral auxiliaries. *Arkivoc* **v**, 375–381.

107. Ammazzalorso, A., Amoroso, R., Bettoni, G., De Fillippis, B., Fantacuzzi, M., Glampietro, L., Maccallini, C., Tricca, M. L. (2006). Asymmetric synthesis of arylpropionic acids and aryloxy acids by using lactamides as chiral auxiliaries. *Eur. J. Org. Chem.* **18**, 4088–4091.

108. Kim, H. J., Shin, E. K., Chang, J. Y., Kim, Y., Park, Y. S. (2005). Dynamic kinetic resolution of α-chloro esters in asymmetric nucleophilic substitution using diacetone-D-glucose as a chiral auxiliary. *Tetrahedron Lett.* **46**, 4115–4117.

109. Kim, H. J., Kim, Y., Choi, E. T., Lee, M. H., No, E. S., Park, Y. S. (2006). Asymmetric syntheses of N-substituted α-amino esters via dynamic kinetic resolution of α-haloacyl diacetone-d-glucose. *Tetrahedron* **62**, 6303–6311.

110. Shin, E. K., Kim, H. J., Kim, Y., Kim, Y., Park, Y. S. (2006). d-Phg-l-Pro dipeptide-derived prolinol ligands for highly enantioselective Reformatsky reactions. *Tetrahedron Lett.* **47**, 1933–1935.

111. Noyori, R., Ikeda, T., Ohkuma, T., Widhalm, M., Kitamura, M., Takaya, H., Akutagawa, S., Sayo, N., Saito, T., Taketomi, T., Kumobayashi, H. (1989). Stereoselective hydrogenation via dynamic kinetic resolution. *J. Am. Chem. Soc.* **111**, 9134–9135.

112. Genet, J. P. (2003). Asymmetric catalytic hydrogenation. Design of new Ru catalysts and chiral ligands: from laboratory to industrial applications. *Acc. Chem. Res.* **36**, 908–918.

113. Mordant, C., Ca no de Andrade, C., Touati, R., Ratovelomanana-Vidal, V., Hassine, B. B., Genet, J. P. (2003). Stereoselective synthesis of diltiazem via dynamic kinetic resolution. *Synthesis* **15**, 2405–2409

114. Berkessel, A., Mukherjee, S., Cleemann, F., Muller, T. N., Lex, J. (2005). Second-generation organocatalysts for the highly enantioselective dynamic kinetic resolution of azlactones. *Chem. Commun.* **42**, 1898–1900.

115. Berkessel, A., Cleemann, F., Mukherjee, S., Muller, T. N., Lex, J. (2005). Highly efficient dynamic kinetic resolution of azlactones by urea-based bifunctional organocatalysts. *Angew. Chem. Int. Ed.* **44**, 807–811.

116. Fülling, G., Sih, C. J. (1987). Enzymatic second-order asymmetric hydrolysis of ketorolac esters: in situ racemization. *J. Am. Chem. Soc.* **109**, 2845–2846.

117. Lin, H. Y., Tsai, S. W. (2003). Dynamic kinetic resolution of (R, S)-naproxen 2, 2, 2- trifluoroethyl ester via lipase-catalyzed hydrolysis in micro-aqueous isooctane. *J. Mol. Catal. B* **24–25**, 111–120.

118. Ong, A. L., Kamaruddin, A. H., Bhatia, S. (2005). Current technologies for the production of (S)-ketoprofen: process perspective. *Process Biochem*, **40**, 3526–3535.

119. Beak, P., Anderson, D. R., Curtis, M. D., Laumer, J. M., Pippel, D. J., Weisenburger, G. A. (2000). Dynamic thermodynamic resolution: control of enantioselectivity through diastereomeric equilibration. *Acc. Chem. Res.* **33**, 715–727.

120. Basu, A., Beak, P. (1996). Control of the enantiochemistry of electrophilic substitutions of N-pivaloyl-R-lithio-o-ethylaniline: stereoinformation transfer based on the method of organolithium formation. *J. Am. Chem. Soc.* **118**, 1575–1576.

121. Basu, A., Gallagher, D. J., Beak, P. (1996). Pathways for stereoinformation transfer: enhanced enantioselectivity via diastereomeric recycling of organolithium/(−)-sparteine complexes. *J. Org. Chem.* **61**, 5718–5719.

122. Thayumanavan, S., Basu, A., Beak, P. (1997). Two different pathways of stereoinformation transfer: asymmetric substitutions in the (−)-sparteine mediated reactions of laterally lithiated N,N-diisopropyl-o-ethylbenzamide and N-pivaloyl-o-ethylaniline. *J. Am. Chem. Soc.* **119**, 8209–8216.

CHAPTER 5

FLUORINE-CONTAINING CHIRAL DRUGS

XIAO-LONG QIU and XUYI YUE

Key Laboratory of Organofluorine Chemistry,Shanghai Institute of Organic Chemistry, Chinese Academy of Sciences, Shanghai, China

FENG-LING QING

Key Laboratory of Organofluorine Chemistry, Shanghai Institute of Organic Chemistry, Chinese Academy of Sciences and College of Chemistry and Chemistry Engineering, Donghua University, Shanghai, China

Chiral Drugs: Chemistry and Biological Action, First Edition. Edited by Guo-Qiang Lin, Qi-Dong You and Jie-Fei Cheng.
© 2011 John Wiley & Sons, Inc. Published 2011 by John Wiley & Sons, Inc.

5.1 INTRODUCTION

Fluorine is one of the most abundant elements on Earth. However, its occurrence is extremely rare in biological compounds. So far, only a dozen organic compounds containing the fluorine atom have been found in nature. It should be noted that all of these naturally occurring fluorinated compounds possess only one fluorine atom (Fig. 5.1), and two of them [nucleocidin, $(2R, 3R)$-2-fluorocitric acid] contain a fluorine atom that links directly to a chiral carbon center [1]. Despite such scarcity, organic chemists, medicinal chemists, and pharmaceutical chemists have synthesized and widely used enormous numbers of fluorine-containing compounds because the introduction of fluorine atom(s) into biologically active molecules can bring about remarkable and profound changes in their physical, chemical, and biological properties [2]. Two of the most notable examples in the field of fluoromedicinal chemistry in the twentieth century are 9α-fluorohydrocortisone (an anti-inflammatory drug) [3] and 5-fluorouracil (an anticancer drug) [4]. In the 1950s, researchers discovered both pharmaceuticals, in which the introduction of just a single fluorine atom to the corresponding natural products brought about some remarkable pharmacological properties. So far, over one hundred fluorine-containing compounds have been approved by the FDA in the United States for treating different diseases. In 2007, nine of the top 20 best-selling drugs worldwide were fluorine-containing compounds or mixtures of fluorine-containing compounds and other compounds (Fig. 5.2). For example, the best-selling and third-best-selling drugs in 2007 were Lipitor and Admair

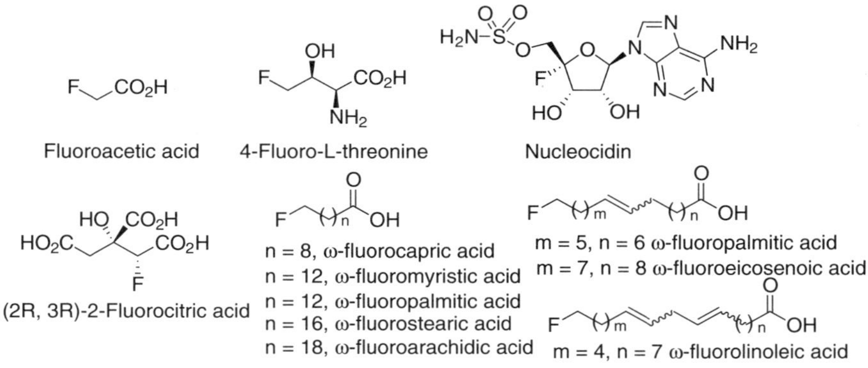

FIGURE 5.1   Structures of naturally occurring fluorine-containing compounds.

Diskus (a mixture of fluticasone propionate and salmeterol), which contain one and three fluorine atoms, respectively.

Incorporation of fluorine into pharmaceutical and veterinary drugs to enhance their pharmacological properties has now become a standard practice. It would not be an exaggeration to say that every drug discovery and development program in the twenty-first century without exception evaluates fluorine-containing compounds as potential drug candidates. Currently, synthesis and biological evaluation of fluorinated prostanoids (for glaucoma), fluorinated conformationally restricted glutamate analogs (for CNS disorders), fluorinated MMP inhibitors (for cancer metastasis intervention), fluoro-taxoids (for cancer), trifluoro-artemisinin (for malaria) and fluorinated nucleosides (for viral infections) are the hot topics in synthetic organic chemistry, medicinal chemistry, and pharmaceutical chemistry. The remarkable change in biological, pharmacological, and pharmaceutical properties achieved through introduction of fluorine atom(s) has intrigued scientists for a long time: The issue has become more and more clear and better understood. All of these results are due to the specific properties of the fluorine atom, including its small steric size, high electronegativity, carbon–fluorine bond strength, capacity to form strong hydrogen bonding, and polar interaction. For example, research has clearly demonstrated the important effects of fluorine substitution on the inter- and intramolecular forces, which affect the binding of ligands, and thus introduce receptor subtype selectivity, at cholinergic and adrenergic receptors [5]. Fluorine substitution can also have a profound effect on drug disposition, in terms of absorption, distribution, drug metabolism [6], and drug clearance. On the other hand, in pharmaceuticals and drug discovery, stereoselective interactions are profoundly important because one or the other enantiomer of a compound can have beneficial or disastrous results. However, although many books (or book chapters) and reviews have described the importance of the incorporation of fluorine atom(s) in medicinal chemistry and chemical biology, few pieces of literature have dealt in detail with the important effect of chiral fluorine [in this chapter, chiral fluorine is defined as a fluorine atom(s) or fluorine-containing

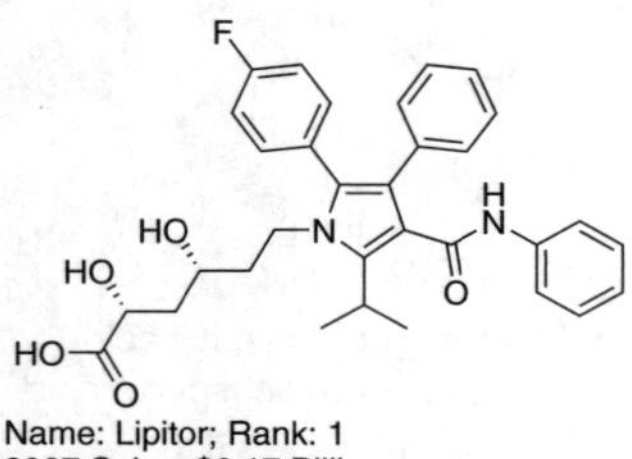

Name: Lipitor; Rank: 1
2007 Sales: $6.17 Billion
Profile: An HMG-Coa reductase inhibitor
used to lower LDL cholesterol levels

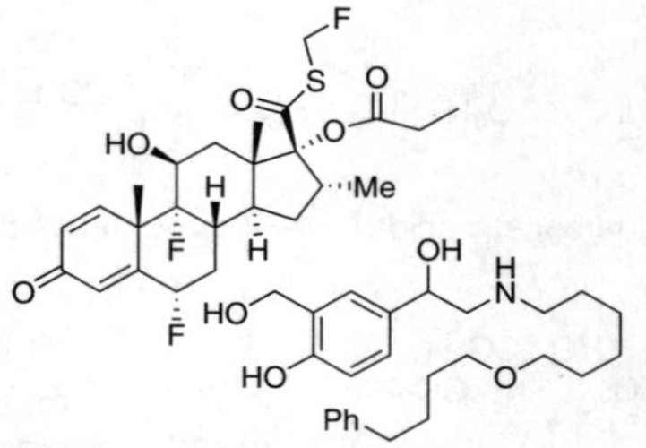

Name: Admair Diskus; Rank: 3
2007 Sales: $3.39 Billion
Profile: An corticosteroid and a bronchodilator
used to treat and prevent asthma

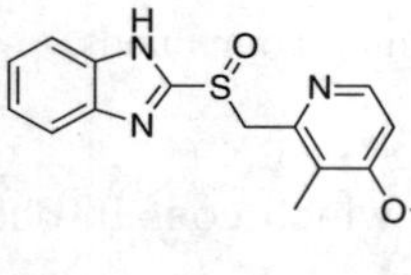

Name: Prevacid; Rank: 4
2007 Sales: $3.32 Billion
Profile: A proton pump inhibitor
used to treat gastric reflux disease

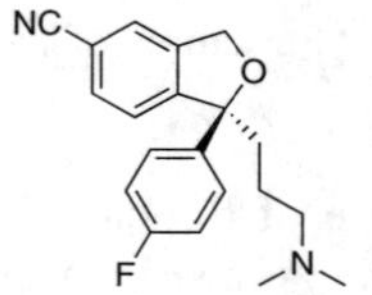

Name: Lexapro; Rank: 9
2007 Sales: $2.30 Billion
Profile: A selective serotonin reuptake
inhibitorused to treat depression
and anxiety

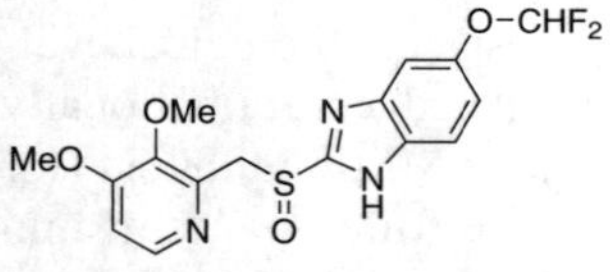

Name: Protonix; Rank: 11
2007 Sales: $2.14 Billion
Profile: A proton pump inhibitor
used to treat esophagus
inflammation and erosion.

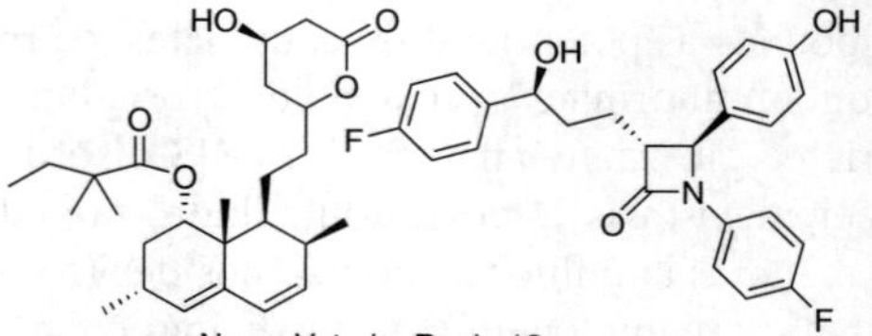

Name:Vytorin; Rank: 12
2007 Sales: $1.94Billion
Profile: A stain and a cholesterol absorption
blocker to treat high cholesterol

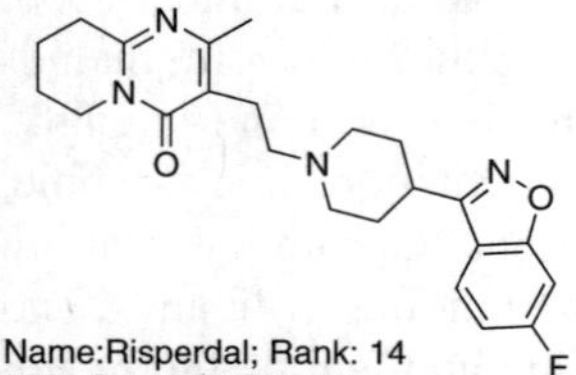

Name:Risperdal; Rank: 14
2007 Sales: $1.79 Billion
Profile: An antipsychotic used to treat
schizophrenia and bipolar mania

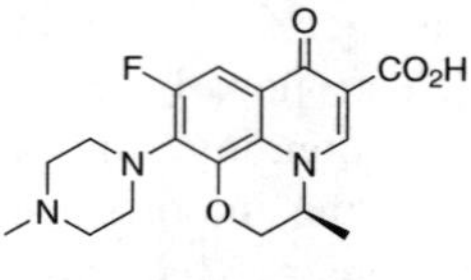

Name:Levaquin; Rank: 19
2007 Sales: $1.43Billion
Profile: A fluoroquinolone antibiotic
used to treat bacterial infections

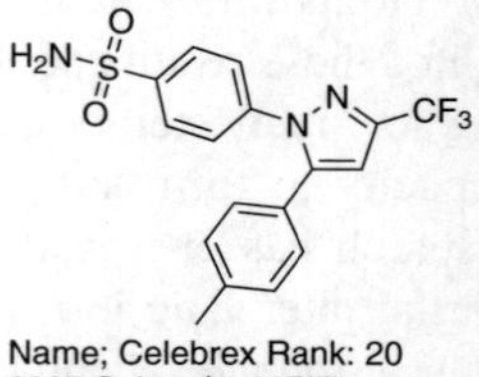

Name; Celebrex Rank: 20
2007 Sales: $1.42Billion
Profile: A COX-2 inhibitor NSAID
used to treat arthritis pain

FIGURE 5.2 Structures and retail dollars of nine fluorine-containing pharmaceuticals that were in the top 20 best-selling drugs in the world in 2007.

group(s) directly linked to a chiral center]. In the first section of this review, some important characteristic property changes associated with the incorporation of fluorine atom(s) or fluorine-containing groups in organic molecules are discussed, based on some classic examples. The second part focuses on synthesis of some chiral-fluorine-containing drugs that have monofluoro, *gem*-difluoromethylene, and trifluoromethyl groups directly linked to a chiral center. The development of procedures for the asymmetric introduction of a fluorine atom or trifluoromethyl group into organic molecules is also reviewed.

5.2 EFFECTS OF FLUORINE ATOM (OR FLUORINE-CONTAINING GROUPS)

5.2.1 Mimic Effect and Block Effect of Fluorine

The van der Waals (vdW) radius of fluorine is 1.35 Å, which is very close to the radius of a hydrogen atom or an oxygen atom. Table 5.1 summarizes some representative physical properties of fluorine compared to some other elements [7]; as shown, the radius of fluorine is only 12.5% larger than that of hydrogen and 3.6% smaller than that of oxygen. In addition, a C–F bond length is 1.39 Å, which is only 0.30 Å longer than a C–H bond. However, it is significantly shorter than a C–Cl bond and a C–Br bond. Because of its similarity to hydrogen, the size of a compound in which one hydrogen atom is replaced with a fluorine atom, would not significantly change compared to its nonfluorinated counterpart. Thus fluorinated compounds often cannot be recognized by some microorganisms and enzymes and easily enter organisms to participate in the metabolic cycle. This observation is the basis for what is regarded as the "mimic effect" of fluorine for hydrogen. Usually, the "mimic effect" occurs when only one hydrogen atom of a compound is replaced with a fluorine atom. It should be noted that the size of a difluoromethyl group or a trifluoromethyl group is bigger than a methyl group [8]. Thus the "mimic effect" seldom occurs when compounds contain difluoromethyl groups and/or trifluoromethyl groups.

TABLE 5.1 Some Physical Data of Selected Elements (Groups)

Element group)	Bond Length C–X/AA	Bond Dissociation Energy (H_3C-X)/kcal/mol	Van der Waals Radius, Å	Electronegativity
H	1.09	103.1	1.20	2.2
F	1.39	108.1	1.35	4.0
O(OH)	1.43	90.2	1.40	3.4
Cl	1.78	81.1	1.80	3.0
S(SH)	1.82	NA	1.85	2.6
Br	1.93	67.9	1.95	2.8
I	2.14	NA	2.15	2.5

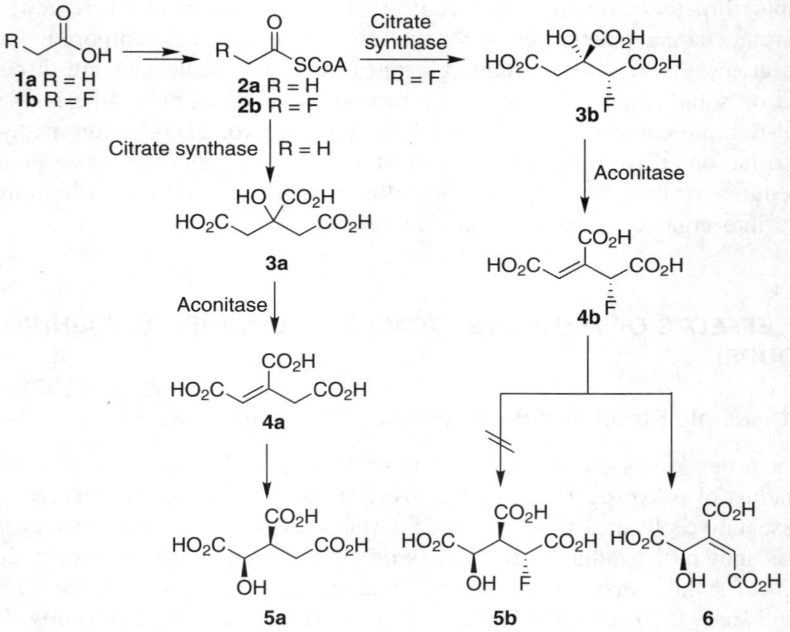

SCHEME 5.1 Metabolism cycle of acetyl acid and fluoroacetyl acid.

One classic example of the "mimic effect" of fluorine is the behavior of fluoroacetic acid in the citric acid metabolism cycle (Scheme 5.1) [9]. Fluoroacetic acid was first found in the South African plant *Dichapetalum cymosum* by Marais et al. in 1943 and was significantly toxic [10]. In organisms, fluoroacetic acid **1b** was recognized by the enzymes because of its size similar to acetic acid and allowed to enter the metabolism cycle **1a**. After **1b** was converted to fluoroacetyl-CoA **2b** by the same enzymatic transformation to acetic acid, $(2R, 3R)$-fluorocitric acid **3b** was produced via citrate synthase since this enzyme could not differentiate **2b** from acetyl-CoA **2a**. In a similar manner, aconitase does not distinguish the citric acid **3a** from **3b** and (R)-fluoro-*cis*-aconitic acid **4b** was obtained via dehydration. In a normal metabolism cycle, **4a** could be further hydroxylated to generate isocitric acid **5a**. However, **4b** is not recognized as a substrate by hydroxylase and therefore could not complete the normal metabolism cycle. Instead,**4b** underwent hydroxylation-defluorination to yield (R)-hydroxy-*trans*-aconitic acid **6**. Thus the "mimic effect" of fluorine allowed fluoroacetic acid **1b** to participate in the citric acid metabolism cycle; however, the final metabolism step was blocked because **4b** could not be distinguished by the enzymes.

In addition, it is clear that replacement of a specific a C—H bond with a C—F bond in some compounds could effectively block the metabolic process, which proceeds via hydroxylation of a C—H bond with the cytochrome P-450 enzyme family. This "blocking" is attributed to the stronger dissociation energy of a C–F bond than that of a C–H bond (by 5.0 kcal/mol) (Table 5.1), and this fluorine-caused "block" function is defined as the "block effect" of fluorine. Since the likely sites of metabolism of molecules by cytochrome P-450-enzymes can now be predicted by computational methods [11], the incorporation of fluorine into a metabolism site(s) has become a useful strategy to prevent the deactivation of a bioactive substance. Of many well-known examples, here this "block effect" will be illustrated with a short description of the development of ezetimibe **8** (brand name: Zetia, $1.40 billion sales in 2007), a drug approved in 2002 that is used to block cholesterol absorption (Fig. 5.3) [12]. The development of ezetimibe began with the search for a novel cholesterol acyl transferase (ACAT) inhibitor. This project finally evolved into a study to optimize in vivo cholesterol-lowering activity, which was elicited by a series of azetidinone analogs. Screening found that the lead compound **7** exhibited poor ACAT activity but had low and consistent in vivo cholesterol-lowering activity, although at that time the molecular target was unknown. Further optimization study established that metabolic oxidation at certain positions of **7** enhanced the bioactivity, while oxidation at other sites decreased the activity. At this stage, strategic replacement of H with F was used successfully to block unwanted metabolic oxidation, which finally led to the discovery of ezetimibe **8**. On the other hand, it was found that the deletion of a fluorine substituent at some special sites would decrease the metabolic stability. For instance, during the discovery of the cyclooxygenase 2 (COX 2) inhibitor Celecoxib **10** (2007 sales: $1.42 billion), a nonsteroidal anti-inflammatory drug

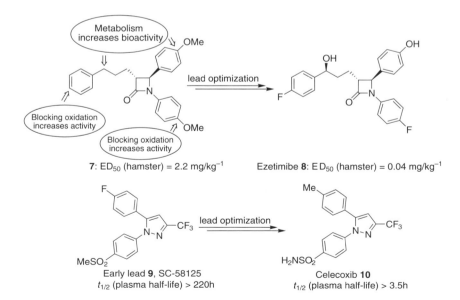

7: ED_{50} (hamster) = 2.2 mg/kg^{-1} Ezetimibe **8**: ED_{50} (hamster) = 0.04 mg/kg^{-1}

Early lead **9**, SC-58125   Celecoxib **10**
$t_{1/2}$ (plasma half-life) > 220h $t_{1/2}$ (plasma half-life) > 3.5h

FIGURE 5.3 Block effect and deblock effect of fluorine successfully used in developing ezetimibe and Celecoxib, respectively.

(NSAID) used in the treatment of osteoarthritis, rheumatoid arthritis, acute pain, painful menstruation, and menstrual symptoms, the extremely high metabolic stability of the lead compound **9**, which resulted in a very long biological half-life ($t_{1/2} > 220h$) in the body, could be reduced to more acceptable levels ($t_{1/2} > 3.5h$) by replacing a fluorine atom with a metabolically labile methyl group [13].

Another classic example of utilizing the "block effect" of fluorine is the development of fluorinated analogs of vitamin D_3 (Fig. 5.4). It has been determined that the hydroxylation of the methylene moiety at the 24-position in the side chain of vitamin D_3 **11** is a critical deactivation step prior to excretion [14]. To block this unwanted metabolism, the "block effect" of fluorine was used. That is, a *gem*-difluoromethyl group was utilized to replace the methylene moiety at the 24-position of **12**, which, as expected, effectively blocked the hydroxylation at this site. The *gem*-difluorinated analog **13** was then smoothly hydroxylated enzymatically at the 1-position to give product **15**. Biological assays demonstrated that **13** was slightly more potent than its nonfluorinated counterpart **12**, while the hydroxylated product **13** displayed 5–10 times higher potency than **14**. In addition, it was also found that 1α, 25-dihydroxy-26,26,26,27,27,27-hexafluorovitamin D_3 **16** (falecalcitrol, marketed in 2001 for treating hyperthyroidism) is several times more potent than its nonfluorinated counterpart **14** [15], which was attributed to the "block effect" of fluorine after trifluoromethyl groups were incorporated.

Interestingly, there are also a few known examples for which the incorporation of fluorine atom(s) or fluorine-containing groups does not prevent metabolism at that site [16]. This phenomenon is observed in particular for phenyl rings with a nitrogen-containing substituent in the *para* position to the fluorine substituent because, during P450-catalyzed metabolism, a rearrangement usually takes place in which the fluorine automatically moves to an adjacent carbon and the phenol metabolite is formed *para* to the nitrogen substituent.

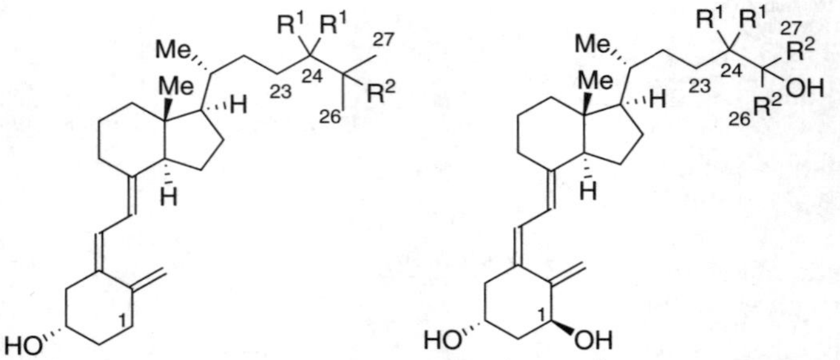

11: $R^1 = R^2 = H$, Vitamin D_3 14: $R^1 = H$, $R^2 = Me$
12: $R^1 = H$, $R^2 = OH$ 15: $R^1 = F$, $R^2 = Me$
13: $R^1 = F$, $R^2 = OH$ 16: $R^1 = H$, $R^2 = CF_3$, falecalcitrol

FIGURE 5.4   Vitamin D_3 and its fluorinated analogs.

5.2.2   Effect of Fluorine on pK_a

It is well-known that fluorine is the most electronegative element (Table 5.1). Thus it is natural that the fluorine atom(s) or fluorine-containing group(s) has some unique inductive effects on the neighboring groups. Once fluorine(s) is incorporated, one of the important inductive changes occurs in the pK_a values of carboxylic acids, alcohols, or protonated amines. As shown in Table 5.2, fluorine substitution can decrease the pK_a values [17]. For example, the acidity of trifluoroacetic acid ($pK_a = 0.52$) is 10^4 times stronger than its nonfluorinated counterpart, acetic acid ($pK_a = 4.76$). After fluorination, weak acid *tert*-butanol ($pK_a = 19.00$) becomes a medium strong acid, $(CF_3)_3COH$ ($pK_a = 5.40$). The inductive effect of fluorine atom(s) or fluorine-containing groups on pK_a could also be observed even when one or more methylene group(s) is inserted between the fluorine moiety and the—CO_2H (–OH,—NH_2, RNH) moiety. The pK_a of $CF_3CH_2CO_2H$ is 3.06, which is still substantially more acidic than $CH_3CH_2CO_2H$ ($pK_a = 4.87$). The pK_a value of CH_3CH_2OH is about 2.6 larger than that of CF_3CH_2OH ($pK_a = 12.39$).

Additionally, the introduction of fluorine(s) into organic amines decreases their amine basicities (increases pK_b or decreases pK_a). For example, the pK_b of 2,2,2-trifluoroethyl amine is about 4.8 times larger than that of ethylamine ($pK_b = 3.3$) and the basicity of benzenyl amine ($pK_b = 9.4$) is significantly stronger than perfluorobenzenyl amine ($pK_b = 14.4$) (Table 5.3). Interestingly, the pK_b values of ethylamines increase almost linearly upon successive fluorine substitutions: $CH_3CH_2NH_2$ ($pK_b = 3.3$), $FCH_2CH_2NH_2$ ($pK_b = 5.0$), $F_2CHCH_2NH_2$ ($pK_b = 6.7$), and $CF_3CH_2NH_2$ ($pK_b = 8.1$) [18]. It should be noted that, based on experimental data, Morgenthaler et al. developed a practical method to predict the pK_a (pK_b) values of alkyl amines through the "σ-transmission effect" of fluorine.

The changes of pK_a (pK_b) that resulted from the introduction of fluorine atom(s) or fluorine-containing groups would improve the bioactivities or stabilities of the compounds in some cases. For example, in the inhibitors of carbonic anhydrase II (a zinc metalloenzyme), a significant difference in potency was observed between $CF_3SO_2NH_2$ ($pK_b = 8.2$, $K_i = 2 \times 10^{-9}$ M) and $CH_3SO_2NH_2$ ($pK_b = 3.5$, K_i in 10^{-4} M range), which is attributed to the substantial increase in the acidity of the sulfonamide functionality. Incorporation

TABLE 5.2 pK_a of Organic Acids and Corresponding Fluorinated Organic Acids

Compound	pK_a	Compound	pK_a	Compound	pK_a
CH_3CO_2H	4.76	$CF_3CH_2CO_2H$	3.06	$(CH_3)_2CHOH$	16.10
CH_2FCO_2H	2.59	$C_6H_5CO_2H$	4.21	$(CF_3)_2CHOH$	9.30
CHF_2CO_2H	1.33	$C_6F_5CO_2H$	1.70	$(CH_3)_3COH$	19.00
CF_3CO_2H	0.52	CH_3CH_2OH	15.93	$(CF_3)_3COH$	5.40
$CH_3CH_2CO_2H$	4.87	CF_3CH_2OH	12.39	————	——

TABLE 5.3   pK_b of Some Amines and Corresponding Fluorinated Amines

Compound	pK_b	Compound	pK_b
$CH_3CH_2NH_2$	3.3	$F_3CCH_2NH_2$	8.1
$FCH_2CH_2NH_2$	5.0	$C_6H_5NH_2$	9.4
$F_2CHCH_2NH_2$	6.7	$C_6F_5NH_2$	14.4

of a CF_3 group could facilitate deprotonation and better binding to the Zn(II) ion in the catalytic domain of the enzyme [19]. If the neutral species of an amine-containing drug is the active form, fluorine-substitution inductive decrease of pK_a can increase activity. This can be illustrated through example of the histidine H2 receptor antagonist mifentidine, whose fluorination was used to probe whether the neutral or the protonated form of this anidine is the active species. Of a series of synthesized fluorinated mifentidine analogs, the N-trifluoroethyl derivative **17** (pK_a = 4.45, K_B = 7.6 pM) was the most active compound, while its nonfluorinated counterpart, N-ethyl derivative **18** (pK_a = 5.57, K_B = 177 pM) displayed low activity (Fig. 5.5) [20]. These data clearly demonstrated that activity correlates with the increased concentration of the neutral species. Additionally, fluorine-substitution inductive decrease of pK_a can improve oral bioavailability in some cases. For instance, van Niel et al. used selective fluorination of 3-(3-(piperidin-1-yl) propyl)indoles and 3-(3-(piperizin-1-yl)propyl)indoles to increase the oral absorption of 5-HT$_{1D}$ versus 5-HT$_{1B}$ selective ligands for treating migraines. As expected, the incorporation of fluorine on the propyl side chain (for example, compound **19**) reduced the pK_a, leading to improved oral bioavailability without the loss of subtype selectivity [21].

Hydrolytic stability may also be successfully addressed by means of fluorine or fluorine-containing group(s) engrafted to the position adjacent to acid-labile functional groups (decrease of pK_a of functional group). For example, the increased hydrolytic stability of the glycosyl bond of a nucleoside could be achieved via the introduction of fluorine into the C−2′ position. In 1959, the Prusoff group reported 1-(2-deoxy-β-D-ribofuranosyl)-5-iodouracil **20** as the first antivirus nucleoside [22]. However, its glycosidic linkage was unstable under acidic conditions. Twenty years later, Watanabe and co-workers found that the incorporation of a fluorine atom into the C-2′ position (**21**) could effectively

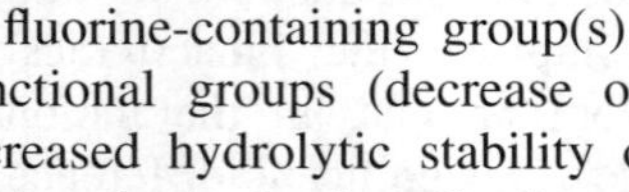

17: R = CH_2CF_3
pK_a = 4.45, K_B = 7.6 picoM
18: R = CH_2CH_3
pK_a = 5.57, K_B = 177 picoM

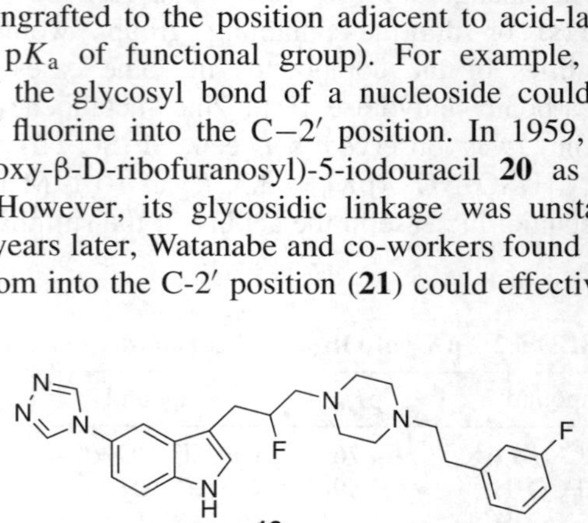

19
fluorinated 3-(3-(piperidin-1-yl)propyl)indole analogue

FIGURE 5.5   Fluorinated amines.

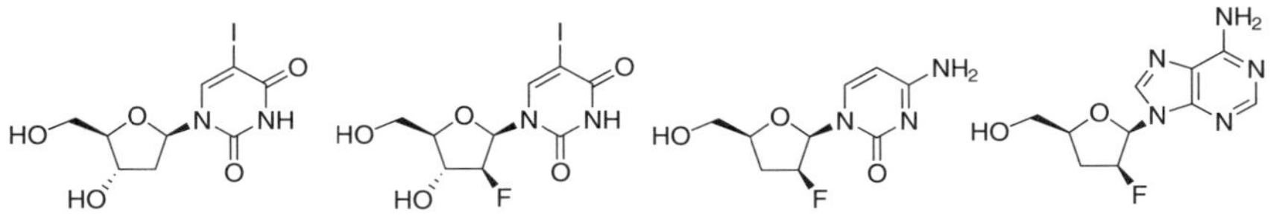

FIGURE 5.6 Improvement of hydrolytic stability after fluorine is incorporated into C-2′ position of nucleosides.

make glycosidic linkage stable and, most delightfully, the resultant fluorinated nucleoside **21** exhibited strong antivirus bioactivities against HSV, HBV, VZV, CMN, and EBV [23]. Similarly, 2,3-dideoxy-2-fluoro nucleosides such as **22** and **23** (FddA, Lodenosine) displayed better hydrolytic stability and stronger bioactivity compared to their parent compounds (Fig. 5.6) [24].

5.2.3 Effect of Fluorine on Lipophilicity

Lipophilicity refers to the ability of a chemical compound to dissolve in fats, oils, lipids, and nonpolar solvents such as hexane or toluene. These nonpolar solvents are themselves lipophilic—the axiom that like dissolves like generally holds true. Thus lipophilic substances tend to dissolve in other lipophilic substances, while hydrophilic (water loving) substances tend to dissolve in water and other hydrophilic substances. In other words, lipophilicity represents the affinity of a molecule or a moiety for a lipophilic environment. For the field of drug discovery and development, this is generally expressed by the partition coefficient ($\log P$ and $\log D$) between water and a water-immiscible solvent, most commonly 1-octanol. $\log P$ is defined as the logarithm of the partition coefficient P, which is defined as the ratio of the concentration of neutral species in octanol divided by the concentration of neutral species in water. $\log D$ refers to the logarithm of the distribution coefficient D, which is defined as the ratio of concentrations of all species (neutral and ionized) in octanol divided by the concentration of all species in water at a given pH (typically 7.4) (Fig. 5.7). In medicinal chemistry, lipophilicity is a key parameter because the absorption and distribution of a drug molecule in vivo are controlled by its balance of lipophilicity as well as ionization. Typically, groups of substantial lipophilicity on the ligand are required to obtain a good binding affinity to the target protein [25]. However, a

$$P = \frac{\text{[neutral species]}_{\text{in octanol}}}{\text{[neutral species]}_{\text{in water}}}$$

$$D = \frac{\text{[neutral species + ionized species]}_{\text{in octanol}}}{\text{[neutral species + ionized species]}_{\text{in water}}}$$

FIGURE 5.7 Definition of $\log P$ and $\log D$.

TABLE 5.4   Lipophilicity Parameters ($\pi_x = \log P_X - \log P_H$, also called Hansch π_x parameters) of Fluorinated Benzene Derivatives Compared with Their Nonfluorinated Counterparts

Entry	X in C_6H_5-X	π_x	Entry	X in C_6H_5-X	π_x	Entry	X in C_6H_5-X	π_x
1	F	0.14	5	OCF_3	1.04	9	$CF_3C(O)NH-$	0.55
2	CH_3	0.56	6	$CH_3(CO)$	-1.27	10	CH_3SO_2-	-1.63
3	CF_3	0.88	7	$CF_3(CO)$	0.08	11	CF_3SO_2-	0.55
4	OCH_3	-0.02	8	$CH_3C(O)NH-$	-1.63			

high lipophilicity typically results in a reduced solubility and a number of other undesirable properties for a compound. Therefore, the right balance between a required lipophilicity and a certain minimal overall polarity of the molecule is one of the recurring challenges for medicinal chemists.

Although some exceptional cases exist, it is generally conceived that the introduction of fluorine or fluorine-containing groups increases the lipophilicity of organic compounds, especially aromatic compounds. For instance, fluorobenzene is slightly more lipophilic than its nonfluorinated analog, benzene (Table 5.4, entry 1). Similarly, trifluoromethyl benzene is 57% more lipophilic than toluene (entry 2 vs. entry 3). In addition, CF_3-Y-Ph (Y = O, CO, CONH, SO_2) are also much more lipophilic than their corresponding counterparts (CH_3-Y-Ph) because the strong electron-withdrawing CF_3 group in the compounds significantly decreases the electron density of the adjacent polar functional groups (Y), thus lowering Y's hydrogen bond-forming capability with water molecules, and hence decreasing hydrophilicity (increasing lipophicity) (Table 5.4, entries 4–11). Along the same line, it is rational that CF_3-substituted benzenes bearing Lewis basic functionalities (such as an amine, alcohol, ether, carbonyl amide et al.) at the *ortho* or *para* position decrease the hydrogen-bond-accepting capability in an aqueous phase, leading to an increase in hydrophobicity and thus lipophicity. Interestingly, the introduction of fluorine into aliphatic compounds resulted in a decrease in lipophilicity, which could be illustrated by some published examples. Pentane ($\log P = 3.11$) is more lipophilic than 1-fluoropentane ($\log P = 2.33$) and 3-fluoropropyl benzene ($\log P = -0.7$); 3,3,3-trifluoropropyl benzene ($\log P = -0.4$) is significantly less lipophilic than its nonfluorinated counterpart [7b].

Changes of the lipophilicity of straight-chain alkanols bearing a terminal CF_3 group would depend on how many—(CH_2)–groups exist between CF_3 and OH (Table 5.5) [26]. CF_3CH_2OH is more lipophilic than ethanol ($\Delta \log P = 0.68$), which can be ascribed to the significant decrease in the basicity of the OH group (decrease of pK_a) by the electron-withdrawing CF_3 group. This through-bond inductive effect extends to three methylene inserts but diminishes beyond four. Thus, compared with butanol and pentanol, both 4-CF_3-butanol and 5-CF_3-pentanol are less lipophilic ($\Delta \log P = -0.25$ and -0.89, respectively). As for alkyl amines, a large enhancement of lipophilicity could be observed when fluorine atom(s) is introduced into position near an amine group, which

TABLE 5.5   Lipophilicity Parameters of Trifluoromethylated Straight-Chain Amines

Entry	Alcohols $CX_3(CH_2)_nOH$	$X = H$ $\log P_H$	$X = F$ $\log P_F$	$\Delta \log P$ $\log P_F - \log P_H$
1	n = 1	−0.32	0.36	0.68
2	n = 2	0.34	0.39	0.05
3	n = 3	0.88	0.90	0.02
4	n = 4	1.40	1.15	−0.25
5	n = 5	2.03	1.14	−0.89

is attributed to the decrease of amine basicity through the inductive effect of fluorine, leading to an increase in the neutral amine component as opposed to the ammonium ion in equilibrium [18].

In 2004, the Böhm group described the statistical effect of replacing a hydrogen by a fluorine atom on the lipophilicity of a compound [25a]. They selected 293 pairs of molecules from the Roche in-house database with measured $\log D$ values that just differed by one fluorine atom. The change in $\log D$ on one H/F exchange is shown in Figure 5.8. A Gaussian distribution with a maximum slightly higher than 0 was observed. On average, the substitution of a hydrogen atom by fluorine increases lipophilicity slightly, by roughly 0.25 log units, which confirms the predictions of atomic increments published by Wildman et al. [27]. They found that the tail of the Gaussian distribution extends to the values below 0, demonstrating that there are quite a number of cases for which an H to F exchange decreases lipophilicity. Furthermore, they found that those compounds (a negative $\log D$ associated with one H/F exchange) are usually characterized by the presence of an oxygen atom close to the fluorine (for selected examples, see Fig. 5.9). Böhm et al. proposed two rationales for this kind of negative $\log D$. One possible explanation is that fluorine's close vicinity to an oxygen atom increased the overall polarity of the molecule, leading to a more pronounced gain in solvation energy in the polar medium relative to the nonpolar solvent. Additionally, it is also possible that the fluorine polarized the neighboring oxygen atoms, leading to stronger hydrogen bonds between the oxygen atom and water molecules. Böhm et al. have also analyzed the other end of the Gaussian distribution shown in Figure 5.8, which contains compounds with a much stronger positive shift in $\log D$ than expected for a single H/F exchange. Most of these compounds contain one or more basic nitrogen atoms. The fluorine substituent reduces the basicity of the nitrogen functionality, leading to an increased $\log D$ that was measured at pH 7.4.

5.2.4   Effect of Fluorine on Protein-Ligand Interactions

It has been known that strategically-placed fluorine can increase the affinity of a molecule for a macromolecular (protein) recognition site through noncovalent bond mechanisms. The effect of the fluorine atom could arise directly from the

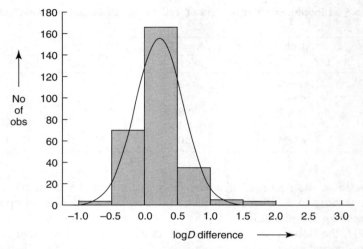

FIGURE 5.8 Histogram of change in $\log D$ observed upon substitution of a hydrogen atom by a fluorine atom. On average, $\log D$ is increased by roughly 0.25.

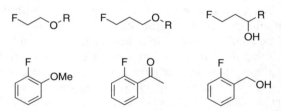

FIGURE 5.9 Chemical substructures with a negative $\log D$ shift associated with one single H/F exchange.

interaction of the fluorine with the protein or indirectly through modulation of the polarity of other groups of the ligand that interact with the protein. It is frequently found that a fluorine substituent leads to a slight enhancement of the binding affinity because of an increased lipophilicity of the molecule, a formed hydrogen bond (F···H) or the change of basicity or acidity resulting from the introduction of fluorine. So far, a large of number of examples for the polar C–F bond-protein interactions were found in the X-ray crystal structure of drug-protein complexes deposited in the Cambridge Structural Database (CSD) and the Protein Data Bank (PDB). These interactions mainly include those between a C–F bond and polar functional groups such as carbonyl and guanidinium ion moieties in the protein side chains, that is, C–F··· C=O and C-F···C(NH$_2$)(= NH) interactions [28]. Of these examples, many indicate that a C–F bond unit serves as a poor hydrogen bond acceptor. Instead of hydrogen bonding, however, a C–F bond was found to form polar interactions with a large number of polarizable bonds in the form of C–F···X(H) (where X = O, N, S) and C–F···C$_\alpha$ –H

($C_\alpha = \alpha$-carbon of α-amino acids) because these C–F···X separations are well beyond hydrogen-bonded contact distance [29]. C–F···H-N (backbone amide) interactions are abundant in the PDB. Out of 788 C–F-containing structures, constraining the F···N separation below the van der Waals contact distance of $d_1 = 3.1$ Å and the angles $a_1 \geq 150°$ and $90° \leq a_2 \leq 150°$ gave 11 structures in which the C–F moiety of the ligand points toward the H–N bond (Fig. 5.10a). In the case of two thrombin inhibitors (compounds **24** and **25**) that differ by only one H/F substitution (Fig. 5.10b) [25a], the F-containing inhibitor **25** is more potent by a factor of 5 and the X-ray crystal structure of the inhibitor-enzyme complexes shows a remarkable conformational difference between the two inhibitors. The crystal structures revealed the existence of a dipolar C–F···H–N interaction with a distance of 3.5 Å, which could be responsible for the observed

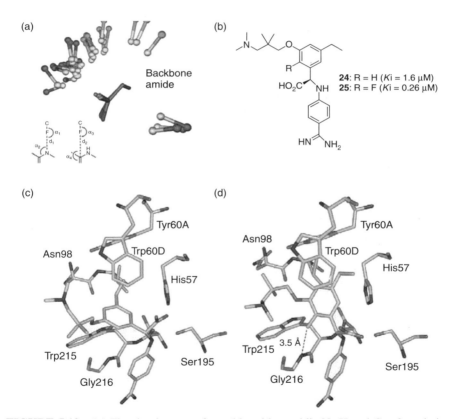

FIGURE 5.10   (a) Fluorine interacts favorably with peptidic N–H and C=O moieties. (b) Structures and binding affinities of a pair of thrombin inhibitors with and without fluorine substitute. (c) and (d) A dramatic change in the conformation and orientation of the thrombin inhibitor is found when hydrogen (c: compound **24**) is replaced by a fluorine atom [d: compound **25**, a dipolar N–H···F–C interaction (3.5 Å) is formed].

change in conformation (Fig. 5.10c and 5.10d). This distance is well beyond H-bonded contact distances, but the conformational change and the resulting gain in bioactivity provide particularly strong evidence for energetically favorable dipolar interactions.

Parlow and co-workers reported a similar observation on the interaction of fluorine with a protein N–H group in the complex of a series of serine protease inhibitors with tissue factor VIIa (TFVIIa) [30]. They described a fluorinated compound **27** (Fig. 5.11) bearing a benzene core, which is a good inhibitor of TFVIIa with K_i = 340 nM. The X-ray structure of the protein-ligand complex reveals a hydrogen bond (3.4 Å) between the fluorine and the N–H group of Gly216 of the protein [30a]. Interestingly, the X-ray structure of the nonfluorinated analog, benzoquinone derivative **26**, also reveals a hydrogen bond (3.3 Å)

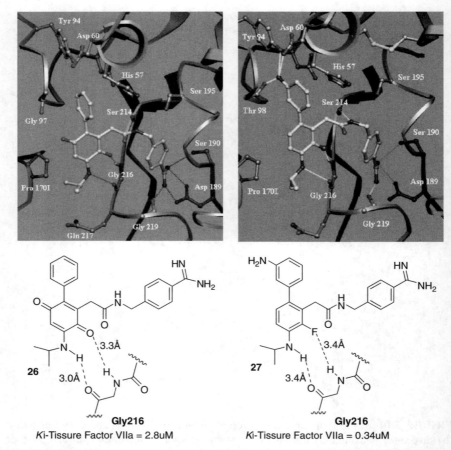

FIGURE 5.11   Crystal structure of **26** (benzoquinone analog) bound in the active site of TFVIIa complex and the crystal structure of **27**(fluorobenzene) bound in the active site of TFVIIa.

between the quinine oxygen atom and peptide nitrogen of Gly216 of the protein, although its bioactivity is much worse ($K_i = 2800$ nM).

Orthogonal multipolar $C-F\cdots C{=}O$ drug-protein interactions are well illustrated by the inhibitor-enzyme complex **28** ($IC_{50} = 19$ nM) with p38 MAP kinase, in which the fluorine atom of the 4-fluorophenyl moiety interacts with the amide carbonyl carbons of Leu104 and Val105 with an equal distance of 3.1 Å (Fig. 5.12a, 5.12b) [31]. In addition, $C-F\cdots C{=}O$ interactions were also seen in the crystal structure of the complex between porcine pancreatic elastase and an inhibitor of the trifluoroacetyl peptide class **29** [32]. X-ray structure demonstrated that the $C-F$ bonds approach the carbonyl C atoms along the pseudotrigonal axis. All three F atoms of the CF_3CO group interacted with backbone $C{=}O$ groups of Cys191 [$d(F\cdots C) = 3.5$ Å, $\alpha(F\cdots C{=}O) = 59°$], Thr213 [$d(F\cdots C) = 3.6$ Å, $\alpha(F\cdots C{=}O) = 87°$], Ser214 [$d(F\cdots C) = 3.0$ Å, $\alpha(F\cdots C{=}O) = 74°$], and Phe215 [$d(F\cdots C) = 3.1$ Å, $\alpha(F\cdots C{=}O) = 103°$] (Fig. 5.12c).

To gain a better understanding of the influence of fluorine substitution on protein-ligand interaction, Müller and co-workers have undertaken a systematic investigation of how protein-ligand interactions are affected by the replacement of H by F in thrombin inhibitors, which contain a rigid tricyclic core [33]. In their study, fluorine was systematically introduced at various positions of the inhibitor skeleton ("fluorine scan") to explore specific interactions of the halogen with active site amino acid residues of the enzyme. The binding mode of the tricyclic inhibitors at the thrombin active site was confirmed by several crystal structures of protein-ligand complexes, which clearly revealed that fluorine atoms reached into the D-pocket of the active site. Among a series of fluorinated analogs, 4-monofluorinated analog rac-**30**, was found to be the most potent inhibitor ($K_i = 0.057$ μM), while the bioactivity of 3-monofluorinated analog **31**, 2-monofluorinated analog **32**, and its nonfluorinated counterpart **33** was 0.36

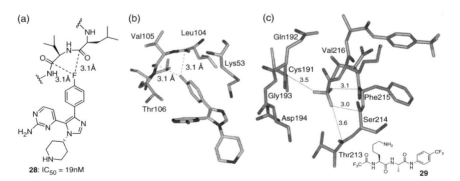

FIGURE 5.12 (a) and (b): Binding of a fluorinated inhibitor **28** to p38 kinase. The fluorine is in close proximity to two carbonyl groups of the protein. (c): $C-F\cdots C{=}O$ interactions seen in the crystal structure of the complex between porcine pancreatic elastase and an fluorinated inhibitor **29**.

μM, 0.50 μM, and 0.31 μM, respectively (Fig.5.13a). The X-ray crystallographic analysis of the enantiopure **30**-thrombin complex indicates that the H–Cα–C=O fragment of the Asn98 residue possesses significant "fluorophilicity." A complex model is outlined in Fig. 5.13b, which illustrates that the C–F residue of enantiopure **30** has strong multipolar H–Cα · · · C=O (3.1 Å) and C–F · · ·C=O (3.5 Å) interactions with the Asn98 residue in the distal hydrophobic pocket ("D-pocket") of thrombin.

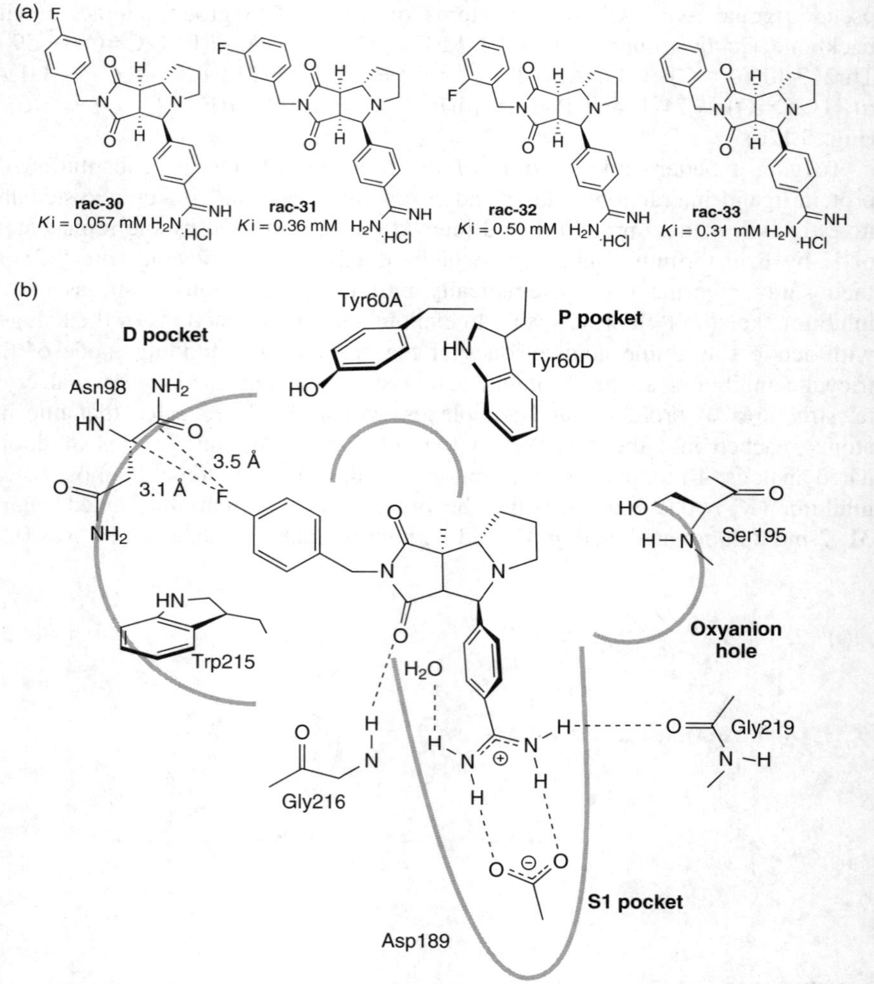

FIGURE 5.13 (a): Structures of fluorinated tricyclic thrombin inhibitors **30–32** and non-fluorinated counterpart **33** and their inhibition activities. (b): Binding model of enantiopure thrombin **30** on the basis of X-ray crystallographic analysis of its protein complex.

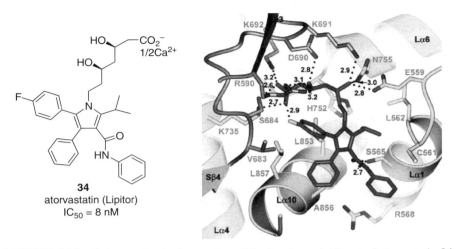

FIGURE 5.14 Orthogonal polar interaction of the fluorine substituent of atorvastatin **34** with Arg590 in HMG-CoA reductase.

Another example of the polar interaction of C–F with the guanidinium carbon of the Arg residue of protein is clearly illustrated by the X-ray crystal structure of complex of HMG-CoA reductase-atorvastatin **34** (brand name: Lipitor, the best-selling cholesterol-lowering drug, which binds to HMG-CoA reductase tightly with $IC_{50} = 8$ nM). Deisenhofer et al. found that a strong C–F···C(NH_2)($=$ NH) polar interaction between the 4-fluorophenyl moiety of atorvastatin and the Arg590 residue of HMG-CoA reductase with a very short F···C distance (2.9 Å) was formed (Fig. 5.14) [34].

5.3 CHIRAL FLUORINE-CONTAINING DRUGS

5.3.1 Monofluorinated Drugs

5.3.1.1 Fluticasone Propionate

Fluticasone propionate **39** [S-(fluoromethyl)-6α, 9-difluoro-11β, 17-dihydroxy-16α-methyl-3-oxoandrosta-1, 4-diene-17β-carbothioate, 17-propionate] is a synthetic glucocorticoid drug that is used for treatment of certain allergic conditions. Currently, GSK markets fluticasone propionate as Flovent (USA/Canada) and Flixotide (EU) for asthma and as Flonase (USA/Canada) and Flixonase (EU/Brazil) for allergic rhinitis, as well as a combination of fluticasone and salmeterol as Advair (USA/Canada) or Seretide (EU). Fluticasone mimics the naturally occurring hormone produced by the adrenal glands, cortisol or hydrocortisone. The exact mechanism of action of fluticasone is as of yet unknown. Fluticasone has potent anti-inflammatory actions. It is believed that fluticasone exerts its beneficial effects by inhibiting several types of signaling pathways involved in allergic, immune, and inflammatory responses. When used as a nasal inhaler or spray, the medication goes directly to the lining within the nose, and very little is absorbed into the rest of the body. The FDA approved fluticasone in October 1994.

The synthesis of fluticasone propionate **39** started from the commercially available material, flumethasone **35** (Scheme 5.2) [35]. The oxidation of **35** with Pd(OAc)$_2$/PPh$_3$/HIO$_4$ in DMA gave the 17β-carboxylic acid **36**, which was further selectively monoacylated with propionyl chloride/Et$_2$NH in acetone, yielding the 17α-propionyloxy derivative **37**. The reaction of **37** with N, N-dimethylthiocarbamoyl chloride/TEA/NaI with 2-butanone as solvent afforded the thioanhydride **38**. Compound **38** was converted to fluticasone propionate **39** by treatment with chlorofluoromethane/NaHS in DMA.

5.3.1.2 Clevudine

Clevudine   **44**   [1-(2′-deoxy-2′-fluoro-β-ʟ-arabinofuranosyl)-5-methyluracil, ʟ-FMAU], is an antiviral drug for the treatment of hepatitis B. Clevudine was approved for HBV in South Korea in 2007 [36]. It is marketed by Bukwang Pharmaceuticals in South Korea under the tradenames Levovir and Revovir. Under license from Bukwang, Pharmasset was developing the drug, but its phase III clinical trial (international, multicenter, randomized, double-blind, 96-week QUASH studies) was terminated because of some myopathy cases in patients.

Chu and co-workers developed an efficient route to 1,3,5-tri-O-benzoyl-2-fluoro-α-ʟ-arabinofuranose **44** starting from the ʟ-xylose **40** via the intermediate **41** and **42** in 12 steps (Scheme 5.3) [37]. Their synthesis highlighted the stereoselective introduction of 2′β-fluorine by treatment of the compound **42** with SOCl$_2$/imidazole followed by KHF$_2$/HF/H$_2$O. Bromination of the compound **43** and subsequent glycosylation with silylated thymine provided ʟ-FMAU **44**.

Recently, the Sznaidman group described a new synthetic route for ʟ-FMAU (Scheme 5.4) [38]. Their synthesis commenced with peracetylation of ʟ-arabinose **45** with Ac$_2$O/pyridine, and the resultant peracetylated arabinose was

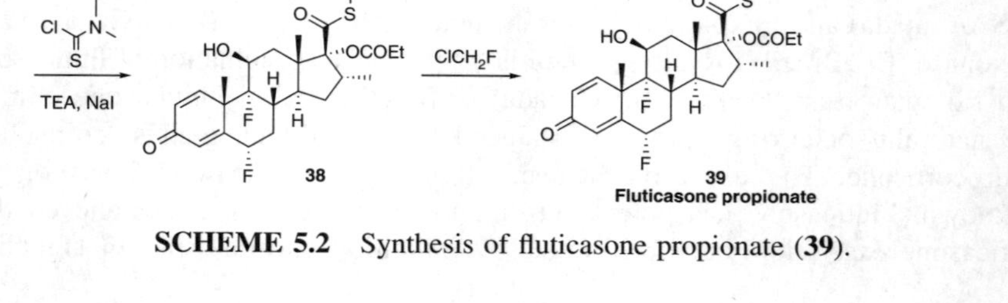

SCHEME 5.2   Synthesis of fluticasone propionate (**39**).

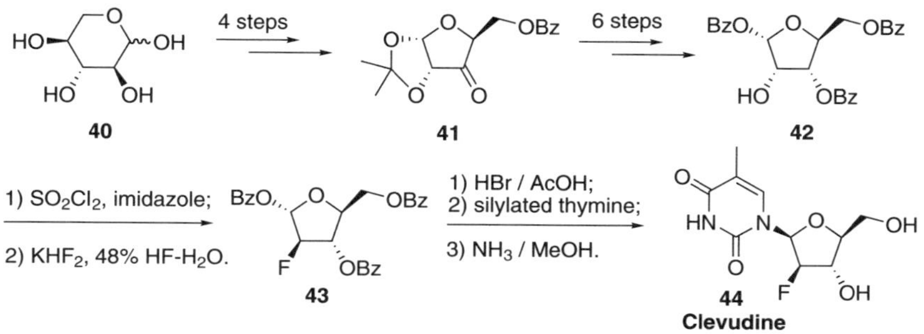

SCHEME 5.3 Synthesis of clevudine (**44**).

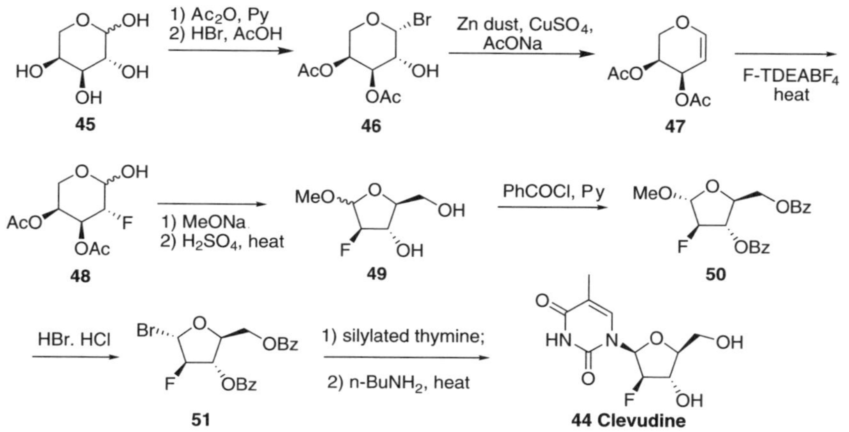

SCHEME 5.4 Synthesis of clevudine (**44**).

converted to the bromo-sugar **46** by treatment with HBr/AcOH/Ac$_2$O. Reaction of **46** with Zn/CuSO$_4$/NaOAc in AcOH/H$_2$O afforded the L-arabinal **47** in 60% yield. Exposure of compound **47** to Selectfluor (F-TEDA-BF$_4$) in refluxing nitromethane/H$_2$O stereoselectively generated the fluoro derivative **48** with 70% yield. Deacetylation of **48** with NaOMe/MeOH provided the hemiacetal, which was then converted into the methyl furanoside **49** by treatment with H$_2$SO$_4$ in MeOH. After benzoylation of the furanoside **49**, the resultant compound **50** was brominated with HBr/AcOH to yield the bromo-sugar **51**. Condensation of the sugar **51** with the silylated thymine followed by deprotection with *n*-BuNH$_2$ afforded the clevudine **44**.

5.3.1.3 Clofarbine

Clofarabine **61** [2-chloro-9-(2'-deoxy-2'-fluoro-β-D-arabinofuranosyl)adenine] is a second-generation purine nucleoside analog that works through incorporation into the DNA molecule and inhibition of further DNA synthesis via a number of different mechanisms [39]. In 2004, Clofarabine was approved by the FDA for the treatment of a type of leukemia known as relapsed or refractory acute lymphoblastic leukemia (ALL) in children, after at least two other types of treatment had failed. This drug is marketed in the U.S. and Canada as Clolar. In Europe and Australia/New Zealand, its brand name is Evoltra. So far, it is not known whether it extends life expectancy. Some investigations of effectiveness in cases of acute myeloid leukemia (AML) and juvenile myelomonocytic leukemia (JMML) have also been carried out. In 2009, Jeha and co-workers reported the phase II study of clofarabine in pediatric patients with refractory or relapsed acute myeloid leukemia [40]. In addition, clofarabine is also being studied in the treatment of other types of cancer.

The synthesis of clofarabine is described as follows (Scheme 5.5) [41]: Protection of the hydroxyl function of 1,2:5,6-di-O-isopropylidene-α-D-allofuranose **52** with TsCl and subsequent treatment of the resultant tosylate with KF in acetamide at 210°C gave 3-deoxy-3-fluoro-1,2:5,6-di-O-isopropylidene-α-D-glucofuranose **53** in medium yield. Exposure of **53** to the mixture of methanol-0.7% aq. H_2SO_4 (1:1) yielded 3-deoxy-3-fluoro-1,2-isopropylidene-α-D-glucofuranose, which was subjected to selective acylation with benzoyl chloride and deisopropylidenation with Amberlite IR-100 (H^+) ion-exchange resin to provide 6-O-benzoyl-3-deoxy-3-fluoro-D-glucofuranose **54**. The KIO_4-mediated oxidative cleavage of glucofuranose **54** resulted in a rearrangement and 5-O-benzoyl-2-deoxy-2-fluoro-3-O-formyl-D-arabinofuranose **55** was obtained. Deformylation of **55** by treatment with NaOMe in methanol provided 5-O-benzoyl-2-deoxy-2-fluoro-D-arabinofuranose **56** in high yield. Acylation of **56** afforded the 1,3-di-O-acetyl derivative **57** in 80% yield, which was treated with HBr/AcOH/CH_2Cl_2 to

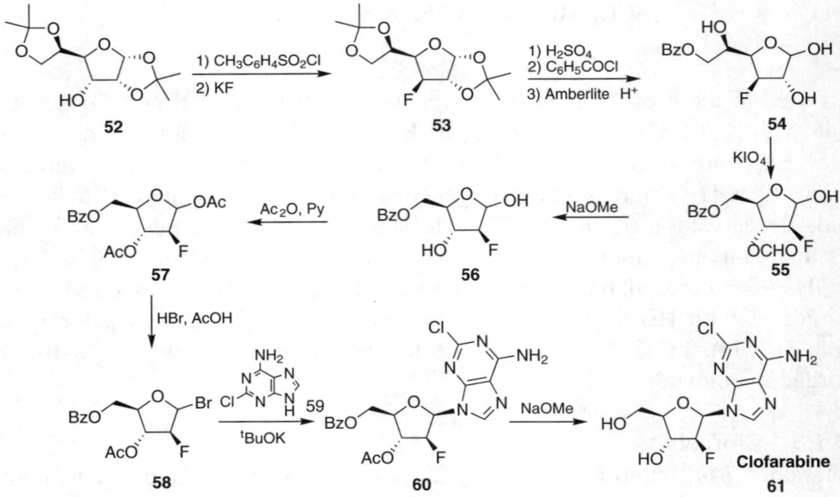

SCHEME 5.5 Synthesis of clofarabine (**61**).

produce 3-*O*-acetyl-5-*O*-benzoyl-2-deoxy-2-fluoro-D-arabinofuranosyl bromide **58**. Condensation of **58** with 2-chloroadenine **59** was investigated in different solvents with BuOK as base, and the acylated clofarabine **60** was afforded. Removal of the protecting groups in **60** gave clofarabine **61**. In 2008, the identification of clofarabine process impurities and their subsequent isolation, synthesis, and characterization was also described in detail [42].

5.3.1.4 Alovudine

Alovudine **66** (3′-deoxy-3′-fluorothymidine), also known as MIV-310 and FLT, first reported by Langen in 1971 [43], is a nucleoside reverse transcriptase inhibitor (NRTI) used for the treatment of HIV patients. Alovudine prevents HIV from entering the nucleus of healthy T cells. This prevents the cells from producing new viruses and decreases the amount of the virus present in the body. Alovudine is being codeveloped by Medivir, a company based in Sweden, and Boehringer Ingelheim. So far, alovudine has not yet been reviewed by the US FDA. In phase II trials of alovudine, side effects appeared to be related to bone marrow damage. Imaging and measurement of proliferation with positron emission tomography (PET) is a commonly used noninvasive staging tool to monitor the response to anticancer drug treatment. [^{18}F]FLT turns out to be a tracer particularly suitable for PET imaging of tumor proliferation because of the lack of degradation in vivo.

The original synthesis of FLT started from readily available thymidine **62**, which was acylated with *p*-chlorobenzoylchloride **63** to yield 5′-*O*-(4-chlorobenzoyl)thymidine **64**. Fluorination of **64** with DAST at −79°C provided the compound **65**, which was converted to **66** via removal of the protecting group with NaHCO₃ [44]. In addition, synthesis of **66** was also accomplished via the opening of the 2,3′-anhydro bond of 2,3′-anhydro-1-(2-deoxy-β-D-threo-pentofuranosyl)thymine with HF-AlF₃ and the reaction of 3′-*O*-mesylthymidine with KHF₂ [45] (Scheme 5.6).

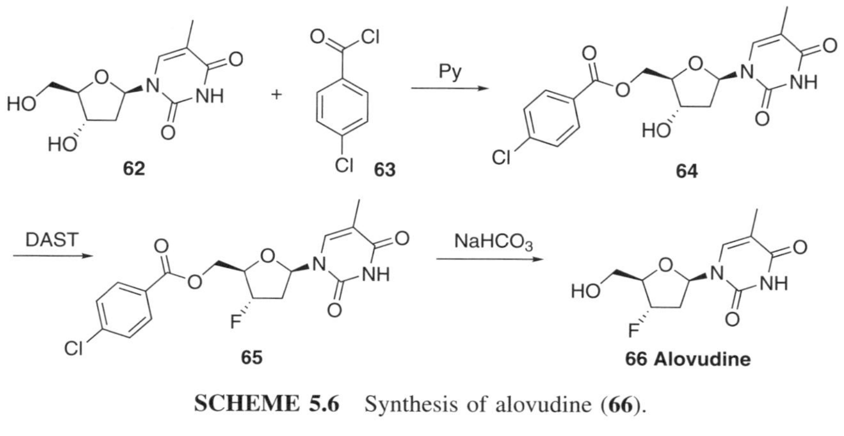

SCHEME 5.6 Synthesis of alovudine (**66**).

5.3.1.5 Fiacitabine

Fiacitabine **74** (2′-fluoro-5-iodo-1-β-D-arabinofuranosylcytosine), also named FIAC, is effective in vitro and in vivo against Herpes simplex viruses types 1 and 2 (HSV-1 and HSV-2) and Varicella zoster virus (VZV). FIAC also inhibits human cytomegalovirus (HCMV) in vitro, as well as Epstein–Barr virus (EBV) [46]. The clinical phase II trials on the safety and effectiveness of FIAC in the treatment of cytomegalovirus (CMV) in patients with AIDS have been completed.

The synthesis of FIAC is described in Scheme 5.7. Tosylation of 1,2:5,6-di-O-isopropylidene-α-D-allofuranose **52** yielded the tosylate **67** in good yield. Fluorination of compound **67** with KF in acetamide at 210°C produced 3-deoxy-3-fluoro-1,2:5,6-di-O-isopropylidene-α-D-glucofuranose, which was treated with MeOH/H$_2$SO$_4$ to generate 3-deoxy-3-fluoro-1,2-isopropylidene-α-D-glucofuranose. Subsequent selective acylation of the sugar with BzCl afforded the 6-O-benzoyl derivative, which was subjected to treatment with Amberlite IR-120 (H$^+$) ion-exchange resin to provide 6-O-benzoyl-3-deoxy-3-fluoro-D-glucofuranose **68**. The oxidative cleavage of glucofuranose **68** with NaIO$_4$ resulted in a rearrangement reaction to give 5-O-benzoyl-2-deoxy-2-fluoro-3-O-formyl-D-arabinofuranose **69**. Deformylation of compound **69** with NaOMe/MeOH gave 5-O-benzoyl- 2-deoxy-2-fluoro-D-arabinofuranose **70**. Acylation of **70** with Ac$_2$O/pyridine afforded the corresponding 1,3-di-O-acetyl derivative, which was further treated with HBr in AcOH/CH$_2$Cl$_2$ to yield 3-O-acetyl-5-O-benzoyl-2-deoxy-2-fluoro-D-arabinofuranosyl bromide **71**. Glycosylation of bromide **71** with tris(trimethylsilyl)cytosine **72** delivered the compound **73**. FIAC was obtained by deprotection of **73** with NH$_3$/MeOH followed by iodination with HIO$_3$/I$_2$ [23, 41a].

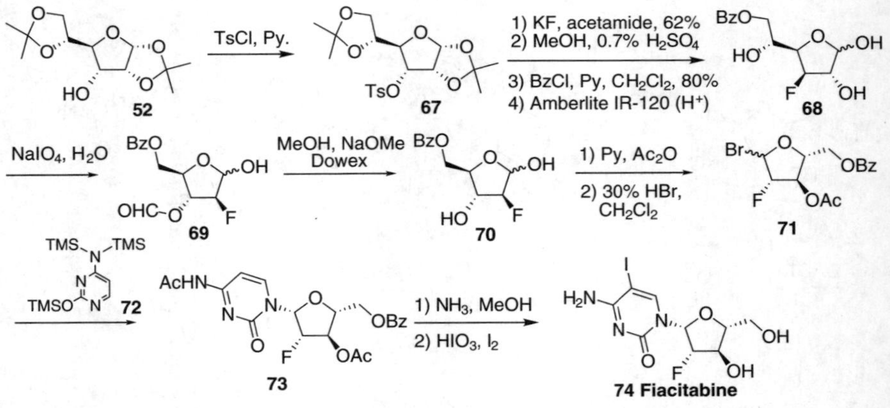

SCHEME 5.7 Synthesis of fiacitabine (**74**).

5.3.1.6 Flindokalner

Flindokalner **82** [(+)-3(*S*)-(5-chloro-2-methoxyphenyl)-3-fluoro-6-(trifluoro-methyl)-2,3-dihydro-1H-indol-2-one, MaxiPost, (*S*)-(+)-BMS-204352], a potassium channel opener for poststroke neuroprotection developed by Bristol-Myers Squibb, is under phase III clinical trial for the treatment or prevention of ischemic stroke.

A chiral process for the synthesis of flindokalner **82** was reported by Starrett and co-workers (Scheme 5.8) [47]. The synthesis started from 2-methoxyphenylacetic acid **75**, which was subjected to chlorination with SO_2Cl_2/AcOH followed by Fischer esterification to give compound **76** in 90% yield. Conversion of **76** to the lithium enolate with LHMDS and subsequent treatment with 4-(trifluoromethyl)-2-nitroflurobenzene **77** resulted in ipso displacement of the fluorine, and the resultant benzylic anion **78** was quenched with *N*-fluorobenzenesulfonimide to provide the racemic ester **79**. Hydrolysis of compound **79** with NaOH afforded the racemic carboxylic acid **80**. Resolution of compound **80** with (*S*)-α-methylbenzylamine was successfully achieved through the formation of a pair of diastereomeric salts in >99% *ee*. Finally, reduction of the nitro group with sodium dithionite followed by cyclization with HCl/MeOH afforded **82** in good yield with an enantiomeric excess of >99% (Scheme 5.8).

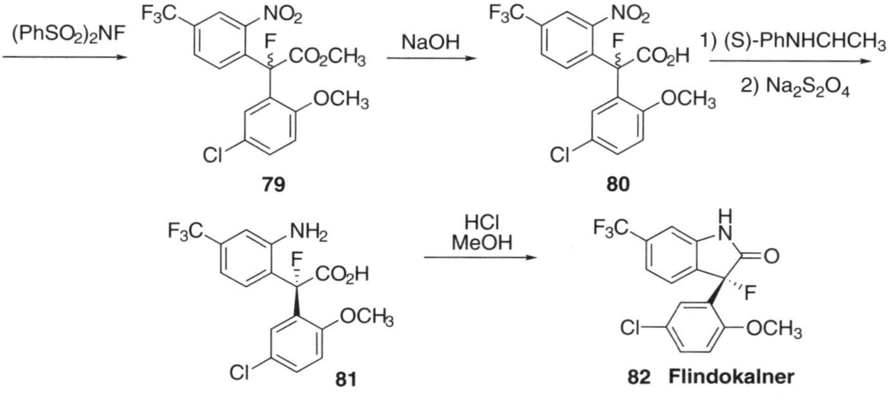

SCHEME 5.8   Synthesis of flindokalner (**82**).

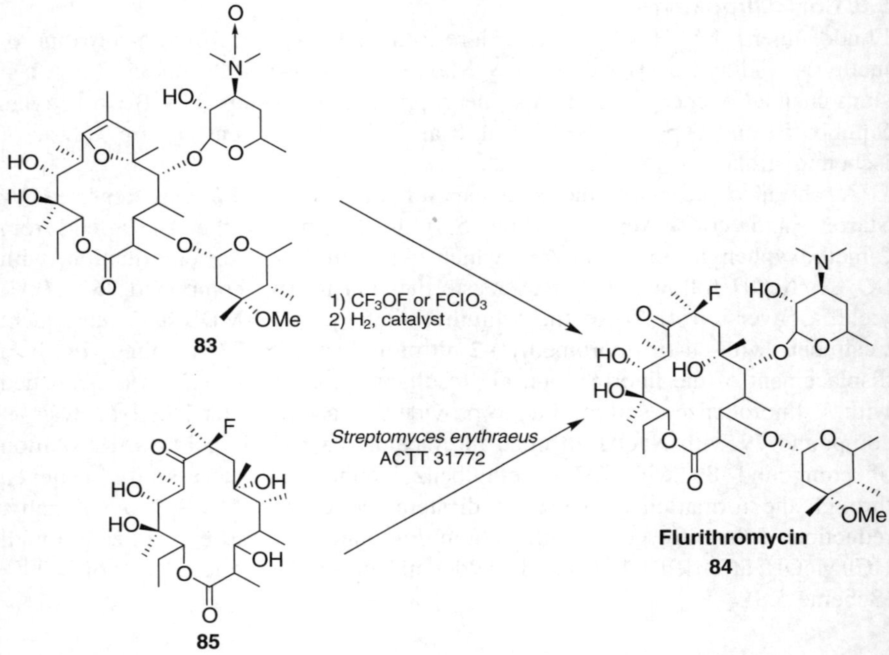

SCHEME 5.9 Synthesis of flurithromycin (**84**).

5.3.1.7 Flurithromycin

Flurithromycin **84**, also known as 8-fluoroerythromycin A, is a macrolide antibiotic. It is a fluorinated derivative of erythromycin. Flurithromycin was marketed in Italy in 1997 by Poli Industria Chimica under the tradename Flurizic.

Flurithromycin can be obtained by two different ways [48]: 1) by fluorination of 8,9-anhydroerythromycin A 6,9-hemiketal *N*-oxide **83** with trifluoromethyl hypofluorite (or perchloryl fluoride), followed by the reduction of the *N*-oxide functionality (Scheme 5.9) and 2) by biological synthesis, starting from (8*S*)-8-fluoroerythronolide A **85** in a culture broth of *Streptomyces erythraeus* (ATCC 31772), a blocked mutant of an erythromycin-producing strain.

5.3.1.8 Sitafloxacin

Sitafloxacin **94** {7-[(4*S*)-4-amino-6-azaspiro[2.4]heptan-6-yl]-8-chloro-6-fluoro-1-[(2*S*)-2-fluorocyclopropyl]-4-oxoquinoline-3-carboxylic acid, also named DU-6859a} is a fluoroquinolone antibiotic with a broad spectrum of activity including Staphylococci, Streptococci, Enterobacteriaceae, and anaerobes. This compound shows promise in the treatment of Buruli ulcers. Sitafloxacin is one of the drugs within the new generation of fluoroquinolones. This compound, together with clinafloxacin, gatifloxacin, gemifloxacin, moxifloxacin, and trovafloxacin,

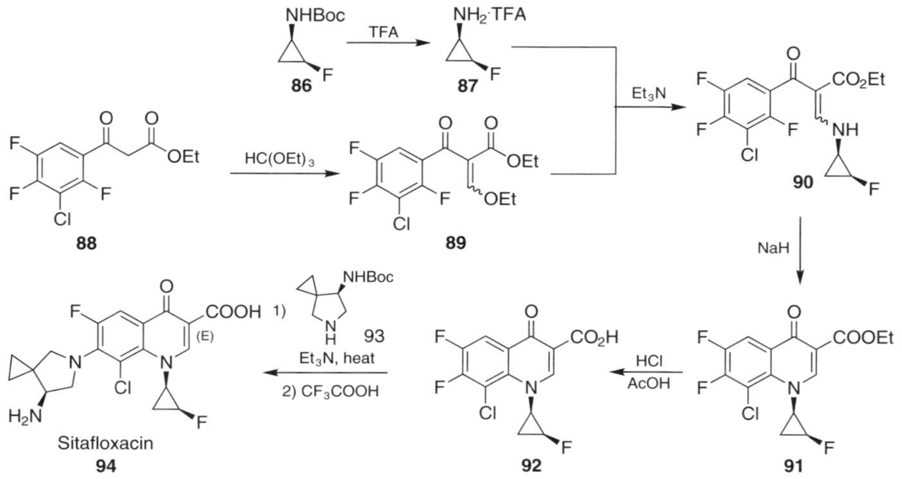

SCHEME 5.10   Synthesis of sitafloxacin (**94**).

constitutes the so-called fourth generation of the fluoroquinolones [49]. The molecule was identified by Daiichi Sankyo Co. Sitafloxacin is currently marketed in Japan by Daiichi Sankyo under the tradename Gracevit.

The synthesis of sitafloxacin **94** is outlined in Scheme 5.10. The reaction of 3-(3-chloro-2,4,5-trifluoro-phenyl)-3-oxo-propionic acid ethyl ester **88** with ethyl orthoformate gave the compound **89**. Condensation of **89** with amine **87**, prepared from (1*R*, 2*S*)-*N*-Boc-2-fluorocyclopropylamine **86** via deprotection with TFA, gave (1*R*, 2*S*)-2-(3-chloro-2,4,5-trifluorobenzoyl)-3-(2-fluorocyclopropylamino) acrylic acid ethyl ester **90**. The cyclization of compound **90** was accomplished by means of NaH, and the quinolone **91** was afforded in good yield. Hydrolysis of compound **91** with HCl/AcOH provided the free acid **92**. Condensation of **92** with (*R*)-7-*tert*-butoxycarbonylamino-5-azaspiro[2.4]heptane **93** and subsequent deprotection with TFA gave sitafloxacin **94** [50].

The chiral intermediate (1*R*, 2*S*)-*N*-Boc-2-fluorocyclopropylamine **86** was prepared in a straightforward fashion (Scheme 5.11). First, cyclization of butadiene **95** with dibromofluoromethane by means of BuONa, followed by oxidation with KMnO$_4$, esterification with EtOH/H$_2$SO$_4$ and reduction with Bu$_3$SnH gave 2-fluorocyclopropanecarboxylic acid ethyl ester **96** as a *cis/trans* mixture, which was crystallized to afford the *cis* isomer **97**. Hydrolysis of **97** with NaOH gave the corresponding acid **98**, which is condensed with (*R*)-α-methylbenzylamine by means of diphenyl chlorophosphate to yield the mixture of diastereomers **99**. Crystallization of **99** provided the optically pure isomer **100**. The amide **100** was hydrolyzed with HCl, followed by treatment with diphenylphosphoryl azide in refluxing BuOH to provide the intermediate **86**.

Intermediate **86** can also be synthesized in six steps from (1*S*, 2*R*)-2-amino-1,2-diphenylethanol **101** (Scheme 5.12). That is, the cyclization of **101** with

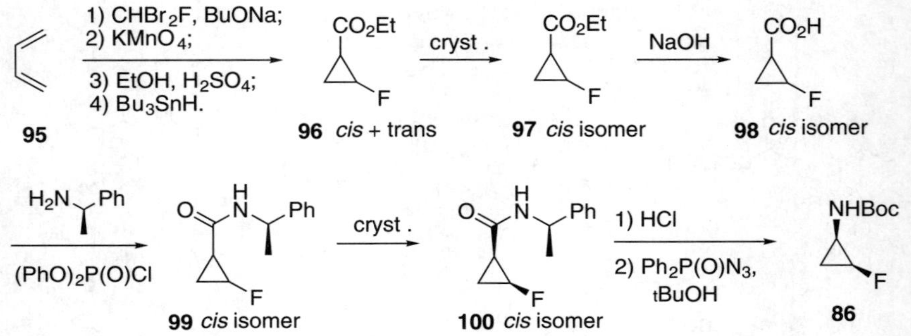

SCHEME 5.11   Synthesis of chiral amine **86**.

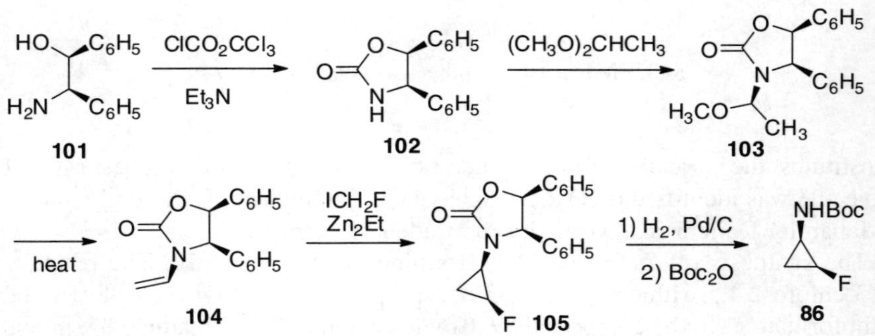

SCHEME 5.12   Synthesis of chiral amine **86**.

$ClCO_2CCl_3$ in Et_3N/CH_2Cl_2 gave $(4R, 5S)$-4,5-diphenyloxazolidin-2-one **102**, which was subjected to treatment with DME and an acid catalyst, yielding the 1-methoxyethyl derivative **103**. The heat treatment ($150°C$) of **103** afforded the vinyl derivative **104**. Cyclization proceeded well via treatment of **104** with ICH_2F/Et_2Zn, and the cyclopropyl-oxazolidinone **105** was generated in a preferentially *cis* way. The hydrogenolysis of **105** over Pd/C in AcOH gave $(1R, 2S)$-2-fluorocyclopropylamine, which was finally converted into **86** by exposure to Boc_2O/Et_3N in THF.

5.3.1.9   6-Fluoro-3-(4-Fluoropiperidin-3-yl)-2-Phenyl-1H-Indole

Schizophrenia is a severe psychiatric illness, and approximately 1% of the world population is affected [51]. It has been suggested [52] that one important feature of atypical antipsychotic drugs is their relative affinities at serotonin 5-HT$_2$ and dopamine D$_2$ receptors, and that affinity higher at the former than the latter led to an atypical profile. Thus developing very high-affinity, selective, and bioavailable h5-HT$_{2A}$ receptor antagonists is an efficient potential approach to the disease.

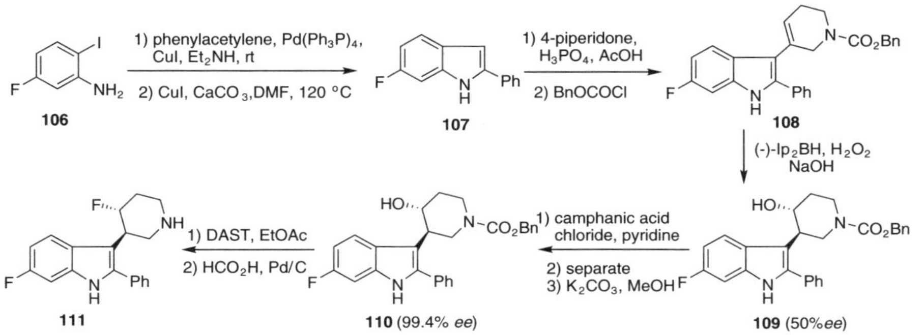

SCHEME 5.13 Synthesis of 3-(4-fluoropiperidin-3-yl)-2-phenylindole **111**.

3-(4-Fluoropiperidin-3-yl)-2-phenylindole **111** was proved to be a high-affinity, selective, and orally bioavailable h5-HT$_{2A}$ receptor antagonist [53] The synthesis of **111** commenced with 2-iodo-5-fluoroaniline **106**, which was condensed with phenylacetylene under palladium catalysis, then cyclized with copper catalysis to give 6-fluoro-2-phenylindole **107** in 76% yield over two steps (Scheme 5.13). Condensation of **107** with 4-piperidone under acidic conditions gave the tetrahydropyridine, which was protected as its benzyloxycarbonyl derivative **108**. Hydroboration of the double bond with bis-isopinocampheylborane ((−)-Ip$_2$BH) gave the *trans* secondary alcohol **109** in 50% *ee*, whose enantiomeric excess was increased to 99.4% *ee* by formation of the camphanate ester using the acid chloride derived from (1R)-(+)-camphanic acid, separation of diastereosiomers by chromatography, and hydrolysis of the ester back to the alcohol. The alcohol having been obtained with high *ee*, **110** was treated with diethylaminosulfur trifluoride (DAST), giving the 3-indolyl-4-fluoropiperidine, which then through deprotection via hydrogenation gave the target molecule **111**.

5.3.1.10 LY-503430

LY-503430 {4′-[(1S)-1-fluoro-1-methyl-2-[(1-methylethyl)sulfonyl]amino]ethyl]-N-methyl-(1,1′-biphenyl)-4-carboxamide, **120**}, a novel α-amino-3-hydroxy-5-methylisoxazole-4-propionic acid (AMPA) receptor potentiator with functional, neuroprotective, and neurotrophic effects in rodent models of Parkinson disease, is an ampakine drug developed by Eli Lilly [54]. LY-503430 produces both nootropic and neuroprotective effects, reducing brain damage caused by 6-hydroxydopamine or MPTP (1-methyl-4-phenyl-1,2,3,6-tetrahydropyridine) and also increasing the levels of the neurotrophic factor brain-derived neurotrophic factor (BDNF) in the brain, particularly in the substantia nigra, hippocampus, and striatum [55]. LY-503430 is orally active, and its main application is the treatment of Parkinson disease, although it has also been proposed in the treatment of Alzheimer disease, depression, and schizophrenia [56].

The first step in the synthesis of LY-503430 was a Wittig olefination of 4-iodoacetophenone **112** with $Ph_3P= CH_2$. The resultant alkene was subjected to the treatment with in situ-generated FBr mixed halogen to afford the fluoro-bromo adduct **113** in 81% yield over two steps (Scheme 5.14). Reaction of **113** with potassium phthalimide **114** in DMF yielded the compound **115** in 85% yield. Cleavage of the phthalimide **115** with methylamine gave the free amine **117** along with the by-product **116**, and the amine **117** was purified by a diastereomeric salt resolution strategy. The resulting diastereomeric salt **118** was converted to its free base via treatment with aqueous NaOH, and the resulting free amine was converted to compound **119** with the following procedure sequences: 1) reaction with $^i PrSO_2Cl/TEA$; 2) Suzuki coupling with 4-carboxybenzeneboronic acid with Pd black as a catalyst to establish the biphenyl linkage; and 3) acidification of the resulting clarified mixture slowly with acetic acid to induce crystallization of a very high purity. Finally, the condensation of the resulting acid **119** with methylamine provided the crude LY-503430, which was purified by recrystallization from acetone-water (1/4) [57].

5.3.1.11   6-FUDCA

Bile acids are steroid acids found predominantly in the bile of mammals, and these acids are amphiphilic compounds synthesized in the liver from cholesterol. They play an important role in solubilizing cholesterol in bile and in the overall digestive process through the formation of micelles. Deconjugation and dehydroxylation of these bile acids lead to the formation of secondary bile acids, which are implicated in colonic carcinogenesis. Fluorination of these bile acids at the 6α-position prevents bacterial dehydroxylation, and this concept led to the identification of 6α-fluoroursodeoxycholic acid **129** (6-FUDCA) as a potential agent for the prevention and treatment of colorectal cancer [58].

The synthesis of 6-FUDCA started from commercially available chenodeoxycholic acid **121** (Scheme 5.15). Oxidation of the C7 hydroxyl group of compound **121** with NaOCl produced the ketone **122**, which was converted to the silyl enol ether intermediate **123** via treatment with TMSCl/NaI. Fluorination of **123** with Selectfluor yielded the fluoroketone **124**, which was isomerized to the thermodynamically more stable 6α isomer **125** by treatment with NaOMe/MeOH. Esterification of acid **125** with TMSCl/MeOH afforded the ester **126**, which was further protected as the acetate with Ac_2O/DMAP to provide compound **127**. Reduction of the ketone **127** was accomplished by means of $PtO_2/H_2/CH_3COOH$ at 40 psi, and compound **128** was obtained in 97% yield. Finally, mesylation of the resulting second hydroxyl in compound **128** with methanesulfonic anhydride produced the corresponding mesylate in high yield, which was subjected to treatment with superoxide anion (KO_2/DMSO) in the presence of the chelator TDA-1 to give the desired displacement product **129** [59].

5.3.1.12   DA-125

Adriamycin is a widely used and highly valued antineoplastic agent, but chronic treatment is limited by cardiotoxicity. Galarubicin hydrochloride **136** (DA-125)

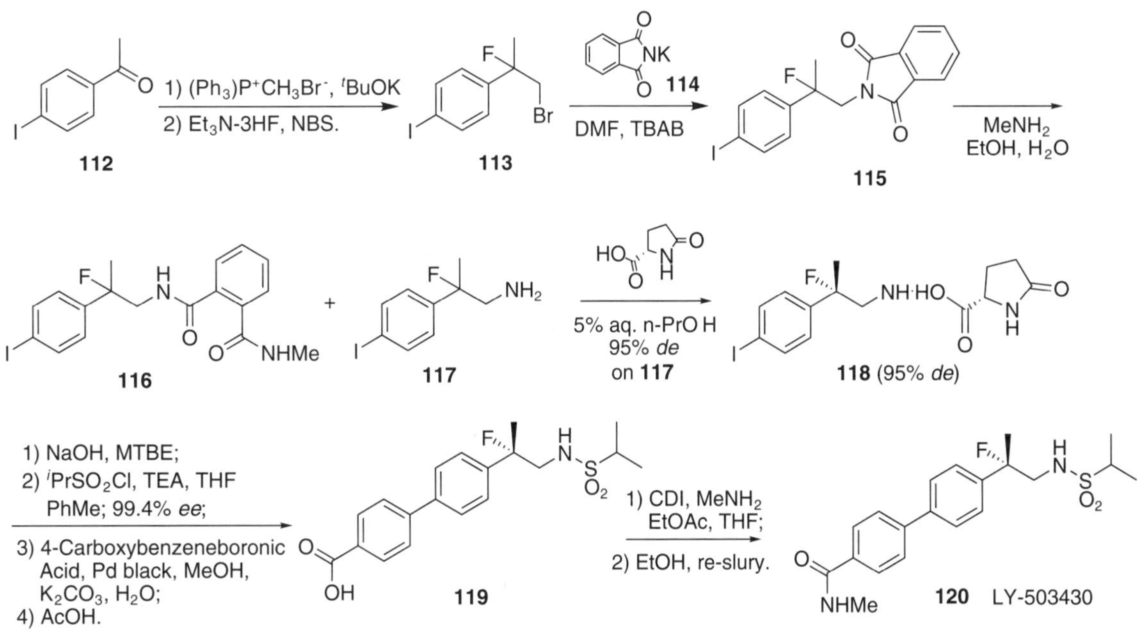

SCHEME 5.14 Synthesis of LY-503430 (**120**).

225

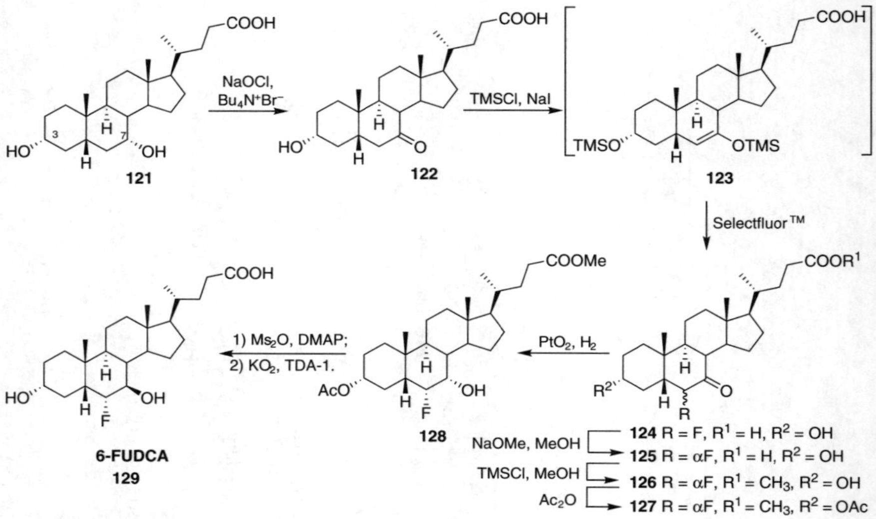

SCHEME 5.15 Synthesis of 6-FUDCA (**129**).

is a new anthracycline antitumor antibiotic derived from adriamycin. It is introduced and currently by Dong-A Pharmaceutical Company in Korea and under clinical phase II trials. The mechanism of action of this new drug lies in inhibition of nucleic acid synthesis through intercalation with DNA. Preclinical studies suggest that DA-125 may have greater pharmacological activity and less cardiac toxicity and skin irritation than adriamycin [60].

The synthesis of DA-**125** is outlined in Scheme 5.16. The coupling of commercially available daunomycinone **132** with 3,4-di-O-acetyl-2,6-dideoxy-2-fluoro-α-L-talopyranosyl bromide **131**, which was obtained from 1,3,4-tri-O-acetyl-derivative **130** via treatment with HBr/HOAc, was performed by a Koenings–Knorr type of reaction to produce the daunomycin derivative **133**. Deprotection of **133** with aqueous NaOH gave the corresponding derivative **134**, which was subsequently treated with Br_2/CH_2Cl_2 and acetone to afford the bromo derivative **135**. Finally, condensation of **135** with N-Boc-β-alanine sodium salt with acetone-water as a solvent gave the protected β-alaninate derivative, which was further treated with hydrochloride in ether to obtain DA-125 **136** [61].

5.3.2 Difluorinated Drugs

5.3.2.1 Gemcitabine

Gemcitabine **141** is a nucleoside analog that exhibits antitumor activity. It was marketed as Gemzar by Eli Lilly in 1996. Chemically, gemcitabine is a

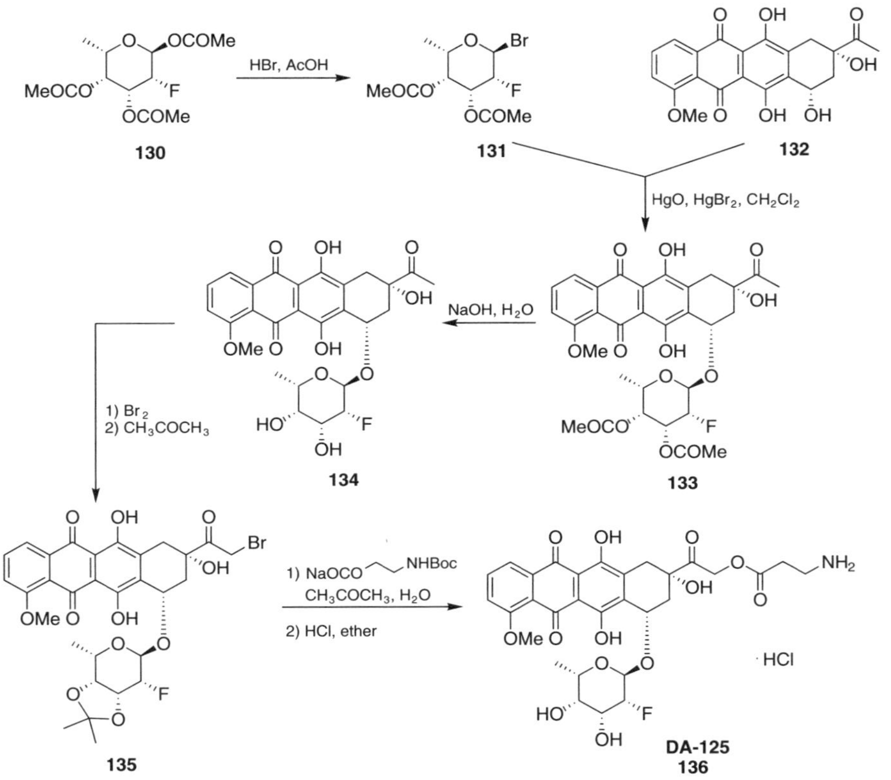

SCHEME 5.16 Synthesis of DA-125 (**136**).

nucleoside analog in which the hydrogen atoms on the 2′ carbons of deoxycytidine are replaced by fluorine atoms. As with fluorouracil and other analogs of pyrimidines, the drug replaces one of the building blocks of nucleic acids, in this case cytidine, during DNA replication. The process arrests tumor growth, as new nucleosides cannot be attached to the "faulty" nucleoside, resulting in apoptosis. Gemcitabine in combination with cisplatin [62], paclitaxel [63], and carboplatin [64] was indicated for the first-line treatment of patients with inoperable, locally advanced (stage IIIA or IIIB), or metastatic (stage IV) non-small cell lung cancer, patients with metastatic breast cancer after failure of prior anthracycline-containing adjuvant chemotherapy, and patients with advanced ovarian cancer that had relapsed at least 6 months after completion of platinum-based therapy, respectively.

The synthesis of gemcitabine started from readily available (R)-glyceraldehyde acetonide **137** [65]. Addition of Reformatsky reagent $BrCF_2CO_2Et$ on **137** in the

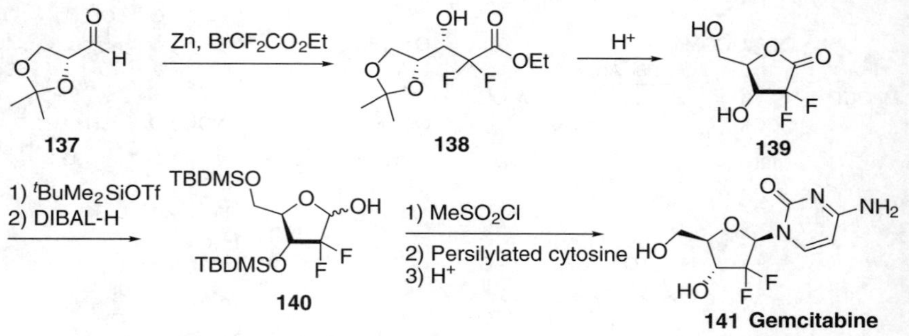

SCHEME 5.17 Synthesis of gemcitabine (**141**).

presence of zinc powder afforded ester **138**, which was hydrolyzed with acid to give the ring-closure product **139**. Protection of the diol with TMSOTf and then reduction of the lactone with DIBAL-H produced the hemiacetal **140**. After mesylation of compound **140** with MsCl, condensation of the resultant mesylate with the persilylated cytosine afforded the fluorinated nucleoside, which was finally deprotected by treatment with acid to give gemcitabine **141**. In 1992, Chou et al. improved the method by utilizing the same synthetic route, but selecting Bz over TBDMS as the protecting group for the hydroxyl groups in the compound **139** [66]. (Scheme 5.17)

5.3.2.2 Lubiprostone

Constipation, defined as infrequent and/or unsatisfactory defecation, is considered secondary in the presence of a recognizable cause or idiopathic or functional when no cause can be identified. Conventional therapeutic approaches for the treatment of constipation include dietary and lifestyle modifications and exercise and, if all else fails, administration of laxatives. Lubiprostone **152**, developed by Sucampo, is a bicyclic fatty acid (prostaglandin E_1) and Cl^- channel opener that increases intestinal water secretion and intestinal fluid Cl^- concentration without altering Na^+ or K^+ concentrations. Lubiprostone was approved by the FDA on January, 2006 for the treatment of constipation [67].

Lubiprostone can be obtained as outlined in Scheme 5.18 [68]. The reaction of aldehyde **142** with fluorinated phosphonate **143** in the presence of thallium ethoxide gave the unsaturated difluoroketone **144**, which was reduced with H_2 over Pd/C to afford the saturated ketone **145**. Reduction of **145** with $NaBH_4$ provided the alcohol **146**, which was further reduced with DIBAL-H to give the lactol **147**. Treatment of **147** with 4-carboxybutyl triphenylphosphonium bromide **148** in the presence of t-BuOK yielded the prostaglandin derivative **149**, which was esterified with benzyl bromide and DBU to give the benzyl ester **150**. Oxidation of **150** with CrO_3/pyridine followed by treatment with AcOH gave compound **151**. Finally, compound **151** was subjected to H_2 over Pd/C to give lubiprostone **152**.

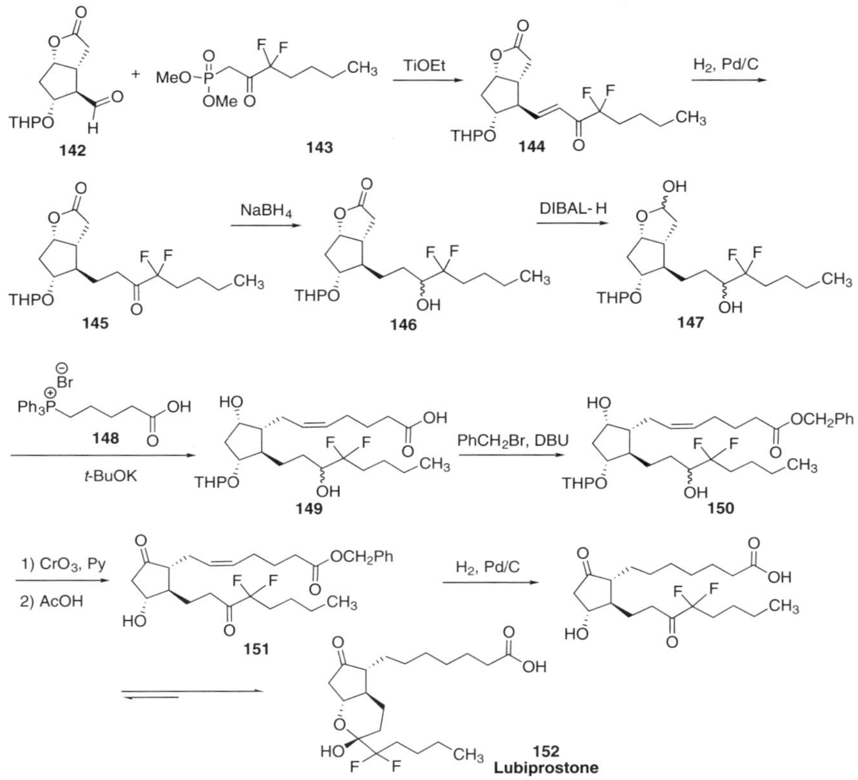

SCHEME 5.18   Synthesis of lubiprostone (**152**).

5.3.3   Trifluoromethylated Drugs

5.3.3.1   Efavirenz

Efavirenz **162** [(*S*)-(−)-6-chloro-4-(cyclopropylethynyl)-4-(trifluoromethyl)-2,4-dihydro-1H-3, 1-benzoxazin-2-one, brand names Sustiva and Stocrin] is a non-nucleoside reverse transcriptase inhibitor (NNRTI) and is used as part of highly active antiretroviral therapy (HAART) for the treatment of human immunodeficiency virus (HIV) type 1. Efavirenz was developed by Merck and approved by the FDA on September 21, 1998, making it the fourteenth approved antiretroviral drug. Efavirenz is also used in combination with other antiretroviral agents as part of an expanded postexposure prophylaxis regimen to reduce the risk of HIV infection in people exposed to a significant risk.

Synthesis of efavirenz is outlined in Scheme 5.19 [69]. The acylation of 4-chloroaniline **153** with pivaloyl chloride (*t*BuCOCl) by means of Na$_2$CO$_3$ in toluene gave the expected anilide **154**, which was acylated with CF$_3$CO$_2$Et by means of *n*-BuLi in THF, yielding, after hydrolysis with HCl, 2′-amino-5′-chloro-2,2,2-trifluoroacetophenone **155**. The benzylation of the amine **155** with

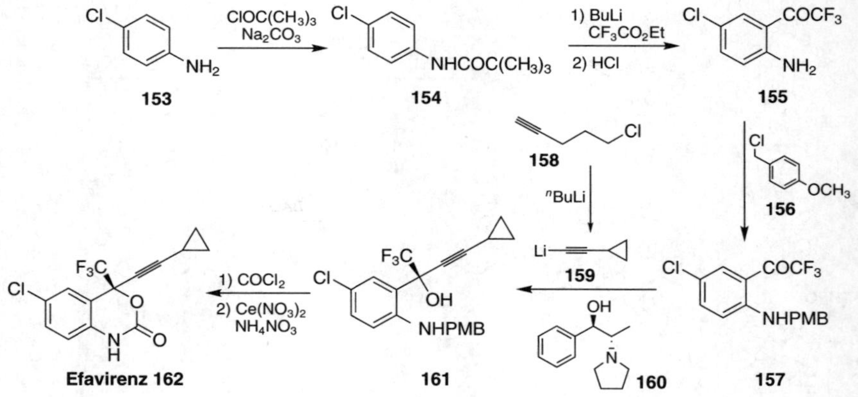

SCHEME 5.19   Synthesis of efavirenz (**162**).

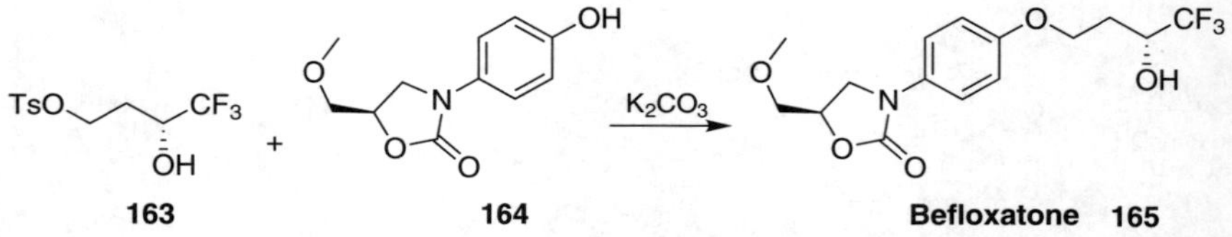

SCHEME 5.20   Synthesis of befloxatone (**165**).

4-methoxybenzyl chloride **156** in basic alumina afforded the protected acetophenone **157** in good yield. Compound **157** was regioselectively condensed with cyclopropylacetylenyl lithium **159** (in situ prepared by cyclization of 5-chloro-1-pentyne **158** with *n*-BuLi in cyclohexane) in the presence of (1*R*, 2*S*)-1-phenyl-2-(1-pyrrolidinyl)-1-propanol **160** to exclusively provide the (*S*)-isomer of the tertiary alcohol **161**. Finally, the cyclization of **161** with phosgene and triethylamine or K_2CO_3 in toluene-THF yielded the benzoxazinone, which was further deprotected with $Ce(NO_3)_2$ in acetonitrile-water to give efavirenz **162**.

5.3.3.2 Befloxatone

Befloxatone **165** (MD-370503), a novel oxazolidinone derivative developed by Sanofi-Aventis, is a potent selective and reversible monoamine oxidase A (MAO-A) inhibitor in vitro ($KiA = 1.9–3.6$ nM) and ex vivo (ED_{50} MAO-A $= 0.02$ mg/kg p.o.). Befloxatone is a relatively selective monoamine oxidase inhibitor and has been shown to have antidepressant activity in various animal models. Befloxatone is obtained by condensation of 1,1,1-trifluoro-4-(tosyloxy)-2(*R*)-butanol **163** with 3-(4-hydroxyphenyl)-5(*R*)-(methoxymethyl)oxazolidin-2-one **164** by means of K_2CO_3 in hot DMF (Scheme 5.20) [70].

Synthesis of trifluoromethyl substituted butanol derivative **163** is described in Scheme 5.21. The reduction of 4,4,4-trifluoro-2-oxobutyric acid ethyl

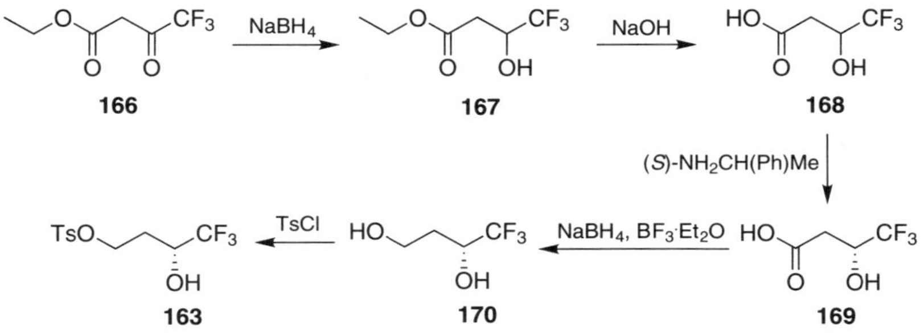

SCHEME 5.21   Synthesis of chiral trifluoromethylated alcohol (**163**).

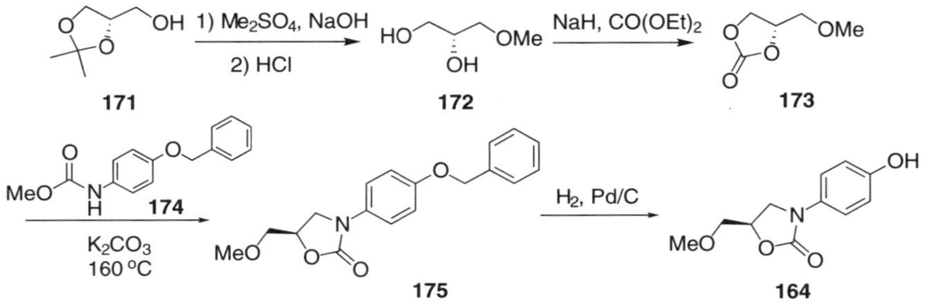

SCHEME 5.22   Synthesis of oxazolidinone (**164**).

ester **166** with NaBH$_4$ gave 4,4,4-trifluoro-2-hydroxybutyric acid ethyl ester **167**, which was hydrolyzed with NaOH/EtOH, yielding the corresponding acid **168**. The optical resolution of **168** was accomplished via treatment with 1-(S)-phenylethylamine in hot ethanol, and the 3-(R)-hydroxy enantiomer **169** was afforded. Reduction of **169** with NaBH$_4$/BF$_3$·Et$_2$O provided 4,4,4-trifluorobutane-1,3-(R)-diol **170**. Finally, this compound was monotosylated by means of TsCl and DMAP in pyridine to afford the intermediate **163**.

The access to segment **164** commenced with methylation of 4-(S)-(hydroxymethyl)-2,2-dimethyl-1,3-dioxolane **171** with Me$_2$SO$_4$/NaOH, and the diol **172** was afforded after hydrolysis with HCl (Scheme 5.22). The cyclization of **172** with CO(OEt)$_2$ by means of NaH yielded 4-(S)-(methoxymethyl)-1,3-dioxolan-2-one **173**. The condensation of **173** with N-(4-benzyloxyphenyl)carbamic acid methyl ester **174** (obtained by reaction of 4-benzyloxyaniline with methyl chloroformate) via treatment with K$_2$CO$_3$ at 160°C provided the protected oxazolidinone **175**, which was finally debenzylated by means of Pd/C-mediated hydrogenation to produce the intermediate **164**.

5.3.3.3 Odanacatib

Osteoporosis is a disease characterized by bone loss causing skeletal fragility and an increased risk of fracture. Cathepsin K is a member of the papain

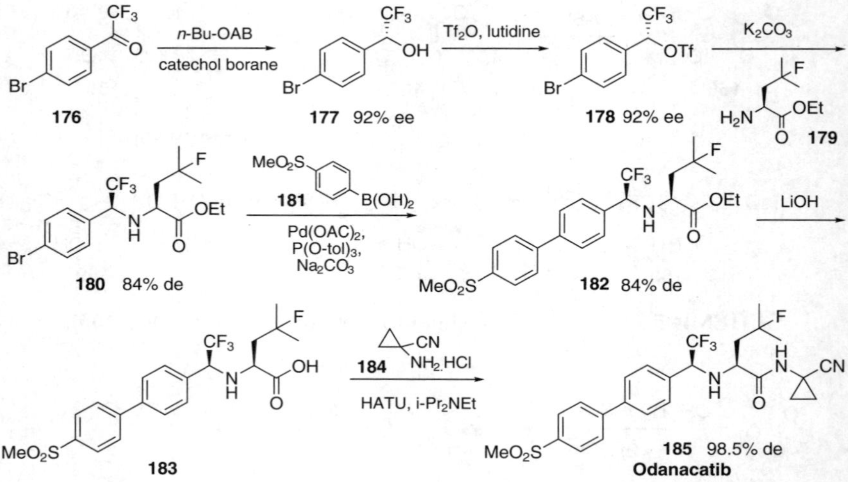

SCHEME 5.23   Synthesis of odanacatib (**185**).

superfamily of cysteine proteases that is abundantly expressed in osteoclasts, the cells responsible for bone resorption. The abundant and selective expression of cathepsin K in osteoclasts has made it an attractive therapeutic target for the treatment of osteoporosis. Odanacatib **185** (MK-0822), developed by Merck, has been identified as a potent and selective inhibitor of cathepsin K and is now in phase III clinical studies for the therapy of osteoporosis [71].

A practical, enantioselective, and chromatography-free synthesis of odanacatib **185** is described in Scheme 5.23 [72]. Chiral alcohol **177** was obtained in a straightforward manner by oxazaborolidine-catalyzed enantioselective reduction of ketone **176**. The key step involves the novel stereospecific S_N2 triflate displacement of a chiral α-trifluoromethylbenzyl triflate **178** with (S)-γ-fluoroleucine ethyl ester **179** to generate the required α-trifluoromethylbenzyl amino stereocenter. The cross-coupling of bromo ester **180** with boronic acid **181** under palladium catalyst gave the desired biaryl **182**. The amidation of acid **183** with cyclopropylaminonitrile **184** in the presence of peptide coupling agents led to odanacatib **185**. The overall synthesis of odanacatib **185** was completed in six steps, in 61% overall yield.

5.4   SYNTHETIC METHODS OF ORGANOFLUORINE COMPOUNDS: ASYMMETRIC ELECTROPHILIC FLUORINATION

The naturally occurring fluorinated organic compounds are rare, and all of the fluorine-containing drugs and polymers are man-made. Promoted by great demand, synthetic methods of organofluorine compounds have made enormous progress. The use of commercially available fluorinated intermediates or building blocks (fluorinated synthons) is the first strategy for the synthesis of fluorinated drugs. As fluorination reagents are readily prepared or are commercially available, recently fluorination reactions to form organofluorine compounds have been used extensively. Fluorination procedures are divided into methods that use electrophilic incorporation of fluorine and those that proceed by nucleophilic attack. The synthetic methods of organofluorine compounds are an extremely broad subject. Because fluorine-containing chiral compounds are mainly prepared by electrophilic fluorination, emphasis will be placed on the development of asymmetric electrophilic fluorination. The fluorinated synthons are not a topic of this chapter.

5.4.1 Nucleophilic Fluorination

The widely used nucleophilic reagents include diethylaminosulfur trifluoride (DAST) and bis(2-methoxyethyl)aminosulfur trifluoride (Deoxofluor) (Fig. 5.15). These reagents transform alcohols into monofluorides and carbonyls into *gem*-difluorides.

5.4.2 Electrophilic Fluorination Agents

Electrophilic fluorination is an important method for incorporation of fluorine atom into organic compounds, and the fluorination mechanism involves the reaction of fluorine cation (F^+) with substrates [73]. Because of the strong electron-withdrawing property of fluorine atom, many practically utilized electrophilic fluorinating reagents (EFRs) are NF-containing compounds.

F_2 was the most early used electrophilic fluorinating reagent for introducing fluorine atom(s) into organic compounds, although its highly poisonous nature and high reactivity significantly limited its utilization [74]. Since the Barton group reported a novel EFR—CF_3OF—in 1968 [75], many EFRs have been successfully developed and applied, for example, $FClO_3$ [76], XeF_2 [77], FNO [78], CF_3OF [79], and NF-containing EFRs. In the 1960s, Banks et al. first proposed that NF-containing compounds could be utilized as EFRs. After that, many NF-containing EFRs have been developed and used because of their advantages (easy preparation, convenient operation, good regio- and stereoselectivity). In the past decades, over 100 NF-containing EFRs have been prepared and utilized. Among these NF-containing EFRs, *N*-fluorobenzene-sulfonimide (NFSI) and 1-chloromethyl-4- fluorodiazoniabicyclo[2.2.2]octane bis(tetrafluoroborate)

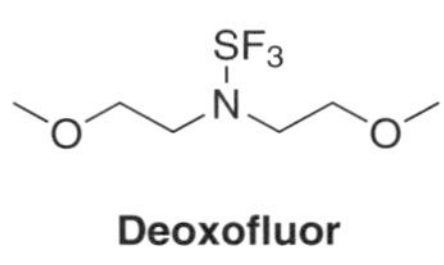

FIGURE 5.15 Two important nucleophilic reagents for fluorination.

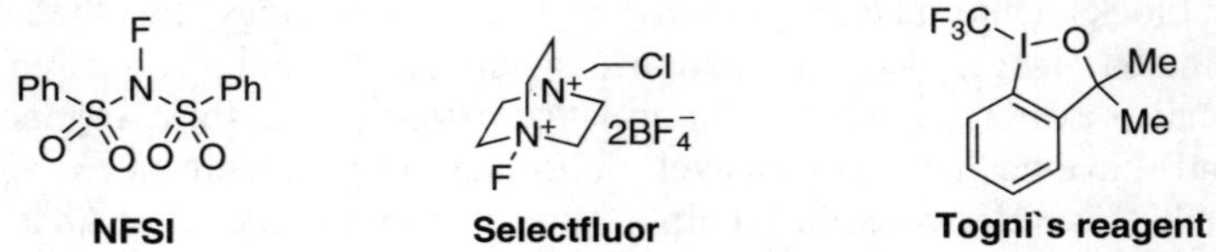

FIGURE 5.16   Three commonly used reagents for fluorination.

(Selectfluor) are commercially available and widely used. Recently, the electrophilic trifluoromethylation agent (Togni's reagent) was reported (Fig. 5.16).

5.4.3 Asymmetric Electrophilic Fluorination with Chiral NF-Containing EFRs

The Lang group, as early as 1988, first synthesized and evaluated the chiral electrophilic N-F fluorinating agents **186a** and **186b** by introducing a fluorine atom into camphorsulfonamide (Scheme 5.24) [80]. They found that asymmetric fluorination of in situ enolated carbanion with **186a** (or **186b**) afforded the chiral α-fluorinated ketone/ester compounds in moderate enantioselectivity. To improve the reactivity and enantioselectivity of camphorsulfonamide-derivated N-F fluorinating agents, Davis and co-workers prepared analogs **186c-d**, which had a chloride atom or methoxy group in the same positions of the camphorsulfonamide skeleton [81]. The reactivity of **186d** was better than **186a/186b**, and reaction with **186d** as fluorinating agent could be performed at −78°C, although enantioselectivity was not significantly increased. Additionally, it should be mentioned that

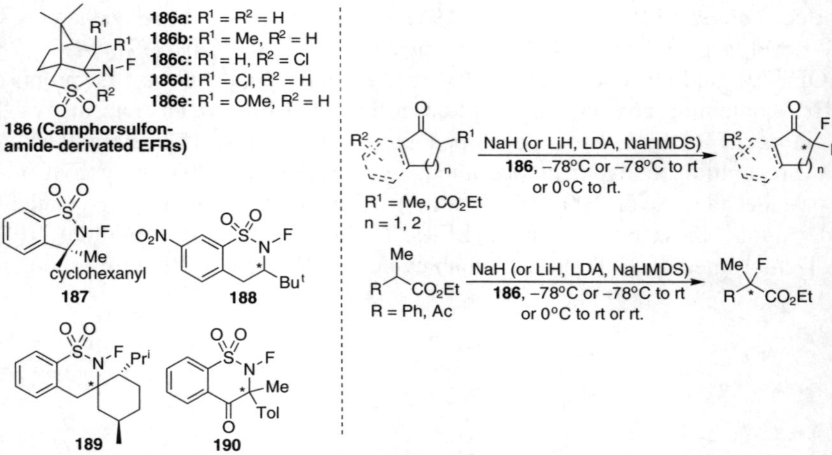

SCHEME 5.24   Examples of chiral sulfonamide-based fluorinating reagents and their applications.

the Takeuchu group and the Liu group also prepared and evaluated other cyclic sulfonamide-containing N-F electrophilic fluorinating agents **187–190** [82].

The Takeuchi group, in 2000, discovered that fluorination of carbanions with Selectfluor occurred in a highly enantioselective manner in the presence of cinchona alkaloid derivatives, such as dihydroquinine 4-chlorobenzoate **191** (DHQB) or dihydroquinidine acetate **192** (DHQDA) (Scheme 5.25) [83]. Cyclic silyl enol ethers and acyclic α-cyano-aryl acetates were the suitable substrates for this enantioselective fluorination, leading to corresponding chiral fluorinated products in moderate to high yields (81%–99%) and in acceptable *ee*. It should be noted that the cinchona alkaloid derivative and Selectfluor must be mixed first for generation of the combination in situ before addition of substrates to achieve enantioselective fluorination. In light of this phenomenon, Takeuchi et al. proposed that this novel enantioselective fluorination was mediated by the chiral N-fluoro species, NF-DHQBBF$_4$ (**193**) and NF-DHQDABF$_4$ (**194**), in situ generated by "fluorine transfer" of cinchona alkaloid by Selectfluor.

Takeuchi's proposed mechanism of asymmetric fluorination mediated by NF-DHQBBF$_4$ (**193**) and NF-DHQDABF$_4$ (**194**) was soon proven by the Cahard group, who designed, synthesized, and evaluated a series of enantioselective electrophilic fluorinating agents: F-QD-BF$_4$ (**195**), F-CN-BF$_4$ (**196**), F-QN-BF$_4$ (**197**), and FCD-BF$_4$ (**198**) (Scheme 5.26) [84]. These fluorinating agents were prepared via a one-step Selectfluor-mediated transfer-fluorination on the naturally occurring cinchona alkaloids, and they exhibited asymmetric induction up to 61% *ee* on fluorination of enolates and silyl enol ether of 2-methyl-1-tetralone. For enolate substrates, Cahard et al. found that F-CN-BF$_4$ appeared less substrate-dependent than other chiral fluorinating agents. Moreover, for silyl enol ether substrate, asymmetric induction was strongly dependent on the reaction conditions, and addition of NaOH to the fluorinating agent considerably improved both the reactivity and the stereoselectivity. Additionally, Cahard and

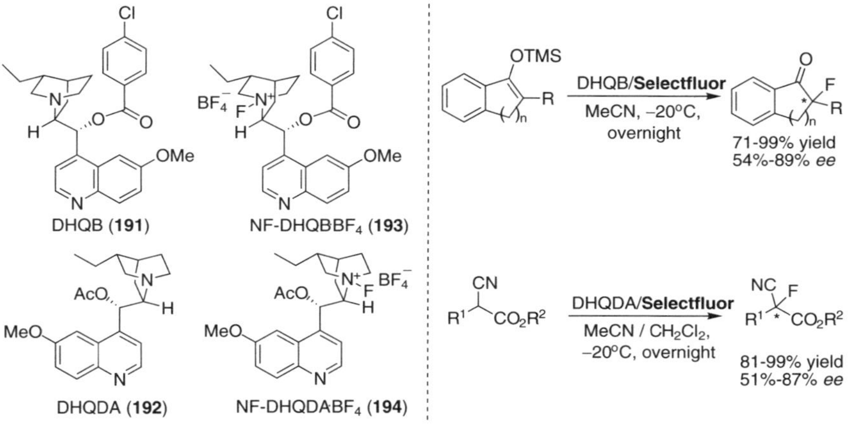

SCHEME 5.25 Cinchona-based fluorinating reagents and their usages.

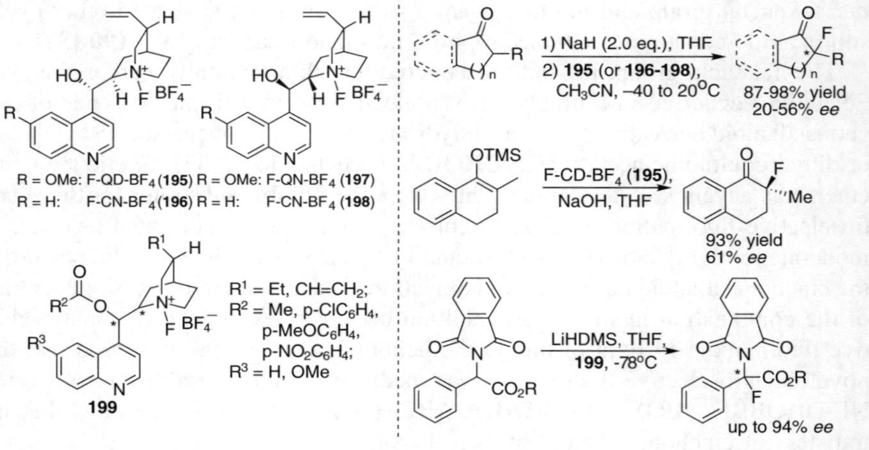

SCHEME 5.26 The electrophilic fluorinating reagents **195–199** and their applications.

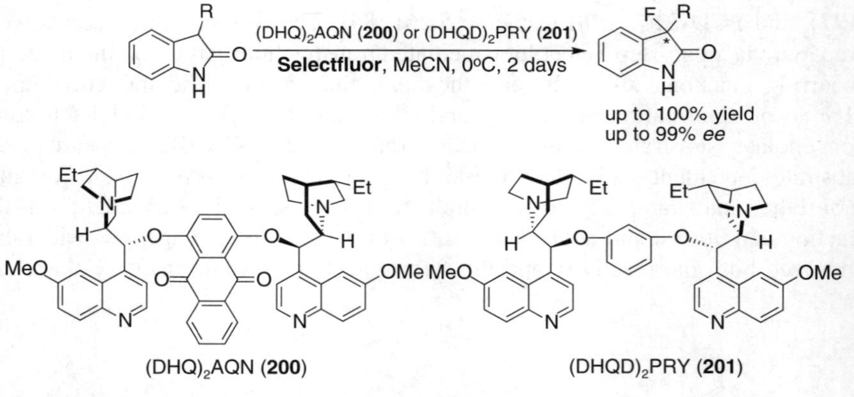

SCHEME 5.27 (DHQ)$_2$AQN (**200**) and (DHQD)$_2$PRY (**201**) with C_2 symmetry.

co-workers also investigated other fluorinating agents **199** derived from some modified cinchona alkaloids [85]. They found that asymmetric fluorination of N-protected-α-phenylaminoglycine derivatives with these N-F fluorinating agents preceded well with an enantiomeric excess up to 94%. Protection of hydroxyl groups in cinchona alkaloids was necessary to obtain high enantioselectivities.

Shibata et al. developed the alternative Selectfluor combinations derived from bis-cinchona alkaloids (DHQ)$_2$AQN (**200**) and (DHQD)$_2$PYR (**201**) (Scheme 5.27) [86]. Fluorination of oxindole derivatives with combinations of **200**-Selectfluor or **201**-Selectfluor furnished the corresponding chiral 3-fluorooxindoles in moderate to excellent *ee* (up to 99% *ee*).

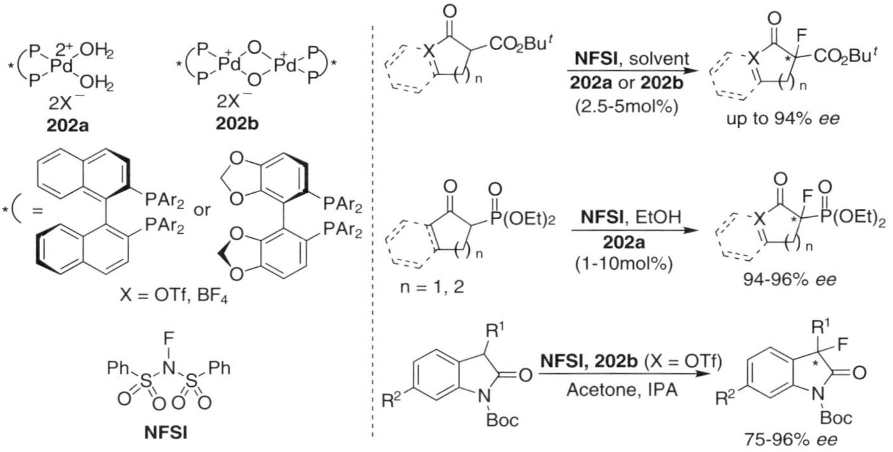

SCHEME 5.28   Chiral ligand-promoted asymmetric electrophilic fluorination.

5.4.4 Asymmetric Electrophilic Fluorination with Chiral Ligands

In the past years, chemists have developed efficient chiral ligands to realize the enantioselective catalytical fluorination mediated by NFSI. Sodeoka and co-workers found that, with NFSI as fluorinating reagent, chiral organic phosphine ligands **202a** and **202b** could efficiently catalyze the enantioselective fluorination of various β-ketoesters, β-ketophosphonates, and oxindoles (Scheme 5.28) [87].

In 2004, the Ma group investigated in detail the Cu(OTf)$_2$-bis(oxazoline)-catalyzed enantioselective electrophilic fluorination of β-ketoester with NFSI as fluorinating reagent (Scheme 5.29) [88]. When chiral complex **203a**/Cu(OTf)$_2$ was used as catalyst, NFSI could enantioselectively fluorinate β-ketoesters, leading to α-fluorinated products in high yields and moderate *ee*. Notably, they found that addition of 1,1,1,3,3,3-hexafluoroisopropanol (HFIP) was crucial for achieving high enantioselectivity. Another interesting report came from Shibata and co-workers, who performed a similar reaction using (S)-Ph-Box **203b** as chiral ligand [89]. Treatment of β-ketoester with NFSI in the presence of complex **203b**/Cu(OTf)$_2$ gave a product of (S)-configuration.

Iwasa [90] designed and prepared the novel optically active N, N, N-tridentate ligands **205** and **206**, which possess both binaphthy axial chirality and carbon-centered chirality, and successfully used the combination of ligands **205** (**206**)-Lewis acids to achieve the catalytic asymmetric α-fluorination of β-keto esters **204** with NFSI as fluorinating reagent (Scheme 5.30). They found that both the yields and enantioselectivities were very high when Ni(ClO$_4$)$_2$ or Mg(ClO$_4$)$_2$ was utilized as the Lewis acid.

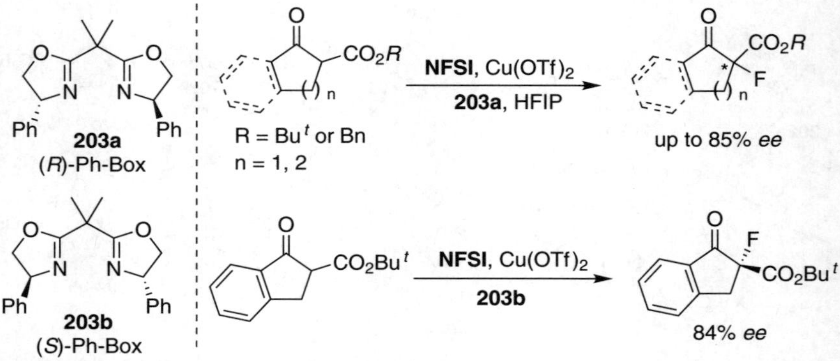

SCHEME 5.29 Asymmetric fluorination catalyzed by Cu(OTf)$_2$-bis(oxazoline).

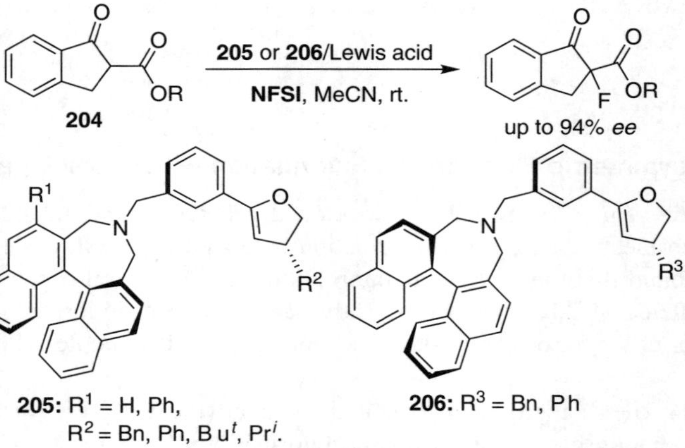

SCHEME 5.30 Combination of ligand **205/206** with Lewis acid in catalytic asymmetric fluorination.

Shibata et al. reported that the catalytic enantioselective fluorination of β-keto esters, oxindoles, and malonates could be realized with extremely high enantiose-lectivities by the use of (R, R)-DBFOX-Ph (**207**)-Lewis acid complex with NFSI utilized as fluorinating reagent (Scheme 5.31) [91].

In 2008, the Lectka group found that catalytic, asymmetric α-fluorination of acid chlorides could be achieved by means of dual metal-ketene enolate acti-vation using the combination of benzoylquinidine (BQd) -NFSI (Scheme 5.32) [92]. (1,3-dppp)NiCl$_2$ and $trans$-(PPh$_3$)PdCl$_2$ were found to be suitable cocat-alysts, efficiently leading to chiral α-fluorinated derivatives. Metal-free reaction also provided the fluorinated compound in high ee, but in low yield. The reaction mechanism involved NFSI-mediated fluorination of the dually activated enolate,

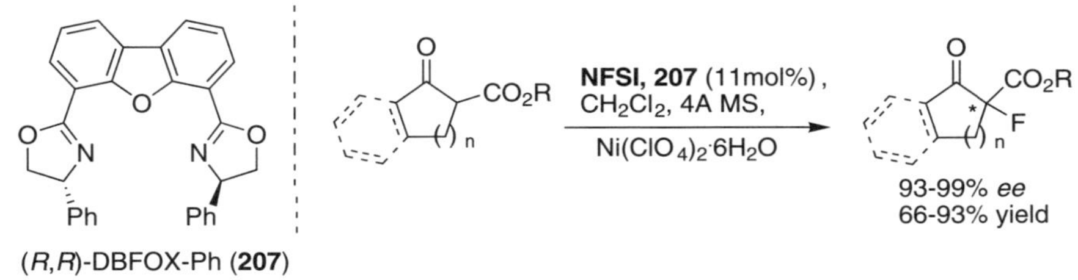

SCHEME 5.31 Combination of DBFOX-Ph with Lewis acid in catalytic asymmetric fluorination.

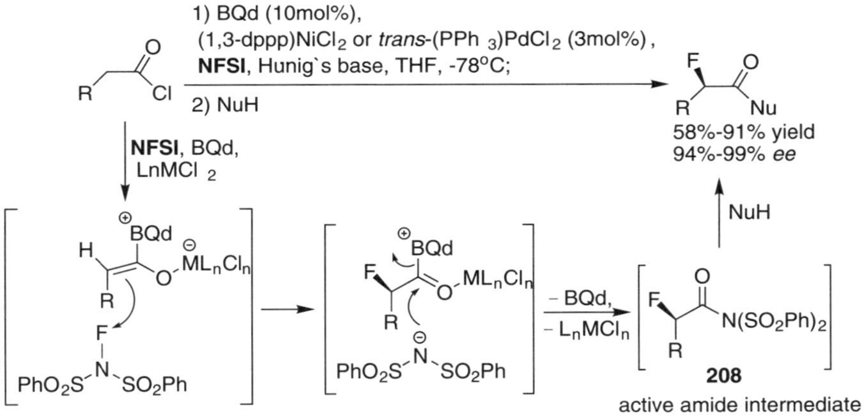

SCHEME 5.32 Fluorination by BQD + NFSI in combination with $LnMnCl_2$.

generating an acyl ammonium salt, which reacted with the liberated dibenzene-sulfonimide anion to form one active amide intermediate **208**. The transacylation of amide intermediate with NuH yielded the desired fluorinated product in high *ee*.

With Selectfluor as fluorinating reagent and chiral titanium complex **209a** (or **209b**) as catalyst, Togni and Hintermann, in 2000, achieved the asymmetric fluorination of β-ketoesters (Scheme 5.33) [93]. Although monosubstituted β-ketoesters did not react with a saturated solution of Selectfluor in CH_3CN at room temperature, addition of Lewis acids (**209a** or **209b**, 5 mol%) efficiently catalyzed the reaction.

Several years later, Togni and co-workers also reported other catalytic enantioselective reactions leading to chiral α-chloro-α-fluoro-β-ketoesters (Scheme 5.34) [94]. The reactions were performed with **209a** (or **209b**) as catalyst, Selectfluor as fluorinating reagent, and NCS as chlorination reagent. The most important finding of this study was that the sequence of addition of the halogenating reagents determined the sense of chiral induction. In the first

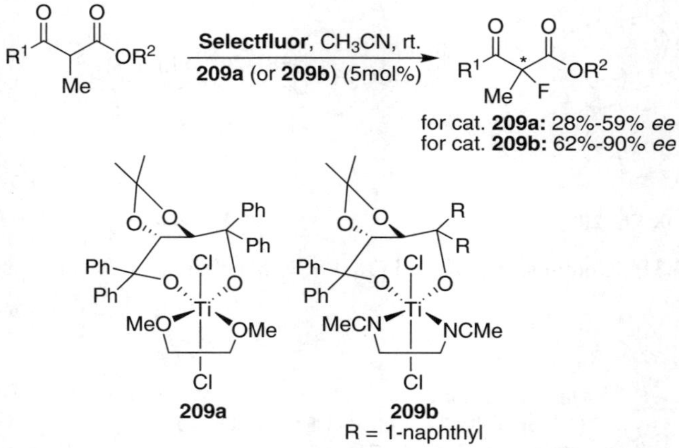

SCHEME 5.33 Use of chiral titanium complex as catalyst.

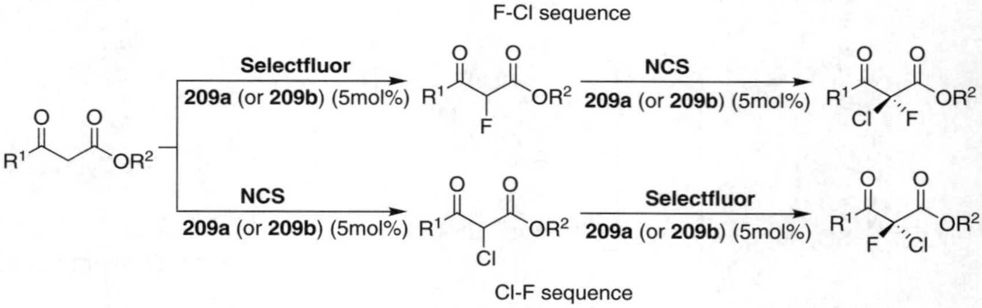

SCHEME 5.34 Togni's system for fluorination.

protocol (F-Cl), a fluorination reaction with Selectfluor in the presence of catalyst **209a** (or **209b**) was carried out. The intermediate was then enantioselectively chlorinated by NCS also in the presence of catalyst **209a** (or **209b**). In the second protocol (Cl-F), the halogenation sequence was inverted. However, the enantioselectivities were less than 65% *ee*.

5.4.5 Asymmetric Electrophilic Fluorination with Organocatalysis

Proline derivative **210** was found to be efficient for the asymmetric catalytic NFSI-mediated fluorination of aldehydes with good enantioselectivities (Scheme 5.35) [95], In addition, the Barbas group screened dozens of catalysts for NFSI-mediated asymmetric fluorination of aldehydes and found that imidazolidinone derivative **211** gave good enantioselectivity [96]. Almost at the same time, MacMillan described that the enantiomer **212** of **211** asymmetrically catalyzed NFSI-mediated fluorination of aldehydes, leading to

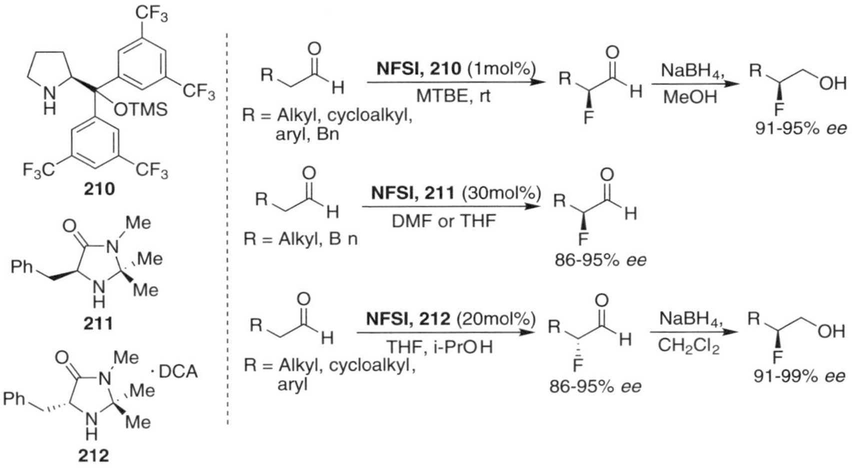

SCHEME 5.35   Organocatalytic fluorination.

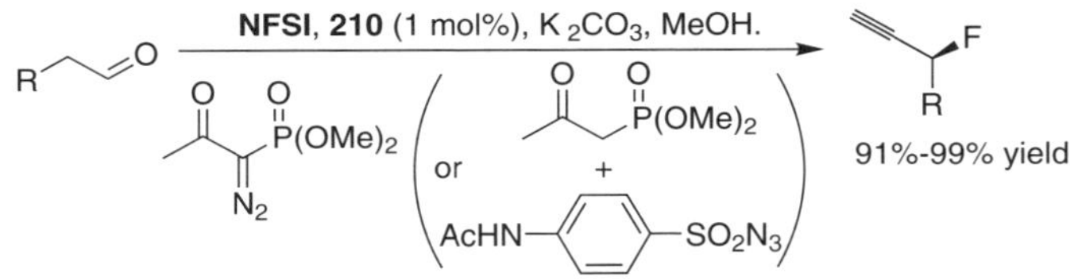

SCHEME 5.36   Organocatalytic fluorination by Jørgensen.

highly enantiomeric excess of 2-fluorinated alcohol derivatives after reduction with NaBH$_4$ [97].

Using the aforementioned proline derivative **210** as the catalyst, Jørgensen and co-workers also presented a simple, direct one-pot organocatalytic approach to the formation of optically active propargylic fluorides (Scheme 5.36) [98]. The approach was based on the organocatalytic α-fluorination of aldehydes and trapping and homologation of the intermediate, providing optically active propargylic fluorides in good yields and enantioselectivities up to 99%.

5.4.6   Asymmetric Trifluoromethylation

In 2009, the MacMillan group described a conceptually new approach to the asymmetric α-trifluoromethylation of aldehydes via the successful merger of enamine and organometallic photoredox [99]. The mechanism of photoredox organocatalysis relied on the propensity of electrophilic radicals (derived from the reduction of an alkyl halide by a photoredox catalyst) to combine with facially biased enamine intermediate (derived from aldehydes and chiral amine catalyst). After screening, they found that the combination of organocatalyst

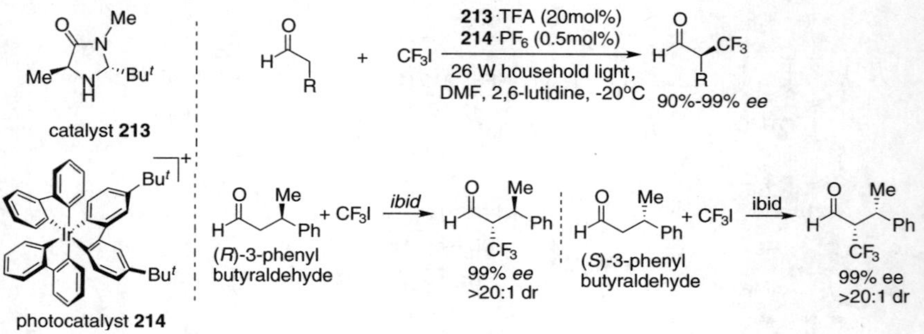

SCHEME 5.37 Photoredox organocatalytic fluorination.

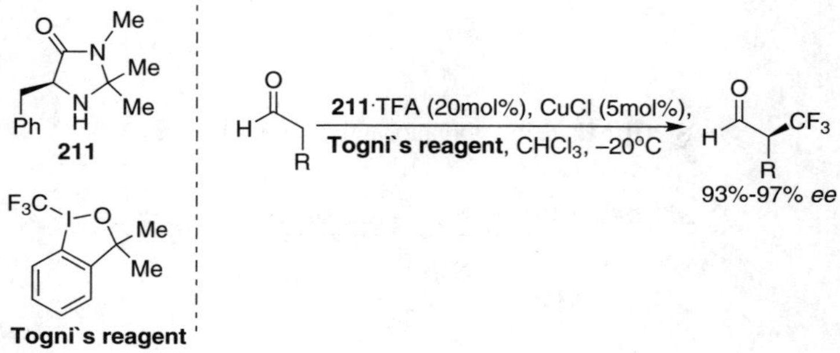

SCHEME 5.38 Fluorination using Togni's reagent.

213 with photoredox catalyst **214** delivered the best enantioselectivities (up to 99% *ee*) for the asymmetric α-trifluoromethylation of aldehydes with CF$_3$I under household light (Scheme 5.37). The mild redox conditions used were compatible with a wide range of functional groups including ethers, esters, amines, carbamates, and aromatic rings. Moreover, significant variation in the steric demand of the aldehyde substituents could be accommodated without loss in enantiocontrol. The most highlighted examples were α-trifluoromethylations of chiral 3-phenyl-butyraldehyde, where exposure of enantiopure (*R*)-3-phenyl-butyraldehyde to catalyst **213** resulted in the diastereoselective production of the *anti*-α,β-disubstituted isomer, while the use of (*S*)-3-phenyl-butyraldehyde with the same amine catalyst afforded the corresponding *syn* adduct with high fidelity.

Very recently, MacMillan and Allen also reported a new mechanism (nonphotolytic) approach to the enantioselective α-trifluoromethylation of aldehydes via the merge of Lewis acid and organocatalysis with an electrophilic trifluoromethylation reagent—Togni's reagent (Scheme 5.38) [100]. Through this methodology, enantioenriched α-trifluoromethylated aldehydes were generated under mild reactions using commercially available, bench-stable reagents and catalyst without the requirement of a light source. Through screening of a series of Lewis acids, FeCl$_2$ and CuCl were found to be both high yielding and enantioselective when the combination of the aforementioned imidazolidinone catalyst **211**-Togni's reagent was used. These mild Lewis acid-organocatalytic conditions tolerated a wide range of functional groups, including aryl rings, ethers, esters, carbamates, and imides. It should be noted that catalyst **211** was ineffective in the aforementioned photolytic trifluoromethylation, which further provided the evidence that this protocol did not involve a radical pathway.

REFERENCES

1. O'Hagan, D., Harper, D. B. (1999). Fluorine-containing natural products. *J. Fluorine Chem.* **100**, 127–133.

2. (a) Banks, R., Smart, B., Tatlow, J. (1994). *Organofluorine Chemistry: Principles and Commercial Applications*. New York: Plenum Press,; (b) Ojima, I., McCarthy, J., Welch, J. (1996). *Biomedical Frontiers of Fluorine Chemistry*. Washington, DC: American Chemical Society; (c) Soloshonok, V. A. (2005). *Fluorine-Containing Synthons*. Washington, DC: American Chemical Society.

3. Fried, J., Sabo, E. F. (1953). Synthesis of 17α-hydroxycorticosterone and its 9α-halo derivatives from 11-epi-17α-hydroxycorticosterone. *J. Am. Chem. Soc.* **75**, 2273–2274.

4. Heidelberger, C., Chaudhuri, N. K., Danneberg, P., Mooren, D., Griesbach, L., Duschinsky, R., Schnitzer, R., Pleven, E., Scheiner, J. (1957). Fluorinated pyrimidines, a new class of tumour-inhibitory compounds. *Nature* **179**, 663–666.

5. (a) Bravo, P., Resnati, G., Angeli, P., Frigerio, M., Viani, F., Arnone, A., Marucci, G., Cantalamessa, F. (1992). Synthesis and pharmacological evaluation of enantiomerically pure 4-deoxy-4-fluoromuscarines. *J. Medicinal Chem.* **35**, 3102–3110; (b) Lu, S. F., Herbert, B., Haufe, G., Laue, K. W., Padgett, W. L., Oshunleti, O., Daly, J. W., Kirk, K. L. (2000). Syntheses of (*R*)- and (*S*)-2- and 6-fluoronorepinephrine and (*R*)- and (*S*)-2- and 6-fluoroepinephrine: effect of stereochemistry on fluorine-induced adrenergic selectivities. *J. Medicinal Chem.* **43**, 1611–1619; (c) Tewson, T. J., Stekhova, S., Kinsey, B., Chen, L., Wiens, L., Barber, R. (1999). Synthesis and biodistribution of *R*- and *S*-isomers of [^{18}F]-fluoropropranolol, a lipophilic ligand for the β-adrenergic receptor. *Nucl. Med. Biol.* **26**, 891–896.

6. Park, B. K., Kitteringham, N. R. (1994). Effects of fluorine substitution on drug metabolism: pharmacological and toxicological implications. *Drug Metab. Rev.* **26**, 43.

7. (a) Bondi, A. (1964). Van Der Waals volumes + radii. *J. Phys. Chem.* **68**, 441–451; (b) Lide, D. R. (2009). *Handbook of Chemistry and Physics*. New York: CRC; (c) Liebman, J. F., Greenberg, A. (1986). *Molecular Structure and Energetics: Vol.3. Studies of Organic Molecules*. Weinheim: VCH.

8. de Riggi, I., Virgili, A., de Moragas, M., Jaime, C. (1995). Restricted rotation and NOE transfer: a conformational study of some substituted (9-anthryl)carbinol derivatives. *J. Org. Chem.* **60**, 27–31.

9. (a) Harper, D. B., O'Hagan, D. (1994). The fluorinated natural products. *Natural Product Rep.* **11**, 123–133; (b) Lauble, H., Kennedy, M. C., Emptage, M. H., Beinert, H., Stout, C. D. 1996). The reaction of fluorocitrate with aconitase and the crystal structure of the enzyme-inhibitor complex. *Proc. Natl. Acad. Sci. U. S. A.* **93**,

13699–13703;   (c) Peters, R., Wakelin, R. W., Buffa, P. (1953). Biochemistry of fluoroacetate poisoning the isolation and some properties of the fluorotricarboxylic acid inhibitor of citrate metabolism. *Proc. R. Soc. B Biol. Sci*. **140**, 497–506;   (d) Tamura, T., Wada, M., Esaki, N., Soda, K. (1995). Synthesis of fluoroacetate from fluoride, glycerol, and β-hydroxypyruvate by *Streptomyces cattleya*. *J. Bacteriol*. **177**, 2265–2269.

10. (a) Marais, J. C. S. (1943). The isolation of the toxic principle "K cymonate" from "Gifblaar", *Dichapetalum cymosum*. *Onderstepoort J. Vet. Res*. **18**, 203; (b) Marais, J. S. C. (1944). Monofluoroacetic acid, the toxic principle of &Gifblaar" *Dichapetalum cymosum* (Hook) Engl. *Onderstepoort J. Vet. Res*. **20**, 67–73.

11. Cruciani, G., Carosati, E., De Boeck, B., Ethirajulu, K., Mackie, C., Howe, T., Vianello, R. (2005). MetaSite: Understanding metabolism in human cytochromes from the perspective of the chemist. *J. Medicinal Chem*. **48**, 6970–6979.

12. Clader, J. W. (2004). The discovery of ezetimibe: a view from outside the receptor. *J. Medicinal Chem*. **47**, 1–9.

13. Penning, T. D., Talley, J. J., Bertenshaw, S. R., Carter, J. S., Collins, P. W., Docter, S., Graneto, M. J., Lee, L. F., Malecha, J. W., Miyashiro, J. M., Rogers, R. S., Rogier, D. J., Yu, S. S., Anderson, G. D., Burton, E. G., Cogburn, J. N., Gregory, S. A., Koboldt, C. M., Perkins, W. E., Seibert, K., Veenhuizen, A. W., Zhang, Y. Y., Isakson, P. C. (1997). Synthesis and biological evaluation of the 1,5-diarylpyrazole class of cyclooxygenase-2 inhibitors: identification of 4-[5-(4-methylphenyl)-3-(trifluoromethyl)-1H-pyrazol-1-yl]benzenesulfonamide (SC-58635, Celecoxib). *J. Medicinal Chem*. **40**, 1347–1365.

14. DeLuca, H. F., Schnoes, H. K. (1984). Vitamin D: metabolism and mechanism of action. In *Annual Reports in Medicinal Chemistry*, Denis, M. B. Ed. Academic Press, vol. 19, p. 179–190.

15. Tanaka, Y., DeLuca, H. F., Kobayashi, Y., Ikekawa, N. (1984). 26,26,26,27,27,27-Hexafluoro-1,25-dihydroxyvitamin D3: a highly potent, long-lasting analog of 1,25-dihydroxyvitamin D3. *Arch. Biochem. Biophys*. **229**, 348–354.

16. (a) Dear, G. J., Ismail, I. M., Mutch, P. J., Plumb, R. S., Davies, L. H., Sweatman, B. C. (2000). Urinary metabolites of a novel quinoxaline non-nucleoside reverse transcriptase inhibitor in rabbit, mouse and human: identification of fluorine NIH shift metabolites using NMR and tandem MS. *Xenobiotica* **30**, 407–426; (b) Koerts, J., Soffers, A. E. M. F., Vervoort, J., De Jager, A., Rietjens, I. M. C. M. (1998). Occurrence of the NIH shift upon the cytochrome P450-catalyzed in vivo and in vitro aromatic ring hydroxylation of fluorobenzenes. *Chem.Res. Toxicol*. **11**, 503–512; (c) Park, B. K., Kitteringham, N. R., O'Neill, P. M. (2001). Metabolism of fluorine-containing drugs. *Annu. Rev. Pharmacol. Toxicol*. **41**, 443–470.

17. (a) Abraham, M. H., Grellier, P. L., Prior, D. V., Duce, P. P., Morris, J. J., Taylor, P. J. (1989). Hydrogen bonding. Part 7. A scale of solute hydrogen-bond acidity based on log K values for complexation in tetrachloromethane. *J. Chem. Soc. Perkin Trans*. **2**, 699–711; (b) Schlosser, M. (1998). Parametrization of substituents: effects of fluorine and other heteroatoms on OH, NH, and CH acidities. *Angew. Chem. Int. Ed*. **37**, 1496–1513; (c) Smart, B. E. (2001). Fluorine substituent effects (on bioactivity). *J. Fluorine Chem*. **109**, 3–11; (d) Speight, J. G. (2005). *Lange's Handbook of Chemistry*. New York: McGraw-Hill.

18. Morgenthaler, M., Schweizer, E., Hoffmann-Röder, A., Benini, F., Martin, Rainer E., Jaeschke, G., Wagner, B., Fischer, H., Bendels, S., Zimmerli, D., Schneider, J.,

Diederich, F., Kansy, M., Müller, K. (2007). Predicting and tuning physicochemical properties in lead optimization: amine basicities. *Chem. Med. Chem.* **2**, 1100–1115.

19. (a) Maren, T. H., Conroy, C. W. (1993). A new class of carbonic anhydrase inhibitor. *J. Biol. Chem.* **268**, 26233–26239; (b) Kim, C. Y., Chang, J. S., Doyon, J. B., Baird, T. T., Fierke, C. A., Jain, A., Christianson, D. W. (2000). Contribution of fluorine to protein-ligand affinity in the binding of fluoroaromatic inhibitors to carbonic anhydrase II. *J. Am. Chem. Soc.* **122**, 12125–12134.

20. Mandell, L. A., Ball, P., Tillotson, G. (2001). Antimicrobial safety and tolerability: differences and dilemmas. *Clin. Infect. Dis.* **32**, S72–S79.

21. Port, R. E., Wolf, W. (2003). Noninvasive methods to study drug distribution. *Invest. New Drugs* **21**, 157–168.

22. Prusoff, W. H. (1959). Synthesis and biological activities of iododeoxyuridine, an analog of thymidine *Biochim. Biophys. Acta* **32**, 295–296.

23. Watanabe, K. A., Reichman, U., Hirota, K., Lopez, C., Fox, J. J. (1979). Nucleosides. 110. Synthesis and antiherpes virus activity of some 2'-fluoro-2'-deoxyarabinofuranosylpyrimidine nucleosides. *J. Medicinal Chem.* **22**, 21–24.

24. (a) Marquez, V. E., Tseng, C. K. H., Kelley, J. A., Mitsuya, H., Broder, S., Roth, J. S., Driscoll, J. S. (1987). 2',3'-Dideoxy-2'-fluoro-ara-a. An acid-stable purine nucleoside active against human immunodeficiency virus (HIV). *Biochem. Pharmacol.* **36**, 2719–2722; (b) Okabe, M., Sun, R. C., Zenchoff, G. B. (1991). Synthesis of 1-(2,3-dideoxy-2-fluoro-beta-D-threo-pentofuranosyl)cytosine (F-ddC). A promising agent for the treatment of acquired immune deficiency syndrome. *J. Org. Chem.* **56**, 4392–4397.

25. (a) Böhm, H. J., Banner, D., Bendels, S., Kansy, M., Kuhn, B., Müller, K., Obst-Sander, U., Stahl, M. (2004). Fluorine in medicinal chemistry. *Chem. Bio. Chem.* **5**, 637–643; (b) Böhm, H. J., Schneider, G. (2003). *Protein-Ligand Interactions: From Molecular Recognition to Drug Design*. Weinheim: Wiley-VCH.

26. Ganguly, T., Mal, S., Mukherjee, S. (1983). Hydrogen bonding ability of fluoroalcohols. *Spectrochim. Acta A Mol. Spectrosc.* **39**, 657–660.

27. Wildman, S. A., Crippen, G. M. (1999). Prediction of physicochemical parameters by atomic contributions. *J. Chem. Inf. Comput. Sci.* **39**, 868–873.

28. Yamazaki, T., Taguchi, T., Ojima, I. (2009). *Unique Properties of Fluorine and Their Relevance to Medicinal Chemistry and Chemical Biology in Fluorine in Medicinal Chemistry and Chemical Biology*. New York: Wiley.

29. (a) Müller, K., Faeh, C., Diederich, F. (2007). Fluorine in pharmaceuticals: looking beyond intuition. *Science* **317**, 1881–1886; (b) Paulini, R., Müller, K., Diederich, F. (2005). Orthogonal multipolar interactions in structural chemistry and biology. *Angew. Chem. Int. Ed.* **44**, 1788–1805.

30. (a) Parlow, J. J., Kurumbail, R. G., Stegeman, R. A., Stevens, A. M., Stallings, W. C., South, M. S. (2003). Synthesis and X-ray crystal structures of substituted fluorobenzene and benzoquinone inhibitors of the tissue factor VIIa complex. *Bioorg. Medicinal Chem. Lett.* **13**, 3721–3725; (b) Parlow, J. J., Kurumbail, R. G., Stegeman, R. A., Stevens, A. M., Stallings, W. C., South, M. S. (2003). Design, synthesis, and crystal structure of selective 2-pyridone tissue factor VIIa inhibitors. *J. Medicinal Chem.* **46**, 4696–4701; (c) Parlow, J. J., Stevens, A. M., Stegeman, R. A., Stallings, W. C., Kurumbail, R. G., South, M. S. (2003). Synthesis and crystal structures of substituted benzenes and benzoquinones as tissue factor VIIa inhibitors. *J. Medicinal Chem.* **46**, 4297–4312.

31. Wang, Z., Canagarajah, B. J., Boehm, J. C., Kassisà, S., Cobb, M. H., Young, P. R., Abdel-Meguid, S., Adams, J. L., Goldsmith, E. J. (1998). Structural basis of inhibitor selectivity in MAP kinases. *Structure* **6**, 1117–1128.

32. Hughes, D. L., Sieker, L. C., Bieth, J., Dimicoli, J. L. (1982). Crystallographic study of the binding of a trifluoroacetyl dipeptide anilide inhibitor with elastase. *J. Mol. Biol.* **162**, 645–658.

33. Olsen, J. A., Banner, D. W., Seiler, P., Wagner, B., Tschopp, T., Obst-Sander, U., Kansy, M., Müller, K., Diederich, F. (2004). Fluorine interactions at the thrombin active site: protein backbone fragments H-C-α-C = O comprise a favorable C-F envionment and interaction of C-F with electrophiles. *Chem. Bio. Chem.* **5**, 666–675.

34. Istvan, E. S., Deisenhofer, J. (2001). Structural mechanism for statin inhibition of HMG-CoA reductase. *Science* **292**, 1160–1164.

35. (a) Kertesz, D. J., Marx, M. (1986). Thiol esters from steroid 17beta.-carboxylic acids: carboxylate activation and internal participation by 17alpha-acylates. *J. Org. Chem.* **51**, 2315–2328; (b) Barkalow, J., Chamberlin, S. A., Cooper, A. J., Hossain, A., Hufnagel, J. J., Langridge, D. C. (2001). Method for the preparation of fluticasone and related 17 beta eta-carbothioic esters using a novel carbothioic acid synthesis and novel purification methods. PCT Int. Appl. WO2001062722; (c) Chu, D., Zhang, D. (2007). Method for the preparation of fluticasone propionate. PCT Int. Appl. WO2007012228.

36. (a) Marcellin, P., Mommeja-Marin, H., Sacks, S. L., Lau, G. K. K., Sereni, D., Bronowicki, J. P., Conway, B., Trepo, C., Blum, M. R., Yoo, B. C., Mondou, E., Sorbel, J., Snow, A., Rousseau, F., Lee, H. S. (2004). A phase II dose-escalating trial of clevudine in patients with chronic hepatitis B. *Hepatology* **40**, 140–148; (b) Yoo, B. C., Kim, J. H., Kim, T. H., Koh, K. C., Um, S. H., Kim, Y. S., Lee, K. S., Han, B. H., Chon, C. Y., Han, J. Y., Ryu, S. H., Kim, H. C., Byun, K. S., Hwang, S. G., Kim, B. I., Cho, M., Yoo, K., Lee, H. J., Hwang, J. S., Kim, Y. S., Lee, Y. S., Choi, S. K., Lee, Y. J., Yang, J. M., Park, J.W., Lee, M. S., Kim, D. G., Chung, Y. H., Cho, S. H., Choi, J. Y., Kweon, Y. O., Lee, H. Y., Jeong, S. H., Yoo, H. W., Lee, H. S. (2007). Clevudine is highly efficacious in hepatitis B e antigen-negative chronic hepatitis B with durable off-therapy viral suppression. *Hepatology* **46**, 1041–1048.

37. (a) Ma, T., Lin, J. S., Newton, M. G., Cheng, Y. C., Chu, C. K. (1997). Synthesis and anti-hepatitis B virus activity of 9-(2-deoxy-2-fluoro-β-l-arabinofuranosyl)purine Nucleosides. *J. Medicinal Chem.* **40**, 2750–2754; (b) Ma, T., Pai, S. B., Zhu, Y. L., Lin, J. S., Shanmuganathan, K., Du, J., Wang, C., Kim, H., Newton, M. G., Cheng, Y. C., Chu, C. K. (1996). Structure-activity relationships of 1-(2-deoxy-2-fluoro-β-l-arabino-furanosyl)pyrimidine nucleosides as anti-hepatitis B virus agents. *J. Medicinal Chem.* **39**, 2835–2843.

38. Sznaidman, M. L., Almond, M. R., Pesyan, A. (2002). New synthesis of l-fmau from l-arabinose. *Nucleosides Nucleotides Nucl. Acids* **21**, 155–163.

39. Chilman-Blair, K., Mealy, N. E., Castaner, J. (2004). Clofarabine. Treatment of acute leukemia. *Drugs Future* **29**, 112–120.

40. Jeha, S., Razzouk, B., Rytting, M., Rheingold, S., Albano, E., Kadota, R., Luchtman-Jones, L., Bomgaars, L., Gaynon, P., Goldman, S., Ritchey, K., Arceci, R., Altman, A., Stine, K., Steinherz, L., Steinherz, P. (2009). Phase II study of clofarabine in

pediatric patients with refractory or relapsed acute myeloid leukemia. *J. Clin. Oncol.* **27**, 4392–4397.

41. (a) Reichman, U., Watanabe, K. A., Fox, J. J. (1975). A practical synthesis of 2-deoxy-2-fluoro-arabinofuranose derivatives. *Carbohydrate Res.* **42**, 233–240; (b) Bauta, W. E., Schulmeier, B. E., Burke, B., Puente, J. F., Cantrell, W. R., Lovett, D., Goebel, J., Anderson, B., Ionescu, D., Guo, R. (2004). A new process for antineoplastic agent clofarabine II. *Org. Process Res. Dev.* **8**, 889–896.

42. Anderson, B. G., Bauta, W. E., Cantrell, J. W. R., Engles, T., Lovett, D. P. (2008). Isolation, synthesis, and characterization of impurities and degradants from the clofarabine process. *Org. Process Res. Dev.* **12**, 1229–1237.

43. Langen, P. (1975). *Antimetabolites of Nucleic Acid Metabolism: The Biochemical Basis of Their Action, with Special Reference to Their Application in Cancer Therapy.* New York: Gordon and Breach.

44. Hoshi, A., Castañer, J. (1994). Alovudine. Antiviral, Anti-HIV. *Drugs Future* **19**, 221.

45. (a) Etzold, G., Hintsche, R., Kowollik, G., Langen, P. (1971). Nucleoside von fluorzuckern.VI: Synthese und reaktivität von 3′-fluor- und 3′-chlor-3′-desoxy-thymidin. *Tetrahedron* **27**, 2463–2472; (b) Kowollik, G., Janta-Lipinski, M., Gaertner, K., Langen, P., Etzold, G. (1973). Nucleoside von Fluorzuckern. XII. Ein neuer Zugang zu 1-(2,3-Didesoxy-3-fluor-beta-D-ribofuranosyl)-pyrimidinen. *J. Praktische Chemie* **315**, 895–900.

46. Watanabe, K. A., Harada, K., Zeidler, J., Matulic-Adamic, J., Takahashi, K., Ren, W. Y., Cheng, L. C., Fox, J. J., Chou, T. C. (1990). Synthesis and anti-HIV-1 activity of 2′-"up"-fluoro analogs of active anti-AIDS nucleosides 3′-azido-3′-deoxythymidine (AZT) and 2′,3′-dideoxycytidine (DDC). *J. Medicinal Chem.* **33**, 2145–2150.

47. Hewawasam, P., Gribkoff, V. K., Pendri, Y., Dworetzky, S. I., Meanwell, N. A., Martinez, E., Boissard, C. G., Post-Munson, D. J., Trojnacki, J. T., Yeleswaram, K., Pajor, L. M., Knipe, J., Gao, Q., Perrone, R., Starrett, J. E. (2002). The synthesis and characterization of BMS-204352 (MaxiPost™) and related 3-fluorooxindoles as openers of maxi-K potassium channels. *Bioorg. Medicinal Chem. Lett.* **12**, 1023–1026.

48. de Angelis, L. (1987). Flurithromycin. *Drugs Future* **12**, 214.

49. Naber, K. G., Adam, D. (1998). The classification of fluoroquinolones. *Chemother. J.* **7**, 66–68.

50. (a) Kimura, Y., Atarashi, S., Kawakami, K., Sato, K., Hayakawa, I. (1994). (Fluorocyclopropyl)quinolones. 2. Synthesis and stereochemical structure-activity relationships of chiral 7-(7-amino-5-azaspiro [2.4] heptan-5-yl)-1-(2-fluorocyclopropyl)quinolone antibacterial agents. *J. Medicinal Chem.* **37**, 3344–3352; (b) Prous, J., Graul, A., Castaner, J. (1994). DU-6859 (sitafloxacin). *Drugs Future* **19**, 827–834.

51. Andreasen, N. C. (1994). *Schizophrenia: from Mind to Molecule.* Washington, DC: American Psychiatric Press.

52. Meltzer, H. Y., Matsubara, S., Lee, J. C. (1989). Classification of typical and atypical antipsychotic drugs on the basis of dopamine D-1, D-2 and serotonin2 pKi values. *J. Pharmacol. Exp. Therapeut.* **251**, 238–246.

53. Rowley, M., Hallett, D. J., Goodacre, S., Moyes, C., Crawforth, J., Sparey, T. J., Patel, S., Marwood, R., Patel, S., Thomas, S., Hitzel, L., O'Connor, D., Szeto, N.,

Castro, J. L., Hutson, P. H., MacLeod, A. M. (2001). 3-(4-Fluoropiperidin-3-yl)-2-phenylindoles as high affinity, selective, and orally bioavailable h5-HT2A receptor antagonists. *J. Medicinal Chem.* **44**, 1603–1614.

54. O'Neill, M. J., Murray, T. K., Clay, M. P., Lindstrom, T., Yang, C. R., Nisenbaum, E. S. (2005). LY503430: pharmacology, pharmacokinetics, and effects in rodent models of Parkinson's disease. *CNS Drug Rev.* **11**, 77–96.

55. (a) Murray, T. K., Whalley, K., Robinson, C. S., Ward, M. A., Hicks, C. A., Lodge, D., Vandergriff, J. L., Baumbarger, P., Siuda, E., Gates, M., Ogden, A. M., Skolnick, P., Zimmerman, D. M., Nisenbaum, E. S., Bleakman, D., O'Neill, M. J. (2003). LY503430, a novel α-amino-3-hydroxy-5-methylisoxazole-4-propionic acid receptor potentiator with functional, neuroprotective and neurotrophic effects in rodent models of Parkinson's disease. *J. Pharmacol. Exp. Therapeut.* **306**, 752–762; (b) Ryder, J. W., Falcone, J. F., Manro, J. R., Svensson, K. A., Merchant, K. M. (2006). Pharmacological characterization of cGMP regulation by the biarylpropylsulfonamide class of positive, allosteric modulators of α-amino-3-hydroxy-5-methyl-4-isoxazolepropionic acid receptors. *J. Pharmacol. Exp. Therapeut.* **319**, 293–298.

56. (a) O'Neill, M. J., Bleakman, D., Zimmerman, D. M., Nisenbaum, E. S. (2004). AMPA receptor potentiators for the treatment of CNS disorders. *Curr. Drug Targets CNS Neurol. Disord.* **3**, 181–194; (b) O'Neill, M. J., Witkin, J. M. (2007). AMPA receptor potentiators: application for depression and Parkinson's disease *Curr. Drug Targets* **8**, 603–620.

57. Magnus, N. A., Aikins, J. A., Cronin, J. S., Diseroad, W. D., Hargis, A. D., LeTourneau, M. E., Parker, B. E., Reutzel-Edens, S. M., Schafer, J. P., Staszak, M. A., Stephenson, G. A., Tameze, S. L., Zollars, L. M. H. (2005). Diastereomeric salt resolution based synthesis of LY503430, an AMPA (α-amino-3-hydroxy- 5-methyl-4-isoxazolepropionic acid) potentiator. *Org. Process Res. Dev.* **9**, 621–628.

58. (a) Nagengast, F. M., Grubben, M. J. A. L., van Munster, I. P. (1995). Role of bile acids in colorectal carcinogenesis. *Eur. J. Cancer* **31**, 1067–1070; (b) Gibson, J. C., Capuano, L. R. (1997). Prevention and Treatment of Colorectal Cancer by 6-Fluoroursodeoxychlolic Acid. WO 97/44043.

59. Konigsberger, K., Chen, G. P., Vivelo, J., Lee, G., Fitt, J., McKenna, J., Jenson, T., Prasad, K., Repic, O. (2002). An expedient synthesis of 6α-fluoroursodeoxycholic acid. *Org. Process Res. Dev.* **6**, 665–669.

60. Chung, M. K., Kim, J. C., Roh, J. K. (1995). Teratogenic effects of DA-125, a new anthracycline anticancer agent, in rats. *Reprod. Toxicol.* **9**, 159–164.

61. Lee, M. G., Yang, J. (1996). DA-125. Antineoplastic agent. *Drugs Future* **21**, 782.

62. Mohedano, N., Sanchez-Rovira, P., Medina, B., Flores-Gonzalez, E., Moreno, M. A., Fernandez, M., Lozano, A. (1997). Phase II trial of gemcitabine in combination with cisplatin and ifosfamide in advanced non-small cell lung cancer (NSCLC). *Lung Cancer* **18**, 43.

63. (a) Colomer, R. (2005). Gemcitabine in combination with paclitaxel for the treatment of metastatic breast cancer. *Women's Health* **1**, 323–329; (b) Moen, M. D., Wellington, K. (2005). Gemcitabine: in combination with paclitaxel in the first-line treatment of metastatic breast cancer. *Am. J. Cancer* **4**, 327–333; (c) Tomao, S., Romiti, A., Tomao, F., Di Seri, M., Caprio, G., Spinelli, G., Terzoli, E., Frati, L. (2006). A phase II trial of a biweekly combination of paclitaxel and gemcitabine in metastatic breast cancer. *BMC Cancer* **6**, 137.

64. (a) Oldfield, V., Wellington, K. (2005). Gemcitabine: in combination with carbo-platin in the treatment of recurrent ovarian cancer. *Am. J. Cancer* **4**, 337–344; (b) Pfisterer, J., Vergote, I., Bois, A. D., Eisenhauer, E. (2005). Combination therapy with gemcitabine and carboplatin in recurrent ovarian cancer. *Intl. J. Gynecol. Cancer* **15**, 36–41.

65. Hertel, L. W., Kroin, J. S., Misner, J. W., Tustin, J. M. (1988). Synthesis of 2-deoxy-2,2-difluoro-D-ribose and 2-deoxy-2,2′-difluoro-D-ribofuranosyl nucleosides. *J. Org. Chem.* **53**, 2406–2409.

66. Chou, T. S., Heath, P. C., Patterson, L. E., Poteet, L. M., Lakin, R. E., Hunt, A. H. (1992). Stereospecific synthesis of 2-deoxy-2,2-difluororibonolactone and its use in the preparation of 2′-deoxy-2′,2′-difluoro-β-D-ribofuranosyl pyrimidine nucleosides: the key role of selective crystallization. *Synthesis* 565–570.

67. Sorbera, L. A., Casta ner, J., Mealy, N. E. (2004). Lubiprostone. Treatment of constipation, treatment of irritable bowel syndrome, treatment of postoperative ileus, ClC-2 channel activator. *Drugs Future* **29**, 336.

68. Ueno, R. (2001). Endothelin antagonist. US6197821.

69. Choudhury, A., Moore, J. R., Pierce, M. E., Fortunak, J. M., Valvis, I., Confalone, P. N. (2003). In situ recycling of chiral ligand and surplus nucleophile for a noncatalytic reaction: amplification of process throughput in the asymmetric addition step of efavirenz (DMP 266). *Org. Process Res. Dev.* **7**, 324–328.

70. Rabasseda, X., Sorbera, L. A., Casta ner, J. (1999). Befloxatone. Antidepressant, MAO-A inhibitor. *Drugs Future* **24**, 1057.

71. Gauthier, J. Y., Chauret, N., Cromlish, W., Desmarais, S., Duong, L. T., Falgueyret, J. P., Kimmel, D. B., Lamontagne, S., Léger, S., LeRiche, T., Li, C. S., Massé, F., McKay, D. J., Nicoll-Griffith, D. A., Oballa, R. M., Palmer, J. T., Percival, M. D., Riendeau, D., Robichaud, J., Rodan, G. A., Rodan, S. B., Seto, C., Thérien, M., Truong, V. L., Venuti, M. C., Wesolowski, G., Young, R. N., Zamboni, R., Black, W. C. (2008). The discovery of odanacatib (MK-0822), a selective inhibitor of cathepsin K. *Bioorg. Medicinal Chem. Lett.* **18**, 923–928.

72. O'Shea, P. D., Chen, C. Y., Gauvreau, D., Gosselin, F., Hughes, G., Nadeau, C., Volante, R. P. (2009). a practical enantioselective synthesis of odanacatib, a potent cathepsin K Inhibitor, via triflate displacement of an α-trifluoromethylbenzyl triflate. *J. Org. Chem.* **74**, 1605–1610.

73. a) Qing, F. L., Qiu, X. L. (2007). *Fluoroorganic Chemistry*. Beijing: Science Press; b) Jiang, Y., Liu, Z. (2009). Progresses in enantioselective electrophilic fluorinations. *Chin. J. Org. Chem.* **29**, 1362–1370; c) Cahard, D., Xu, X., Couve-Bonnaire, S., Pannecoucke, X. (2010). Fluorine & chirality: how to create a nonracemic stereogenic carbon-fluorine centre? *Chem. Soc. Rev.* **39**, 558–568.

74. a) Rozen, S. (1988). Elemental fluorine as a legitimate reagent for selective fluorination of organic-compounds. *Acc. Chem. Res.* **21**, 307–312; b) Rozen, S. (1996). Elemental fluorine: not only for fluoroorganic chemistry. *Acc. Chem. Res.* **29**, 243–248.

75. Barton, D. H. R., Godinho, L. S., Hesse, R. H., Pechet, M. M. (1968). Organic reactions of fluoroxy-compounds—electrophilic fluorination of activated olefins. *Chem. Commun.* 804.

76. a) Barton, D. H. R., Ganguly, A. K., Hesse, R. H., Loo, S. N., Pechet, M. M. (1968). Organic reactions of fluoroxy-compounds—electrophilic fluorination of aromatic

rings. *Chem. Commun*. 806; b) Schack, C. J. Christe, K. O. (1979). Reactions of fluorine perchlorate with fluorocarbons and the polarity of the O-F bond in covalent hypofluorites. *Inorg. Chem*. **18**, 2619–2620.

77. Tius, M. A. (1995). Xenon difluoride in synthesis. *Tetrahedron* **51**, 6605–6634.

78. Schmutzler, R. (1968). Nitrogen oxide fluorides. *Angew. Chem. Int. Engl*. **7**, 440–455.

79. Rozen, S. (1996). Selective fluorinations by reagents containing the OF group. *Chem. Rev*. **96**, 1717–1736.

80. Differding, E., Lang, R. W. (1988). New fluorinating reagents. 1. The 1st enantios-elective fluorination reaction. *Tetrahedron Lett*. **29**, 6087–6090.

81. Davis, F. A., Zhou, P., Murphy, C. K., Sundarababu, G.., Qi, H. Y., Han, W., Przes-lawski, R. M., Chen, B. C., Carrol, P. J. (1998). Asymmetric fluorination of enolates with nonracemic N-fluoro-2,10-camphorsultams. *J. Org. Chem*. **63**, 2273–2280.

82. a) Takeuchi, Y., Suzuki, T., Satoh, A., Shiragami, T., Shibata, N. (1999). N-fluoro-3-cyclohexyl-3-methyl-2,3-dihydrobenzo [1,2-d] isothiazole 1,1-dioxide: an efficient agent for electrophilic asymmetric fluorination of enolates. *J. Org. Chem*. **64**, 5708–5711; b) Shibata, N., Liu, Z. P., Takeuchi, Y. (2000). Novel enantios-elective fluorinating agents. (R)- and (S)-N-fluoro-3-tert-butyl-7-niitro-3,4-dihydro-2H-benzo[e][1,2]thiazine 1,1-dioxides. *Chem. Pharm. Bull*. **48**, 1954–1958; c) Liu, Z. P., Shibata, N., Takeuchi, Y. (2000). Novel methods for the facile construction of 3,3-disubstituted and 3,3-spiro-2H, 4H-benzo[e] 1,2-thiazine-1,1-diones: synthe-sis of (11S, 12R, 14R)-2-fluoro-14-methyl-11(methylethyl) spiro [4H-benzo[e]-1,2-thiazine-3,2′-cyclohexane]-1,1-dione, an agent for the electrophilic asymmetric fluo-rination of aryl ketone enolates. *J. Org. Chem*. **65**, 7583–7587; d) Sun, H. M., Liu, Z. P., Tang, L. Q. (2008). Synthesis of novel chiral fluorinating agents. (+)- and (−)-N-fluoro-3-methyl-3-(4-methylphenyl)-2H- benzo[e][1,2] thiazine-1,1,4-triones. *Chin. Chem. Lett*. **19**, 907–910.

83. Shibata, N., Suzuki, E., Takeuchi, Y. (2000). A fundamentally new approach to enantioselective fluorination based on cinchona alkaloid derivatives/selectfluor com-bination. *J. Am. Chem. Soc*. **122**, 10728–10729.

84. Cahard, D., Audouard, C., Plaquevent, J. C., Roques, N. (2000). Design, synthesis and evaluation of a novel class of enantioseletive electrophilic fluorinating agents: N-fluoro ammonium salts of cinchona alkaloids (F-CA-BF₄). *Org. Lett*. **2**, 3699–3701.

85. Mohar, B., Baudoux, J., Plaquevent, J. C., Cahard, D. (2001). Electrophilic fluori-nation mediated by cinchona alkalodis: Highly enantioselective synthesis of alpha-fluoro-alpha-phenylglycine derivatives. *Angew. Chem. Int. Ed*. **40**, 4214–4216.

86. Shibata, N., Suzuki, E., Asahi, T., Shiro, M. (2001). Enantioselective fluorination mediated by cinchona alkaloid derivatives/Selectfluor combinations: reaction scope and structural information for N-fluorocinchona alkaloids. *J. Am. Chem. Soc*. **123**, 7001–7009.

87. a) Hamashima, Y., Yagi, K., Takano, H., Tamas, L., Sodeoka, M. (2002). An efficient enantioselective fluorination of various beta-ketoesters catalyed chiral palladium complexes. *J. Am. Chem. Soc*. **124**, 14530–14531; b) Hamashima, Y., Suzuki, T., Takano, H., Shimura, Y., Sodeoka, M. (2005). Catalytic enantioselective fluorination of oxindoles. *J. Am. Chem. Soc*. **127**, 10164–10165; c) Hamashima, Y., Suzuki, T., Shimura, Y., Shimizu, T., Umebayashi, N., Tamura, T., Sasamoto, N., Sodeoka, M. (2005). An efficient catalytic enantioselective fluorination of beta-ketophosphonates

using chiral palladium complexes. *Tetrahedron Lett*. **46**, 1447–1450; d) Suzuki, T., Goto, T., Hamashima, Y., Sodeoka, M. (2007). Enantioselective fluorination of tert-butoxycarbonyl lactones and lactams catalyzed by chiral Pd(II)-bisphosphine complexes. *J. Org. Chem*. **72**, 246–250; e) Hamashima, Y., Suzuki, T., Takano, H., Shimura, Y., Tsuchiya, Y., Moriya, K., Goto. T., Sodeoka, M. (2006). Highly enantioselective fluorination reactions of beta-ketoesters and beta-ketophosphonates catalyzed by chiral palladium complexes. *Tetrahedron* **62**, 7168–7179.

88. Ma, J. A., Cahard, D. (2004). Copper(II) triflate-bis(oxazoline)-catalysed enantioselective electrophilic fluorination of beta-ketoesters. *Tetrahedron Asymmetry* **15**, 1007–1011.

89. Shibata, N., Ishimaru, T., Nagai, T., Kohno, J., Toru, T. (2004). First enantio-flexible fluorination reaction using metal-bis(oxazoline) complexes. *Synlett* **10**, 1703–1706.

90. Shibatomi, K., Tsuzuki, Y., Nakata, S., Sumikawa, Y., Iwasa, S. (2007). Synthesis of axially chiral cyclic amino-substituted 2-(oxazolinyl)pyridine ligands for catalytic asymmetric fluorination of beta-keto esters. *Synlett* 551–554.

91. a) Shibata, N., Kohno, J., Takai, K., Ishimura, T. S., Nakamura, T., Toru, S. K. (2005). Highly enantioselective catalytic fluorination and chlorination reactions of carbonyl compounds of two-point binding. *Angew. Chem. Int. Ed*. **44**, 4204–4207; b) Reddy, D. S., Shibata, N., Nagai, J., Nakamura, S., Toru, T., Kanemasa, S. (2008). Desymmetrization-like catalytic enantioselective fluorination of malonates and its application to pharmaceutically attractive molecules. *Angew. Chem. Int. Ed*. **47**, 164–168.

92. Paull, D. H., Scerba, M. T., Alden-Danforth, E., Widger, L. R., Lectka, T. (2008). Catalytic, asymmetric alpha-fluorination of acid chlorides: dual metal-ketone enolate activation. *J. Am. Chem. Soc*. **130**, 17260.

93. Hintermann, L., Togni, A. (2000). Catalytic enantioselective fluorination of beta-ketoesters. *Angew. Chem. Int. Ed*. **39**, 4359.

94. Frantz, R., Hintermann, L., Perseghini, M., Broggini, D., Togni, A. (2003). Titanium-catalyzed stereoselective geminal heterodihalogenation of beta-ketoesters. *Org. Lett*. **5**, 1709–1712.

95. Marigo, M., Fielenbach, D., Braunton, A., Kjoersgaard, A., Jørgensen, K. A. (2005). Enantioselective formation of stereogenic carbon-fluorine centers by a simple catalytic method. *Angew. Chem. Int. Ed*. **44**, 3703–3706.

96. Steiner, D. D., Mase, N., Barbas, C. F. III. (2005). Direct asymmetric aipha-fluorination of aldehydes. *Angew. Chem. Int. Ed*. **44**, 3706–3710.

97. Beeson, T. D., MacMillan, D. W. C. (2005). Enantioselective organocatalytic alpha-fluorination of aldehydes. *J. Am. Chem. Soc*. **127**, 8826–8828.

98. Jiang, H., Falcicchio, A., Jensen, K. L., Paix ao, M. W., Bertelsen, S., Jørgensen, K. A. (2009). Target-directed organocatalysis: a direct asymmetric catalytic approach to chiral propargylic and allylic fluorides. *J. Am. Chem. Soc*. **131**, 7153–7157.

99. Nagib, D. A., Scott. M. E., MacMillan, D. W. C. (2009). Enantioselective alpha-trifluoromethylation of aldehydes via photoredox organocatalysis. *J. Am. Chem. Soc*. **131**, 10875.

100. Allen, A. E., MacMillan, D. W. C. (2010). The productive menger of iodonium salts and organocatalysis: A non-photolytic approach to enantioselective alpha-trifluoromethylation of aldehydes. *J. Am. Chem. Soc*. **132**, 4986.

CHAPTER 6

INDUSTRIAL APPLICATION OF CHIRAL TECHNOLOGIES

HUI-YIN (HARRY) LI, RUI LIU, CARL BEHRENS, and CHAO-YING NI

Wilmington PharmaTech Company LLC, Newark, DE and
University of Delaware, Newark, DE

Chiral Drugs: Chemistry and Biological Action, First Edition. Edited by Guo-Qiang Lin,
Qi-Dong You and Jie-Fei Cheng.
© 2011 John Wiley & Sons, Inc. Published 2011 by John Wiley & Sons, Inc.

6.1 INTRODUCTION

The field of asymmetric synthesis continues to be a very fertile area for new discoveries that advance the state of the art of organic synthesis as a whole. Year after year, new asymmetric reactions are discovered and added to the chiral toolbox that is now available to chemists as they design ways to synthesize a particular target molecule. However, relatively few asymmetric methods have demonstrated the unique combination of effectiveness, practicality, and economy required to make the transition from a laboratory tool that is useful for the synthesis of limited quantities of material under a specialized set of circumstances to an industrially useful process that enables the efficient synthesis of enantiopure compounds on a commercial scale.

There have been many excellent and comprehensive reviews of the applications of asymmetric synthesis on an industrial scale in the chemical literature, and the interested reader is referred to these sources and the references and examples cited therein [1]. The intention of this chapter is simply to provide an overview of a few selected examples of asymmetric synthetic methods that have made the leap from being simply a clever discovery to becoming a truly useful method on an industrial scale. This is by no means an exhaustive survey of the field. Instead, this chapter is designed to give the reader an overview of some of the asymmetric syntheses that have been utilized by the industry to manufacture high-profile pharmaceutical and agricultural products on a commercial scale.

The asymmetric syntheses in this survey can be grouped into several major categories. Classical methods such as chiral resolution (amlodipine, duloxetine, and sertraline hydrochloride), hydrolytic kinetic resolution (linezolid), and the use of starting materials from the chiral pool or commercially available chiral starting materials (oseltamivir phosphate, ezetimibe, and atorvastatin) are still very desirable because of the proven utility of such processes and the ready availability of materials from the chiral pool. Homogeneous asymmetric hydrogenation of olefins and ketones (e.g., L-DOPA, metolachlor, biotin, sitagliptin, and aprepitant) is a mainstay of industrial synthesis because of the atom economy of the reduction process. Asymmetric isomerization is a very useful technique, as illustrated by the synthesis of menthol. The industrial application of chiral oxidation technology is found in the asymmetric epoxidation of an olefin (indinavir sulfate) and the asymmetric oxidation of a sulfide (esomeprazole). Examples of specialized strategies such as crystallization-induced diastereoselective transformation (aprepitant) and asymmetric desymmetrization (nelfinavir) are also provided. Finally, in some cases, enzymatic methods are challenging conventional asymmetric synthesis.

For example, a highly efficient nitrilase-catalyzed asymmetric synthesis of the atorvastatin side chain precursor has been developed.

6.2   CHIRAL RESOLUTION

Chiral resolution has traditionally been a very powerful tool for asymmetric synthesis. In a typical resolution, a racemic compound is converted to the corresponding diastereomeric salt by reaction with an enantiomerically pure acid (e.g., tartaric acid) or base (e.g., α-methylbenzylamine) as appropriate. The diastereomeric salts often have sufficiently different physical properties to enable a preparatively useful separation by crystallization. Resolution is advantageous because recrystallization is a very well-known and robust procedure for large-scale use, and the resolving agents are often readily available and highly economical materials that are available from the chiral pool. One disadvantage is that half of the material is necessarily lost in the resolution process, unless the undesired enantiomer can be racemized and recycled. Classical resolution procedures are illustrated by amlodipine (tartaric acid) and sertraline (mandelic acid), while duloxetine provides an example of the resolve-racemize-recycle (RRR) approach. Kinetic resolution is a related variation in which a racemic compound is enantiomerically enriched by the selective reaction of one isomer. Linezolid provides an example in which chiral glycidyl butyrate derived by hydrolytic kinetic resolution was used in an industrial synthesis.

6.2.1   Amlodipine Besylate (Norvasc®)

Amlodipine besylate (**6**) is a long-acting calcium channel blocker belonging to the dihydropyridine class of compounds [2]. It is useful as an antihypertensive and in the treatment of angina. It can be administered as the besylate salt (branded) or the maleate salt (generic). Most of the desirable pharmacological activity can be attributed to the (S)-$(-)$-enantiomer.

The original preparation of amlodipine relies upon the Hantzsch synthesis to construct the core dihydropyridine ring system of the molecule as shown in Scheme 6.1 [3]. The synthesis starts with the reaction of 2-azidoethanol sodium salt with ethyl 4-chloroacetoacetate **1** to afford the keto ester **2**. Using Hantzsch conditions, treatment of **2** with ammonium acetate gives **3**, which is followed by reaction with 2-(2-chlorobenzylidene)acetoacetate **4** to give the azido dihydropyridine **5**. Catalytic hydrogenation of **5** with Pd on $CaCO_3$ and salt formation with benzenesulfonic acid delivers the racemic amlodipine besylate **6** [4] (Scheme 6.1).

Racemic amlodipine can be resolved by crystallization with either D-tartaric acid or L-tartaric acid [5]. Also, a novel bis((S)-mandelic acid)-3-nitrophthalate resolving agent (**7**) in Scheme 6.2 has also been reported to produce the desired amlodipine enantiomer in over 99% *ee* [6].

An alternative patented procedure in Scheme 6.3 describes the synthesis of the amlodipine maleate salt (**6**) by a Hantzsch reaction involving

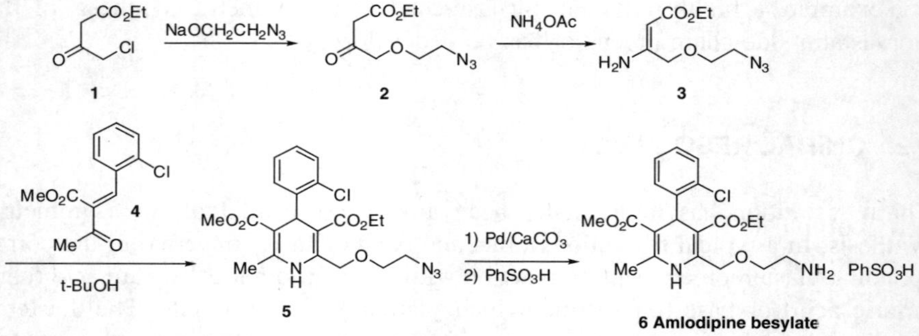

SCHEME 6.1 Synthesis of amlodipine besylate (**6**).

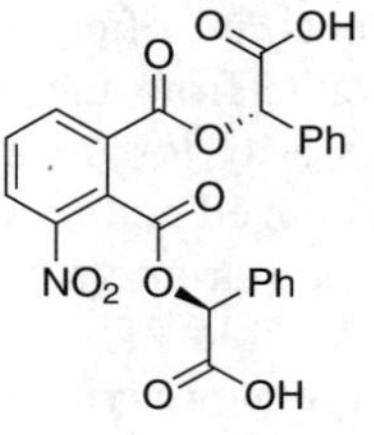

SCHEME 6.2 A novel bis((*S*)-mandelic acid)-3-nitrophthalate resolving agent (**7**).

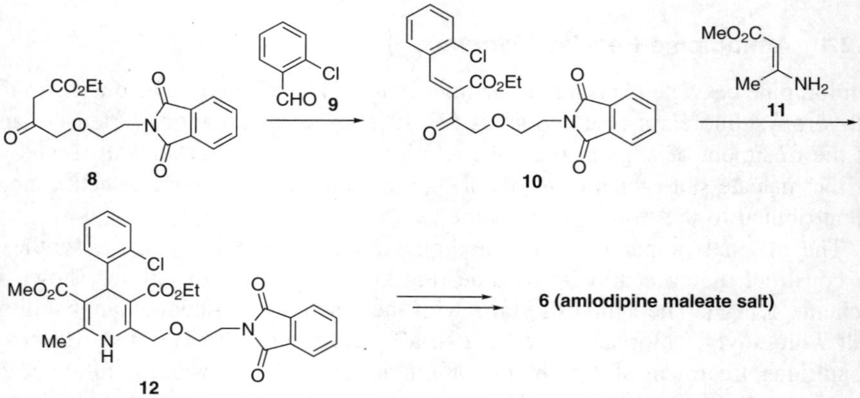

SCHEME 6.3 Synthesis of amlodipine maleate salt (**6**).

2-chlorobenzaldehyde (**9**), methyl aminocrotonate (**11**), and a phthalimide-protected acetoacetate derivative **8** [which is structurally similar to compound (**2**) in the previous synthesis] to afford racemic "phthalimidoamlodipine" (**12**) in >98% purity after a single recrystallization [7]. The phthalimide protecting group can be removed by treatment with methylamine, and the amlodipine is then isolated by crystallization as the maleate salt.

6.2.2 Duloxetine (Cymbalta®)

Duloxetine is a dual inhibitor of serotonin and norepinephrine uptake that is useful in the treatment of major depressive disorders and generalized anxiety disorders [8].

An efficient enantioselective synthesis of duloxetine has been developed [9,10]. As shown in Scheme6.4, 2-acetylthiophene **13** was subjected to Mannich reaction conditions (Me$_2$NH·HCl, CH$_2$O, HCl, IPA) to afford the adduct **14** in 91% yield. Enantioselective reduction of **14** was accomplished by using the Li(ent-Chirald®)$_2$AlH$_2$ complex [11] to afford (S)-**15** in 90% yield and 85–90% ee. Conversion of (S)-**15** to the corresponding sodium salt with NaH followed by reaction with 1-fluoronaphthalene gave the aryl ether **16** in 91% yield. Finally, N-demethylation of **16** with 2,2,2-trichloroethyl chloroformate delivered duloxetine **17** in 90% yield. According to the authors, this process was used to make large amounts of duloxetine, but the process was not taken to production because the overall efficiency was at best only comparable to a classical resolution [12].

Another asymmetric synthesis that was investigated by Eli Lilly is shown in Scheme 6.5 in which the key step is a CBS reduction of the acyl thiophene **19** to give the chiral alcohol **20** [13,14].

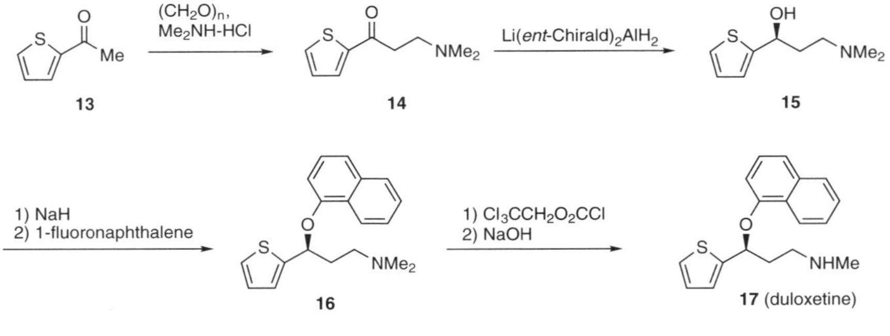

SCHEME 6.4   Pilot asymmetric synthesis of duloxetine (**17**).

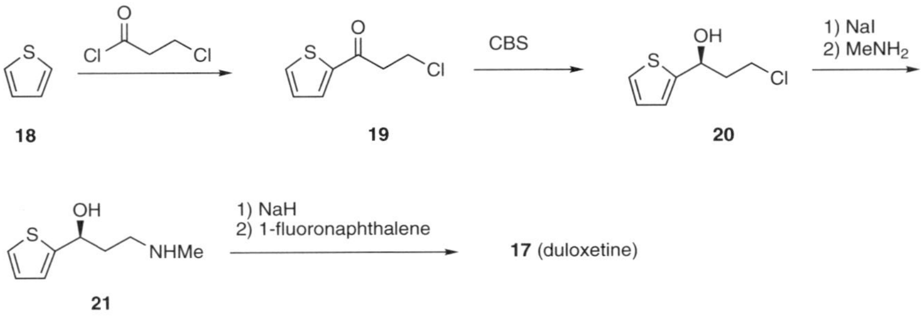

SCHEME 6.5   Alternative asymmetric synthesis of duloxetine (**17**).

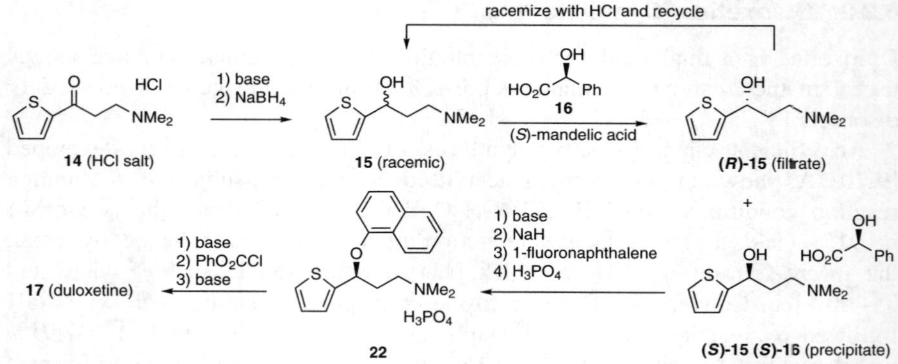

SCHEME 6.6 Industrial asymmetric synthesis of duloxetine.

The industrial synthesis of duloxetine in Scheme 6.6 developed at Eli Lilly is an example of a resolution-racemize-recycle (RRR) process [15]. The process begins similarly to the enantioselective synthesis. Reaction of 2-acetylthiophene with Me$_2$NH·HCl and paraformaldehyde and HCl in isopropanol gave the Mannich adduct, which was isolated as the HCl salt **14**. This HCl salt **14** in ethanol was neutralized with aqueous NaOH and reduced with NaBH$_4$ to give the racemic alcohol (±)-**15**. The product was not isolated, but instead extracted after work-up into MTBE. The MTBE solution of (±)-**15** was treated with a solution of (S)-mandelic acid **16** (0.45 eq.) in ethanol at 50°C. The precipitate initially formed had only a 20–60% diastereomeric excess, but after the mixture was refluxed and cooled to ambient temperature, the desired (S)-**15**·(S)-**16** salt was isolated in up to 98% diastereomeric excess and 43% yield.

The undesired (R)-**15** enantiomer, which remained in solution in MTBE, could be racemized by treatment with HCl and neutralized with NaOH to regenerate a mixture of the racemic alcohol (±)-**15** in MTBE that would be suitable for NaBH$_4$ reduction. After four reduction-racemization cycles, (S)-**15** was isolated in 81% yield [16].

Treatment of (S)-**15** with NaH and potassium benzoate in DMSO followed by 1-fluoronaphthalene gave the aryl ether, which was isolated as the corresponding phosphate salt **22** in 91% ee and 79.6% yield. Compound **22** was converted to the free base with aqueous ammonia and toluene, N-demethylated by reaction with phenyl chloroformate and i-Pr$_2$NEt in toluene followed by hydrolysis with NaOH, and acidified with HCl to afford duloxetine HCl (**17** HCl salt) in 32% yield from (S)-**15**.

Overall, the RRR process based on classical resolution methodology was found to be a more robust and cost-effective method than enantioselective synthesis for commercial use.

Recently, several publications have appeared describing potential improvements to the synthesis of duloxetine [17–19]. Also, a potential enzymatic route to duloxetine has been reported [20].

6.2.3   Sertraline Hydrochloride (Zoloft®)

Sertraline hydrochloride **27** is an inhibitor of serotonin uptake and is useful for the treatment of depression. In the original commercial synthesis in Scheme 6.7, the sertraline ketone **23** was treated with methylamine in the presence of $TiCl_4$ as a dehydrating agent to drive the reaction to form the imine **24** to completion [21]. The imine **24** was then selectively reduced with Pd/C to afford **25** as a 6:1 *cis:trans* isomer mixture that was resolved by crystallization with (D)-mandelic acid to deliver the corresponding (D)-(−)-mandelic acid salt **26**, which is a key intermediate in the synthesis of **27**. A significant drawback to this procedure is the use of $TiCl_4$, which results in the generation of hazardous waste containing TiO_2.

Pfizer researchers sought to improve the overall synthesis by focusing on the key steps of converting the sertraline ketone **23** to the imine **24** and converting the imine **24** to the sertraline amine **25** [22]. The first advance came when the researchers discovered that the imine **24** had limited solubility in alkanol solvents such as methanol and ethanol. This was crucial because the limited solubility of the product in the reaction solvent enabled the equilibrium between **23** and **24** to be shifted toward the product side (e.g., >95% of **24** in MeOH, EtOH, IPA, or *n*-PrOH) without the need to utilize a conventional dehydrating agent.

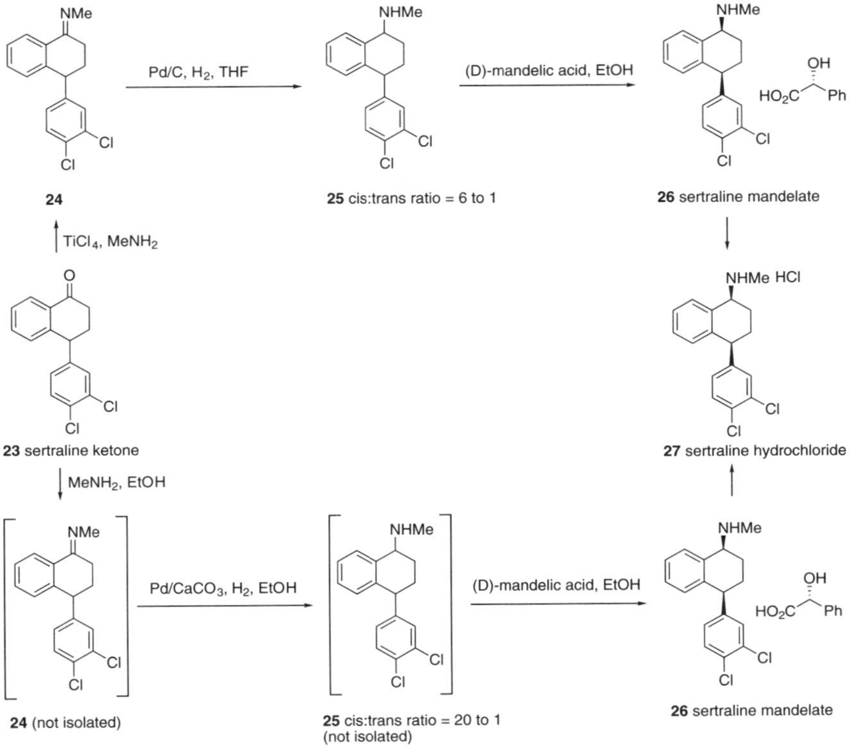

SCHEME 6.7   Synthesis of sertraline hydrochloride (**27**).

The next improvement was to avoid isolating the imine **24** by telescoping the hydrogenation step with the reduction step. To accomplish this, a design of experiments (DoE) approach was applied to the selection of the proper catalyst and hydrogenation conditions for the conversion of **24** to **25**. From this work it was found that 1% (w/w) of Pd/CaCO$_3$ in alkanol solvents provided the best overall results, with a *cis:trans* ratio of 20:1 of **25** and excellent quality and yield.

The final improvement involved telescoping the resolution of the racemic mixture by crystallization of the mandelic acid salt with the rest of the procedure. Since the previous process used ethanol as the crystallization solvent, it was relatively straightforward to avoid the need to isolate the sertraline amine **25**.

In the improved overall process, the sertraline ketone **23** is converted to the imine **24** with 3.1 eq. of methylamine in anhydrous ethanol at 50–55°C. After the reaction is complete, the reaction mixture is cooled to room temperature, treated with Pd/CaCO$_3$ and decolorizing carbon, and hydrogenated at up to 40°C. After the hydrogenation is complete, the catalyst and decolorizing carbon are removed by filtration and excess methylamine is removed by distillation with replenishment of anhydrous ethanol during the distillation. Finally, (D)-mandelic acid (0.9 eq.) is added and the mixture is refluxed and then cooled slowly to less than 25°C to afford sertraline mandelate **26** in 40% yield with respect to **23**.

Since more than 100 metric tons of sertraline hydrochloride is produced annually, the impact of improving the efficiency of the manufacturing process is quite significant. Pfizer researchers estimated that the new process would eliminate 440 tons per year of TiO$_2$-containing waste and over 40 tons of undesired isomer waste. The number of solvents was reduced from five to two, and the total volume was just 24% of the original process. For these accomplishments, Pfizer received a Presidential Green Challenge Award in 2002 [23].

6.2.4   Linezolid (Zyvox®)

Linezolid **36** is an oxazolidinone antibiotic with a novel mechanism of action involving the inhibition of protein synthesis [24,25]. It was launched in 2000, and it has proven useful against gram-positive bacterial infections.

The original synthesis of linezolid in Scheme 6.8 began with the selective reaction of 3,4-difluoronitrobenzene **28** with morpholine to afford **29**. The product was then reduced with ammonium formate and Pd/C under transfer hydrogenation conditions to give the intermediate aniline **30**, which was protected by reaction with benzyl chloroformate to deliver the carbamate **31**. The carbamate was deprotonated with *n*-BuLi in THF and then reacted with the chiral (*R*)-glycidyl butyrate **32** to produce the oxazlidinone **33**. The oxazolidinone **33** was mesylated, reacted with sodium azide, reduced with Pd/C and H$_2$, and acetylated with acetic anhydride to from linezolid **36** [26,27].

In a more recent version of the synthesis in Scheme 6.8, it was discovered that the functionalized epichlorohydrin derivative **35** can be reacted with compound **31** to provide **36** directly, which shortens the synthesis by four steps [28,29]. Thus

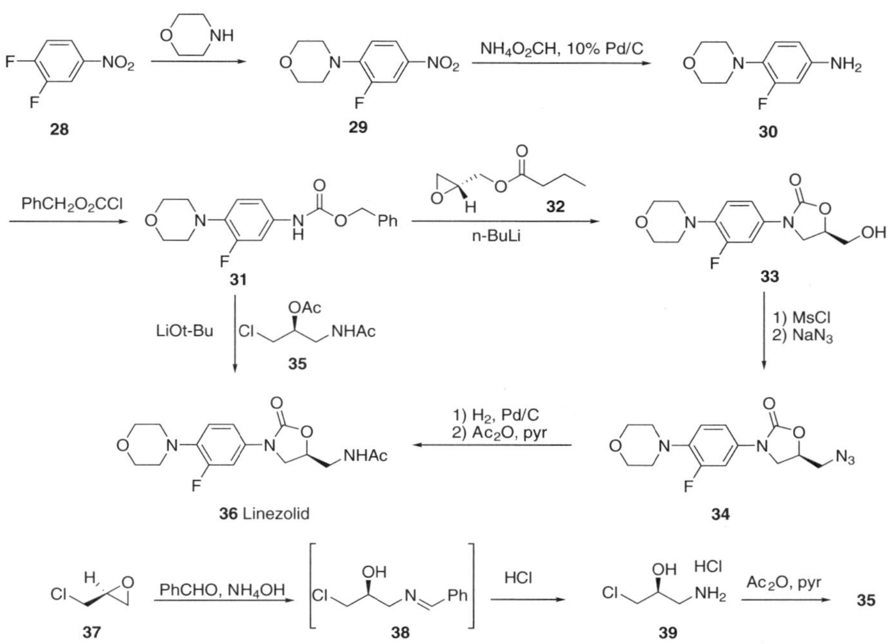

SCHEME 6.8 Synthesis of linezolid (**36**).

(S)-epichlorohydrin, which was commercially available on a large scale in an optically pure form by using Jacobsen's hydrolytic kinetic resolution technology (HKR) with racemic epichlorohydrin, was added to a mixture of benzaldehyde and aqueous ammonia in ethanol to afford the intermediate benzylidene **38**. After complete reaction, the Schiff's base was not isolated but instead was directly hydrolyzed with aq. HCl to afford (2S)-1-amino-3-chloro- propanol hydrochloride **39** as a solid in 77% yield. This material was successfully diacetylated with acetic anhydride and pyridine in methylene chloride to afford **35** as a white crystalline solid in 83% yield. The carbamate **31**, which was an intermediate in the previous synthesis, was reacted with lithium *t*-butoxide and **35** to afford linezolid **36** directly.

6.3 CHIRAL POOL SYNTHESIS

Another time-honored method to make enantiomerically pure compounds is to design the synthesis so that it incorporates or utilizes the chirality of the economical and abundant materials that are found in the naturally occurring chiral pool of natural compounds. This practice is illustrated by the synthesis of oseltamivir from quinic acid and ezetimibe from (S)-3-hydroxy-γ-lactone. The chiral side chain of atorvastatin was first made by synthesis from isoascorbic acid.

6.3.1 Oseltamivir Phosphate (Tamiflu®)

Oseltamivir phosphate **58** is a pro-drug of Ro 64-0802 **59**, which is a potent competitive inhibitor of influenza A and B neuraminidase. It is used in the treatment and prevention of influenza infections, and it has the advantage of being orally administered. The compound was developed by Gilead and is marketed by Hoffman-La Roche.

Many interesting synthetic routes to oseltamivir phosphate have been published, including those by Karpf and Trussardi [30], Corey and co-workers [31], Shibasaki and co-workers [32], Fukiyama and co-workers [33], and Trost and Zhang [34], and various patents [35,36]. A review of this area has been published [37].

The Roche synthesis of oseltamivir phosphate **58** in Scheme 6.9 proceeds via the key epoxide intermediate **51**. Compound **51** is structurally related to the naturally-occurring (−)-shikimic acid, but because of the limited availability of (−)-shikimic acid, an alternative route was developed from (−)-quinic acid **40** by Rohloff et al. [38] and improved by a team from Roche [39]. The synthesis begins with the selective protection of the *cis* vicinal diol moiety of **40** as the acetonide by reaction with 2,2-dimethoxypropane and TsOH in EtOAc to afford the lactone **41**. After a distillative workup, the lactone **41** was selectively hydrolyzed with NaOEt in EtOH to give the ester diol **42**. The hydrolysis of the lactone is reversible under the reaction conditions, and a 5:1 mixture of **42:41** was obtained at ambient temperature. However, it was discovered that a 13:1 mixture in favor of the desired compound **42** could be obtained at −20°C. Workup was accomplished by neutralization with HOAc at −20°C to prevent relactonization followed by distillation, resulting in about 90–93% yield from **40**.

The mixture of **42** and **41** was selectively mono-mesylated (<3% dimesylate) at the 5-position with MsCl and Et$_3$N in methylene chloride at 0–5°C to give **43**. The workup was done by acidification to pH 7.5 to 7.8 with 0.5% aq. HCl, extraction, and evaporation. Most of the corresponding mesyl lactone was removed by crystallization from EtOAc. The next step was the selective dehydration of **43** to afford the shikimic acid derivative **44**. Although this transformation had previously been accomplished by treatment with thionyl chloride, the procedure was not particularly selective and was furthermore expensive (because of the use of a Pd catalyst for the selective removal of an impurity) and cumbersome, so an alternative procedure was sought. It was then discovered that reaction of **43** with Vilsmeier's salt in methylene chloride followed by warming to about 75°C was highly selective for the desired unsaturated ester **44** with only a trace of the isomeric unsaturated ester **45** (ratio of **44** to **45** was ∼35–50:1) being observed. The drawback to this procedure is that about 14–18% of compound **46** is formed as an unavoidable side-product of the reaction, as all attempts to use chloride-free Vilsmeier's reagents were unsuccessful. During the workup, the undesired chloro compound **46** and the isomeric **45** were removed by crystallization from methanol to afford **44** in 65% yield and >98% purity.

Compound **44** was transketalized with 3-pentanone and catalytic MsOH to afford, after neutralization with Et$_3$N to prevent hydrolysis, the pentylideneketal

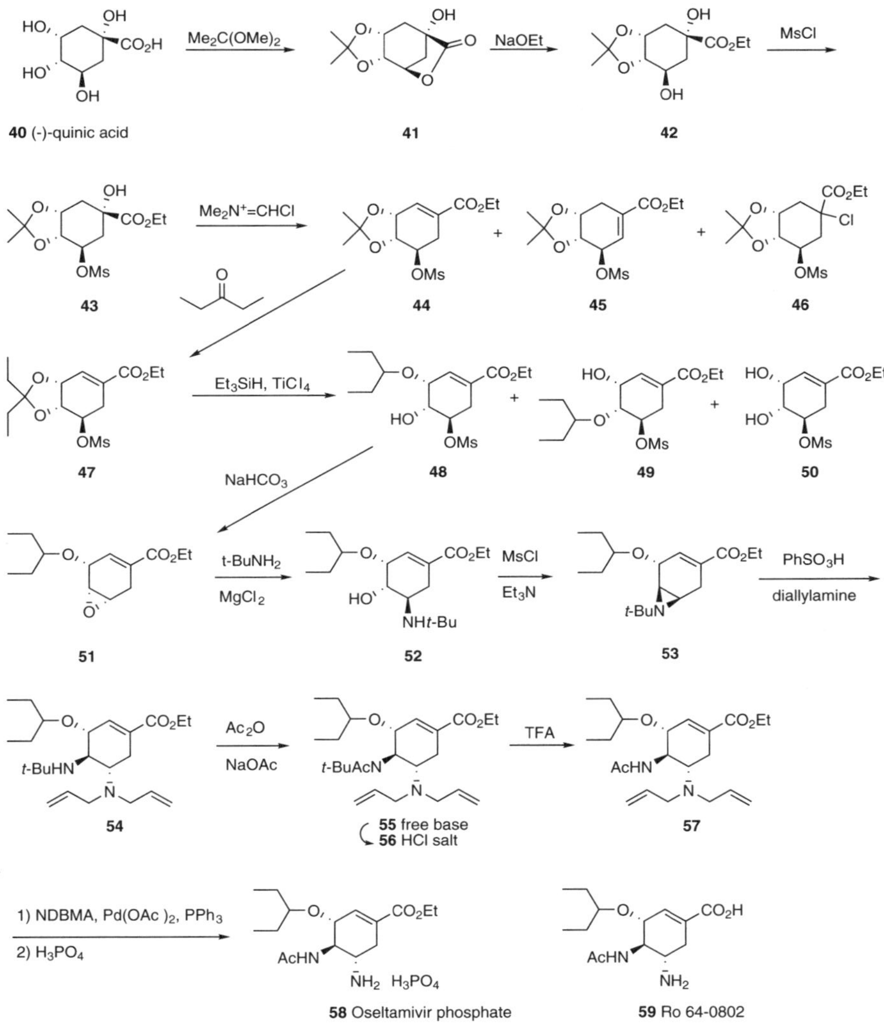

SCHEME 6.9 The Roche synthesis of oseltamivir phosphate (**58**).

47. This material can be selectively reduced to the pentyl ether **48**. The use of BH$_3$ · SMe$_2$ and TMSOTf in methylene chloride gives the desired product **48** in a 10:1:1 ratio with the isomeric pentyl ether **49** and the diol **50**. However, the procedure was not robust enough for plant use. After extensive screening it was discovered that the use of Et$_3$SiH and TiCl$_4$ in methylene chloride at about −34°C delivered a 32:1 ratio of **48** to **49** along with about 2–4% of the diol **50**. After workup, the mesylate **48** was cyclized to the key intermediate epoxide **51** by reaction with sodium bicarbonate in aqueous ethanol. The overall yield was about 68% from compound **44**.

The epoxide **51** can be transformed into oseltamivir phosphate **58** in a process described by Karpf and Trussardi, or in an improved second-generation synthesis described by Harrington et al. [40]. Epoxide **51** participates in a ring-opening reaction with inversion of configuration at the less hindered epoxide terminus with t-BuNH$_2$ in the presence of MgCl$_2$ in toluene to give the trans amino alcohol **52**. Selective O-sulfonation of **52** with MsCl in toluene (the presence of the bulky t-Bu substituent on the nitrogen ensured selective mesylation of the alcohol [41]) followed by cyclization with Et$_3$N afforded the aziridine **53** in 93% yield. Compound **53** undergoes a selective aziridine ring-opening reaction with diallylamine and benzenesulfonic acid, again with inversion of configuration, at the less hindered aziridine terminus to afford **54**, which was then acetylated with Ac$_2$O and NaOAc to give **55** and then converted to the HCl salt **56** with HCl in ethanol. The synthesis was completed by reaction of the HCl salt **56** with TFA to remove the t-Bu protecting group to give **57**, which was followed by treatment with 1,3-dimethylbarbituric acid (NDBMA) in the presence of catalytic Pd(OAc)$_2$ and PPh$_3$ in absolute ethanol to remove both allyl protecting groups and finally acidification with phosphoric acid to give oseltamivir phosphate **58** in 61% overall yield from epoxide **51**.

Recently, a chemoenzymatic approach was reported (Scheme 6.10) in which a genetically engineered *Escherichia coli* strain that is deficient in both shikimate kinase isoenzymes delivered shikimic acid **61** with a titer of 84 g/l and in 33% yield from glucose **60** [42]. Shikimic acid is an alternative to quinic acid as a starting material for the synthesis of oseltamivir phosphate **58**. An alternative chemoenzymatic approach involves the synthesis of kanosamine **62** from glucose by *Bacillus pumilus* ATCC 21143, followed by the conversion of kanosamine by a specialized *E. coli* strain to aminoshikimic acid **63** [43], which can then be converted to oseltamivir phosphate **58** by conventional chemistry.

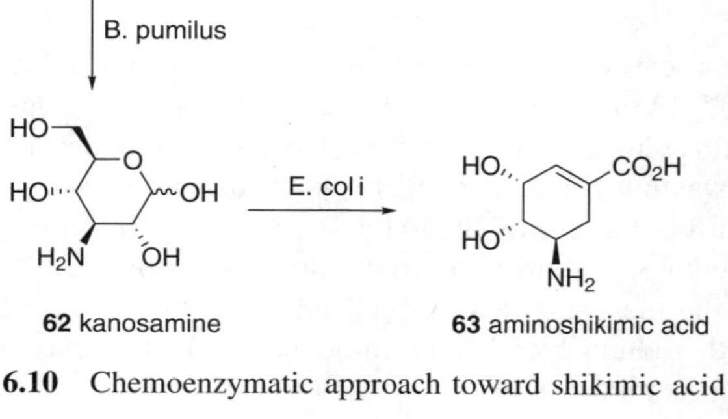

SCHEME 6.10   Chemoenzymatic approach toward shikimic acid synthesis.

6.3.2 Ezetimibe (Zetia®)

Ezetimibe is a potent intestinal cholesterol absorption inhibitor. It has been marketed as a single agent (Zetia®) and as a combination therapy with statins, which have a different mechanism of action. The combination of ezetimibe with simvastatin (Zocor®) is marketed as Vytorin®.

Several synthetic procedures that are useful for the synthesis of azetidinones have been reviewed [44,45]. The original commercial synthesis of ezetimbe **68** in Scheme 6.11 utilizes a CBS asymmetric reduction of the aryl ketone (*S*)-**64** to afford the chiral alcohol **65**. Compound **65** is then protected as a TMS ether with 2 eq. of TMS-Cl, enolized with TiCl$_4$, and reacted with 4-hydroxybenzylidene(4-fluoro)aniline **66** to afford an intermediate phenol that is protected as the TMS ether in situ to afford **67** as a crystalline solid. Compound **67** was cyclized to the azetidinone with *N,O*-bistrimethylsilylacetamide, and the TMS protecting groups were simultaneously removed with 2 eq. of TBAF in a one-pot reaction to afford ezetimibe **68**.

A potential alternative enantio- and diastereoselective synthesis of ezetimibe **68** utilized the chiral starting material (*S*)-3-hydroxy-γ-lactone **69** to generate the required stereochemistry of the final compound and is shown in Scheme 6.12 [46]. Reaction of **69** with 2 eq. of LDA generated the dilithium salt, which undergoes 1,2-addition with 4-benzyloxybenzylidene(4-fluoro)aniline followed by in situ formation of the γ-lactam by intramolecular ring opening of the γ-lactone to afford, after crystallization, a 64% yield of a 95:5 mixture of *trans*-**70** and *cis*-**70** with the desired absolute configuration resulting from the directing effect of the chiral center in the starting material **69**. Oxidative cleavage of the side chain diol **70** with NaIO$_4$ in acetonitrile delivered the aldehyde **71**. Interestingly, the *cis*-aldehyde-**71** was not stable under the reaction conditions and epimerized entirely to the desired *trans*-aldehyde-**71** in 90% *ee*. Reaction of **71** with the pre-formed TMS enolate of 4-fluorobenzaldehyde in the presence of TiCl$_4$ afforded

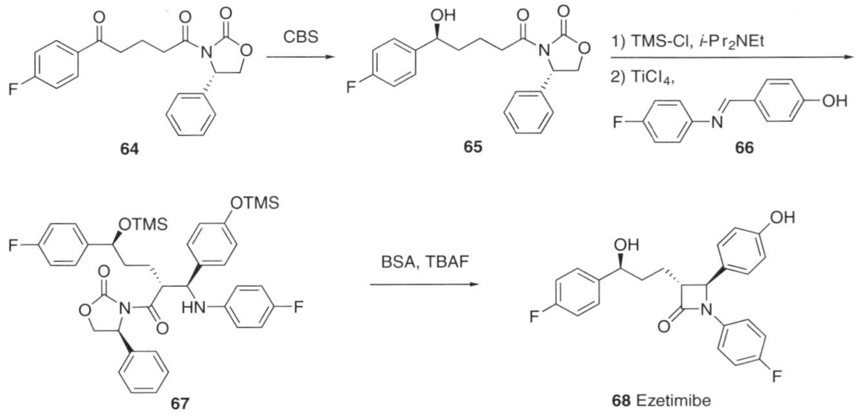

SCHEME 6.11   Synthesis of ezetimibe (**68**).

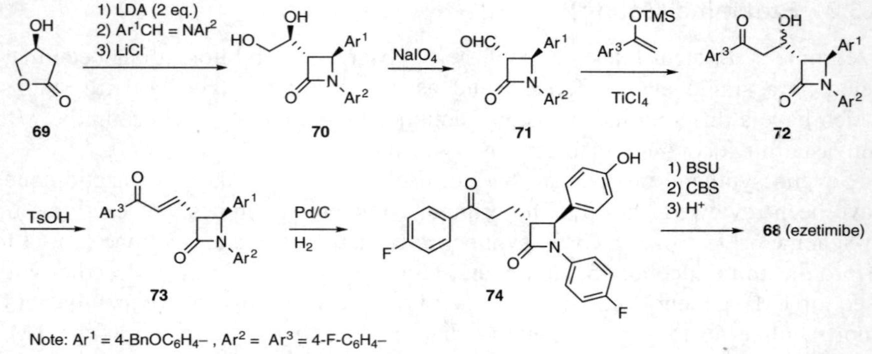

SCHEME 6.12 Alternative synthesis of ezetimibe (**68**).

a diastereomeric mixture of alcohols **72** that were directly dehydrated with TsOH and molecular sieves to afford the unsaturated ketone **73** in 75% yield from **71**. Both the unsaturated ketone and the benzyl protecting group of **73** were reduced with Pd/C in EtOH to afford **74** in 90% yield. The synthesis was completed by protecting the free phenol as the TMS ether with bis(trimethylsilyl)urea, CBS asymmetric reduction of the aryl ketone with a chiral oxazaborolidine in 97% *ee*, and finally deprotection of the TMS ether upon aqueous workup afforded, after crystallization, ezetimibe **68** in 99.7% *ee* and 79% yield from **73**.

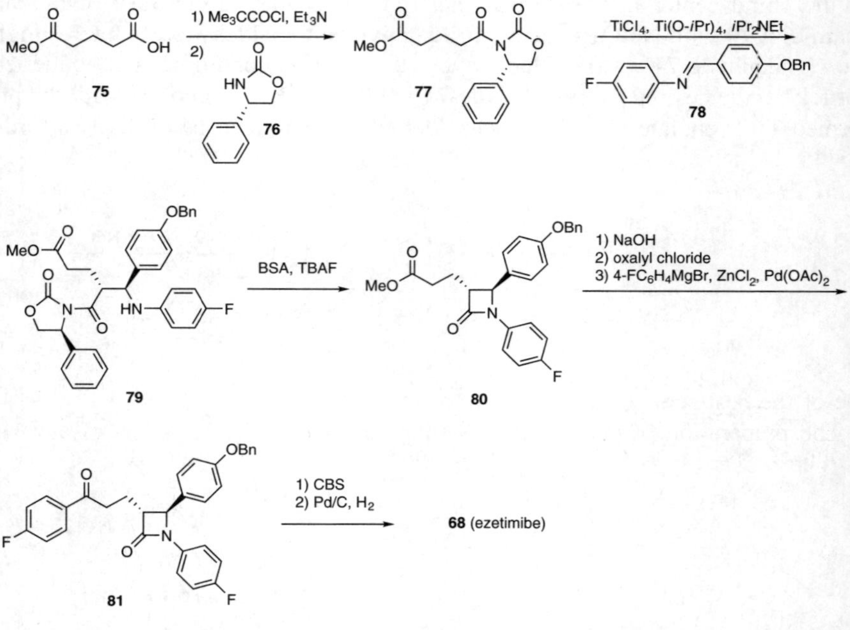

SCHEME 6.13 An improved synthesis of ezetimibe (**68**).

More recently, an alternative scaleable synthesis was reported as shown in Scheme 6.13 [47]. Reaction of monomethyl glutarate **75** with pivaloyl chloride and Et$_3$N in methylene chloride gives the mixed anhydride, which then reacts with (S)-4-phenyl-2-oxazolidinone **76** to give the chiral oxazolidinone **77**. Enolization of **77** with TiCl$_4$ and Ti(i-OPr)$_4$ in methylene chloride followed by addition to 4-benzyloxybenzylidene(4-fluoro)aniline **78** affords the chiral oxazoline **79** with a 97:3 diastereomeric ratio. Thus, the chirality of **76** was transmitted to two additional stereochemical centers with this diastereoselective conjugate addition. Compound **79** was cyclized to the corresponding *trans*-β-lactam **80** by treatment with *N,O*-bis(trimethylsilyl)acetamide and TBAF in toluene at 60°C. The pendant methyl ester group of **80** was hydrolyzed with aqueous NaOH and the converted to the acid chloride with oxalyl chloride and catalytic DMF in methylene chloride. Coupling of the acid chloride with 4-fluorophenylmagnesium bromide in the presence of ZnCl$_2$ and Pd(OAc)$_2$ in THF affords the keto lactam **81**. CBS asymmetric reduction of the arylketone moiety of **81** with borane dimethyl sulfide and (R)-methyloxazaborolidine provided the secondary alcohol with the desired (S)-configuration. The synthesis of ezetimibe **68** was completed by removal of the benzyl protecting group with 5% Pd/C in aqueous methanol.

6.3.3   Atorvastatin (Lipitor®)

Hypercholesterolemia is characterized by the presence of high cholesterol levels in the blood, which if left untreated may contribute to heart disease and other problems [48]. HMG-CoA reductase (HMGR), which converts 3-hydroxy-3-methylglutaryl-CoA to mevalonic acid, is the rate-limiting enzyme in cholesterol biosynthesis. HMGR inhibitors (or "statins") have become prevalent in the treatment of hypercholesterolemia because of their typically favorable efficacy and safety profiles [49].

A number of compounds from this class have been approved for marketing in the US. Three compounds are natural products or semisynthetic derivatives of natural products including lovastatin, simvastatin, and pravastatin. An additional three compounds are synthetic inhibitors of HMGR including fluvastatin, rosuvastatin, and atorvastatin. Furthermore, a combination of simvastatin and ezetimibe (an intestinal cholesterol absorption inhibitor) is being marketed for the treatment of hypercholesterolemia [50].

Atorvastatin (Lipitor®) **100** had 2008 sales of about US$12 billion, and it is one of the best selling branded pharmaceuticals in the world [51].

The preparation of one of the key intermediates in the synthesis of atorvastatin is shown in Scheme 6.14 [52]. The synthesis begins with the Knoevenagel condensation of isobutyrylacetanilide **82** with benzaldehyde using β-alanine and acetic acid in hexane to afford the enone **83** in 85% yield. Next, the enone was subjected to a Stetter reaction with 4-fluorobenzaldehyde with *N*-ethylthiazolium bromide **84** as the catalyst to deliver the 1,4-diketone **85** in 80% yield. This material, when cyclized with 3-aminopropionaldehyde diethyl acetal (a model amine that can also be elaborated into the fully functionalized atorvastatin side chain) by

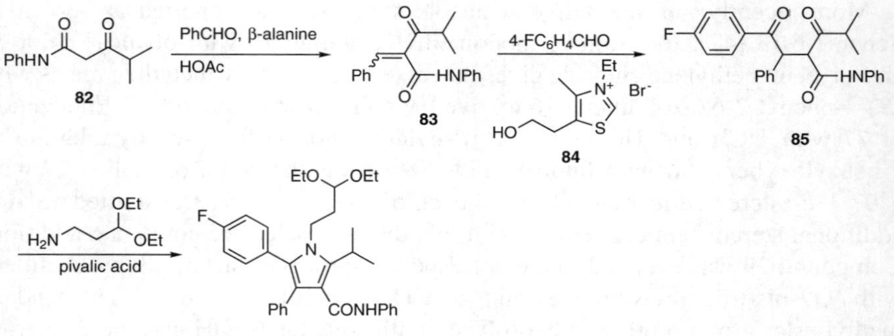

SCHEME 6.14 Synthesis of the key intermediate (**86**) for atorvastatin.

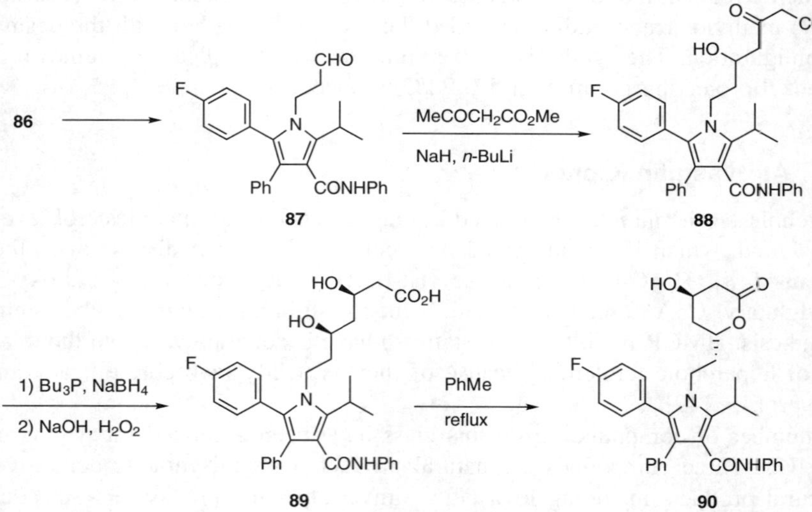

SCHEME 6.15 Synthesis of the racemic atorvastatin lactone (**90**).

the Paal–Knorr reaction, affords **86**, which contains the penta-substituted pyrrole ring system of atorvastatin, in 43% yield.

Compound **86** is an intermediate in the synthesis of atorvastatin. However, the synthesis was rather lengthy because **86** was first converted into the racemic atorvastatin lactone **90** as shown in Scheme 6.15. The enantiomers were then separated by reacting the racemic atorvastatin lactone with the readily available chiral (*R*)-α-methylbenzylamine to form the corresponding amide, separating the resulting diastereomers by crystallization, and cleaving the amide with lactonization to afford the optically pure atorvastatin lactone [53–55].

The next challenge was to synthesize the stereochemically pure amine **98** corresponding to side chain of atorvastatin to enable a convergent synthesis. An

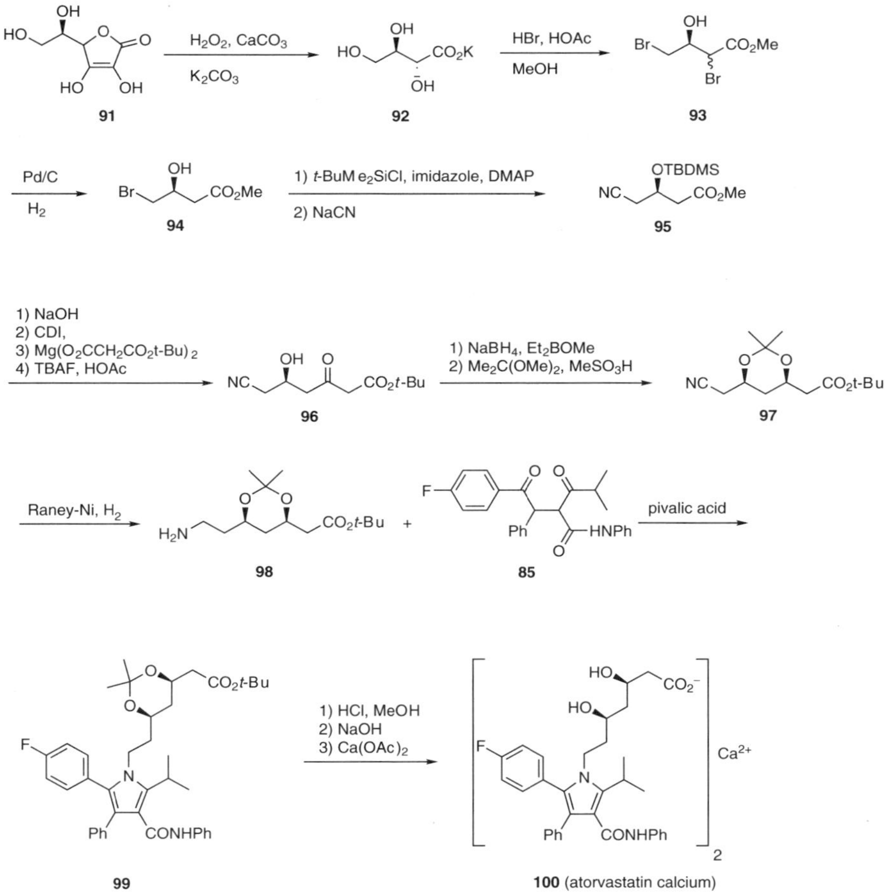

SCHEME 6.16   Synthesis of atorvastatin calcium (**100**).

efficient synthesis (Scheme 6.16) utilized isoascorbic acid, a member of the chiral pool, to derive the desired side chain stereochemistry. Thus, oxidative cleavage of isoascorbic acid **91** followed by dibromination with concomitant esterification, selective reduction of the 2-bromo group, protection of the alcohol, and finally cyanation afforded **95** [56].

The chiral compound **95** was then elaborated into the fully functionalized, stereochemically pure side chain **98** [57,58]. First, the methyl ester was homologated to the β-ketoester and deprotected to afford **96**. The key step was the stereoselective reduction of the β-ketoester **96** with NaBH$_4$ and Et$_2$BOMe at low temperatures to deliver the *syn*-diol, which was converted into **98** by conventional means.

Finally, the tetra-substituted ketone **85** and the amine **98** were reacted in refluxing heptane-toluene with 1 eq. of pivalic acid to afford the protected atorvastatin intermediate **99**, which after removal of the acetonide protecting group, hydrolysis of the *t*-butyl ester, and conversion to the calcium salt yielded atorvastatin calcium **100**.

The synthesis of atorvastatin is fairly lengthy, and therefore a considerable amount of effort has been directed toward improving the synthesis. Codexis received a Presidential Green Challenge Award in 2006 for improvements to the synthesis of the atorvastatin side chain [59]. Codexis developed specific enzymes using recombinant-based directed evolution technology that were used for the selective enzymatic reduction of ethyl 4-chloroacetoacetate to (*S*)-4-chloro-3-hydroxybutyrate followed by an enzymatic cyanation to give ethyl (*R*)-4-cyano-3-hydroxybutyrate **105**, which is a key intermediate in the synthesis of the atorvastatin side chain [60].

More recently, a newer enzymatic method has been reported for the synthesis of **105** (Scheme 6.17) in which the desired stereochemistry of the side chain is created in a highly efficient way from an achiral starting material [61]. The key step is the nitrilase-catalyzed desymmetrization of the *meso* compound 3-hydroxyglutaronitrile (**103**) by the highly selective hydrolysis of one of the nitrile groups [62].

The synthesis begins with the inexpensive industrial compound epichlorohydrin **101**. In principle, it is possible to obtain **103** in a one-pot reaction with epichlorohydrin with the use of excess NaCN. However, this process was not considered safe for scale-up because it was anticipated that the reaction would be very difficult to control. Therefore, a stepwise approach was used in which epichlorohydrin was first converted into 4-chloro-3-hydroxybutyronitrile **102** with a slight excess of cyanide and then converted into **103** in a subsequent reaction with additional cyanide. The synthesis of **102** was accomplished by cofeeding HCN (1.1 eq.) and epichlorohydrin (1.0 eq.) over 4 hours into a mixture of triethanolamine and tetrabutylammonium bromide in water at 45–50°C and stirring for about 8 hours until the reaction was completed by GC. The crude **102** was added to a mixture of aq. NaCN and HCl (pH = 10) at 50–55°C over 4 hours and heated at 50–55°C for about 6 hours. After the reaction was completed by GC, the reaction was worked up and purified by wiped film evaporation to afford **103**.

The *meso* compound 3-hydroxyglutaronitrile **103** was desymmetrized with Nitrilase BD9570 (Diversa) at pH 7.5 and 27°C to afford (*R*)-4-cyano-3-hydroxybutyric acid **104** in 81% yield and 98.8% *ee*. The acid **104** was converted

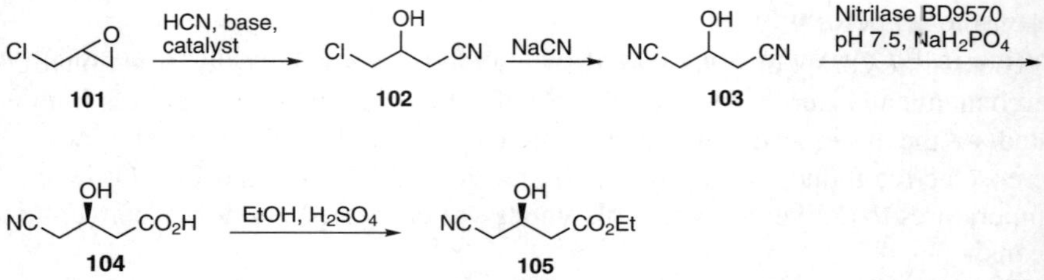

SCHEME 6.17 Chemoenzymatic synthesis of the side chain (**105**) for atorvastatin.

to the atorvastatin side chain precursor ethyl (*R*)-4-cyano-3-hydroxybutyrate **105** with ethanol and sulfuric acid in 99% yield and 98.7% *ee*.

6.4   ASYMMETRIC DESYMMETRIZATION

Asymmetric desymmetrization is a process in which a *meso* compound is converted into an optically active product. This can be done by enzymes, as illustrated in the biocatalytic synthesis of the atorvastatin side chain. The synthesis of nelfinavir illustrates the asymmetric desymmetrization of a *meso*-epoxide.

6.4.1   Nelfinavir (Viracept®)

Nelfinavir **116** is a member of a family of HIV protease inhibitors that are useful for the treatment of HIV. The compound was initially developed by Agouron.

The Agouron synthesis in Scheme 6.18 starts with the Jacobsen asymmetric desymmetrization of *meso*-3,4-epoxytetrahydrofuran **106** with azidotrimethylsilane to afford **107** in 96% yield and 99% *ee*, followed by azide reduction (with Pd/C and PtO$_2$ and H$_2$) and hydrolysis to afford the free amino alcohol [63]. The hygroscopic free amino alcohol was then converted to the tosylate salt **108** for convenience. The tosylate salt **108** was then acylated with the commercially available 3-acetoxy-2-methylbenzoyl chloride **109** in ethyl acetate at room temperature and neutralized with triethylamine to afford the intermediate **110**.

A key element of this synthesis is the selective rearrangement of the tetrahydrofuran **110** to the oxazoline **113** without the formation of the undesired oxazoline isomer **111**. During process development, it was discovered that with normal addition of MsCl to **110** in the presence of excess TEA the undesired oxazoline **111** was formed exclusively because of base-induced intramolecular mesylate displacement of **112** by the amide to generate the *cis*-fused oxazoline ring system. However, the use of reverse addition of a deficiency of TEA to a mixture of **110** and MsCl permitted the intermediate mesylate **112** to be formed exclusively without formation of the undesired product **111**. Treatment of the mesylate **112** with acetic anhydride followed by sulfuric acid caused rearrangement of the tetrahydrofuran ring system, resulting predominantly in the formation of the desired oxazoline **113** with <5% of the undesired *cis*-fused oxazoline **111** [64].

Therefore, a highly telescoped process was developed in which a slurry of **110** (prepared as before by the reaction of **108** with **109** in the presence of triethylamine in EtOAc) was not isolated but instead cooled with an ice acetone bath and treated with MsCl followed by slow addition of triethylamine to generate the mesylate **112**, which was also not isolated but was treated with acetic anhydride followed by sulfuric acid to afford, after work-up, the oxazoline **113** as an oil. The oxazoline **113** was then dissolved in methanol and treated with potassium carbonate and the commercially available perhydroisoquinoline **114** at 50°C to afford, via the corresponding epoxide, the crude adduct **115**. After reslurrying in methyl isobutylketone, the product **115** was obtained in >95% purity and in

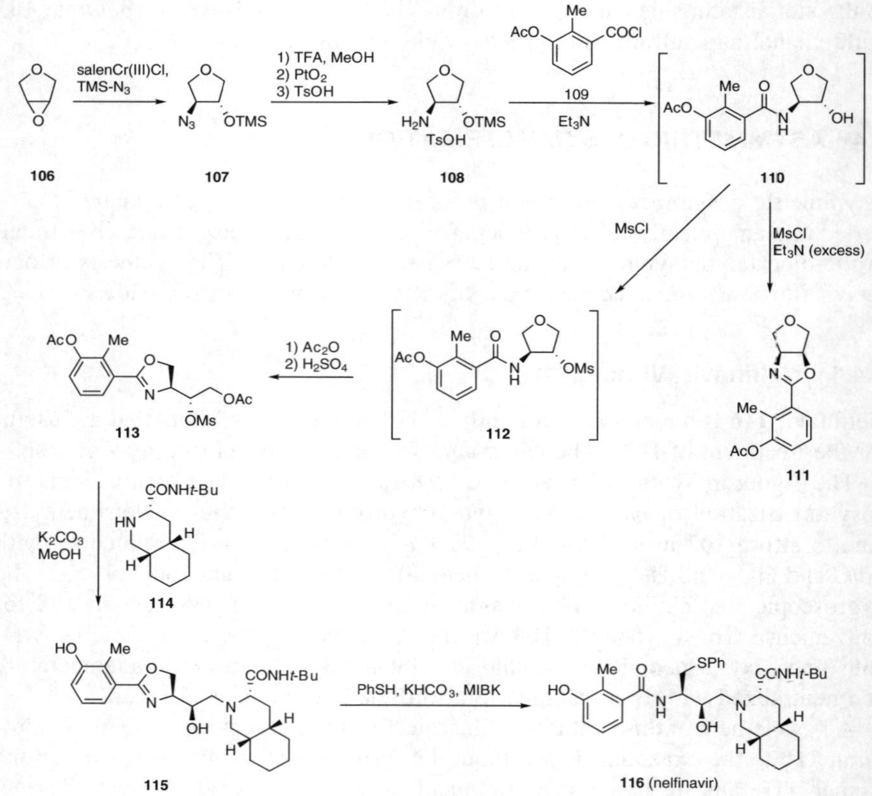

SCHEME 6.18   Synthesis of nelfinavir (**116**).

72% overall yield from the tosylate salt **108**. Finally, the oxazoline ring system of **115** can be cleaved by the reaction with thiophenol and potassium bicarbonate in MIBK to afford nelfinavir **116** in 82% yield.

6.5   STEREOSELECTIVE ISOMERIZATION

The famous synthesis of L-menthol illustrates the utility of asymmetric isomerization in which an allyl amine undergoes a selective allylic rearrangement to furnish the optically active enamine in 96–98% *ee*.

6.5.1   L-Menthol

The synthesis of L-menthol is an example of the synthesis of a chiral compound from a prochiral substrate via a stereoselective isomerization. Menthol is used in a variety of consumer and pharmaceutical products, and the demand is estimated at 4,500 tons per year.

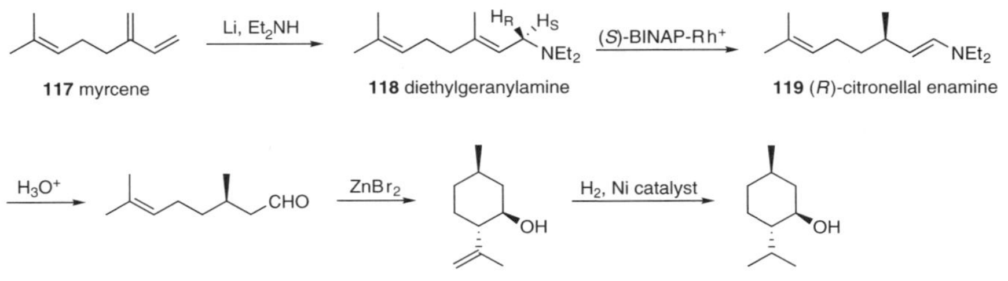

SCHEME 6.19 Synthesis of L-menthol (**122**).

Takasago Co. has developed a commercial process for the synthesis of $(-)$-menthol based on the catalytic asymmetric isomerization of an allylic amine as shown in Scheme 6.19. The synthesis begins by the reaction of myrcene **117** with diethylamine in the presence of lithium diethylamide to afford N,N-diethylgeranylamine **118**. This material is subjected to an asymmetric isomerization process mediated by Rh(I) and (S)-BINAP to afford the (R)-diethyl enamine **119** in 97% ee. The synthesis is completed by aqueous acidic hydrolysis of the enamine to $(+)$-citronellal **120**, Lewis acid-mediated cyclization of the aldehyde into $(-)$-isopulegol **121** ($\sim$100% ee after crystallization), and finally hydrogenation with a nickel catalyst to produce $(-)$-menthol **122** [65].

N,N-diethylgeranylamine **118** is a prochiral compound, and the allylic H_R and H_S protons are enantiotopic. When this allylic amine is exposed to the Rh(I) and (S)-BINAP catalyst, a [1,3]-suprafacial hydrogen shift occurs to form the (R)-enamine. The catalyst differentiates between the H_R and H_S protons in order to deliver the desired product in high enantiomeric excess. This asymmetric reaction to manufacture $(-)$-citronellal is run at a 9 ton per batch scale, and $(-)$-citronellal is converted into $(-)$-menthol at a scale of about 1,500 tons per year.

6.6 ASYMMETRIC REDUCTION

The asymmetric reduction of olefins and ketones has been widely used in the synthesis of organic synthesis. Industrially useful asymmetric reductions are exemplified by L-DOPA (enamide), metolachlor (imine), biotin (olefin), and sitagliptin (enamine), while the synthesis of aprepitant involves the asymmetric reduction of a ketone.

6.6.1 L-DOPA

In principle, one of the most straightforward and attractive chiral technologies is asymmetric hydrogenation, in which hydrogen is selectively added to one side of an unsaturated molecule to create the chiral product. Researchers have investigated asymmetric heterogeneous catalysis for many years, but in general it

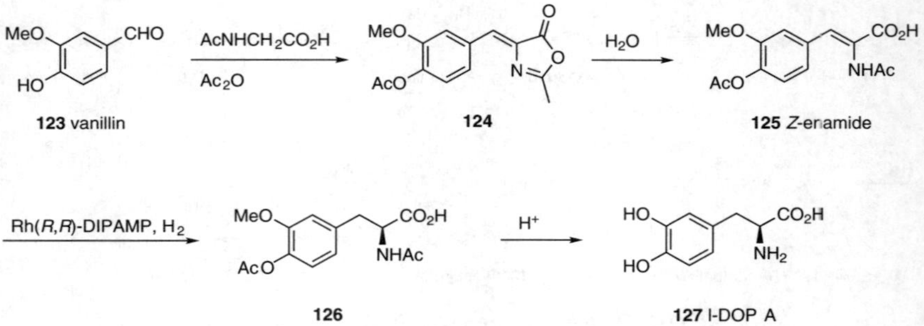

SCHEME 6.20 Synthesis of L-DOPA (**127**).

has been of limited commercial utility [66]. Homogeneous asymmetric catalysis, however, has become a very successful commercial technology. A major breakthrough occurred with the discovery by Wilkinson of RhCl(PPh₃)₃ as an efficient achiral homogeneous hydrogenation catalyst [67]. The utility of this reaction was extended into the realm of asymmetric synthesis by the use of chiral phosphine ligands in place of the simple triphenylphosphine ligand [68].

Knowles discovered that enamides successfully undergo asymmetric hydrogenation. A series of monodentate chiral phosphines were investigated as ligands in the reduction of an enamide with results up to about 88% *ee*. The key advance came when Knowles, inspired by the discovery of Kagan's bidentate DIOP ligand, discovered that the phosphine-based bidentate chiral ligand DIPAMP was highly efficient at chirality transfer during enamide hydrogenation reactions [69]. One of the best examples of the commercialization of this technology is Monsanto's synthesis of the anti-Parkinson's compound L-DOPA (Scheme 6.20), a material that was produced on a ton scale. Thus the Z-enamide **125** (prepared by the reaction of vanillin **123** wth acetylglycine followed by hydrolysis of the adduct **124**) was asymmetrically hydrogenated with Rh(I) and (*R,R*)-DIPAMP to afford the desired product **126** in 95% *ee*. The product was then deprotected under acidic conditions to afford L-DOPA **127**.

6.6.2 Metolachlor (Dual Magnum®)

Metolachlor is a grass herbicide that is used for maize. It is known commercially as Dual Magnum® [70]. Even though it contains just one stereochemical center, there are four possible stereoisomers of this compound because of the presence of atropisomers resulting from restricted rotation about the phenyl-nitrogen bond. Initially the product was sold as a mixture of all four stereoisomers when it was introduced to the market in 1976. However, by 1982 it was known that most of the herbicidal activity of metolachlor was found in the (*S*)-isomers and that the atropisomerism was not important for biological activity [71].

Because of the immense scale at which this compound was being manufactured (in excess of 20,000 tons per year), an enormous economy could be realized if an asymmetric synthesis could be developed. As a result a multiyear research campaign was launched to solve the asymmetric synthesis of metolachlor, which resulted in the development of a highly successful commercial process [72].

The racemic synthesis of metolachlor involves reductive amination of 2-ethyl-6-methyaniline with methoxyacetone followed by acylation with chloroacteyl chloride to give the desired product.

The asymmetric synthesis of metolachlor (Scheme 6.21) focused on the selective reduction of the imine intermediate **129** formed by the reaction of 2-ethyl-6-methyaniline **128** with methoxyacetone. Asymmetric hydrogenation of the imine **129** was accomplished with [Ir(COD)$_2$Cl]$_2$ and the xyliphos ligand **130** at high pressure (80 bar) at 50°C in the presence of acetic acid and iodide as additives to afford **131** in about 79% *ee* [73]. The synthesis was completed by the reaction of **131** with chloroacetyl chloride to furnish the metolachlor **132**. The final optimized process was exceedingly efficient, with a substrate-to-catalyst ratio of >1,000,000 and a reaction time of about 4 hours. This particular procedure is regarded as the largest-scale example of an industrial catalytic asymmetric process.

6.6.3 Biotin

Biotin is a water soluble B-complex vitamin. It is involved in important metabolic pathways such as gluconeogenesis, fatty acid synthesis, and amino acid catabolism [74]. Biotin also functions as a cofactor that aids in the transfer of carbon dioxide to various target macromolecules. In 2003, the world demand for synthetic biotin was about 35 metric tons per year [75].

Biotin has been a target for total synthesis [76]. However, an efficient industrial synthesis of biotin as shown in Scheme 6.22 was patented by Lonza [77]. The

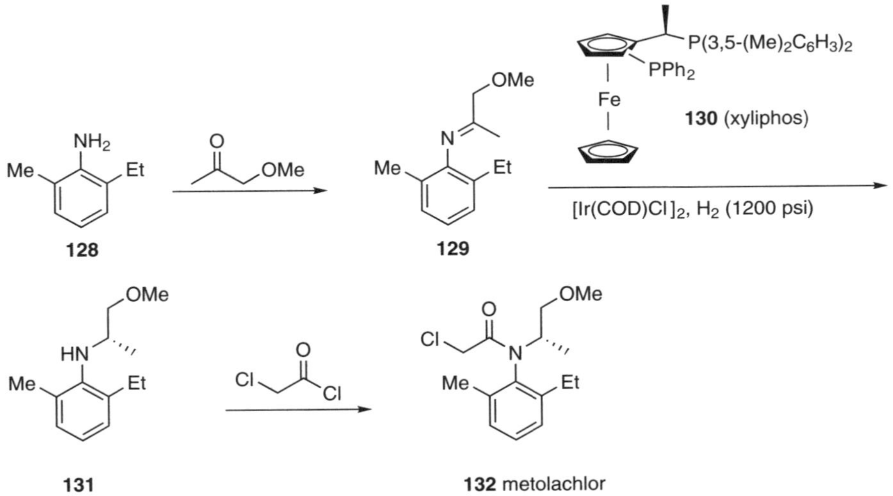

SCHEME 6.21 Synthesis of metolachlor (**132**).

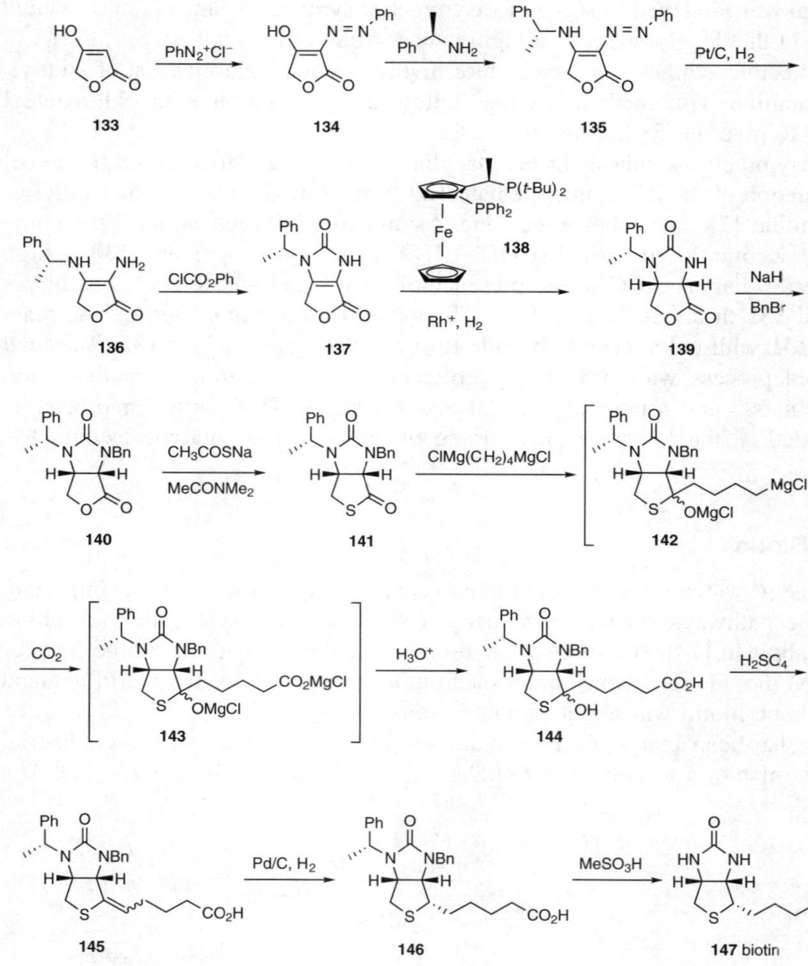

SCHEME 6.22   Synthesis of biotin (**147**).

synthesis begins with the electrophilic amination of tetronic acid **133** with phenyl-diazonium chloride to give **134** and condensation of the ketone moiety with the readily available (R)-α-methylbenzylamine to afford **135** [78]. Subsequent reduction to **136** with Pt/C and H_2 followed by cyclization with phenyl chloroformate produces the first critical intermediate **137**. Heterogeneous hydrogenation of the tetrasubstituted olefin under standard conditions with Rh/Al$_2$O$_3$ provided only a 70:30 isomer ratio because of the directing effect of the chiral auxillary. It was possible to isolate the pure desired diastereomer by crystallization, but the 58% yield was unacceptably low. After an extensive exploratory research effort to examine various metals (e.g., Rh, Ir, etc.) and ligands (e.g., diop, bppf, etc.) it was discovered that hydrogenation with H_2 using Rh with **138**, a member of the josiphos family of ferrocenyldiphosphine ligands, delivered compound **139** in 95% yield and 99:1 diastereomeric ratio [79].

With the desired absolute stereochemistry of the *cis*-fused ring system secured by the homogeneous asymmetric hydrogenation, compound **139** was protected by *N*-alkylation with NaH and benzyl bromide to afford **140**. Treatment of **140** with potassium thioacetate in dimethylacetamide at 150°C in the presence of catalytic hydroquinone and with the rigorous exclusion of atmospheric oxygen resulted in clean conversion to the thiolactone **141**.

The valeric acid side chain of biotin was introduced in a stepwise, telescoped procedure. Thiolactone **141** was first reacted with the di-Grignard reagent to afford the adduct **142**. Treatment of **142** with carbon dioxide led to a one-carbon homologation to afford the carboxylate **143**, which was quenched with aqueous acid to give **144** as a pair of diastereomers. Acidification of **144** with concentrated sulfuric acid in toluene at 65°C dehydrated the tertiary alcohol to afford **145** as a mixture of isomers.

The synthesis of D-(+)-biotin **147** was completed by selective hydrogenation of **145** from the less-hindered side of the trisubstituted olefin with Pd/C to set the desired side chain configuration in **146** followed by the removal of both of the benzylic protecting groups by treatment with methanesulfonic acid to furnish **147**.

6.6.4   Sitagliptin (Januvia®)

Type 2 diabetes mellitus (T2DM) is a major epidemic. Formerly known as non-insulin-dependent diabetes mellitus (NIDDM) or adult-onset diabetes, it is characterized by high blood glucose [80]. It is estimated that approximately 150 million people worldwide suffer from diabetes, and this value is expected to double by 2025 [81]. In the United States there are roughly 24 million cases of diabetes, and typically about 90% of these cases are type 2 [82].

Dipeptidyl peptidase IV (DPP-4) inhibitors have recently emerged as a new class of antihyperglycemic agents for the treatment of T2DM [83]. Sitagliptin phosphate **154** is a selective and potent inhibitor of DPP-4 developed by Merck [84]. The molecule contains just one chiral center, but it is very interesting to observe how the process to make the drug substance was dramatically refined and improved as a result of innovative process chemistry research.

The first-generation large-scale synthesis reported by Merck began with the prochiral β-ketoester as shown in Scheme 6.23 [85]. This procedure was used to make >100 kg of sitagliptin. The stereochemistry of the final product was introduced by the asymmetric reduction of the β-ketoester **148** using (*S*)-BinapRuCl$_2$-trimethylamine complex in methanol at 90 psi H$_2$ and 80°C in the presence of catalytic HBr to afford the (*S*)-hydroxy ester **149** which was 90.8% *ee* by chiral HPLC. The (*S*)-hydroxy ester **149** was not isolated but was instead hydrolyzed with NaOH in aqueous methanol to afford the isolated (*S*)-hydroxy carboxylic acid in 83% yield and 94% *ee*. The stereochemical center was then inverted by a two-step procedure involving a Mitsunobu reaction. The carboxylic acid was coupled with *O*-benzylhydroxylamine hydrochloride using EDC-HCl and

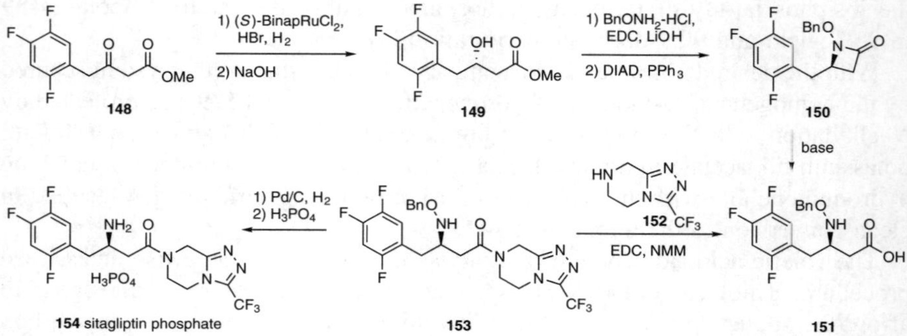

SCHEME 6.23 First First-generation synthesis of sitagliptin phosphate (**154**).

LiOH in aqueous THF to afford the hydroxamate. Without isolation, the secondary alcohol was activated with DIAD and PPh$_3$, whereupon intramolecular cyclization delivered the β-lactam **150** in 82% yield and 99.7% *ee*. The β-lactam was hydrolyzed with LiOH in aqueous THF to the protected β-amino acid, coupled with the triazolopiperazine **152** using EDC-HCl and *N*-methylmorpholine in acetonitrile to afford **153**, deprotected by catalytic hydrogenation (Pd/C, H$_2$), and acidified with phosphoric acid to deliver sitagliptin phosphate **154** in 52% overall yield. The synthesis was very successful, and it delivered the necessary product for the initial development of the compound, but it was not without drawbacks. Although the yield was high, there were multiple steps and isolations and waste associated with such steps. Furthermore, creating the stereocenter and then immediately inverting it by a Mitsunobu reaction was not particularly efficient.

A highly efficient second-generation synthesis was recently reported by Merck as shown in Scheme 6.24 that features the synthesis of the key dehydrositagliptin intermediate **161** in one pot. Asymmetric hydrogenation of this material leads to the sitagliptin free base **162** in >95% *ee*. The free base was converted into the desired final pharmaceutical form (i.e., the phosphate monohydrate salt) in >99.9% purity and >99.9% *ee* [86,87].

The one-pot preparation of **161** begins by treating 2,4,5-triphenylacetic acid **155**, Meldrum's acid **156**, and DMAP in acetonitrile sequentially with *i*-Pr$_2$NEt and *t*-BuCOCl, to afford the Hunig's base salt **157**. The triazolopiperazine hydrochloride salt **152** (HCl salt) is added, followed by CF$_3$CO$_2$H to afford the ketoamide **160**. A solution of **160** in acetonitrile is added to NH$_4$OAc in methanol to deliver dehydrositagliptin **161** in 82% overall yield. A considerable effort was made to understand the details of the reaction mechanism and optimize the process. In summary, under basic conditions the Hunig's base salt **157** is fairly stable. However, upon acidification of **157** with CF$_3$CO$_2$H the free acid **158** degraded to the oxo-ketene **159** and reacted with the amine hydrochloride **152** to afford the ketoamide **160**. The ketoamide **160** was converted exclusively to the *Z*-isomer of dehydrositagliptin **161** under the reaction conditions.

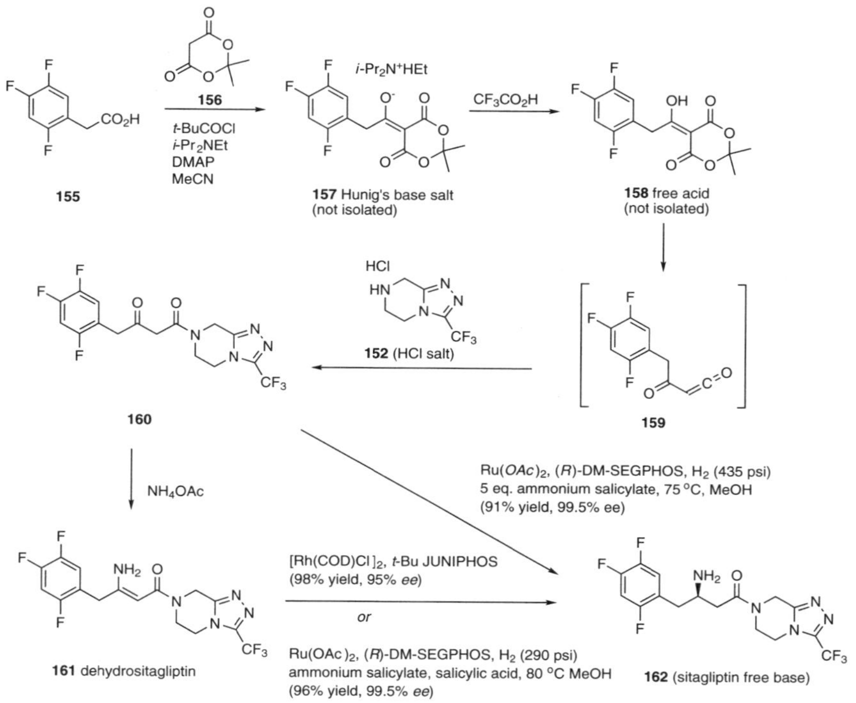

SCHEME 6.24   Second-generation synthesis of sitagliptin free base (**162**).

Next, enantioselective hydrogenation of the dehydrositagliptin **161** was accomplished with [Rh(COD)Cl]$_2$ and t-Bu JUNIPHOS in 98% yield and 95% ee. This was noteworthy because the direct hydrogenation of an unprotected enamine was unprecedented. Residual rhodium was removed and recovered with Ecosorb C-941, and the product was crystallized as the free base from IPA in 84% yield and >99% ee. The free base was converted to the phosphate monohydrate form in aqueous IPA to afford sitagliptin phosphate in 96% yield, 99.9 A% purity, and >99.9% ee. This is the current manufacturing process for sitagliptin phosphate.

Overall, this new process increased the overall yield of sitagliptin by nearly 50% and reduced the waste generated per kilogram of sitagliptin by 80%, from 250 kg to 50 kg. Furthermore, 95% of the rhodium catalyst was recovered and recycled. Merck earned a Presidential Green Challenge Award in 2005 for this improved synthesis of sitagliptin [88].

More recently, further process improvements were discovered that may ultimately supercede the current process [89]. An improved catalyst for the asymmetric reduction of the dehydrositagliptin **161** was discovered. Reduction of **161** with Ru(OAc)$_2$ and (R)-DM-SEGPHOS with 290 psi H$_2$ in the presence of salicylic acid and ammonium salicylate at 80°C delivers sitagliptin **162** in 96% yield and 99.5% ee.

The Merck researchers have also discovered that it is possible to perform an asymmetric reductive amination directly on **160** without the need to first convert it to dehydrositagliptin 161. Thus treatment of 160 at 435 psi H_2 with Ru(OAc)$_2$ and (R)-DM-SEGPHOS in the presence of 5 eq. of ammonium salicylate at 75°C affords sitagliptin **162** in 91% yield and 99.5% *ee*.

6.6.5   Aprepitant (Emend®)

Aprepitant (**177**) is a potent, orally active NK$_1$ receptor antagonist that is useful for the treatment of chemotherapy-induced emesis [90]. Aprepitant is a structurally interesting molecule with a *cis*-configuration of substituents on the morpholine ring and three stereocenters in close proximity. A practical and efficient synthesis of this material was required to support the commercialization of this compound [91].

The synthesis in Scheme 6.25 began with the conversion of N-benzylethanolamine **163** into the lactam lactol **164**. This can be readily accomplished by a two-step procedure involving cyclization with diethyl oxalate and reduction with lithium tri(*sec*-butyl)borohydride. However, a more direct route that avoided the need for the reduction of the intermediate lactam lactone was discovered. Treatment of **163** with 2.3 eq. of aqueous glyoxylic acid in aqueous THF with heating afforded **164** in 76% yield after crystallization directly from the reaction mixture.

Separately, enantiomerically pure (R)-3,5-bis(trifluorophenyl)-*sec*-phenethyl alcohol **168** was synthesized either by the selective reduction of 3,5-bis(trifluorophenyl)-acetophenone **167** with a ruthenium-catalyzed transfer hydrogenation using (1S, 2R)-*cis*-amino-2-indanol as the chiral ligand (91% *ee* and 92% yield) or with an oxazaborolidine-catalyzed (CBS) borane reduction (93% *ee* and 97% yield). In either case the chiral alcohol **168** could be recrystallized to >99% *ee*.

The most critical part of the whole process was the coupling of racemic **164** with enantiopure (R)-**168** to afford the desired (R,R)-**169**. This was envisioned to occur in several distinct steps including 1) activation of **164**, 2) reaction of activated **164** with chiral alcohol **168** to afford a mixture of **169** and **170**, and 3) equilibration of **169** and **170** and the selective crystallization of **169** from the reaction mixture via a crystallization-induced diastereoselective transformation [92]. Each of these independent steps was optimized in great detail.

Initially, lactol **164** was activated for nucleophilic substitution by conversion to the imidate **165** by stirring with trichloroacetonitrile and potassium carbonate in methylene chloride. The imidate **165** reacted with **168** in the presence of catalytic TMSOTf to produce a 55:45 mixture of the diastereomers **169** and **170** in 90% yield, but **165** was not an ideal intermediate because it was an unstable, moisture-sensitive, and difficult to handle material, along with the fact that it also produced significant amounts of a dimeric species. It was ultimately discovered that the trifluoroacetate derivative **166** (prepared by the addition of 1.0 eq. of TFAA to **164**) offered the best combination of activation of **164** toward nucleophilic

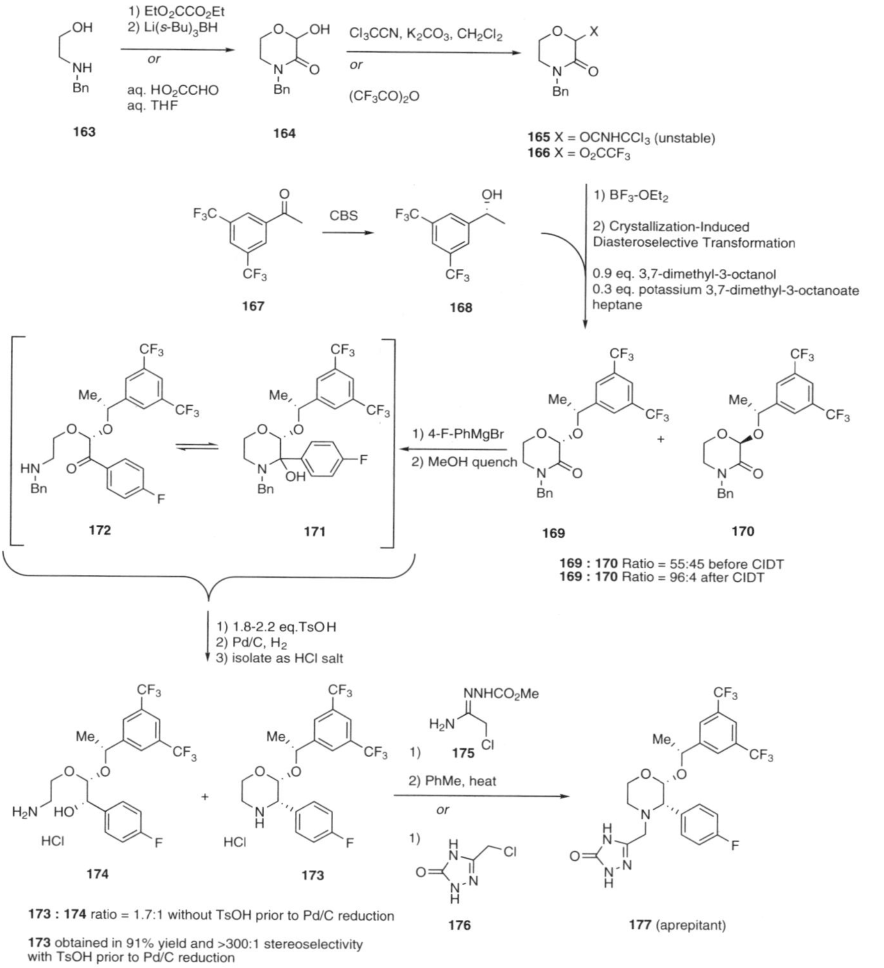

SCHEME 6.25   Synthesis of aprepitant (**177**).

substitution with a minimal tendency toward transesterification. Treatment of the trifluoroacetate ester **166** with **168** catalyzed by $BF_3 \cdot Et_2O$ affords a 55:45 mixture of **169** and **170** in 95–98% yield.

The isomers **169** and **170** were separated chromatographically in order to examine their properties individually. Compound **169** (the desired isomer) was found to be a highly crystalline compound, whereas compound **170** was found to be a much lower melting solid. During crystallization studies, it was observed that seeding a mixture of the isomers in solution with **169** led to selective crystallization of **169**, setting the stage for a possible crystallization-induced diastereoselective transformation provided that a equilibrium between **169** and **170** could be established in a suitable solvent.

After careful process research it was determined that the optimum procedure is to dissolve the crude 55:45 mixture of **169** and **170** and 0.9 eq. of 3,7-dimethyl-3-octanol in heptane, cool to −10 to −5°C, and seed with **169**. Upon addition of 0.3 eq. of potassium 3,7-dimethyl-3-octanoate the base-catalyzed crystallization-driven epimerization is initiated, and over a period of 5 hours the 55:45 mixture is converted into the equivalent of a 96:4 mixture of **169** to **170** (homogenized basis) because the desired diastereomer crystallizes with high recovery and the undesired diastereomer is continuously epimerized. The desired product **169** can be isolated in 83–85% yield (based on chiral alcohol **168**) and 99% *ee*.

The next step is the conversion of the lactam **169** to the morpholine **171**. It was not necessary to activate **169** toward nucleophilic addition because it reacted readily with 4-fluorophenylmagnesium bromide in a variety of solvents to give the presumed equilibrium mixture of **171** and **172**. This mixture could be reduced to afford **173** as the major product in a complex mixture. The issue was that during the workup of the initial Grignard reaction a mixture of **171** and **172** as well as other side-products were observed. Therefore, a telescoped procedure that avoided isolation of the Grignard adduct was sought. After careful investigation it was determined that quenching the Grignard reaction mixture into methanol followed by hydrogenation in the presence of Pd/C successfully reduced the formation of the side-products but led to just a 1.7 to 1.0 mixture of **173** to **174**. Fortunately, it was determined that acidification of the quenched Grignard reaction mixture prior to hydrogenation dramatically shifted the hydrogenation selectivity toward the desired product **173**. In the optimized procedure, **169** was reacted with 1.3 eq. of 4-fluorophenylmagnesium bromide in THF, quenched with methanol, treated with 1.8–2.2 eq. of 4-toluenesulfonic acid, and hydrogenated with 5% Pd/C to afford **173** as the HCl salt in 91% yield and >300:1 stereoselectivity in the reduction [93].

Finally, **173**-HCl could be converted to aprepitant in two steps by the reaction with **175** followed by thermolysis in toluene in 85% yield, or in one step by the reaction with **176**.

Merck received a Presidential Green Challenge Award in 2005 for this redesigned, efficient synthesis of aprepitant that nearly doubled the yield of the first-generation synthesis [94].

6.7 ASYMMETRIC OXIDATION

Asymmetric oxidation is also widely practiced in organic synthesis. The industrially useful asymmetric oxidation of a sulfide is illustrated by the synthesis of esopremazole, and the asymmetric oxidation of indene is demonstrated in the synthesis of indinavir sulfate.

6.7.1 Esomeprazole (Nexium®)

Esomeprazole **179** (Scheme 6.26) is a proton pump inhibitor that is useful in the treatment of heartburn and acid reflux disease [95]. Esomeprazole is

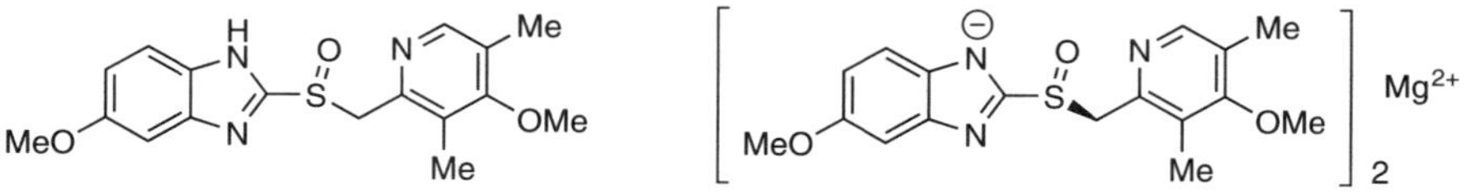

SCHEME 6.26   Omeprazole (**178**) and esomeprazole (**179**).

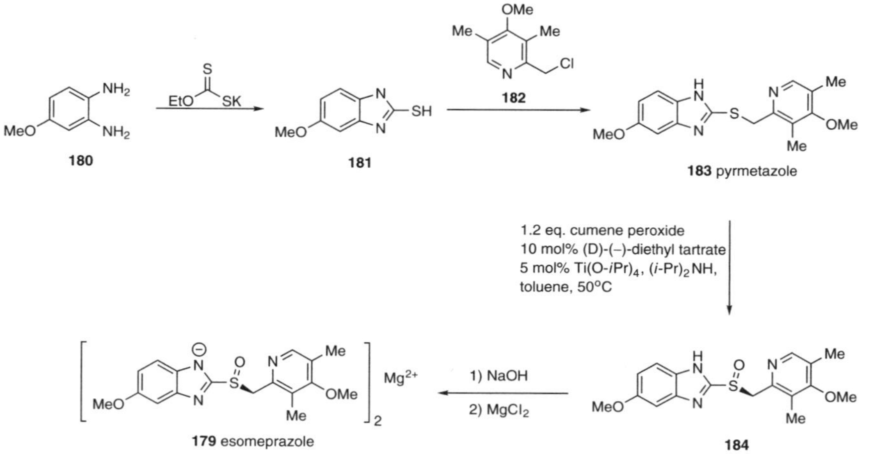

SCHEME 6.27   Synthesis of esomeprazole (**179**).

the magnesium salt of the (*S*)-enantiomer of racemic omeprazole (Prilosec®) **178**. The first development supply of esomeprazole was provided by covalently linking omeprazole with L-mandelic acid and separating the resulting diastereomers by HPLC [96]. However, because of the unacceptable volume and time requirements of such a procedure at large scale, an asymmetric synthesis was required.

The synthesis of omeprazole, the achiral precursor of esomeprazole, is fairly straightforward in Scheme 6.27. Treatment of 4-methoxybenzene-1,2-diamine **180** with potassium ethyl xanthate gives thiobenzimidazole **181**. Compound **181** was coupled with the 2-chloromethylpyridine derivative **182** to afford the adduct pyrmetazole **183**, which was nonselectively oxidized by *m*-chloroperbenzoic acid to give omeprazole **178** [97,98].

Attempts to synthesize esomeprazole **179** by asymmetric oxidation of **183** to the chiral sulfoxide **184** following Kagan's original procedure were unsuccessful. However, after considerable experimentation it was found that asymmetric oxidation with a modified Sharpless reagent was indeed successful. A mixture of **183** with catalytic Ti(O-*i*Pr)$_4$ (0.3 eq.), (D)-(−)-diethyl tartrate (0.6 eq.), and water (0.1 eq.) in toluene was heated at 54°C for about 50 minutes, followed by treatment with Hunig's base (0.3 eq.) and stoichiometric cumene hydroperoxide to afford **184** in 94% *ee* [99]. The purity was upgraded to >99.5% *ee* by crystallization of the **184** sodium salt from MIBK and acetonitrile. Conversion of the sodium salt to the magnesium salt completed the synthesis of esomeprazole **179**.

6.7.2 Indinavir Sulfate (Crixivan®)

HIV protease inhibitors are useful for the treatment of HIV [100]. Indinavir sulfate **199** (Crixivan®, Merck) is one member of a family of transition state mimic HIV protease inhibitors, including compounds such as saquinavir (Invirase®, Roche), ritonavir (Norvir®, Abbott), and nelfinavir (Viracept®, Agouron).

Indinavir sulfate contains five chiral centers, so for a synthesis of this material to be truly practical, it must demonstrate a very high degree of stereochemical control. The Merck synthetic strategy (Schemes 6.28–6.30) begins with the asymmetric epoxidation of indene **185** using a chiral manganese-salen catalyst. The chirality of the optically active indene oxide **186** is then used to induce the desired stereochemistry of two additional stereochemical centers of a synthetic intermediate **191**. The fifth stereochemical center is set by a chiral resolution in the synthesis of a second key intermediate, **197**. Finally, these optically active fragments are joined and further elaborated to deliver indinavir [101].

The synthesis of the key intermediate **191** in Scheme 6.28 begins with the asymmetric epoxidation of indene **185** using a manganese-salen catalyst in methylene chloride with catalytic 4-phenyl propyl pyridine-*N*-oxide and sodium hypochlorite as the stoichiometric oxidant to afford (1*S*,2*R*)-indene oxide **186** in 89% yield and 88% *ee* [102,103]. The epoxide **186** is treated with fuming sulfuric acid in acetonitrile to afford the intermediate oxazoline, which is then hydrolyzed to afford (1*S*,2*R*)-1-amino-2-indanol as the sulfate salt. The amino alcohol sulfate salt is neutralized with sodium hydroxide, crystallized with L-tartaric acid, and neutralized with sodium hydroxide to afford (1*S*,2*R*)-1-amino-2-indanol **187** in >99% *ee* [104].

Reaction of **187** with 3-phenylpropionyl chloride followed by protection with 2-methoxypropene in the presence of catalytic acid affords the acetonide **188**,

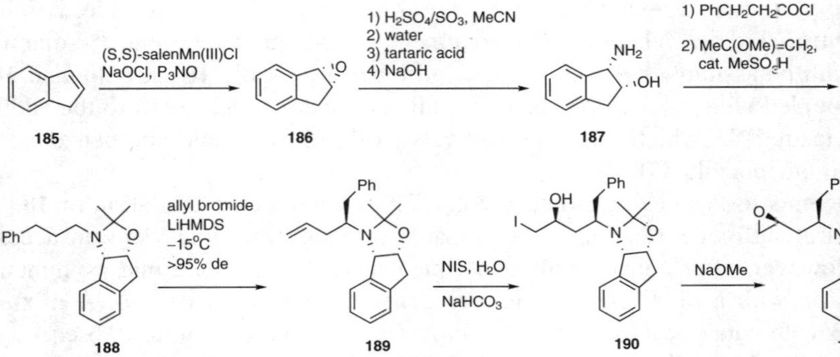

SCHEME 6.28 Synthesis of the key intermediate **191** for indinavir.

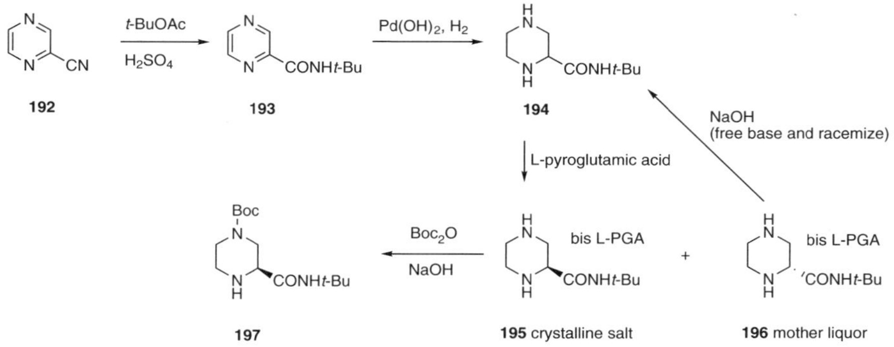

SCHEME 6.29   Synthesis of key intermediate piperazinecarboxamide **197**.

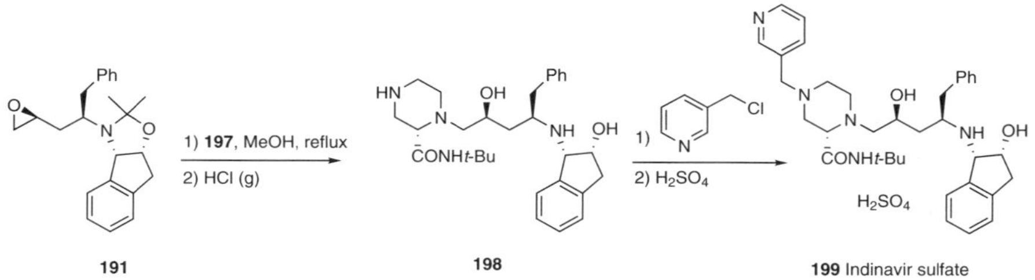

SCHEME 6.30   Synthesis of indinavir sulfate (**199**).

which is then treated with LiHMDS and allyl bromide at $-30°C$ to afford **189** with very high diastereoselectivity (96:4) in favor of the desired (R)-isomer. The olefin **189** is then subjected to selective iodohydroxylation with NIS in aq. NaHCO$_3$ to afford the iodo alcohol **190** in >97% de [105]. (An alternative diastereoselective iodohydroxylation procedure was published in which NaI and NaOCl in an IPAC/NaHCO$_3$ biphasic system at pH 8–9.5 was employed to synthesize the iodoalcohol **190** [106].) The iodo alcohol **190** is converted to the epoxide **191** with retention of configuration by reaction with sodium methoxide to deliver the first key intermediate in the synthesis of indinavir.

The second key intermediate in the synthesis is the differentially protected chiral piperazinecarboxamide (S)-**197**. The racemic form of this material is conveniently available as shown in Scheme 6.29 from cyanopyrazine **192** by reaction with t-BuOAc and H$_2$SO$_4$ to make the t-Bu-pyrazinecarboxamide **193** followed by reduction of the pyrazine ring with Pd(OH)$_2$ and H$_2$ to afford **194**. The racemic **194** can be resolved by crystallization with L-pyroglutamic acid (L-PGA) to deliver the desired (S)-isomer as the bis salt **195**. The undesired (R)-isomer **196** can be racemized with a base and recycled. The (S)-isomer salt **195** is then selectively protected as the mono-Boc derivative with Boc$_2$O and NaOH to afford **197**.

The epoxide **191** is joined by a ring-opening reaction with piperazine (S)-**197** in refluxing methanol (Scheme 6.30) followed by cleavage of the Boc and acetonide protecting groups with HCl to afford the adduct **198**. Finally, the synthesis is completed by alkylation of **198** with 3-picolyl chloride followed by sulfate salt formation to afford indinavir sulfate **199**.

CITED BY URL

Pedersen J, Wallace M. 1999. Wiley Journals DTD: Guidelines for reference tagging. Available at http://jws-edcd.wiley.com:8255/refguide.html. Accessed 2002 Feb 4.

PATENTS

With Inventors

Harred JF, Knight AR, McIntyre JS, inventors; Dow Chemical Company, assignee. Epoxidation process. US patent 3,654,317. 1972 Apr 4.

Citation by Patent Number (more detailed information is not available)

US Patent 3,654,317. 1972 Apr 4.

REFERENCES

1. a) Federsel, H. J. (2005). Asymmetry on large scale: the roadmap to stereoselective processes. *Nat. Rev. Drug Discov.* **4**, 685–697;   b) Li, J. J., Johnson, D. S., Sliskovic, D. R., Roth, B. D. (2004). *Contemporary Drug Synthesis*. New York: John Wiley & Sons, Inc.;   c) Gadamasetti, K. G., editor. (1999). *Process Chemistry in the Pharmaceutical Industry*. CRC Press. d) Gadamasetti, K., Braish, T., editors. (2008). *Process Chemistry in the Pharmaceutical Industry*, Volume 2, CRC Press;   e) Sheldon, R. A. (1993). *Chirotechnology*. Marcel Dekker;   f) Walker, D. (2008). *The Management of Chemical Process Development in the Pharmaceutical Industry*. John Wiley & Sons, Inc..   g) Collins, A. N., Sheldrake, G. N., Crosby J., editors. (1992). *Chirality in Industry*. John Wiley & Sons, Inc.;   h) Collins, A. N., Sheldrake, G. N., Crosby J., editors. (1997). *Chirality in Industry II*. John Wiley & Sons, Inc.;   i) Ager, D. J., editor. (1999). *Handbook of Chiral Chemicals*. Marcel Dekker;   j) Seyden-Penne, J. (1995). *Chiral Auxillaries and Ligands in Asymmetric Synthesis*. John Wiley & Sons, Inc.. k) Ojima, I., editor. (2000). *Catalytic Asymmetric Synthesis*, 2nd ed. John Wiley & Sons, Inc.;   l) Johnson, D. S., Li, J. J., editors. (2007). *The Art of Drug Synthesis*. John Wiley and Sons, Inc..   m) Blaser, H. U., Schmidt, E., editors. (2004). *Asymmetric Catalysis on an Industrial Scale*. Wiley-VCH.

2. Arrowsmith, J. E., Campbell, S. F., Cross, P. E., Stubbs, J. K., Burges, R. A., Gardinier, D. G., Blackburn, K. J. (1986). Long-acting dihydropyridine calcium antagonists. 1,2-alkoxymethyl derivatives incorporating basic substituents. *J. Med. Chem.* **29**, 1696–1702.

3. Christen, D. P. (2007). *The Art of Drug Synthesis.* Johnson, D. S., Li, J. J. editors. John Wiley & Sons, Inc., p. 159–167.

4. Campbell, S. F., Cross, P. E., Stubbs, J. K. (1986). 2-(Secondary aminoalkoxymethyl) dihydropyridine derivatives as anti-ischaemic and antihypertensive agents. Patent No. US 4,572,909.

5. a) Joshi, R. R., Joshi, R. A., Karade, N. B., Gurjar, M. K. (2005). Process for preparation of chiral amlodipine salts. Patent No. U.S. 6,846,932; b) Chung, Y. S., Ha, M. C. (2007). Processes for the preparation of *S*-(−)-amlodipine. Patent No. US 7,202,365.

6. Lee, H. W., Shin, S. J., Yu, H., Kang, S. K., Yoo, C. L. (2009). A novel chiral resolving reagent, bis((*S*)-mandelic acid)-3-nitrophthalate, for amlodipine racemate resolution: scalable synthesis and resolution process. *Org. Process Res. Dev.* **13**, 1382–1386.

7. Peters, T. H. A., Benneker, F. B. G., Slanina, P., Bartl, J. (2005). Process for making amlodipine, derivatives thereof, and precursors therefor. Patent No. U.S. 6,858,738.

8. a) Wong, D. T., Robertson. D. W., Bymaster, F. P., Krushinski, J. H., Reid, L. R. (1988). LY 227942. An inhibitor of serotonin and norepinephrine uptake-biochemical pharmacology of a potential antidepressant drug. *Life Sci.* **43**, 2049–2057; b) Deeter, J., Frazier, J., Staten, G., Staszak, M. Weigel, L. (1986). Asymmetric synthesis and absolute stereochemistry of LY248686. *Drugs Future* **11**, 134–135; c) Ankier, S. I. (1986). Recent progress in the development of new antidepressant drugs. *Prog. Med. Chem.* **23**, 121–185; d) Robertson, D. W., Krushinski, J. H., Fuller, R. W., Leander, J. D. (1988). The absolute configurations and pharmacological activities of the optical isomers of fluoxetine, a selective serotonin-uptake inhibitor. *J. Med. Chem.* **31**, 1412–1417; e) Fuller. R. W. (1986). Pharmacological modification of serotonergic function-drugs for the study and treatment of psychiatric and other disorders. *J. Clin. Psychiatry* **47**, 4–8.

9. Deeter J., Frazier J., Staten G., Staszak M., Weigel L. (1990). Asymmetric synthesis and absolute stereochemistry of LY248686. *Tetrahedron Lett.* **31**, 7101–7104.

10. Piñiero-Núñez, M. (2007). *The Art of Drug Synthesis.* Johnson, D. S., Li, J. J., editors. John Wiley & Sons, Inc., chapter 14, p. 199–213.

11. a) Yamaguchi, S., Mosher, H. S., Pohland, A. (1972). Reversal in stereoselectivity depending upon the age of a chiral lithium alkoxyaluminohydride reducing agent. *J. Am. Chem. Soc.* **94**, 9254–9255; b) Yamaguchi, S., Mosher, H. S. (1973). Asymmetric reductions with chiral reagents from lithium aluminum hydride and (+)-(2*S*,3*R*)-4-dimethylamino-3-methyl-1,2-diphenyl-2-butanol. *J. Org. Chem.* **38**, 1870–1877; c) Brinkmeyer R.

S., Kapoor. V. M. (1977). Asymmetric reduction. Reduction of acetylenic ketones with chiral hydride agent. *J. Am Chem. Soc.* **99**, 8339–8341.

12. Astleford B. A., Weigel, L. O. (1997). *Chirality in Industry II*, Collins, A. N., Sheldrake, G. N., Crosby J., editors. John Wiley & Sons, Inc., chapter 6, p. 99–117.

13. Li, J., Liu, K., Sakya, S. (2005). Synthetic approaches to the 2004 new drugs. *Mini-Reviews Medicinal Chem.* **5**, 1133–1144.

14. Bymaster, F. P., Beedle, E. E., Findlay, J., Gallagher, P. T., Krushinski, J. H., Mitchell, S., Robertson, D. W., Thompson, D. C., Wallace, L. Wong, D. T. (2003). Duloxetine (Cymbalta™), a dual inhibitor of serotonin and norepinephrine reuptake. *Bioorg. Med. Chem. Lett.* **13**, 4477–4480.

15. Berglund, R. A., (1994). Asymmetric synthesis. Patent No. US 5,362,886.

16. Astleford B. A., Weigel, L. O. (1997). *Chirality in Industry II*, Collins, A. N., Sheldrake, G. N., Crosby, J., editors, John Wiley & Sons, Inc., chapter 6, p. 99–117.

17. Fujima, Y., Ikunaka, M., Inoue, T., Matsumoto, J. (2006). Synthesis of (*S*)-3-(*N*-methylamino)-1-(2-thienyl)propan-1-ol: revisiting Eli Lilly's resolution-racemization-recycle synthetic of duloxetine for its robust processes. *Org. Process Res. Dev.* **10**, 905–913.

18. Ini, S., Abramov, M., Liberman, A. (2009). Process for the Purification of Duloxetine Hydrochloride. Patent No. U.S. 7,534,900.

19. Butchko, M. A., Merschaert, A., Moder, K. P. (2009). Process for the Asymmetric Synthesis of Duloxetine. Patent No. US 7,538,232.

20. Sturmer, R. (2008). Method for the Production of (*S*)-3-Methylamino-1-(thieny-2-yl) propan-1-ol, Patent No. U.S. 7,435,563.

21. Li, J. J., Johnson, D. S., Sliskovic, D. R., Roth, B. D. (2004). *Contemporary Drug Synthesis*. Wiley-Interscience, chapter 10, p. 125–147.

22. Taber, G. P., Pfisterer, D. M., Colberg, J. C. (2004). A new and simplified process for preparing *N*-[4-(3,4-dichlorophenyl)-3,4-dihydro-1(2H)-naphthalenylidene] methanamine and a telescoped process for the synthesis of (1*S*-*cis*)-4-(3,4-dichlorophenol)-1,2,3,4-tetrahydro-*N*-methyl-1-naphthalenamine mandelate: Key Intermediates in the synthesis of sertraline hydrochloride. *Org. Process Res. Dev.* **8**, 385–388.

23. *The Presidential Green Chemistry Challenge*, Award Recipients 1996–2009, U.S. EPA.

24. a) Diekema., D. J., Jones, R. N. (2001). Oxazolidinone antibiotics. *Lancet* **358**, 1975–1982; b) Aoki, H., Ke, L., Poppe, S. M., Poel, T. J., Weaver, E. A., Gadwood, R. C., Thomas, R. C., Shinabarger, D. L., Ganoza, M. C. (2002). Oxazolidinone antibiotics target the P site on *escherichia coli* ribosomes. *Antimicrob. Agent Chemother.* **46**, 1080–1085.

25. Shinabarger, D. (1999). Mechanism of action of the oxazolidinone antibacterial agents. *Expert Opin. Invest. Drugs* **8**, 1195.

26. a) Barbachyn, M. R., Brickner, S. J., Hutchingson, D. K. (1995). 3-Substedphenyl-5-amino-methyl-2-oxo-oxazolidine(s) are antimicrobial agents, useful against e.g. staphylococci, streptococci, enterococci, Bacteroides spp.

Mycobacterium tuberculosis. WO 95/07271; b) Brickner, S. J., Hutch-ingson, D. K., Barbachyn, M. R., Manninen, P. R., Ulanowicz, D. A., Garmon, S. A., Crega, K. C., Hendges, S. K., Toops, D. S., Ford, C. W., Zurenko, G. E. (1996). Synthesis and autibacterial activity of U-100592 and U-100766, two oxazolidinone antibacterial agents for the potential treatment of multidrug-resistant gram-positive bacterial infections. *J. Med. Chem.* **39**, 673–679; c). Barbachyn, M. R., Brickner. S. J., Hutchingson. D. K. (1997). Substituted oxazine and thiazine oxazolidinone antimicrobials. US 5,688,792; d) Pearlman, B. A., Perrault, W. R., Barbachyn, M. R., Man-ninen. P. R., Toops, D. S., Houser, D. J., Fleck, T. J. (1997). Process to prepare oxazolidinones. WO 97/37980; e) Pearlman. B. A. (1999). New intermediates of pharmacologically active enantiomeric 5-acylamino-3-phenyloxazolidinones. WO 99/24393.

27. Li, J. J., Johnson, D. S., Sliskovic, D. R., Roth, B. D. (2004). *Contemporary Drug Synthesis*. John Wiley & Sons, Inc. p. 84.

28. Perrault, W. R., Pearlman, B. A., Godrej, D. B. (2002). Preparation of pharmaceutically active (*S*)-2-oxo-5-oxazolidinylmethylacetamides via a one step process from (*S*)-acetamidoacetoxypropanes and *N*-aryl-*O*-alkylcarbamates. WO02/085849.

29. Perrault W. R., Pearlman, B. A., Godrej, D. B., Jeganathan, A., Yamagata, K., Chen, J. J., Lu, C. V., Herrinton, P. M., Gadwood, R. C., Chan, L., Lyster, M. A., Maloney, M. T., Moeslein, J. A., Greene, M. L., Barbachyn, M. R. (2003). The synthesis of *N*-aryl-5(*S*)-aminomethyl-2-oxazolidinone antibacterials and derivatives in one step from aryl carbamates. *Org. Process Res. Dev.* **7**, 533–546.

30. Karpf, M., Trussardi, R. (2001). New, azide-free transformation of epoxides into 1,2-diamino compounds: synthesis of the anti-influenza neuraminidase inhibitor oseltamivir phosphate (Tamiflu). *J. Org. Chem.* **66**, 2044–2051.

31. Yeung, Y. Y., Hong, S., Corey, E. J. (2006). A Short enantioselective path-way for the synthesis of the anti-influenza neuramidase inhibitor oseltamivir from 1,3-butadiene and acrylic acid. *J. Am. Chem. Soc.* **128**, 6310–6311.

32. a) Fukuta, Y., Mita, T., Fukuda, N., Kanai, M., Shibasaki, M. (2006). De novo synthesis of tamiflu via a catalytic asymmetric ring-opening of *meso*-aziridines with $TMSN_3$. *J. Am. Chem. Soc.* **128**, 6312–6313; b) Mita, T., Fukuda, N., Roca, F. X., Kanai, M., Shibasaki, M. (2007). Second generation catalytic asymmetric synthesis of tamiflu: allylic substitution route. *Org. Lett.* **9**, 259–262.

33. Satoh, N., Akiba, T., Yokoshima, S., Fukuyama, T. (2007). A practical synthesis of (−)-oseltamivir. *Angew. Chem. Int. Ed.* **46**, 5734–5736.

34. Trost, B. M., Zhang, T. (2008). A concise synthesis of (−)-oseltamivir. *Angew. Chem. Int. Ed.* **47**, 1–4.

35. Karpf, M., Trussardi, R. (2009). Process for the preparation of 4,5-diamino shikimic acid derivatives. Patent No. U.S. 7,514,580.

36. Trussardi, R. (2009). Preparation of oseltamivir phosphate. Patent No. U.S. 7,531,687.

37. Johnson, D. S., Li, J. J., editors. (2007). *The Art of Drug Synthesis*. John Wiley & Sons, Inc., chapter 7, p. 95–114.

38. Rohloff, J. C., Kent, K. M., Postich, M. J., Becker, M. W., Chapman, H. H., Kelly, D. E., Lew, W., Louie, M. S., McGee, L. R., Prisbe, E. J., Schultze, L. M., Yu, R. H., Zhang, L. (1998). Practical total synthesis of the anti-iufluenza drug GS-4104. *J. Org. Chem*. **63**, 4545–4550.

39. Federspiel, M., Fischer, R., Hennig, M., Mair, H. J., Oberhauser, T., Rimmler, G., Albiez, T., Bruhin, J., Estermann, H., Gandert, C., Göckel, V., Götzö, S., Hoffmann, U., Huber, G., Janatsch, G., Lauper, S., Röckel-Stäbler, O., Trussardi, R., Zwahlen, A. G. (1999). Industrial synthesis of the key precursor in the synthesis of the anti-influenza drug oseltamivir phosphate (Ro 64-0796/002, GS-4104-02): ethyl (3*R*,4*S*,5*S*)-4,5-epoxy-3-(1-ethyl-propoxy)-cyclohex-1-ene-1-carboxylate. *Org. Process Res. Dev*. **3**, 266–274.

40. Harrington, P. J., Brown, J. D., Foderaro, T., Hughes, R. C. (2004). Research and development of a second-generation process for oseltamivir phosphate, prodrug for a neuraminidase inhibitor. *Org. Process Res. Dev*. **8**, 86–91.

41. Poch, M., Verdaguer, X., Moyano, A., Pericàs, M. A., Riera, A. (1991). A versatile enantiospecific approach to 3-azetidinols and aziridines. *Tetrahedron Lett*. **32**, 6935–6938.

42. a) Tao, J., Zhao, L., Ran, N. (2007). Recent advances in developing chemoenzymatic processes for active pharmaceutical ingredients. *Org. Process Res. Dev*. **11**, 259–267; b) Chandran, S. S., Yi, J., Draths, K. M., von Daeniken, R., Weber, W., Frost, J. W. (2003). Phosphoenolpyrurate availability and the biosynthsis of shikimic acid. *Biotechnol. Prog*. **19**, 808–814.

43. Guo, J., Frost, J. W. (2004). Synthesis of aminoshikimic acid. *Org. Lett*. **6**, 1585–1588.

44. Rosenblum, S. B. (2007). *The Art of Drug Synthesis*. Johnson, D. S., Li, J. J., editors. Wiley Interscience, chapter 13, p. 183–196.

45. Thiruvengadam, T. K., Sudhakar, A. R., Wu, G. (1999). *Practical Enantio- and Diastereo-selective Processes for Azetidinones, Process Chemistry in the Pharmaceutical Industry*. Gadamasetti, K. G., editor. Marcel Dekker, p. 221–242.

46. Wu, G., Wong, Y., Chen, X., Ding, Z. (1999). A Novel one-step diastereo- and enantioselective formation of trans-azetidinones and its application to the total synthesis of cholesterol absorption inhibitors. *J. Org. Chem*. **64**, 3714–3718.

47. Sasikala, C., Padi, P., Sunkara, V., Ramayya, P., Dubey, P. K., Uppala, V., Praveen, C. (2009). An improved and scalable process for the synthesis of ezetimibe, an antihypercholesterolemia drug. *Org. Process Res. Dev*. **13**, 907–910.

48. Durrington, P. (2003). Dyslipidaemia. *Lancet* **362**, 717–731.

49. Farmer, J. A. (1998). Aggressive lipid therapy in the statin era. *Prog. Cardiovasc. Dis*. **41**, 71–94.

50. Flores, N. A. (2004). Ezetimibe + simvastatin (Merck/Schering-Plough). *Curr. Opin. Investig. Drugs* **5**, 984–992.
51. Pfizer. (2009). *2008 Annual Report*. Pfizer 2009-04-23.
52. Li, J. J., Johnson, D. S., Sliskovic D. R., Roth, B. D. (2004). *Contemporary Drug Synthesis*. John Wiley & Sons, Inc., p. 113–124.
53. Wang, N. Y., Hsu, C. T., Sih, C. J. (1981). Total Synthesis of (+)-compactin (ML-236B). *J. Am. Chem. Soc.* **103**, 6538–6539.
54. Narasaka, K., Pai, H. C. (1980). Stereoselective synthesis of meso (or erythro) 1,3-diols from β-hydroxyketones. *Chem. Lett.* **9**, 1415–1418.
55. Lynch, J. E., Volante, R. P., Wattley, R. V., Shinkai, I. (1987). Synthesis of an HMG-CoA reductase inhibitor: a diastereoselective aldol approach. *Tetrahedron Lett.* **28**, 1385–1387.
56. Sletzinger, M., Verhoeven, T. R., Volante, R. P., McNamara, J. M., Corley, E. G., Liu, T. M. (1985). A diastereospecific, non-racemic synthesis of a novel β-hydroxy-δ-lactone HMG-CoA reductase inhibitor. *Tetrahedron Lett.* **26**, 2951–2954.
57. Chen, K. M., Hardtmann, G. E., Prasad, K., Repic, O., Shapiro, M. J. (1987). 1,3-*syn* Diastereoselective reduction of β-hydroxyketones utilizing alkoxydialkylboranes. *Tetrahedron Lett.* **28**, 155–158.
58. Brower, P. L., Butler, D. E., Deering, C. F., Le, T. V., Millar, A., Nanninga, T. N., Roth, B. D. (1992). The synthesis of (4*R-cis*)-1,1-dimethylethyl 6-cyanomethyl-2,2-dimethyl-1,3-dioxane-4-acetate, a key intermediate for the preparation of CI-981, a highly potent, tissue selective inhibitor of HMG-CoA reductase. *Tetrahedron Lett.* **33**, 2279–2282.
59. http://www.epa.gov/gcc/pubs/docs/award_recipients_1996_2009.pdf.
60. a) Davis, S. C., Grate, J. H., Gray, D. R., Gruber, J. M., Huisman, G. W., Ma, S. K., Newman, L. M. (2004). Producing 4-cyano-3-hydroxybutyric acid ester from 4-halo-3-hydroxybutyric acid ester, comprises contacting 4-halo-3-hydroxybutyric acid ester with halohydrin dehalogenase and cyanide. WO 04015132;   b) Davis, S. C., Jenne, S. J., Krebber. A., Huisman, G. W., Newman, L. M. (2005). Novel ketoreductase polypeptide having enhanced activity than wild-type polypeptide, useful for stereospecific reduction of ketones. WO 0501713;   c) Davis, S. C., Fox, R. J., Gavrilovic, V., Huisman, G. W., Newman, L. M. (2005). New isolated polypeptide having halohydrin dehalogenase activity, useful for catalyzing conversion of 4-halo-3-hydroxybutyric acid derivatives to 4-substituted-3-hydroxybutyric acid derivatives. WO 05017141;   d) Davis, S. C., Grate, J. H., Gray, D. R., Gruber, J. M., Huisman, G. W., Ma. S. K., Newman, L. M., Sheldon, R., Wang. L. A. (2005). Producing vicinal cyano hydroxy substituted carboxylic acid ester useful in synthesis of pharmaceuticals, from vicinal halo hydroxy substituted carboxylic acid ester, by contacting with halohydrin dehalogenase and cyanide. WO 05018579.
61. Bergeron, S., Chaplin, D. A., Edwards, J. H., Ellis, B. S. W., Hill, C. L., Holt-Tiffin, K., Knight, J. R., Mahoney, T., Osborne, A. P., Ruecroft, G. (2006). Nitrilase-catalysed desymmetrisation of 3-hydroxyglutaronitrile:

preparation of a statin side-chain intermediate. *Org. Process Res. Dev.* **10**, 661–665.

62. a) DeSantis., G., Zhu, Z., Greenberg, W., Wong, K., Chaplin, J., Hanson, S., Farwell, B., Nicholson, L., Rand, C., Weiner, D., Robertson, D., Burk, M. (2002). An enzyme library approach to biocatalysis: development of nitrilases for enantioselective production of carboxylic acid derivative. *J. Am. Chem. Soc.* **124**, 9024–9025;
    b) DeSantis, G., Wong, K., Farwell, B., Chatman, K., Zhu, Z., Tomlinson, G., Huang, H., Tan, X., Bibbs, L., Chen, P., Kretz, K. Burk, M. (2003). Creation of a productive, highly enantioselective nitrilase through gene site saturation mutagenesis (GSSM). *J. Am. Chem. Soc.* **125**, 11476–11477.

63. Schaus, S. E., Larrow, J. F., Jacobsen, E. N. (1997). Practical synthesis of enantiopure cyclic 1,2-amino alcohols via catalytic asymmetric ring opening of meso epoxides. *J. Org. Chem.* **62**, 4197–4199.

64. Zook, S. E., Busse, J. K., Borer B. C. (2000). A concise synthesis of the HIV-protease inhibitor nelfinavir via an unusual tetrahydrofuran rearrangement. *Tetrahedron Lett.* **41**, 7017–7021.

65. a) Akutagawa, S., Tani, K. (2000). *Catalytic Asymmetric Synthesis*, 2nd ed., Ojima, I. editor. John Wiley & Sons, Inc., chapter 3, p. 145–161;
    b) Akutagawa, S. (1992). *Chirality in Industry*, Collins, A. N., Sheldrake, G. N., Crosby, J., editors. John Wiley & Sons, Inc., chapter 16, p. 313–323;
    c) Sheldon, R. A. (1993). *Chirotechnology*. Marcel Dekker, p. 276–277, p. 344–346.

66. Federsel, H.-J. (2005). Asymmetry on a large scale: the roadmap to stereoselective progress. *Nat. Rev. Drug Discov.* **4**, 685–697.

67. Osborn, J. A., Jardine, F. H., Young, J. F., Wilkinson, G. (1966). The preparation and properties of tris(triphenylphosphine) halogenorhodium (I) and some reactions thereof including catalytic homogeneous hydrogenation of olefins and acetylenes and their derivatives. *J. Chem. Soc. A.* **12**, 1711–1732.

68. Knowles, W. S., Sabacky, M. J. (1968). Catalytic asymmetric hydrogenation employing a soluble, optically active, rhodium complex. *J. Chem. Soc. Chem. Commun.* 1445–1446.

69. Knowles, W. S. (2004). Asymmetric hydrogenations–the Monsanto L-Dopa process. In *Asymmetric Catalysis on an Industrial Scale, Challenges, Approaches, and Solutions*, Blaser, H. U., Schmidt, E., editors. Wiley-VCH, p. 23–38. See also Selke, R. (2004). The other L-Dopa process. In *Asymmetric Catalysis on an Industrial Scale, Challenges, Approaches, and Solutions*, Blaser, H. U., Schmidt, E., editors. Wiley-VCH, p. 39–53.

70. Blaser, H. U., Spindler, F. (1997). Enantioselective catalysis for agrochemicals: the case history of the DUAL MAGNUM® herbicide. *Chimia* **51**, 297–299.

71. Spindler, F., Pugin, B., Jalett, H. P., Buser, H. P., Pittelkow, U., Blaser, H. U. (1996). A technically useful catalyst for the homogeneous enantiose-lective hydrogenation of *N*-aryl imines: a case study. *Catal. Org. Reaction* **68**, 153–166.

72. a) Blaser H. U., Hanreich, R., Schneider, H.-D., Spindler, F., Steinacher, B. (2002). The chiral switch of (*S*)-metolachlor: a personal account of an industrial odyssey in asymmetric catalysis. *Adv. Synth. Catal*. **344**, 17–31; b) (2004). The chiral switch of metalochlor: the development of a large-scale enantioselective catalytic process In: *Asymmetric Catalysis on an Industrial Scale, Challenges, Approaches, and Solutions*, Blaser, H. U., Schmidt, E., editors. Wiley-VCH, p. 55–70.

73. Blaser, H. U., Breiden, W., Pugin, B., Spindler, F., Studer, M., Togni, A. (2002). Solvias Josiphos ligands: from discovery to technical applications. *Topics Catal*. **19**, 3–16.

74. Laneman, S. A. (1999). *Handbook of Chiral Chemicals*, Ager, D. J., editor. Marcel Dekker, chapter 9, p. 143–176.

75. Warm, A., Naughton, A. B., Saikali, E. A. (2003). Process development implications of biotin production scale-up. *Org. Process Res. Dev*. **7**, 272–284.

76. a) De Clercq, P. J. (1997). Biotin: a timeless challenge for total synthesis. *Chem. Rev*. **97**, 1755–1792;   b) Deroose, F. D., DeClercq, P. J. (1995). Novel enantioselective synthesis of (+)-biotin. *J. Org. Chem*. **60**, 321–330; c) Moolenaar, M. J., Speckamp, W. N., Hiemstra, H., Poetsch, E., Casutt. M. (1995). Synthesis of D-(+)-biotin through selective ring-closure of *N*-acyliminium silyl enol ethers. *Angew. Chem. Int. Ed*. **107**, 2391–2393; d) Fujisawa, T., Nagai, M., Koike, Y., Shimizu, M. (1994). Diastereoface discrimination in the addition of acetylide to a chiral aldehyde, leading to a synthesis of (+)-deoxybiotin in enantiomerically pure form starting from L-cysteine. *J. Org. Chem*. **59**, 5865–5867.

77. (a) McGarrity. J., Tenud. L., Meul, T. (1988). New N-protected furo- or thieno-imidazolo-di: one derives—useful as intermediates for optically active biotin. Eur. Pat. Appl. EP 0270076 A1;   b) McGarrty, J., Tenud. L. (1988). Optically active biotin prodn. From furo-imidazole–di: one deriv.—by hydrogenation, conversion to thio-lactone, reacting with Grignard reagent and hydrogenating, used for treating dermatitis, ect. Eur. Pat Appl. EP 0273270 A1.

78. Imwinkelried, R. (1997). Catalytic asymmetric hydrogenation in the manu-facture of d-biotin and dextromethorphan. *Chimia* **51**, 300–302.

79. Bader, R. R., Baumeister, P., Blaser, H. U. (1996). Catalysis at Ciba-Geigy. *Chimia* **50**, 99–105.

80. Kumar, V., Abbas, A., Fausto, N. (2004). *Robbins and Cotran Pathologic Basis of Disease*, 7th ed. Elsevier, p. 1194–1195.

81. Zimmet, P., Alberti, K. G., Shaw, J. (2001). Global and societal implications of the diabetes epidemic. *Nature* **414**, 782–787.

82. a) American Diabetes Association title = Total Prevalence of Diabetes and Pre-diabetes url = http://www.diabetes.org/diabetes-statistics/prevalence. jsp|accessdate = 2008-11-29; b) Inzucchi, S. E., Sherwin, R. S. (2005). The prevention of type 2 diabetes mellitus. *Endocrinol. Metab. Clin. N. Am.* **34**, 199–219.

83. a) Thornberry, N. A., Weber, A. E. (2007). Discovery of JANUVIA™ (sitagliptin), a selective dipeptidyl peptidase iv inhibitor for the treatment of type 2 diabetes. *Curr. Topics Med. Chem.* **7**, 557–568; b) Gwaltney, S. L, Stafford, J. A. (2005). Inhibitors of dipeptidyl peptidase 4. *Annu. Rep. Med. Chem.* **40**, 149–165; c) Ahren, B., Landin-Olsson, M., Jansson, P. A., Svenson, M., Holmes, D., Schewizer, A. (2004). Inhibition of dipeptidyl peptidase-4 reduces glycemia, sustains insulin levels, and reduces glucagon levels in type 2 diabetes. *J. Clin. Endocrinol. Metab.* **89**, 2078–2084; d) Weber, A. E. (2004). Dipeptidyl peptidase IV inhibitors for the treatment of diabetes. *J. Med. Chem.* **48**, 4135–4141.

84. Weber A. E., Thornberry, N. (2007). Case History; JANUNIA™(Sitagliptin), a selective dipeptidy peptidase IV Inhibitor for the treatment of type 2 diabetes. *Annu. Rep. Med. Chem.* **42**, 95–109.

85. Hansen, K. B., Balsells, J., Dreher, S., Hsiao, Y., Kubryk, M., Palucki, M., Rivera, N., Steinhuebel, D., Armstrong, J. D., Askin, D., Grabowski, E. J. J. (2005). First generation process for the preparation of the DPP-IV inhibitor sitagliptin. *Org. Process Res. Dev.* **9**, 634–639.

86. Hansen, K. B., Hsiao, Y., Xu, F., Rivera, N., Clausen, A., Kubryk, M., Krska, S., Rosner, T., Simmons, B., Balsells, J., Ikemoto, N., Sun, Y., Spindler, F., Malan, C., Grabowski, E. J. J., Armstrong, J. D. (2009). Highly efficient asymmetric synthesis of sitagliptin. *J. Am. Chem. Soc.* **131**, 8798–8804.

87. Xu, F. (2008). *Process Chemistry in the Pharmaceutical Industry*, vol. 2. Gadamasetti, K., Braish, T. editors. CRC Press, chapter 21, p. 333–347.

88. http://www.epa.gov/gcc/pubs/docs/award_recipients_1996_2009.pdf.

89. Steinhuebel, D., Sun, Y., Matsumura, K., Sayo, N., Saito, T. (2009). Direct asymmetric reductive amination. *J. Am. Chem. Soc.* **131**(32), 11316–11317.

90. a) Navari, R. M., Reinhardt, R. R., Gralla, R. J., Kris, M. G., Hesketh, P. J., Khojasteh, A., Kindler, H., Grote, T. H., Pendergrass, K., Grunberg, S. M., Carides, A. D., Gertz, B. J. (1999). Reduction of cisplatin-induced emesis by a selective neurokinin-1-receptor antagonist. *N. Engl. J. Med.* **340**, 190–195; b) Hale, J. J., Mills, S. G., MacCoes, M., Finke, P. E., Cascieri, M. A., Sadowski, S., Ber, E., Chicchi, G. G., Kurtz, M., Metzger, J., Eiermann, G., Tsou, N. N., Taltersall, D., Rupniak, N. M. J., Williams, A. R., Rycroft, W., Hargreaves, R., MacIntyre, D. E. (1997). Structural optimization affording 2-(*R*)-(1-(*R*)-3,5-bis(trifluoromethyl) phenylethoxy)-3-(*S*)-(4-fluoro)phenyl-4-(3-oxo-1,2,4-triazol-5-yl) methylmorpholine, a potent, orally active, long-acting morpholine acetal human NK-1 receptor antagonist. *J. Med. Chem.* **41**, 4607–4614; c) Rupniak, N. M. J., Tattersall, F. D., Willams, A. R., Rycroft, W., Carlson, E. J., Cascieri,

M. A., Sadowski, S., Ber, E., Hale, J. J., Mils, S. G., MacCoss, M., Seward, E., Huscroft, I., Owen, S., Swain, C. J., Hill, R. G., Hargreaves, R. J. (1997). In vitro and in vivo predictors of the anti-emetic activity of tachykinin NK1 receptor antagonists. *Eur. J. Pharmacol*. **326**, 201–209.

91. Brans, K. M. J., Payack, J. F., Rosen, J. D., Nelson, T. D., Candelario, A., Huffman, M. A., Zhao, M. M., Li, J., Craig, B., Song, Z. J., Tschaen, D. M., Hansen, K., Devine, P. N., Pye, P. J., Rossen, K., Dormer, P. G., Reamer, R. A., Welch, C. J., Mathre, D. J., Tsou, N. N., McNamara, J. M., Reider, P. J. (2003). Efficient synthesis of NK1 receptor antagonist aprepitant using a crystallization-induced diastereoselective transformation. *J. Am. Chem. Soc*. **125**, 2129–2135.

92. Anderson, N. G. (2005). Developing processes for crystallization-induced asymmetric transformation. *Org. Process Res. Dev*. **9**, 800–813.

93. Brands, K. M., Krska, S. W., Rosner, T., Conrad, K. M., Corley, E. G., Kaba, M., Larsen, R. D., Reamer, R. A., Sun, Y., Tsay, F. R. (2006). Understanding the origin of unusual stepwise hydrogenation kinetics in the synthesis of the 3-(4-fluoropheyl)morpholine moiety of NK1 receptor antagonist aprepitant. *Org. Process Res. Dev*. **10**, 109–117.

94. http://www.epa.gov/gcc/pubs/docs/award_recipients_1996_2009.pdf.

95. Federsel, H. J., Larsson, M. (2004). An innovative asymmetric sulfide oxidation: the process development history behind the new antiulcer agent esomeprazole In: *Asymmetric Catalysis on an Industrial Scale, Challenges, Approaches, and Solutions*, Blaser, H. U., Schmidt, E., editors. Wiley-VCH, p. 413–436.

96. Butters, M., Catterick, D., Craig, A., Curzons, A., Dale, D., Gillmore, A., Green, S. P., Marziano, I., Sherlock, J. P., White, W. (2006). Critical assessment of pharmaceutical processes—a rationale for changing the synthetic route. *Chem. Rev*. **106**, 3002–3027.

97. Li, J. J., Johnson, D. S., Sliskovic, D. R., Roth, B. D. (2004). *Contemporary Drug Synthesis*. John Wiley & Sons, Inc., p. 21–27.

98. Kohl, B., Sturm, E., Senn-Bilfinger, J., Simon, W. A., Kruger, U., Schaefer, H., Rainer, G., Figala, V., Klemm, K. (1992). (H^+, K^+)-ATPase inhibiting 2-[(2-pyridy(methyl)sulfinyl) benzimidazoles. 4. A novel series of dimethoxypyridyl-substituted inhibitors with enhanced selectivity, the selection of pantoprazole as a clinical candidate. *J. Med. Chem*. **35**, 1049–1057.

99. Cotton, H., Elebring, T., Larsson, M., Li, L., Sörenson, H., von Unge, S. (2000). Asymmetric synthesis of esomeprazole. *Tetrahedron Asymmetry* **11**, 3819–3825.

100. a) Vacca, J. P., Dorsey, B. D., Schleif, W. A., Levin, R. B., McDaniel, S. L., Darke, P. L., Zugay, J., Quintero, J. C., Blahy, O. M., Roth, E., Sardana, V. V., Schlabach, A. J., Graham, P. I., Condra, J. H., Gotlib, L., Holloway, M. K., Lin, J., Chen, I.-W., Vastag, K., Ostovic, D., Anderson, P. S., Emini, E. A., Huff, J. R. (1994). L-735,524: An orally bioavailable human immunodeficiency virus type 1 protease inhibitor. *Proc. Natl. Acad. Sci. U. S. A.* **91**, 4096–4100; b) Dorsey, B. D., Levin, R. B., McDaniel,

S. L., Vacca, J. P., Guare, J. P., Darke, P. L., Zugay, J. A., Emini, E. A., Schleif, W. A. (1994). L-735,524: the design of a potent and orally bioavailable HIV protease inhibitor. *J. Med. Chem.* **37**, 3443–3451.

101. Reider, P. J. (1997). Advances in AIDS Chemotherapy: the asymmetric synthesis of CRIXIVAN®. *Chimia* **51**, 306–308.

102. a) Senanayake, C. H., Roberts, F. E., DiMichele, L. M., Ryan, K. M., Liu, J., Fredenburgh, L. E., Foster, B. S., Douglas, A. W., Larsen, R. D., Verhoeven, T. R., Reider, P. J. (1995). The behavior of indene oxide in the ritter reaction: A simple route to *cis*-aminoindanol. *Tetrahedron Lett.* **36**, 3993–3996;   b) Senanayake, C. H., Smith, G. B., Ryan, K. M., Fredenburgh, L. E., Liu, J., Roberts, F. E., Larsen, R. D., Verhoeven, T. R., Reider, P. J. (1996). The role of 4(3-phenylpropyl)pyridine *N*-oxide (P₃NO) in the manganese-salen-catalyzed asymmetric epoxidation of indene. *Tetrahedron Lett.* **37**, 3271–3274.

103. Senanayake, C. H., Jacobsen, E. N. (1999). *Process Chemistry in the Pharmaceutical Industry*, Gadamasetti, K. G., editor. Marcel Dekker, p. 347–368.

104. Larrow, J. F., Roberts, E., Verhoeven, T. R., Ryan, K. M., Senanayake, C. H., Reider, P. J., Eric N. Jacobsen, E. N. (2004). *Org. Synthesis* **10**, 29–34.

105. a) Maligres, P. E., Upadhyay, V., Rossen, K., Cianciosi, S. J., Purick, R. M., Eng, K. K., Reamer, R. A., Askin, D., Volante, R. P., Reider, P. J. (1995). Diastereoselective *syn*-epoxidation of 2-alkyl-4-enamides to epoxyamides: synthesis of the Merck HIV-1 protease inhibitor epoxide intermediate. *Tetrahedron Lett.* **36**, 2195–2198;   b) Maligres, P. E., Weissman, S. A., Upadhyay, V., Cianciosi, S. J., Reamer, R. A., Purick, R. M., Sager, J., Rossen, K., Eng, K. K., Askin, D., Volante, R. P., Reider, P. J. (1996). Cyclic imidate salts in acyclic stereochemistry: diastereoselective *syn*-epoxidation of 2-alkyl-4-enamides to epoxyamides. *Tetrahedron* **52**, 3327–3338.

106. LeBlond, C. R., Rossen, K. R., Gortsema, F. P., Zavialov, I. A., Cianciosi, S. J., Andrews, A. T., Sun, Y. (2001). Harvesting short-lived hypoiodous acid for efficient diastereoselective iodohydroxylation in Crixivan® synthesis. *Tetrahedron Lett.* **42**, 8603–8606.

CHAPTER 7

STRUCTURAL BASIS AND COMPUTATIONAL MODELING OF CHIRAL DRUGS

DEPING WANG
Biogen IDEC Inc., Cambridge, MA

ERIC HU
Gilead Sciences Inc., Foster City, CA

7.1 INTRODUCTION

The word "chirality" is derived from the Greek word χειρ (cheir), meaning "hand." A person's two hands are mirror images of each other, and they are

Chiral Drugs: Chemistry and Biological Action, First Edition. Edited by Guo-Qiang Lin, Qi-Dong You and Jie-Fei Cheng.
© 2011 John Wiley & Sons, Inc. Published 2011 by John Wiley & Sons, Inc.

said to have chirality. Lord Kelvin defines chirality as the nonsuperimposable characteristics of two mirror objects: "I call any geometrical figure, or group of points, chiral, and say it has chirality, if its image in a plane mirror, ideally realized, cannot be brought to coincide with itself" [1]. A chiral object does not carry a center of symmetry or a plane of symmetry and cannot be mapped to its mirror image by rotations and translations alone. A nonchiral object is called achiral or amphichiral. Achiral molecules normally carry more than one rotation or translation symmetry, such as benzene (C_6H_6).

Any isometry or symmetry operator can be written as $(v \rightarrow Av + b)$, in which (A) is an orthogonal matrix and (b) is a vector. The determinant of (A) is either 1 (orientation-preserving) or -1 (orientation-reversing). An object is achiral if and only if there is at least one orientation-reversing isometry. Any object possessing a plane of symmetry $(x,y,z) \rightarrow (x,y,-z)$ or a center of symmetry $(x,y,z) \rightarrow (-x,-y,-z)$ is achiral. Notably, there are also achiral objects lacking both plane and center of symmetry such as:

$$F_0 = \{(1,0,0),(0,1,0),(-1,0,0),(0,-1,0),(2,1,1),$$
$$(-1,2,-1),(-2,-1,1),(1,-2,-1)\}$$

Chirality is different from symmetry, which has five elements: symmetry axis, plane of symmetry, center of symmetry, rotation-reflection axis, and identity. An asymmetric object can have chirality, but the opposite is not always true. Most chiral molecules have point chirality at a single stereogenic atom, normally carbon, which has four different substituents. It is possible for a molecule to be chiral without a point chirality. For example, triphenylphosphine has three phenyl substituents attached to the phosphorus in a propeller-like fashion (Fig. 7.1a). 1,1'-Bi-2-naphthol (BINOL) (Fig. 7.1b) has axial symmetry but without a stereogenic center. (E)-cyclooctene (Fig. 7.1c) has two non-coplanar and dissymmetric rings. Some macromolecules such as fullerenes (C76, C78, and C84) (Fig. 7.1d) have inherent chiralities.

Most biological active molecules are chiral. Nearly all amino acids are in the L-form, and biologically relevant glucoses are in the D-form. Macromolecules

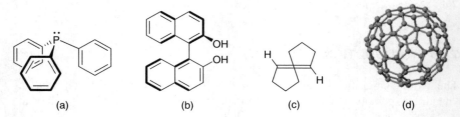

| (a) | (b) | (c) | (d) |

FIGURE 7.1 Chiral molecules do not always have stereogenic centers. (a) Propellor-like triphenylphosphine; (b) BINOL has axial chirality; (c) (E)-cyclooctene has planar chirality; (d) C76 has inherent chirality.

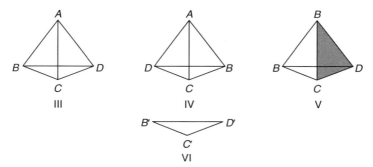

FIGURE 7.2 Easson–Stedman model depicts the planar recognition between a drug and a receptor.

such as protein α-helices and DNA have a right-handed turn and are also chiral. Enzymes also exhibit a stereochemical preference for many natural substrates such as hormones, neurotransmitters, and endogenous opioids. The alternative forms are normally inactive and sometimes even toxic. The purpose of homochirality remains unclear and is the subject of a lot of debate [2]. It is generally believed to serve as an efficient way to reduce entropy barriers in the formation of large organisms mainly via chiral amplification and transmission [3].

The three-dimensional nature of drug molecules tends to be neglected, even though stereoselectivity has been known since the early years of the twentieth century [4]. Mainly because of the advances in asymmetric synthesis of chiral drug molecules and increasing appreciation and understanding of the potential biological significance administered with enantiomers, both regulatory agencies and pharmaceutical industries have shown great advancement toward the development of single stereoisomer products [5,6].

Easson and Stedman proposed in the 1930s that the spatial arrangement of the functionalities in the drug and complementary sites on the receptor is responsible for the drug-receptor recognition (Fig. 7.2) [7]. Arrangement enantiomer **IV** rather than **III** or **V** achieves the best binding with **VI**, the surface of a tissue receptor.

However, the observations that some achiral molecules were more potent than the "less" active enantiomers make the Easson–Stedman hypothesis true only in sites of direct action [8]. Mesecar and Koshland identified four different sites based on the X-ray structures of dehydrogenase with L- and D-isocitrate [9]. The three-point Easson–Stedman model assumes that the substrate approaches a planar surface from one direction (Fig. 7.3). A fourth location becomes essential to discriminate the two enantiomers either serving as the directionality or due to an additional binding site.

The enantiomers with greater and weaker affinity are called the eutomer and the distomer, respectively. The ratio of affinities of eutomer to distomer is described as the eudismic ratio or the eudismic index for its logarithmic form.

The first oral treatment for male erectile dysfunction (MED), sildenafil (Viagra™, Fig. 7.4), is a cGMP phosphodiesterase enzyme PDE5 inhibitor. It prolongs the effect of cGMP, and hence the erection, by preventing cGMP

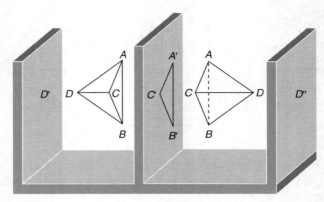

FIGURE 7.3   The four-point model illustrates the directionality of the three-point planar recognition.

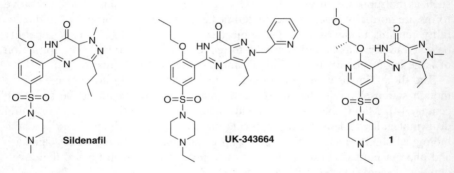

FIGURE 7.4   Selective PDE5/6 inhibitors reflect the enantioselective interactions between ligand and receptor.

breakdown in the event of nitric oxide release. Selective PDE5 inhibitors such as sildenafil, also exhibits efficacy in pulmonary arterial hypertension. Compound **1** (Fig. 7.4) was associated with significant nonlinearity in its pharmacokinetics (PK) and dose-dependent T_{max} was observed, indicating P-glycoprotein binding [10]. The slope of the eudismic indexes vs. pIC_{50} values of the high-affinity stereoisomers (eutomers) is called the eudismic affinity quotient (EAQ). A steeper slope strongly indicates an enantioselective interaction between receptor and ligand that was first proposed by Pfeiffer in the 1950s [11]. Chirality provides another optimizing variable independent of molecular weight. Taking advantage of information of chirality, Pfizer scientists identified highly efficient inhibitors to have excellent selectivity between PDE5 and PDE6 and greatly improved PK linearity and compound **1** (Fig. 7.4) has been advanced to clinical development [10].

More than one-half of marketed drugs are chiral [12], and most of the chiral drugs are administered as racemic largely because of limited synthetic and chiral resolution techniques or high manufacturing costs for single enantiomers. However, the importance of chirality of drugs has been increasingly recognized in the wake of the thalidomide tragedy [13]. Thalidomide is a sedative drug that was prescribed to pregnant women from 1957 to the early 1960s in at least 46 countries under different brand names, and it was withdrawn from the market after being found to be teratogenic, a cause of birth defects. The thalidomide molecule is chiral and was administered as a racemic compound, in which the enantiomers had equal sedating effect. After its withdrawal, studies in pregnant mice and rats showed that the S enantiomer of thalidomide was teratogenic, whereas the R enantiomer was devoid of any such symptoms. On the other hand, both R and S enantiomers were equally teratogenic in rabbits [14]. FDA won great praise in the 1960s for not allowing thalidomide to be marketed in the United States.

Over the last decade, advances in new chemical and analytical technologies have facilitated stereoselective synthetic processes and separation of individual enantiomers from racemic mixtures. More enantiopure drugs have been brought to the market in response to FDA's new guidelines and have demonstrated their advantages in terms of efficacy and tolerability. As a result, the worldwide market of single-enantiomer drugs has been increasing dramatically and the annual sales exceeded $225 billion in 2005 [15]. However, much of the information regarding one stereoisomer showing better therapeutic benefits than the other was obtained not until the drugs had advanced into clinical trials or been on the market for years. For example, (S)-omeprazole, a proton pump inhibitor used for treatment of gastroesophageal reflux, has been shown to be superior to its racemic formulations in clinical trials [16]. On the other hand, enantiopure drugs do not always perform better in the clinical setting than their racemic counterparts because of systemic exposure and pharmacokinetics-related issues. Patients will benefit most, in addition to significant drug developmental cost reduction, if we have a better understanding of drug chirality and even predict the pharmacodynamic effects of stereoisomers in the human body.

Structure-based computational drug design is one of the most reliable, mature, and extensively utilized approaches in modern drug discovery. This chapter reviews the structural basis and computational modeling of chiral drugs.

7.2 STRUCTURAL BASIS OF CHIRAL DRUGS

Enantiomers of a chiral drug have identical physical and chemical properties in an achiral environment. However, they often present different chemical and pharmacological behaviors in a chiral environment such as the human body. Because our living systems are mostly chiral, enantiomers interact with biological targets differently and, therefore, may exhibit widely different pharmacokinetic properties. In other words, each enantiomer of a chiral drug could behave very differently in the human body. When one enantiomer is responsible for the activity of interest, its counterpart could be inactive, or have totally different activities that are either desirable or undesirable. The final chirality of a drug is largely determined by

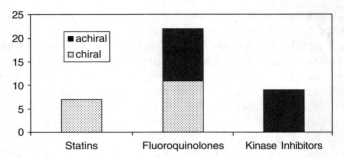

FIGURE 7.5 Comparison of chiral and achiral drugs on the market: statins, fluoro-quinolines, and kinase inhibitors.

the binding pocket of its target. Some targets require proper chiral compounds to maintain potency, some only prefer chiral compounds slightly, and some do not need chiral centers at all. For example, Figure 7.5 shows the comparison of three well-studied drug targets: statins [3-hydroxy-3-methylglutaryl-coenzyme A (HMG-CoA) reductase (HMGR) inhibitors], fluoroquinolones (DNA topoisomerase IV inhibitors), and kinase inhibitors reported in the MDL Drug Data Report database (MDDR) in 2009. As shown in Figure 7.5, all the drugs on the market for HMGR are chiral compounds and almost half of DNA topoisomerase IV inhibitors are chiral, while none of the kinase inhibitors is chiral. The structural basis behind this is discussed in the following sections.

7.2.1 Stereoselective Binding Pockets

Most of the therapeutical targets consist of either chiral amino acids (L-amino acids) or nucleotides (DNA or RNA); therefore, chiral drugs are normally required to achieve desired in vitro and in vivo activities. For example, HMGR catalyzes the rate-limiting step in cholesterol biosynthesis where a high level of cholesterol poses a critical risk for coronary heart disease. An effective clinical treatment is to block HMGR with small-molecule inhibitors such as statins. So far, seven drugs (Fig. 7.6) have been approved for this purpose. All statins share an HMG-like moiety, which may be present in an inactive lactone form like that in simvastatin and lovastatin. This lactone form is enzymatically hydrolyzed to its active hydroxyl acid form. Statins have rigid, hydrophobic groups that covalently bond to the HMG-like moiety, which carries two chiral centers, C3 and C5 of the dihydroxyheptanoic acid segment.

Recently the structures of the catalytic portion of human HMGR complexed with six different statins have been determined by a delicate crystallography study [17]. Figure 7.7a shows the crystal structure of atorvastatin (Lipitor) with HMGR. The HMG moiety of the statin occupies the active site of the HMGR enzyme, while the bulky hydrophobic groups of the statin sit in a mostly greasy enzyme pocket (Fig. 7.7a). The HMG moiety interacts with several polar residues around the pocket (Fig. 7.7b). K691 participates in a hydrogen-bonding network with E559 and the O5-hydroxyl of the statins. The terminal carboxylate of the

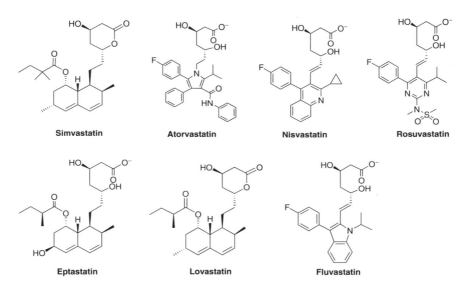

FIGURE 7.6 Chemical structures of FDA-approved statins.

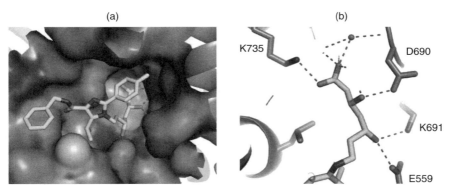

FIGURE 7.7 Binding mode of atorvastatin (pdb code: 1HWK). (a) Surface representation; (b) Interactions between the HMG moieties and the protein.

HMG moiety forms a salt bridge with K735. Apparently, the HMGR enzyme is stereoselective because of the asymmetric binding pocket around dihydroxyheptanoic acid. A number of hydrogen bonds and ion pairs help to achieve charge and shape complementarity between the protein HMGR and the HMG-like moiety of the statins. Both chiral centers are critical to maintain potency. Significant efforts have been made to modify the HMG moiety during the early drug discovery process. For example, a series of 5-substituted 3,5-dihydroxypentanoic acids and their derivatives have been prepared and tested for inhibition of HMG-CoA reductase in vitro [18]. In general, unless a carboxylate anion is formed and the hydroxyl groups remain unsubstituted in an erythro relationship, inhibitory activity is greatly reduced.

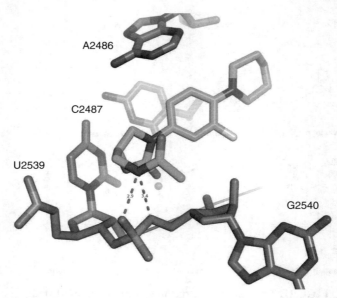

FIGURE 7.8 Binding mode of linezolid in ribosome 50S subunit (pdb code: 3CPW).

Similar to proteins, some DNA or RNA inhibitor targets are also stereos-elective and require chiral inhibitors. For example, linezolid (Fig. 7.8) is an oxazolidinone antibacterial targeting the 50S subunit of prokaryotic ribosomes. To gain insight into its mechanism of action, a crystal structure of linezolid bound to the *Haloarcula marismortui* 50S subunit has been determined [19]. As shown in Figure 7.8, linezolid binds to the 50S A-site near the catalytic center and competes with incoming A-site substrates. The fluorophenyl ring is sandwiched by A2486 and C2487. The oxazolidinone ring is stacked against the base moiety of U2539, where it makes favorable van der Waals interactions. In addition, both the oxazolidinone ring and the C5 acetamide arm show good shape complementarity with a portion of the ribosomal A-site surface residues located near the mouth of the exit tunnel. Furthermore, the acetamide NH participates in a hydrogen bond with the phosphate group of G2540. The (S)-configuration of C5 in oxazolidinone is necessary to entail proper interactions with key pharmacophores in the ribosome 50S subunit. Alternating the configuration from (S)- to (R)- completely demolishes compound potency [20].

7.2.2 Nonstereoselective Binding Pockets

Protein kinases catalyze key phosphorylation reactions in signaling transduction that affect every aspect of cell growth, differentiation, and metabolism. Kinases have become increasingly important anticancer targets. Currently, there are nine ATP-competitive (Fig. 7.9) kinase inhibitors approved for clinical use and many more in clinical trials. Of the eight drugs on the market, four belong to the class of type I inhibitors that target the ATP binding site of a kinase domain. The highly

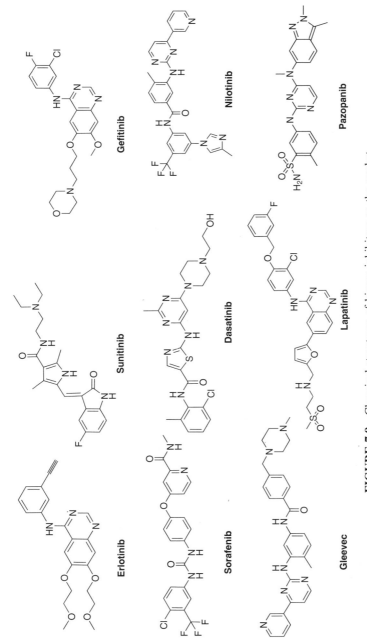

FIGURE 7.9 Chemical structures of kinase inhibitors on the market.

conserved kinase domain consists of a bilobed structure, with Mg-ATP situated in a deep cleft located between the N- and C-terminal lobes. Type I inhibitors bind to the "hinge" residues through the formation of hydrogen bonds and hydrophobic interactions in the region occupied by the ATP adenine ring. Type II inhibitors typically use ATP binding sites, but they also exploit unique hydrogen bonding patterns resulting from activation loop (DFG) movement. Several residues of the activation loop are able to fold away from the canonical conformation required for ATP phosphate transfer and open up the back channel for extra binding. Erlotinib is used here as an example to illustrate the general binding pocket for type I kinase inhibitors.

4-Anilinoquinazoline is a common hinge binder in many kinase inhibitors or drugs, such as erlobtinib and gefetinib. These inhibitors bind at the site normally occupied by ATP during a phosphorus transfer. A crystal structure of EFGR kinase with erlotinib was reported at 2.6 Å resolution [21]. As shown in Figure 7.10, the N1–C8 bond in erlotinib quinazoline faces the "hinge," with the ether substituent extended into the solvent and the aniline substituent on the opposite end sequestered in a hydrophobic pocket. The quinazoline N1 acts as a hydrogen bond acceptor coupled with M769 amide nitrogen. The other quinazoline nitrogen atom (N3) is 4.1 Å away from the T766 side chain, which is able to form a water-bridged hydrogen bond as indicated in the X-ray crystal structure. In comparison, anilinoquinazoline shares two common features with ATP: forming a hydrogen bond with hinge residue M769 and aromatic rings occupying the same hydrophobic space. So far, these two features are preserved in most kinase-small molecule complexes without the involvement of any chiral centers [22]. Many kinase drugs or inhibitors were designed based on these two features, and some of them also occupy a back pocket or DFG-out pocket for selectivity, which is beyond the scope of this discussion.

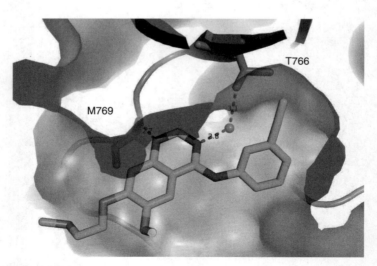

FIGURE 7.10   Binding pocket of erlotinib in EGFR kinase (pdb code: 1M17).

(a)   (b)

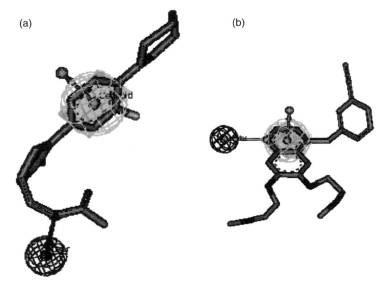

FIGURE 7.11 Comparison of the pharmacophore features of a chiral and an achiral drug. (a) Key pharmacophores in linezolid: aromatic ring and hydrogen bond donor; (b) Key pharmacophores in erlotinib: aromatic ring and hydrogen bond acceptor. (Graphical displays generated with the Discovery Studio Visualizer).

Comparison of HMGR, bacterial 50S, and kinase inhibitor binding pockets shows some distinct characteristics. As shown in Figure 7.11a, in the bacterial ribosome 50S inhibitor complex, one or more sp^3 carbon linker atoms are necessary to link hydrogen bond acceptor/donor and aromatic ring. In drug targets such as EGFR kinase (Fig. 7.11b), however, hydrogen bond acceptors/donors align along the edge of the aromatic ring and no sp^3 atom is required. In summary, both hydrophobic and polar interactions with the target contribute to potency for most inhibitors or drugs; the relative position of these key pharmacophore points determines the chirality of drugs.

In many chiral drugs, a chiral center is not critical for binding potencies. Instead, it modulates the molecular properties. For instance, ibuprofen is sold as a mixture of isomers. Although the (S)-isomer is most effective, the (R)-isomer can be converted to the (S)-isomer in the human body and thus is considered a prodrug [23]. By contrast, alclofenac is a nonchiral drug. In the crystal structure of COX-1 with ibuprofen, the propionate inhibitor binds in the cyclooxygenase active site, forming hydrogen bonds between the carboxylate of the inhibitor and R120 and Y355 in the binding pocket (Fig 7.12a) [24]. The structurally similar alclofenac binds in a nearly identical manner in the COX-1 active site, forming an ion pair with R120 and a hydrogen bond with Y355 (Fig. 7.12b). The binding pocket is made up of mainly hydrophobic residues including L352, Y385, W387, Y348, F518, G526, and S530. This cavity is not nearly filled by either of the two ligands. As a result, none of the ibuprofen atoms has effective van der Waals

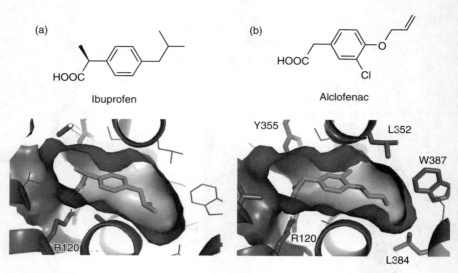

FIGURE 7.12 (a) Chemical structure of ibuprofen and its binding site (pdb code: 1EQG); (b) Chemical structure of alclofenac and its binding site (pdb code: 1HT8).

contact with the enzyme. Alclofenac is slightly bigger than ibuprofen, and the peripheral hydrocarbon atoms are within van der Waals distances with residue W387. Interestingly, the methyl group in the ibuprofen chiral center is at least 3.7 Å away from the nearest protein atom without forming effective van der Waals interactions. It is likely that the methyl group enhances binding affinity through implementing torsional constraints and reduces entropy loss upon protein binding. Although the two drugs bind similarly in the pocket, they show distinct kinetics. Ibuprofen is a classical time-independent competitive inhibitor, while alclofenac is a time-dependent and slow tight-binding inhibitor. The difference between time-dependent and time-independent inhibition is clinically relevant [25].

Chiral centers in some drugs were introduced late in the lead optimization stage and selected for clinical trials because of better cellular potencies or superior drug properties. Generally, these chiral functional groups bind in a solvent-accessible area. For example, fluoroquinolones are one of the important antibiotics groups in the clinic that target topoisomerase IV. A structure-activity relationship (SAR) study has shown that substituents at the C7 position of the quinolone core greatly influence their antibacterial spectrum and safety. As discussed above, about half of the fluoroquinolone antibiotics drugs are chiral. Except for levofloxacin, the chiral center in the fluoroquinolone antibiotics, if any, is located at the C7 substituent in most cases. Clinafloxacin (Fig. 7.13) is one of the chiral compounds still in development that show broad antibacterial activities. Recently, the crystal structure of clinafloxacin bound in the enzyme was solved [26]. The C7 substituent of the quinolone forms stacks with DNA bases and reaches into a large solvent-accessible area above the cleaved DNA. It is assumed that no significant contribution to binding affinity is gained from the stacking interactions. In fact,

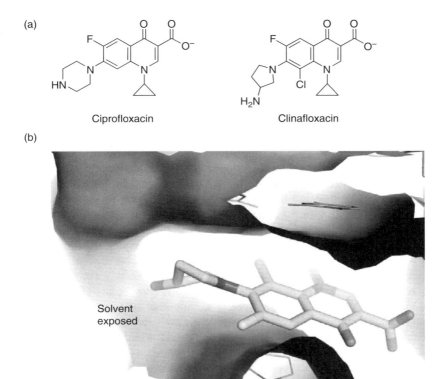

FIGURE 7.13   (a) Chemical structures of ciprofloxacin and clinafloxacin; (b) Binding pocket of clinafloxacin in topoisomerase IV (pdb code: 3FOE).

in vitro susceptibility testing showed that (*R*)- and (*S*)-isomers were comparable in their inhibitory activities based on the fact that all MICs were within twofold for each organism tested [27].

As mentioned above, enantiomers can have quite distinct PK profiles even though the chiralities themselves have nothing to do with binding affinities to the targets. The drug omeprazole contains a sulfoxide group and is a racemate (Fig. 7.14). As the two isomers of omeprazole drugs have the same physicochemical properties, they both undergo a nonenzymatic, proton-catalyzed transformation to the active intermediate. The two isomers show identical dose-response

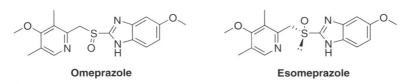

FIGURE 7.14   Chemical structure of omeprazole and esomeprazole.

curves when tested in vitro for inhibition of acid production in isolated gastric gland. Nevertheless, it has been shown that the (S)-isomer has better bioavailability and oral potency in inhibiting gastric acid secretion because of stereoselective metabolism of omeprazole [28].

7.3 MOLECULAR MODELING IN CHIRAL DRUG DESIGN

7.3.1 Library Design

As mentioned above, the use of enantiopure drugs has continued to grow in the pharmaceutical industry, and nearly half of all marketed drugs are enantiopure. However, combinatorial and high-throughput chemistry efforts have continued to focus on achiral molecules for two reasons. First, the technology to prepare chiral molecules in quantities and formats required for combinatorial chemistry is not widely available. Second, the significance of addressing stereospecificity in the early stages of drug discovery is not widely appreciated because of difficulty in synthesis and relevant costs. However, addressing chirality in the early stage of drug discovery offers several benefits. Enantiopure compounds can reduce false negative lead identifications since chiral compounds that are active as single isomers can exhibit little or no activity as racemic mixtures. In addition, the screening results are more easily analyzed and used for building a predictive SAR model. As discussed elsewhere in this book, the synthetic techniques for preparing and purifying chiral compounds in automated parallel reactions have been greatly improved. The design and preparation of large-quantity chiral libraries also have become feasible. Furthermore, new computational tools are improved to better assist library design and result interpretation.

One challenge of library design is to select an optimally sized library for synthesis, because the size of a combinatorial library based on a chiral template can be astronomical. The use of computational methods can help to design libraries that maximize the three-dimensional (3D) structure-activity information. Generally, 3D descriptors are used to encode bit strings that describe a molecule's 3D chemical profile. These bit strings are then used to select the subset of products for which the informational entropy of the bit in the strings is maximized [29]. The process of computer-aided library design generally begins with a collection of computer-proposed combinatorial syntheses, the so-called virtual library that essentially is an enumeration of different reagents and chemical reactions from the chiral original templates. The reaction can be filtered to remove undesirable compounds. In addition, drug-property filters, such as Lipinski's rule of five [30], can be applied. After the virtual library is constructed, a set of representative low-energy conformations is generated for each molecule. These conformations illustrate which region of chemical space each molecule can explore. An abstract representation for each conformation of each molecule is created with topological queries. Atom types, 3D coordinates, and bond types are mapped onto chemical features, which include hydrogen bond donors and acceptors, charged groups, hydrophobic groups, and aromatic rings. The collection of relative feature positions of all searched conformations of a given molecule is used to construct a molecule's 3D descriptors (also called signatures or fingerprints).

Pharmacophore fingerprints consist of sets of features and their associated interfeature distances and angles [31], which are discussed below. For each molecular conformer, distances between all feature pairs are calculated and binned so that all possible combinations of two, three, and four features are considered. Different feature combinations are mapped to a particular bit of string in the signature that is set previously. All possible two-, three-, and four-point pharmacophore feature combinations present in the molecule's conformers are mapped into a single bit string that identifies pharmacophore features present in all the conformers. For a typical pharmacophore fingerprint, the length is about 10 million bits. It should be mentioned that pharmacophore fingerprints are 3D fingerprints. In general, enantiomers are more similar than the diastereomers. This is impossible for two-dimensional (2D) fingerprints.

In addition to pharmacophore fingerprints, shape-feature signatures or shape-fingerprints have also been introduced by different groups. Heigh et al. designed shape-fingerprints, in which binary bit strings encode molecular shape [32]. In the fingerprints, individual bits of shape-fingerprints represent "reference shapes." Bits are set when the shape similarity measured to the reference shapes is greater than a certain value. The association between a bit and a reference shape is analogous to "keyed" representations of fingerprints, as opposed to "hashed" representations in which bit positions cannot be associated with specific features. Shape-fingerprints can be used for clustering of large data sets, evaluating the diversity of compound libraries. Independently, Beroza et al. introduced shape-feature signatures [33]. Unlike pharmacophore fingerprints, which consider only features and their related positions, shape-feature signatures consider the steric volume of the molecule. The shape-feature signature space is defined by a shape catalog and a set of positions within the shape that might contain a chemical feature. Each such signature consists of molecular shapes, represented by an approximate steric envelope, and locations of chemical features (e.g., hydrogen-bond acceptor).

Once a set of molecule signatures is complete, a matrix is constructed in which each row corresponds to a particular feature. The designed molecules from the combinatorial synthesis are ranked by analyzing this matrix. A number of rows are chosen that have the highest informational entropies. Maximizing the informational entropies ensures higher likelihoods that assay results for the set of molecules will most efficiently identify which components in the 3D molecular descriptors are correlated with activity.

7.3.2 Virtual Screening

As the frontrunner of high-throughput technology in modern drug discovery, high-throughput screening (HTS) has become an integral part of early-stage discovery efforts in all large pharmaceutical companies. Even though compound collections continue to grow in number, making literally millions of compounds available for biological evaluation, more sophisticated and efficient screening methods are still necessary and are being pursued. In recent years computer-aided virtual screening (VS) has become more and more popular because of increasing enrichment of biological and target structural information and improved computational methods. The goal of VS is to identify a small number of candidate compounds from a vast collection of compound databases that are predicted to be active. VS is often the starting point when assays are too complex or time-consuming for HTS. VS methods are categorized into ligand-based and receptor-based methods. Ligand-based VS includes 2D or 3D fingerprints, pharmacophore, and shape-based methods. Receptor-based methods require 3D information of the targets, and docking and scoring are normally employed. The details and comparison of these methods are reviewed elsewhere [34]. Among these methods, both shape- and structure-based methods are able to distinguish chiral compounds from achiral compounds. However, 2D fingerprints fail to do so, although they are extremely fast. Recently, Mason et al. developed a 3D four-point pharmacophore fingerprints method to solve this problem. As shown in Figure 7.15, 3D pharmacophore fingerprints consist of multiple three- and four- point pharmacophores and can be calculated systematically. The six pharmacophoric features defined in the software package Catalyst [35] (hydrogen bond donors, hydrogen bond acceptors, acidic centers, basic centers, hydrophobic regions, and aromatic ring centroids) are automatically assigned to atoms or dummy atom centroids. A significant increase in the amount of shape information was found with the use of four-point pharmacophores, including the ability to distinguish chirality.

Receptor-based VS using a protein target with an experimentally determined structure has become an established method for lead discovery and for potency improvement in lead identification and lead optimization. The molecular docking process consists of sampling the coordinate space of the binding site and scoring each possible ligand pose to identify the predicted binding mode for that ligand. A large number of docking programs are available for use in VS that differ in the sampling algorithms used, the handling of ligand and protein flexibility, and the

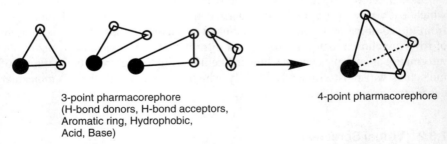

3-point pharmacorephore
(H-bond donors, H-bond acceptors,
Aromatic ring, Hydrophobic,
Acid, Base)

4-point pharmacorephore

All combinations of the 6 features and 7 or 10 distance ranges
For 7 ranges: 0-2.5, 2.5-4, 4-6, 6-9, 9-13, 13-18, >18 Å

FIGURE 7.15   Definition of multiple potential 3D pharmacophore fingerprints.

scoring functions [36]. Docking of chiral compounds is same as that of achiral compounds using any standard automated or manual docking procedures. One caveat is the treatment of pseudochiral pyramidal sp^3 amine nitrogen atom. In most docking protocols, conformational sampling is restricted in torsional space and the sp^3 nitrogen atom will not be inverted. Some docking programs are able to invert this amine to the pseudoinverted position, such as FlexX [37] and Glide [38].

The success of a docking algorithm depends on comprehensive sampling methods and good scoring functions. Over the years, great progress has been made to enhance sampling during docking by taking into account protein flexibility, while the individual scoring function is still not robust enough to predict protein-ligand binding affinity accurately. Thus the application of the scoring function is quite limited during lead optimization [39]. Nevertheless, a number of success stories show that docking protocols are able to distinguish enantiomers case by case with different scoring functions. For example, the papain/CLIK-148 coordinate system was employed as a model to study the interactions of a nonpeptide thiocarbazate inhibitor with cathepsin L [40]. This small-molecule inhibitor, a thiol ester containing a diacyl hydrazine functionality and one stereogenic center, was most active as the (S)-enantiomer, with an IC_{50} of 56 nM; the (R)-enantiomer displayed only weak activity (33 μM). Correspondingly, molecular docking studies with Extra Precision Glide revealed good correlation between scores and biological activities for the two thiocarbazate enantiomers. The highest-ranked pose for the (R)-enantiomer obtained from the XP Glide docking study (Fig. 7.16) has a docking score of −7.0 kcal/mol, 2 kcal/mol lower than the (S)-enantiomer. In addition, key hydrogen-bonding and hydrophobic contacts present in the complexes of both the active (S)-enantiomer with papain and CLIK-148 with papain are completely disrupted in the (R)-enantiomer binding mode. While the (S)-enantiomer makes at least six hydrogen-bonding contacts with the active site residues and large hydrophobic contacts within the S1′, S2, and S3 subpockets, the (R)-enantiomer forms only two hydrogen bonds to papain and does not occupy the S1′ subsite at all.

In addition to these VS methods, another critical issue is the quality of the virtual libraries used for screening. Although plenty of 3D databases are available,

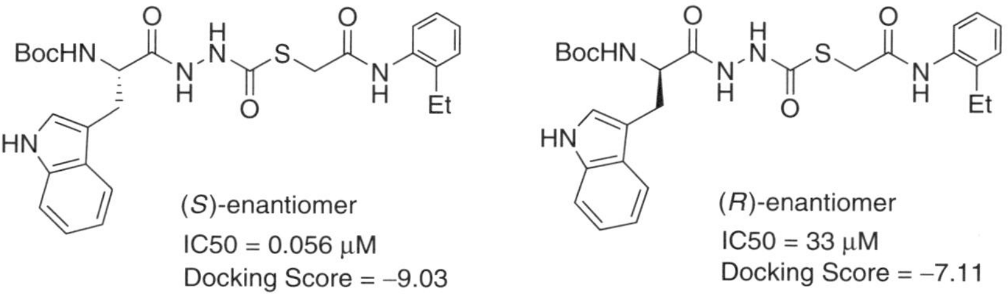

(S)-enantiomer
IC50 = 0.056 μM
Docking Score = −9.03

(R)-enantiomer
IC50 = 33 μM
Docking Score = −7.11

FIGURE 7.16 Cathepsin L inhibitory activities and Glide docking scores for (R)- and (S)-enantiomers.

their stereochemistry was often randomly assigned in the process of 2D-to-3D conversion. A virtual screening library that correctly registers accurate stereochemical information can, in theory, be utilized to provide information about the relative binding affinities of enantiomers. Recently, Brooks et al. conducted an interesting virtual screening study using the NCI Diversity Set 3D database [41]. The database represents a collection of highly diverse small molecules that contain both augmented enantiomers. Their results achieved significant ranking separation based on docking scores of the individual enantiomers in the database. Furthermore, their study demonstrates that potential lead candidates may be overlooked in databases containing 3D structures representing only one single enantiomer of a racemic chiral compound. In other words, potential improvements for virtual screening in computational lead generation entail a comprehensive approach to the inclusion of correct and complete stereoisomers.

7.3.3 Critical Assessment of Chiral Preference

Scoring functions are not optimal mainly because of the lack of treatment of protein flexibility and the solvation of the system. To overcome these issues, Monte Carlo (MC) statistical mechanics or molecular dynamics (MD) are typically applied to predict relative binding affinity of stereoisomers. During the simulation, classical force fields are used, and the methodology allows extensive sampling of all degrees of freedom for the complexes and representations of the aqueous surroundings with discrete water molecules. Free energy perturbation (FEP) and thermodynamic integration (TI) calculations then provide rigorous means to compute free energy changes. To assess the stereoisomer preference of chiral inhibitors, perturbations are made to convert one chiral isomer to the other chiral isomer using thermodynamic cycles. The conversions involve a coupling parameter that causes one molecule to be smoothly mutated to the other by changing the force field parameters and geometry. For example, Rao and Murcko successfully predicted binding preference of stereoisomers of Ro 31–8959 by FEP analysis [42]. Ro 31–8959 (Fig. 7.17a; the carbon atom with the star is the chiral center) is a highly potent inhibitor of HIV proteinase, and the (R)-configuration of hydroxyl is at least 250-fold better than the (S)-configuration. To compute the relative binding affinities of (R)- and (S)-isomers, a thermodynamic cycle (Fig. 7.17b) was used by converting both hydroxyl groups to hydrogen (des-OH). As shown in Figure 7.17b, the average changes of free energy (G_{bind}) when mutating the hydroxyl in both the (R)-configuration and the (S)-configuration to a hydrogen atom are 4.01 and 0.61 kcal/mol, respectively. The difference between these two values is 3.4 kcal/mol, in agreement with the observed 250-fold difference in binding.

In the above example, the crystal structures are critical for FEP calculations. In many cases, the crystal structures of the complexes of inhibitors and their targets are not available or they are extremely difficult to crystallize. These factors make it very challenging to accurately predict the binding modes and their chiral preference for compounds with chiral centers with the use of FEP. Instead,

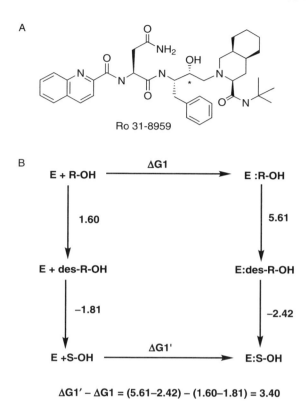

FIGURE 7.17   (a) Chemical structure of Ro 31–8959, a hydroxyethylamine inhibitor. The chiral center is highlighted with a star. (b) The thermodynamic cycle showing results obtained with the (R)-OH and (S)-OH model.

a combination of homology modeling, induced-fit-docking (IFD), and experimental mutation studies is useful. For example, recently Koldsø et al. proposed reversed binding modes for (S)- and (R)-citalopram binding to human serotonin transporters (hSERT), using homology models based on a homologous protein [43]. The predicted binding modes are also experimentally validated with human wild type and 15 mutants. These findings can help us understand the difference in binding of (S)- and (R)-citalopram, and they emphasize the importance of developing homochiral drugs.

7.3.4   QSAR Modeling with Chirality Descriptors

The relationship between chemical or biological activity and molecular structure and properties, namely, the quantitative structure-activity relationship (QSAR), structure-activity relationship (SAR), or quantitative structure-property relationship (QSPR), is a fundamental problem in medicinal chemistry and biochemistry.

QSAR is a well-established technique that has been applied broadly throughout different stages of drug discovery. 2D-QSAR uses molecular properties or topological descriptors of chemical structures (such as molecular connectivity indices) and is the most widely used QSAR technique because of its simplicity and efficiency in practice [44]. Unfortunately, these descriptors lack the ability to discriminate between stereoisomers, which limits their application in QSAR. To circumvent this problem, chirality descriptors derived from molecular graphs were introduced and applied to QSAR studies of ecdysteroids [45]. In this study, a combination of chirality descriptors and conventional 2D-QSAR (chirality insensitive) topological descriptors was applied to all data sets. The results are similar or better in comparison to models generated with 3D-QSAR approaches. The multidimensional QSAR techniques, such as 3D- and 4D-QSAR, are devised to handle flexible molecules by incorporating conformational sampling in the analysis. These techniques are able to successfully generate indirect spatial descriptors independent of translational and rotational states of the molecules, but an assumed active conformation is always required. As a result, they are inherently unable to describe the chiral arrangement of the pharmacophores. Recently, a chirality-sensitive flexibility (CSF) [46] descriptor was introduced. CSF is defined by the distance between a pharmacophore point and a plane defined by three pharmacophore points. This descriptor resulted in improved QSAR models on 38 PGF2 analogs in terms of both prediction accuracy and precision of the chiral geometric features of the predicted active conformations.

7.3.5 ADMET Modeling of Chiral Drugs

Drug enantiomers can have dramatically different pharmacokinetic and pharmacodynamic profiles since the human biological system is chiral. Drugs have to interact with all different kinds of proteins before being delivered to their targets. For instance, thiamylal is reported to have a higher concentration of the (S)-enantiomer than of the (R)-enantiomer in humans. The total clearance and volume distribution of the (R)-enantiomer were significantly greater than those of the (S)-enantiomer mainly because of enantioselective binding to serum protein [47].

Absorption, distribution, metabolism, excretion, and toxicology (ADMET) studies are widely used in drug research and development to help obtain the optimal balance of properties necessary to convert lead compounds into safe and effective drugs in humans. Recently, drug discovery efforts have been aimed at identifying and addressing metabolism issues at the earliest possible stage by developing and applying innovative *in silico* models. Many QSAR models have been developed for ADMET property predictions based on large corporate databases or published data. These models have been successfully used to understand SAR and prioritize compounds. For example, QSAR and pharmacophore modeling were employed to investigate stereoselective interactions between transporter (hOCT1) and both (R)- and (S)-verapamil. Verapamil is a chiral drug (Fig. 7.18) administered as a racemic mixture. Clinical and in vitro

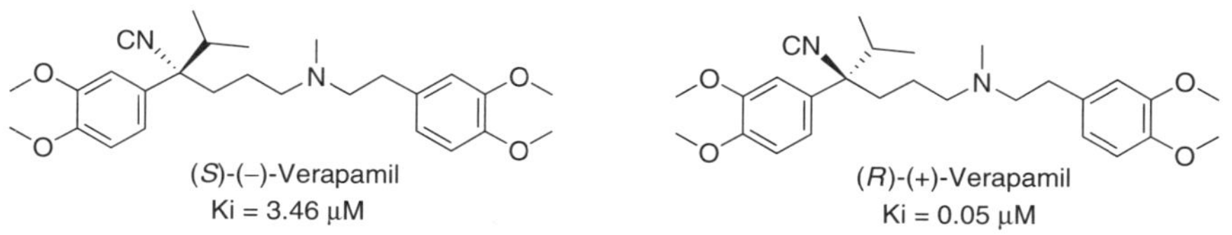

FIGURE 7.18 Chemical structure of (R)- and (S)-verapamil and experimentally determined K_i value using the CMAC (hOCT1) column.

studies have established that the cardiovascular effects of the S form are 10 times greater than R form [48]. One of the proposed pharmacophores was able to explain the source of the observed enantioselectivity. When (R)-verapamil was fitted to the proposed pharmacophore, all the relevant functional groups of the molecule matched the hypothesis, while (S)-verapamil could be mapped to only three of the model feature sites.

Protein structural information relevant to ADMET has been growing quickly in recent years, including the ligand-binding domain of PXR [49], transcriptional regulator of CYP3A4, human serum albumin [50,51], and the human P450 2C9 [52]. Structure-based approaches have been used successfully to build predictive models. It will be important to see more protein structures, especially bound with relevant cocrystals, which should continue to revolutionize *in silico* ADMET work.

7.3.6 Molecular Modeling Study of Chiral Drug Crystals

Application of molecular modeling in the solid state ranges from investigation of the influence of crystal packing on molecular structures to the determination of thermodynamic and dynamic properties of crystals. Over the last 10 years, modeling techniques have been applied in pharmaceutical research to polymorphism prediction, morphology modification, and crystal engineering. Differences in intermolecular interactions between homochiral and racemic crystals are subtle and complex. Accurate prediction of lattice energies of chiral drugs is desired to optimize resolution techniques, which are playing a major role in the industrial production of pure enantiomers. As an example, Crystal Packer in Cerius [53] using the Dreiding II force field was used to calculate the lattice energies of a number of chiral drugs with known crystals [54]. Comparison of the calculated energies among ephedrine derivatives reveals that a greater coulombic energy corresponds to a higher melting temperature, while greater van der Waals energy correlates with a higher enthalpy of fusion. Between the homochiral and racemic crystals, van der Waals forces appear to contribute significantly to the differences in both melting behavior and enthalpy of fusion. The shape complementarity of the paired enantiomers results in more effective van der Waals interactions, and therefore greater discrimination between homochiral and racemic crystals. Although this calculation method was based on available crystal structures, the results also offer critical insight into molecular interactions in certain chiral drugs in crystalline and salt formations.

7.4 PERSPECTIVES

With goals of improving drug efficacy and cutting development cost in the face of regulatory pressures, the pharmaceutical industry will continue to invest in chiral drugs. Advances in structural biology and pharmacology have provided us with great opportunities to understand how enantiomers work differently in the human body. Molecular modeling is able to provide guided design of chiral compounds through structure- and ligand-based design, virtual screening, and informatics-based data mining. With further improvement of computational methodologies and computer power, computer-aided drug discovery will be an increasingly valuable tool for improved design and characterization of chiral drugs along with the advances in asymmetric synthetic techniques and chiral separation.

REFERENCES

1. Kelvin, L. (1904). *Baltimore Lectures on Molecular Dynamics and the Wave Theory of Light*. London: C. J. Clay and sons; Baltimore, Publication agency of the Johns Hopkins University.
2. Meierhenrich, U. J. (2008). *Amino Acids and the Asymmetry of Life*. Heidelberg, Berlin, New York: Springer.
3. Julian, R. R. Myung, S., Clemmer, D. E. (2005) Do homochiral aggregates have an entropic advantage? *J. Phys. Chem. B* **109**, 440–444.
4. Cushny, A. R. (1926). *Biological Relations of Optically Isomeric Substances*. London: Bailliere, Thindalland and Cox.
5. Maureen, R. A. (2002). Chiral roundup. *Chem. Eng. News* **80**, 43–50.
6. Reddy, I. K., Mehvar, R. (Eds). (2004). *Chirality in Drug Design and Development*. New York: Marcel Dekker, Inc.
7. Easson, L. H., Stedman, E. (1933). Studies on the relationship between chemical constitution and physiological action: molecular dissymmetry and physiological activity. *Biochem. J*. **27**, 1257–1266.
8. Patil, P. N., LaPidus, J. B., Tye, A. (1970). Steric aspects of adrenergic drugs. *J. Pharm. Sci*. **59**, 1205–1234.
9. Mesecar, A. D., Koshland, D. E. (2000). A new model for protein stereospecificity. *Nature* **403**, 614–615.
10. Bunnage, M. E., Mathias, J. P., Wood, A., Miller, D., Street, S. D. A. (2008). Highly potent and selective chiral inhibitors of PDE5: an illustration of Pfeiffer's rule. *Bioorg. Med. Chem. Lett*. **18**, 6033–6036.
11. Pfeiffer, C. C. (1956). Optical isomerism and pharmacological action, a generalization. *Science* **124**, 29–31.
12. Millership, J. S., Fitzpatrick, A. (1993). Commonly used chiral drugs: a survey. *Chirality* **5**, 573–576.

13. Smithells, D. (1998). Was the thalidomide tragedy preventable? *Lancet* **351**, 1591.

14. Singh, S. S. (2008). Chapter 14: Toxicokinetics: an integral component of preclinical toxicity studies. Gad, S. C. Ed. In *Preclinical Development Handbook: Toxicology*. Wiley-Interscience, Wiley-Blackwell. p. 459–507.

15. Erb, S. (2006). Single-enantiomer drugs poised for further market growth. *Pharmaceut. Technol*. Oct. 3.

16. Flockhart, D. A., Nelson, H. S. (2002). Single isomer versus racemate: is there a difference? Clinical comparisons in allergy and gastroenterology. *CNS Spectr*. **7**, 23–27.

17. Istvan, E. S., Deisenhofer, J. (2001). Structural mechanism for statin inhibition of HMG-CoA reductase. *Science* **292**, 1160–1164.

18. Stokker, G. E., Hoffman, W. F., Alberts, A. W., Cragoe, E. J., Deana, A. A., Gilfillan, J. L., Huff, J. W., Novello, F. C., Prugh, J. D., Smith, R. L. (1985). 3-Hydroxy-3-methylglutaryl-coenzyme A reductase inhibitors. 1. Structural modification of 5-substituted 3,5-dihydroxypentanoic acids and their lactone derivatives. *J. Med. Chem*. **28**, 347–358.

19. Ippolito, J. A., Kanyo, Z. F., Wang, D., Franceschi, F. J., Moore, P. B., Steitz, T. A., Duffy, E. M. (2008). Crystal structure of the oxazolidinone antibiotic linezolid bound to the 50S ribosomal subunit. *J. Med. Chem*. **51**, 3353–3356.

20. Zhou, C. C., Swaney, S. M., Shinabarger, D. L., Stockman, B. J. (2002). ^{1}H nuclear magnetic resonance study of oxazolidinone binding to bacterial ribosomes. *Antimicrob. Agents Chemother*. **46**, 625–629.

21. Stamos, J., Sliwkowski, M. X., Eigenbrot, C. (2002). Structure of the epidermal growth factor receptor kinase domain alone and in complex with a 4-anilinoquinazoline inhibitor. *J. Biol. Chem*. **277**, 46265–46272.

22. Aronov, A. M., McClain, B., Moody, C. S., Murcko, M. A. (2008). Kinase-likeness and kinase-privileged fragments: toward virtual polypharmacology. *J. Med. Chem*. **51**, 1214–1222.

23. Bonabello, A., Galmozzi, M. R., Canaparo, R., Isaia, G. C., Serpe, L., Muntoni, E., Zara, G. P. (2003). Dexibuprofen (S+-isomer ibuprofen) reduces gastric damage and improves analgesic and antiinflammatory effects in rodents. *Anesth. Analg*. **97**, 402–408.

24. Selinsky, B. S., Gupta, K., Sharkey, C. T., Loll, P. J. (2001). Structural analysis of NSAID binding by prostaglandin H_2 synthase: time-dependent and time-independent inhibitors elicit identical enzyme conformations. *Biochemistry* **40**, 5172–5180.

25. Copeland, R. A., Williams, J. M., Giannaras, J., Nurnberg, S., Covington, M., Pinto, D., Pick, S., Trzaskos, J. M. (1994). Mechanism of selective inhibition of the inducible isoform of prostaglandin G/H synthase. *Proc. Natl. Acad. Sci. U. S. A.* **91**, 11202–11206.

26. Laponogov, I., Sohi, M. K., Veselkov, D. A., Pan, X., Sawhney, R., Thompson, A. W., McAuley, K. E., Fisher, L. M., Sanderson, M. R. (2009). Structural insight into the quinolone-DNA cleavage complex of type IIA topoisomerases. *Nat. Struct. Mol. Biol*. **16**, 667–669.

27. Humphrey, G. H., Shapiro, M. A., Randinitis, E. J., Guttendorf, R. J., Brodfuehrer, J. I. (1999). Pharmacokinetics of clinafloxacin enantiomers in humans. *J. Clin. Pharmacol*. **39**, 1143–1150.

28. Olbe, L., Carlsson, E., Lindberg, P. (2003). A proton-pump inhibitor expedition: the case histories of omeprazole and esomeprazole. *Nat. Rev. Drug Discov*. **2**, 132–139.

29. Miller, J. L., Bradley, E. K., Teig, S. L. (2003). Luddite: an information-theoretic library design tool. *J. Chem. Inf. Comput. Sci*. **43**, 47–54.

30. Lipinski, C. A., Lombardo, F., Dominy, B. W., Feeney, P. J. (2001). Experimental and computational approaches to estimate solubility and permeability in drug discovery and development settings. *Adv. Drug Deliv. Rev*. **46**, 3–26.

31. McGregor, M. J., Muskal, S. M. (1999). Pharmacophore fingerprinting. 1. Application to QSAR and focused library design. *J. Chem. Inf. Comput. Sci*. **39**, 569–574.

32. Haigh, J. A., Pickup, B. T., Grant, J. A., Nicholls, A. (2005). Small molecule shape-fingerprints. *J. Chem. Inf. Model*. **45**, 673–684.

33. Beroza, P., Suto, M. J. (2000). Designing chiral libraries for drug discovery. *Drug Discov. Today* **5**, 364–372.

34. Lyne, P. D. (2002). Structure-based virtual screening: an overview. *Drug Discov. Today* **7**, 1047–1055.

35. Catalyst, Accelrys, 9685 North Scranton Road, San Diego, CA. http://www.accelrys.com/.

36. Kroemer, R. T. (2007). Structure-based drug design: docking and scoring. *Curr. Protein Pept. Sci*. **8**, 312–328.

37. Rarey, M., Kramer, B., Lengauer, T., Klebe, G. (1996). A fast flexible docking method using an incremental construction algorithm. *J. Mol. Biol*. **261**, 470–489.

38. Friesner, R. A., Murphy, R. B., Repasky, M. P., Frye, L. L., Greenwood, J. R., Halgren, T. A., Sanschagrin, P. C., Mainz, D. T. (2006). Extra precision glide: docking and scoring incorporating a model of hydrophobic enclosure for protein-ligand complexes. *J. Med. Chem*. **49**, 6177–6196.

39. Wang, R., Lu, Y., Fang, X., Wang, S. (2004). An extensive test of 14 scoring functions using the PDBbind refined set of 800 protein-ligand complexes. *J. Chem. Inform. Comput. Sci*. **44**, 2114–2125.

40. Beavers, M. P., Myers, M. C., Shah, P. P., Purvis, J. E., Diamond, S. L., Cooperman, B. S., Huryn, D. M., Smith, A. B. (2008). Molecular docking of cathepsin L inhibitors in the binding site of papain. *J. Chem. Inf. Model*. **48**, 1464–1472.

41. Brooks, W. H., Daniel, K. G., Sung, S., Guida, W. C. (2008). Computational validation of the importance of absolute stereochemistry in virtual screening. *J. Chem. Inf. Model*. **48**, 639–645.

42. Rao, B. G., Murcko, M. A. (1994). Reversed stereochemical preference in binding of Ro 31-8959 to HIV-1 proteinase: a free energy perturbation analysis. *J. Comput.Chem*. **15**, 1241–1253.

43. Koldsø, H., Severinsen, K., Tran, T. T., Celik, L., Jensen, H. H., Wiborg, O., Schiøtt, B., Sinning, S. (2010). The two enantiomers of citalopram bind to the human serotonin transporter in reversed orientations. *J. Am. Chem. Soc*. **132**, 1311–1322.

44. Willett, P. (2006). Similarity-based virtual screening using 2D fingerprints. *Drug Discov. Today* **11**, 1046–1053.

45. Golbraikh, A., Tropsha, A. (2003). QSAR modeling using chirality descriptors derived from molecular topology. *J. Chem. Inf. Comput. Sci*. **43**, 144–154.

46. Dervarics, M., Otvös, F., Martinek, T. A. (2006). Development of a chirality-sensitive flexibility descriptor for 3+3D-QSAR. *J. Chem. Inf. Model*. **46**, 1431–1438.

47. Sueyasu, M., Ikeda, T., Taniyama, T., Futugami, K., Kataoka, Y., Oishi, R. (1997). Pharmacokinetics of thiamylal enantiomers in humans. *Int. J. Clin. Pharmacol. Ther*. **35**, 128–132.

48. Longstreth, J. A. (1993). Verapamil: a chiral challenge to the pharmacokinetic and pharmacodynamic assessment of bioavailability and bioequivalence. In *Drug Stereochemistry: Analytical Methods and Pharmacology*. *2nd ed*. New York: Marcel Dekker. Inc., p. 315–335.

49. Watkins, R. E., Wisely, G. B., Moore, L. B., Collins, J. L., Lambert, M. H., Williams, S. P., Willson, T. M., Kliewer, S. A., Redinbo, M. R. (2001). The human nuclear xenobiotic receptor PXR: structural determinants of directed promiscuity. *Science* **292**, 2329–2333.

50. Petitpas, I., Grüne, T., Bhattacharya, A. A., Curry, S. (2001). Crystal structures of human serum albumin complexed with monounsaturated and polyunsaturated fatty acids. *J. Mol. Biol*. **314**, 955–960.

51. Petitpas, I., Bhattacharya, A. A., Twine, S., East, M., Curry, S. (2001). Crystal structure analysis of warfarin binding to human serum albumin: anatomy of drug site I. *J. Biol. Chem*. **276**, 22804–22809.

52. Williams, P. A., Cosme, J., Ward, A., Angove, H. C., Matak, V. D., Jhoti, H. (2003). Crystal structure of human cytochrome P450 2C9 with bound warfarin. *Nature* **424**, 464–468.

53. Krieger, J. H. (1995). New software expands role of molecular modeling technology. *Chem. Eng. News* **27**, 30–40.

54. Li, Z. J., Ojala, W. H., Grant, D. J. (2001). Molecular modeling study of chiral drug crystals: lattice energy calculations. *J. Pharm. Sci*. **90**, 1523–1539.

CHAPTER 8

PHARMACOLOGY OF CHIRAL DRUGS

YONGGE LIU and XIAO-HUI GU

Otsuka Maryland Medicinal Laboratories, Inc., Rockville, MD

Chiral Drugs: Chemistry and Biological Action, First Edition. Edited by Guo-Qiang Lin, Qi-Dong You and Jie-Fei Cheng.
© 2011 John Wiley & Sons, Inc. Published 2011 by John Wiley & Sons, Inc.

8.1 INTRODUCTION

Pharmacology is the study of substances that interact with living organisms through binding to regulatory molecules (receptors) that activate or inhibit normal biochemical functions. Substances that are administered to the body to achieve medical purposes (therapeutic, diagnostic, and preventive) are considered pharmaceuticals or drugs. The scope of pharmacology is very extensive, embracing physical and chemical properties of drugs, drug-body interactions, and therapeutic and other uses of the drugs, as well as associated undesirable or toxic effects. The effects of drugs on the body are called pharmacodynamics, including biochemical and physiological effects and mechanisms of actions. The effects of the body on the drugs are termed pharmacokinetics, including absorption, distribution, metabolism, and excretion (often referred to as ADME). The undesirable and/or toxic effects of the drugs are studied by another branch of pharmacology, called toxicology.

The use of exogenous substances to alter biological functions of human beings dates back thousands of years to Chinese and Egyptian use of natural substances as remedies, largely based on systems of thought without experimentation and observation. Only since the eighteenth and nineteenth centuries, with the advances of chemistry and biology, has pharmacology become a biochemical science that involves experimentation to understand how drugs work at the organ and tissue level. The advance of modern pharmacology came in the mid-twentieth century with the arrival of molecular biology and the availability of new technologies, which helped identify the molecular mechanisms of many drugs. With the accompanying expansion of knowledge of the underlying pathology of the diseases, we are now able to better tailor treatments and design drugs to target individual diseases. In fact, drugs are becoming one of the major tools to improve the quality and duration of life and prevent epidemics of deadly diseases. However, with the desire to live longer and better, and the emergence of new diseases, we need to meet the challenges by improving current and discovering new treatments.

The majority of drugs are synthesized chemically as small molecules. However, new biotechnology has permitted the development of therapeutic proteins through biological processes. These biological drugs range from small-sized peptides such as insulin to large-sized antibodies. Discussion of the biological agents is beyond the scope of this chapter and can be found elsewhere [1–3]. Here, we first discuss general principles of pharmacodynamics of small-chemical drugs, followed by a discussion of effects of chirality, an optical property of the drug, on the pharmacodynamics. For information about other branches of pharmacology, pharmacokinetics and toxicology, readers are recommended to read the relevant chapters in this book.

8.2 BASIC PRINCIPLES OF PHARMACODYNAMICS

8.2.1 Physical Properties of the Drug

To interact with their receptors, drugs must have appropriate physicochemical properties such as sizes, shapes, and electrical charges. Drug-receptor interactions are often described as a "key and lock" mechanism, in which a drug (key) must fit in and bind to a particular receptor (lock). The specific binding of a drug to its receptor is critical for intended biological effects of the drug while limiting its interaction with other types of receptors. The majority of small-molecule drugs have a size ranging from 100 to 500 Daltons. If too small, a drug may lack the specific properties to fit into the receptor uniquely, thus lacking specificity. If too large, a molecule may limit its ability to reach the targeted tissue (pharmacokinetic issues). Another factor affecting the drug-receptor interaction is the shape of the drug, as the three-dimensional structure of the drug must be complementary to that of the receptor. This includes the chirality of the drug, as more than half of the small-molecule drugs exist as enantiomeric pairs. We discuss this issue in great detail below in this chapter.

Even if the drug fits well with the receptor, forces are needed to keep the drug on the receptor. This is accomplished mainly by three types of chemical forces or bonds. The first is a very strong but uncommon covalent bond, often resulting in an "irreversible" biological reaction. The second is an electrostatic bond that is weaker than the covalent bond but is more common in the drug-receptor interaction. The strength of electrostatic bonds ranges from strong electrostatic interactions between permanently charged ionic molecules to hydrogen bonds and weaker dipole interactions such as van der Waals forces. The third is the hydrophobic bond, which is the weakest of the three but is important for interactions between drugs and lipid cell membranes.

8.2.2 Receptors

The best-characterized receptors are transmembrane proteins that mediate the effects of neurotransmitters (such as acetylcholine, dopamine, and norepinephrine), autacoids (such as serotonin, histamine, and prostaglandins), and hormones. Effects mediated by these receptors are responsible for the majority of drug actions. A common mechanism of transferring the signal from extracellular receptor binding to inside the cell is the seven-transmembrane domain G proteins [4,5]. Binding to a receptor by a ligand causes the activation of G proteins, which then generate an intracellular regulatory molecule known as a second messenger. The second messenger is so called because it follows the arrival of the first messenger, the activation of a cell surface receptor by a ligand. Three major second messengers are cyclic nucleotides (cAMP or cGMP), Ca^{2+}, and phosphatidylinositol. Another major class of receptors is intracellular proteins. One well-known example is the steroid receptors that enter the nucleus and modulate protein expression. Intracellular receptors can also be enzymes, such as phosphodiesterase type 5 (PDE5), the target of sildenafil (Viaga) for the treatment of male erectile dysfunction. Other receptors include transport proteins (for example, Na^+/K^+-ATPase, a receptor for digitalis glycosides) and occasionally structural proteins such as tubulin, a receptor for antiinflammatory and anticancer drugs.

8.2.3 Drug-Receptor Interactions

8.2.3.1 Relationship Between Drug Concentration and Response

Drug-receptor binding can be divided into two steps: binding to the receptor and subsequent responses from the receptor. The first step can be described by the mass action law, with the following mathematic equation (often called the drug binding curve): $B = (B_{max} \times C)/(C + K_D)$, where B is the concentration of receptors occupied by the drug at concentration C, B_{max} is the maximal concentration of receptors that can be bound to the drug (usually determined as the maximal number of receptors occupied at saturating concentrations of the drug), and K_D is the concentration at which 50% of the receptors are occupied by the drug. K_D, also called the equilibrium dissociation constant, represents the affinity of the drug to the receptor. The higher the affinity (as thus the lower the K_D value), the more tightly the drug binds to the receptor.

Depending on the type of drug, receptors can take different actions after the drug binding. If a drug activates the receptor, then the drug is called an agonist of the receptor. The response from an agonist binding to the receptor (often called the drug-response curve) can be described similarly as the drug binding curve: $E = (E_{max} \times C)/(C + EC_{50})$, where E is the effect at drug concentration C, E_{max} is the maximal effect that can be generated by the drug, and EC_{50} is the drug concentration that produces half-maximal effect. ED_{50} is often used to compare the potency of the drug. Graphically, these two equations are hyperbolic curves, as shown in Figure 8.1.

It is apparent that drug binding or receptor response increases quickly at low concentrations of the drug but slowly at high concentrations (Fig. 8.1). To make the visual comparison easier over large concentration ranges, pharmacologists often transform the hyperbolic curve into a sigmoid curve by plotting against the logarithm of the drug concentrations (Fig. 8.2). These curves demonstrating the relationship between drug binding and receptor responses are often used by pharmacologists as concentration-response or dose-response curves.

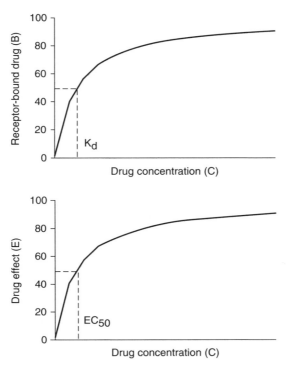

FIGURE 8.1   The top panel plots drug concentration vs. percentage of receptors bound by the drug. As the drug concentration increases, more and more receptors are occupied by the drug until near 100% occupancy. K_D represents the drug concentration that produces 50% of maximal drug-receptor occupancy, and indicates the affinity of the drug to the receptor. The bottom panel plots drug concentration vs. cellular response to the drug binding to the receptor. Potency of the drug is represented by EC_{50}, the drug concentration that produces 50% of the maximal responses. Note that both curves are hyperbolic.

8.2.3.2 Agonists, Partial Agonists, and Inverse Agonists

As discussed above, an agonist is a drug that mimics the full effects of endogenous regulatory ligands (such as neurotransmitters and hormones) for the receptor. Thus, when the receptors are fully occupied by the drug, a maximal and full response is reached. However, there are drugs that produce only a partial effect even with the ability to occupy all the receptors. These drugs are termed *partial agonists*. In contrast to the agonist, if drugs produce the effects opposite to those of an agonist after occupying the receptors, these type of drugs are called *inverse agonists*.

8.2.3.3 Antagonists

An antagonist is a compound that binds to the receptor but has no intrinsic regulatory activity. However, antagonists prevent the binding of an agonist to

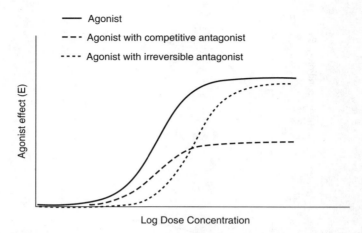

FIGURE 8.2 Pharmacologists usually use the logarithm of the concentration or dose to transform the hyperbolic into sigmoid curves to expand the scales of the concentrations at low end. In the presence of reversible competitive antagonist, an agonist drug effect curve is shifted to the right, but maximal effect of the agonist can still be achieved, although a higher concentration is needed. In the presence of an irreversible antagonist, however, the agonist's potency is not affected but maximal effect can no longer be achieved.

the receptor. Antagonists can be divided into two classes, competitive and irreversible. A competitive antagonist competes with the binding of an agonist reversibly, and therefore the inhibition can be overcome by increasing the concentration of the agonist. This effect can be demonstrated by the right-shifting of the concentration-response curve (Fig. 8.2). The maximal effect is unaltered, but higher concentrations of the agonist are needed to achieve the same response in the absence of the antagonist. On the other hand, an irreversible antagonist binds to the active site of the receptor irreversibly (covalently, for example), thus preventing the agonist from producing any effect. The possible maximal effect is reduced, but the agonist potency is unchanged so long as receptors are still available for the agonists to bind (Fig. 8.2).

8.2.4 Signaling Mechanisms in Drug Action

Thus far, we have discussed drug-receptor interactions and drug effects using drug concentration-response curves. Biological and cellular responses are much more complex than simply the binding of a drug to a receptor, which only starts the chain reactions of signal transduction. Below, we discuss the five basic strategies that the cell uses to transduce the transmembrane signals.

8.2.4.1 Intracellular Receptors for Lipid-Soluble Drugs
Drugs in this class can cross the cell membrane (lipophilic) and activate an intracellular receptor. One example is the nuclear receptors that modulate gene

transcription. Once inside the cell, these drugs or endogenous ligands bind to soluble DNA-binding proteins (DNA response elements) in the nucleus, stimulating the transcription of genes. Examples of this type of drugs include steroid and thyroid hormones, corticosteroids, mineralocorticoids, sex hormones, and vitamin D. Additional examples include intracellular enzymes, for instance, guanylyl cyclase that can be activated by nitric oxide and phosphodiesterase that regulates the level of intracellular cAMP (an important second messenger as discussed below).

8.2.4.2 *Ligand-Regulated Transmembrane Enzymes*
These receptors are polypeptides consisting of an extracellular ligand-binding domain and a cytosolic enzyme domain, with the two domains linked by a hydrophobic segment of the polypeptide that crosses the lipid bilayer of the plasma membrane. Cytosolic enzyme activity is increased by the binding of the ligand on the extracellular binding site. One important member of this class of receptors is the receptor tyrosine kinases, high-affinity cell surface receptors for many polypeptide growth factors, cytokines, and hormones. Many of these receptors exist as monomers. Once the ligand binds to the extracellular domain of these receptors, they form dimers, which lead to the activation of the intracellular kinase domain of the receptors. Subsequently, tyrosine residues in the cytosolic domain are phosphorylated, which in turn regulate downstream signaling proteins that regulate transcription and cell proliferation. One example is the epidermal growth factor receptor. Studies have shown its involvement in tumor growth, and several anticancer drugs have successfully been developed recently to inhibit the activity of this receptor [6].

8.2.4.3 *Cytokine Receptors*
Cytokines are small proteins that are secreted by certain cells to regulate immunity and inflammation within the body. They include a diverse group of peptide ligands comprising growth hormone, erythropoietin, interferon, as well as other growth and differentiation regulators. The signaling pathways for the cytokines are initiated by binding to their receptors on the cell surface. Unlike the receptor tyrosine kinase, cytokine receptors do not have intrinsic kinase activity. Instead, cytokine receptors couple to other protein kinases that modulate cellular functions. One common kinase is the Janus kinase family, a separate tyrosine kinase.

8.2.4.4 *Ligand-Gated Channels*
These receptors are transmembrane ion channels that open and close according to the binding of a ligand. These ion channels are usually very selective to allow only one type of ions (Na^+, Ca^{2+}, K^+, Cl^-, and Mg^{2+}) to pass in response to the ligand binding. The most-studied endogenous ligands for the ion channels are synaptic transmitters, including acetylcholine, γ-aminobutyric acid, and the excitatory amino acids (glycine, aspartate, and glutamate). The receptors are located at synapses to convert the chemical signals of presynaptically released neurotransmitters directly and very quickly into postsynaptic electrical signals. Many agonists or antagonists of ion channels are important drugs for the treatment of central nerve diseases.

8.2.4.5   G Protein-Coupled Receptors

G protein-coupled receptors (GPCRs) are the most well-known and well-studied receptors. They comprise a large group of transmembrane proteins that sense the binding of extracellular ligands to the receptors and respond with the activation of a G protein. The activated G protein then changes the activity of an effector, usually an ion channel, which can alter membrane potential, or enzyme, which consequently changes the concentrations of intracellular second messengers [4,5]. Important second messengers are cAMP, calcium, and phosphoinositides, which are discussed in detail below in this chapter.

G proteins are comprised of two subunits, G_α and $G_{\beta\gamma}$, and bound to the intracellular side of the membrane with the receptor. When a GPCR is activated by a ligand, the receptor changes its conformation to release GDP from the G_α subunit. This allows the binding of GTP to the G_α, which triggers the dissociation of G protein from the receptor and $G_{\beta\gamma}$ from GTP-bound G_α. Both G_α and $G_{\beta\gamma}$ subunits can regulate the activity of an effector (an ion channel or an intracellular enzyme). The G_α subunit containing intrinsic GTPase activity then hydrolyzes GTP to GDP, allowing it to reassociate with $G_{\beta\gamma}$ and the receptor, and returns to the unstimulated state, ready to start the next cycle.

The G proteins include a diverse group of subfamilies, as listed in Table 8.1, and can be generally divided into two groups, one affecting the intracellular cAMP level and the other modulating the intracellular Ca^{2+} and phosphoinositide level.

8.2.4.6   Second Messenger Signal Transduction Pathways

Cells use the second messenger system to relay the signals from cell surface receptors to target molecules in the cytosol or nucleus. In addition, the second messengers help amplify the signals so that a limited number of ligands can produce large cellular responses. The common change at the molecular level is the reversible phosphorylation (transfer of phosphate from ATP) of target proteins by kinases. Two major protein kinases are the cAMP-dependent protein kinase, PKA, and the phosphoinositide-sensitive protein kinase, PKC. Other common

TABLE 8.1   G Proteins and Their Receptors and Effectors

G- Protein	Receptors for	Effectors
G	Catecholamines, histamine, serotonin, vasopressin, glucagon, and many other hormones	Stimulate adenylyl cyclase to increase cAMP
G_i	Acetylcholine, opoidsopioids, serotonin, and many other hormones	Inhibit adenylyl cyclase to decrease cAMP
G_q	Acetylcholine, serotonin, and many other hormones	Stimulate phospholipase C to generate DAG and IP_3
G_{olf}	Olfactory epithelium	Increase cAMP

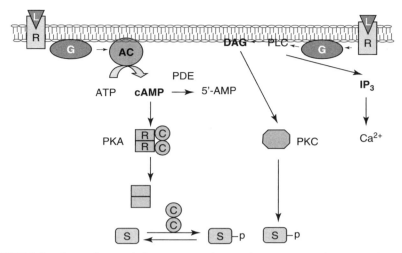

FIGURE 8.3   G protein coupled receptors and second messenger pathways. In the cAMP pathway, an agonist ligand (L) binds to a receptor (R), activating a stimulatory G- protein (G_s), which activates adenylyl cyclase (AC). AC then converts ATP to cAMP. The next step is the binding of cAMP to the regulatory domain (r) of protein kinase A (PKA), resulting in the dissociation of catalytic domain (c) of PKA. The final result of this pathway is the transfer of a phosphate (p) from ATP to the substrate (S) protein (phosphorylation) by the catalytic domain of the PKA. In the phospholipase C (PLC) pathway, after ligand-receptor binding and subsequent activation of a G- protein, PLC hydrolyzes membrane-bound phosphatidylinositol 4,5 bisphosphate (PIP_2) to yield diacylglycerol (DAG) and inositol- 1,4,5 trisphosphate (IP_3). DAG is confined to the membrane and activates PKC, while IP_3 diffuses through the cytosol and triggers the release of Ca^{2+} from intracellular stores.

second messengers include Ca^{2+} and cGMP. Figure 8.3 summarizes the major pathways through GPCRs to second messengers.

8.2.4.6.1   cAMP.   cAMP is produced from ATP by adenylyl cyclase in response to receptor-mediated stimulation of a stimulatory G protein (G_s). Most of the cAMP effects are mediated by protein phosphorylation through PKA, which can exert both immediate effects, such as the increase of heart rate in the "fight or flight" response, and chronic effects, such as gene regulation.

PKA is composed of a cAMP-binding regulatory dimer and two catalytic domains. Once cAMP binds to the regulatory sites, catalytic domains are activated and released through the cytosol and to the nucleus, where they transfer phosphate from ATP to substrate proteins.

8.2.4.6.2 Ca^{2+} and Phosphoinositides.   Two important second messengers are produced by the actions of phospholipase C (PLC). After the activation of

PLC through GPCR stimulation, it produces diacylglycerol (DAG) and inositol 1,4,5-trisphosphate (IP$_3$) through the hydrolysis of membrane-bound phosphatidylinositol 4,5-bisphosphate (PIP$_2$). DAG is confined to the membrane and activates PKC, while IP$_3$ diffuses through the cytosol and triggers the release of Ca^{2+} from intracellular stores. Ca^{2+} can bind to calmodulin and regulate other protein kinase activities.

8.2.4.6.3 cGMP. In comparison to cAMP, cGMP-based signal transduction is more limited. The best-illustrated role of cGMP is the regulation of smooth muscle tone. The mechanism to generate cGMP is quite similar to that for cAMP, namely, guanylyl cyclase converting GTP to cGMP. One major effect of cGMP is to relax smooth muscle cells through cGMP-dependent protein kinase (PKG).

8.2.4.7 Interplay Among Signal Transduction Pathways

The estimated number of receptors is enormous [7], yet there are only a few primary second messengers (cAMP, Ca^{2+}, phosphoinositides, and cGMP). These second messengers impact nearly every aspect of cellular life from microbes to human beings. But how do the cells relay extracellular signals and respond to them with a high degree of specificity with only a few second messengers? Several layers of complexity make this possible [4,5,8,9]. Cells express different isoforms for receptors and G proteins to allow different responses to the same ligand. In addition, ligands for the same receptor have the ability to induce different receptor conformations that can activate different downstream signaling pathways (called biased agonism) [9,10]. Receptors may also form dimers with the same type (homodimerization) or completely different types (heterodimerization) that can alter pharmacology and signaling pathways [8,11]. The compartmentation of signaling molecules contributes to the specificity as well: Second messengers are often localized within the cells to precise subcellular domains (plasma membrane or intracellular organelles) to allow selective activation of a subset of targets. Furthermore, cross talk between different signal messengers is very common to allow fine-tuning of cellular responses. For example, second messengers often oppose or complement each other. An increase of intracellular Ca^{2+} causes the contraction of smooth muscle cells, while cAMP and cGMP have the opposite effects. cAMP also reduces intracellular Ca^{2+} concentration by increasing the uptake to the intracellular stores. Thus the interplay among the second messengers significantly increases the complexity of signal transduction pathways and presents intriguing ways to achieve signaling specificity.

8.2.5 Receptor Desensitization and Signal Termination

To maintain a relatively stable condition, biological systems often utilize negative feedback mechanisms to prevent overstimulation. This is true for the receptors. After sustained activation by a ligand, receptor responses often decrease or "desensitize" with time in a reversible fashion, but the full response can be achieved again if the ligand has been removed for a certain period of time. One common mechanism is the receptor phosphorylation by the G protein-coupled receptor kinase and the binding of β-arrestin [12–14], resulting in uncoupling of the G protein from the receptor. Additionally, β-arrestin causes receptor internalization, in which receptors are taken up from the plasma membrane and internalized into intracellular lysosome vesicles for degradation. Finally, downregulation of the receptor may occur after prolonged stimulation (hours to days) that reduces receptor biosynthesis and increases receptor internalization and degradation.

To maintain dynamic regulation, cells have various ways to quickly terminate the second messengers. Phosphorylation can be rapidly reversed by specific and nonspecific phosphatases. cAMP and cGMP are hydrolyzed by a group of enzymes called phosphodiesterases, while IP$_3$ is dephosphorylated and DAG can either be dephosphorylated or deacylated. Ca^{2+} is actively removed from the cytosol to intracellular stores and pumped out of the cells.

8.3 EFFECT OF CHIRAL PROPERTY ON DRUGS

The chiral property of a drug refers to a center of three-dimensional asymmetry. Thus a chiral drug has two non-superimposable mirror-image forms or enantiomers, similar to the right and left hands of a person. Enantiomers are formed around a chiral center, a carbon atom that possesses asymmetric spatial arrangement. If a molecule has two or more chiral centers, its chiral isomers are called diastereomers. As we discussed above, the binding of a drug to a biomolecular receptor in the body is the first and critical step to initiate pharmacological effects. To ensure a specific binding, the drug and receptor must have a perfect three-dimensional fit, often referred to as the unique "key (drug) and lock (receptor)" configuration. Because stereoselectivity plays an important role in biosynthesis of drug targets, it is not surprising that chiral property affects drug binding. This is easily understandable if we use the analogy of hand and glove: A right hand would not fit into a left-hand grove, and vice versa. Therefore, the enantiomers of drugs may have different effects on the receptors, which can be reflected in pharmacodynamics, pharmacokinetics, and toxicology.

Although the concept of "chirality" has been known since the 1870s, it was about a century later that chemists began to recognize the influences of chiral characteristics on a drug and its interaction with the body. Before the 1980s, the majority of drugs were either achirals or racemates, an equimolar mixture of the two enantiomers. But since that time, with the advance of new technologies to separate and produce single enantiomers, it has become possible to develop stereochemically pure drugs. Because of the potential different pharmacology and toxicology of each enantiomer, the single enantiomer has become the preferred chemical identity for drug companies and regulatory authorities [15,16]. In fact, the proportion of new drugs that are single enantiomers, versus racemates and diastereomers, increased from ∼20% in 1990 to almost 90% in 2006 (http://www.fda.gov/downloads/AboutFDA/CentersOffices/CDER/ucm10353 2.pdf; accessed in July, 2010). Four of the current (2010) top-selling drugs, atorvastatin (Lipitor), clopidogrel (Plavix), valsartan (Diovan), and rosuvastatin

(Crestor) are single enantiomer drugs (http://www.reuters.com/article/idUSLDE63 C0BC20100413; accessed in July, 2010).

Because of the potential to have vast pharmacological differences, each enantiomer from a racemic mixture can be considered as an individual drug. The pharmacodynamics of racemic drugs can be divided into five major categories. The majority of racemic drugs belong to the first category, in which one enantiomer (eutomer) has the major therapeutic effects while the other is inactive or has negligible activity (distomer). In this case, the eutomer binds to the intended receptor while the distomer has no or only minimal binding. The second category is chiral drugs with two enantiomers of equal activity. This type of chiral drugs is unsurprisingly rare, as dictated by the three-dimensional requirement of drug-receptor binding. If we use the hand-glove analogy again: it is difficult to imagine a glove that can fit both left and right hands. The third category is drugs that are racemic mixtures, but with one enantiomer being rapidly converted to the other enantiomer in vivo and the biological activities residing in the second enantiomer (chiral inversion). Thus the first enantiomer is essentially the pro-drug (a drug that is inactive when administered but converted to an active form in vivo), and all of the biological activities are induced by the second enantiomer. The fourth category is chiral drugs that have enantiomers exhibiting separate biological activities, sometimes with one enantiomer having desirable and the other having undesirable effects. This is usually caused by the binding of different enantiomers to different receptors. The final category of chiral drugs has one enantiomer antagonizing the effects of another enantiomer. Obviously, a racemic mixture as a drug in this class is not ideal, and instead an enantiomer with the intended biological activity (not the antagonizing enantiomer) should be developed as a drug. In the following sections, we discuss each category of chiral drugs.

8.3.1 Drugs with One Major Bioactive Enantiomer

The majority of racemic drugs have one major bioactive enantiomer. Pharmacologically this is expected, as drug-receptor binding requires three-dimensional conformation specificity; therefore often only one stereoisomer has biological activity. This type of drugs extends to almost all of the therapeutic areas. Although only one enantiomer is pharmacologicaly active in this category, drugs developed before the 1980s are usually racemates because of the difficulties of separating and producing large quantities of the chemicals. Since then, more and more chiral drugs have been developed as enantiomers because of the ever-increasing recognition of the advantages of enantiomers versus racemates by the pharmaceutical industry, as discussed below in this chapter.

In the cardiovascular arena, drugs in this class include β-adrenergic receptor antagonists, calcium channel antagonists, and angiotensin-converting enzyme (ACE) inhibitors that are used for the treatment of hypertension, heart failure, arrhythmias, etc.

β-Adrenergic receptor antagonists, first developed in the late 1950s, are still widely used for the treatment of angina, myocardial infarction, hypertension,

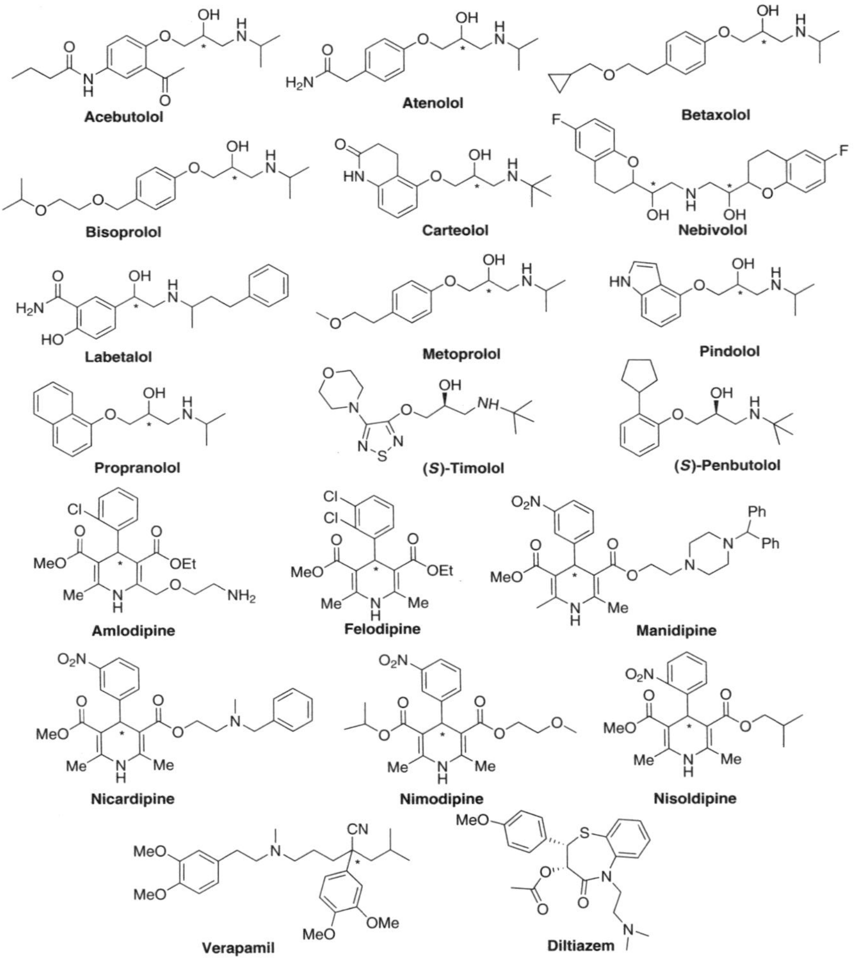

FIGURE 8.4 Chemical structures of β-adrenergic and calcium channel blocking drugs.

and various other conditions. Many β-blockers are still marketed as racemic drugs, of which nearly all the (S)-enantiomers are more potent than the (R)-enantiomers in blocking β-adrenergic receptors [17]. This group of β-blockers includes acebutolol, atenolol, betaxolol, bisoprolol, carteolol, labetalol, metoprolol, pindolol, and propranolol (Fig. 8.4). One exception is nebivolvol, which has the (R)-enantiomer being more potent in inhibiting the β-receptors [18]. Newer β-blockers were developed as single (S)-enantiomers, such as timolol and penbutolol. Many calcium channel blockers are also commonly marketed as racemates. These include amlodopine, felodipine, manidipine, nicardipine, nimodipine, nisoldipine, and verapamil. Diltiazem is a diastereoisomer with two pairs of enantiomers. Similar to the β-blockers, the (S)-enantiomers dominate calcium channel blocking activity [17].

Drugs with one eutomer are also common in other therapeutic areas, including neurological diseases, local anesthetics, antibiotics, antihistaminics, and proton pump inhibitors. However, because of reasons similar to those for cardiovascular drugs (lack of technology for separation during the development period), many of the older drugs were developed and marketed as racemates. Because of space limitations these drugs are not discussed here, and interested readers can find the relevant information elsewhere [17,19–26].

In contrast to older drugs, many chiral drugs developed in the last three decades are marketed as enantiomers. This includes one of the successful classes of cardiovascular drugs, the ACE (angiotensin-converting enzyme) inhibitors [27]. The first ACE inhibitor, captopril (Fig. 8.5), was marketed in 1981 for the treatment of hypertension and heart failure. The (S, S)-captopril is more than 100 times more potent in the inhibitory activity than the (R, R)-isomer [28]. Thus only

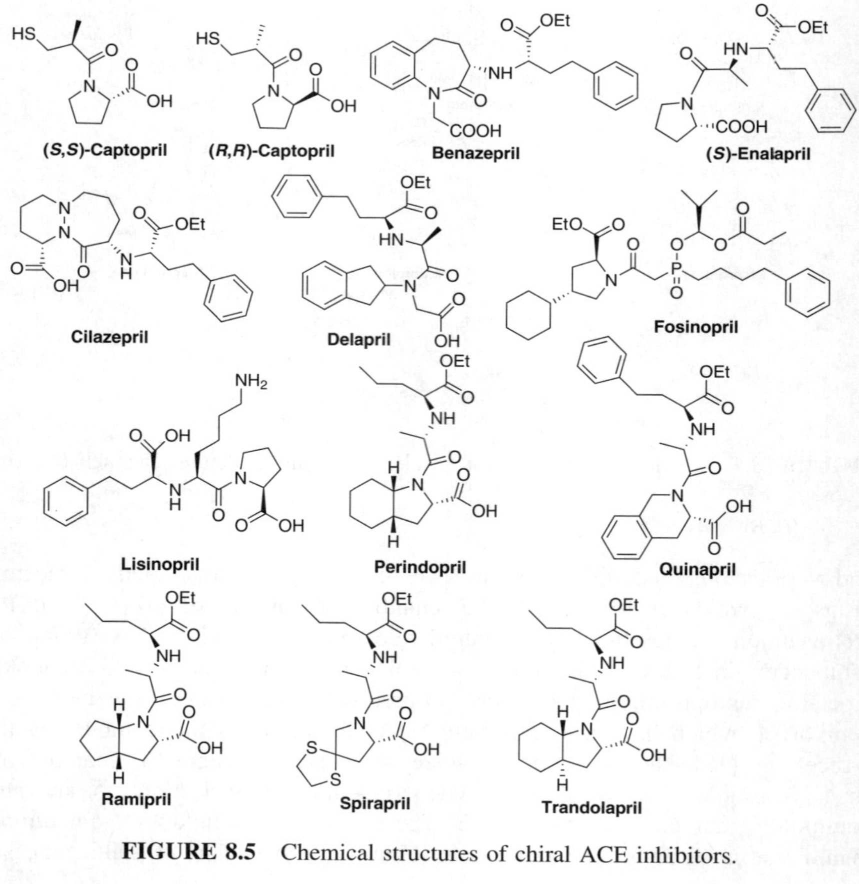

FIGURE 8.5 Chemical structures of chiral ACE inhibitors.

the (S, S)-enantiomer was developed and marketed. Other drugs in this class include benazepril, cilazepril, delapril, enalapril, fosinopril, lisinopril, perindopril, quinapril, ramipril, spirapril, and trandolapril.

Another major breakthrough of modern medicine was the arrival of statins, that is, the oral HMG-CoA reductase inhibitors, around the middle of the 1990s [29]. The statins are indicated for the treatment of hypercholesterolemia, mixed dyslipidemia, and dysbetalipoproteinemia, and to prevent cardiovascular disease in patients with multiple risk factors [30]. Atorvastatin (Lipitor) is the best-known statin. The racemate of atorvastatin is a diastereoisomer with two pairs of enantiomers (Fig. 8.6), and the marketed drug is the (R, R)-atorvastatin. Preclinical studies demonstrated that the (R, R)-isomer is about 65-fold more potent in the inhibition of HMG-CoA reductase activity than the (S, S)-isomer [31]. Other commonly available stains are also marketed as enantiomers, such as simvastatin and pravastatin.

8.3.2   Drugs with Equally Active Enantiomers

Drugs in this category are in fact very rare. Since both enantiomers show equal pharmacological activity, there is no need to separate them. Thus most of these chiral drugs are marketed as racemic mixtures (Fig. 8.7). One example is flecainide, an antiarrhythmic. There was no significant difference between the effects of flecainide enantiomers on basic electrophysiological parameters of canine

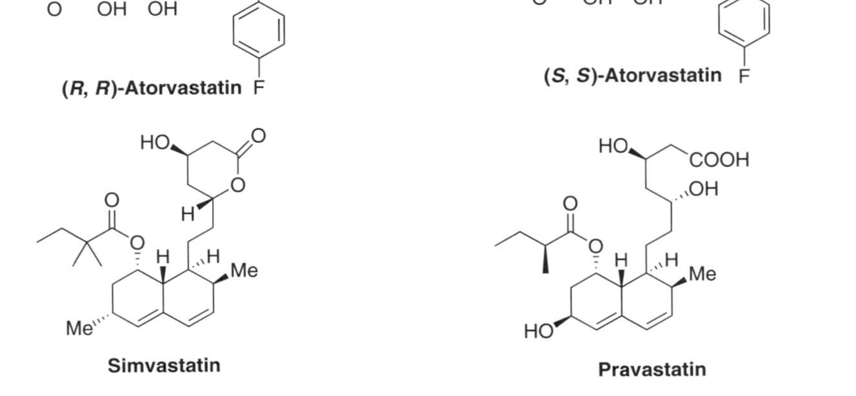

FIGURE 8.6   Chemical structures of statins.

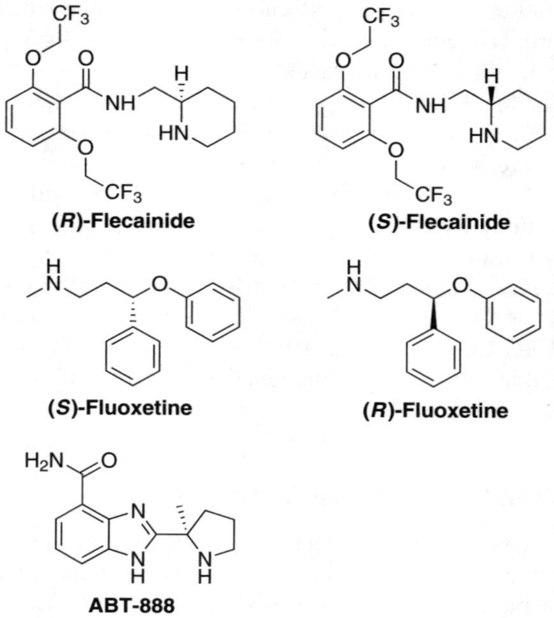

FIGURE 8.7 Chemical structures of the enantiomers of flecainide, fluoxetine, and ABT-888.

Purkinje fibers and in vivo efficacy [32,33]. Fluoxetine, a selective serotonin reuptake inhibitor, is an antidepressant. Preclinical studies demonstrated equal pharmacological activities of fluoxetine enantiomers [34,35]. ABT-888 (Veliparib) is a poly(ADP-ribose) polymerase (PARP)-1 inhibitor that is being developed as a novel anticancer drug. Although both enantiomers have identical enzymatic ($K_i = 5$ nM) and cellular potency ($EC_{50} = 3$ nM) [36], the (R)-enantiomer has better oral bioavailability. This led to the selection of the (R)-enantiomer as the clinical candidate. Thus not only do the pharmacodynamic properties of the chiral drug influence the selection of a drug for development, but pharmacokinetics and toxicity also play important roles, as discussed in other chapters.

8.3.3 Racemic Drugs with Chiral Inversion

This class of drugs are racemic mixtures, but with one enantiomer being rapidly converted to the other enantiomer in vivo. The largest class of drugs in this category is the nonsteroidal antiinflammatory drugs (NSAIDs). They inhibit the cyclooxygenases (COX), the enzymes responsible for the synthesis of prostaglandins and other mediators from arachidonic acid, and thus are used for the treatment of inflammatory diseases and as analgesics and antipyretics. Most of the NSAIDs are 2-aryl-propionic acids or "profens" and are racemic mixtures [37,38]. Of the two enantiomers, the (S)-isomer is the primary

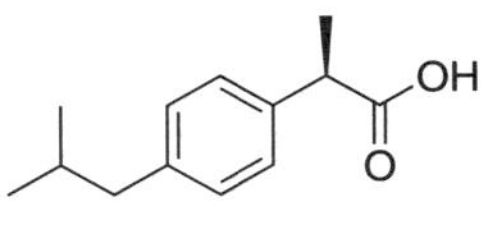

(*S*)-Ibuprofen

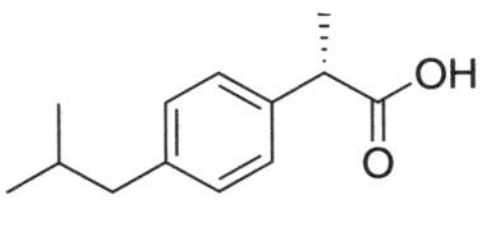

(*R*)-Ibuprofen

FIGURE 8.8   Chemical structures of the two ibuprofen enantiomers.

inhibitor of COX. However, the activities of the two enantiomers are essentially indistinguishable in vivo, because of the unidirectional metabolic bioconversion of the (*R*)-enantiomer to the (*S*)-enantiomer. The most well-known compound in this class is ibuprofen (Fig. 8.8) [39,40].

8.3.4   Drugs with One Active Enantiomer and One Enantiomer with Undesirable Effects

Obviously, only the active enantiomer can be marketed from this class of chiral drugs. One of the well-known drugs in this class is clopidogrel (Fig. 8.9), a thienopyridine derivative that inhibits the ADP P2Y$_{12}$ receptor. As a leading antiplatelet agent, clopidogrel reduces cardiovascular events associated with atherosclerotic diseases. Clopidogrel is a racemic mixture, with (*S*)-clopidogrel providing all of the favorable antiplatelet activity but with no significant neurotoxicity while the (*R*)-enantiomer contributes no inhibition of platelet activity but virtually all of the neurotoxicity (U.S. Patent No. 4,847,265). Thus only the (*S*)-clopidogrel was developed and eventually marketed. Dopa (3,4-dihydroxyphenylalanine, Fig. 8.9) is a precursor of dopamine that is used in the clinical treatment of Parkinson disease. Dopa is a racemic mixture, but only (*S*)-Dopa is clinically useful because (*R*)-Dopa causes agranulocytosis [41]. The use of (*S*)-Dopa resulted in a reduced dose and a reduction of adverse effects [42,43]. Another example is ofloxacin, a racemic mixture used as an antibiotic. However, the antimicrobial activity mainly resides in the (*S*)-enantiomer. Importantly, the (*S*)-enantiomer has improved bioavailability and reduced acute toxicity [44,45]. These facts led to the development of the (*S*)-enantiomer as a new antibiotic, levofloxacin (Fig. 8.9) [46].

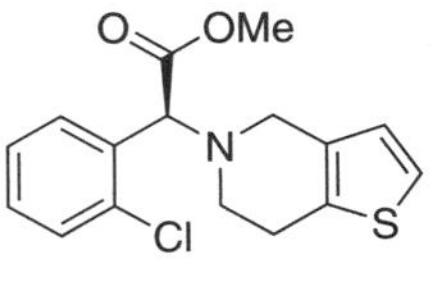

(*S*)-Clopidogrel

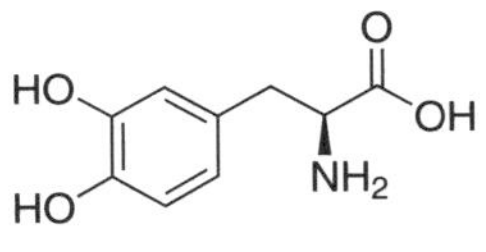

(*S*)-Dopa

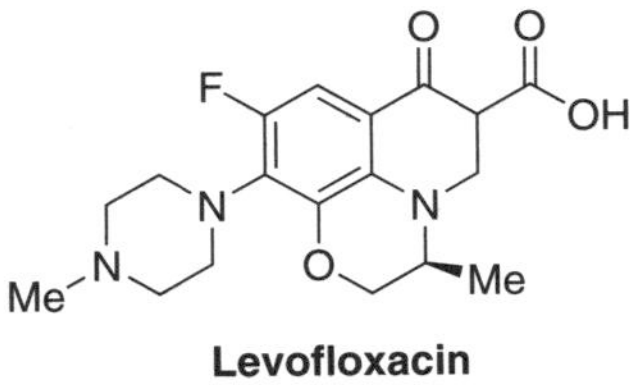

Levofloxacin

FIGURE 8.9   Chemical structures of clopidogrel, (*S*)-dopa, and levofloxacin.

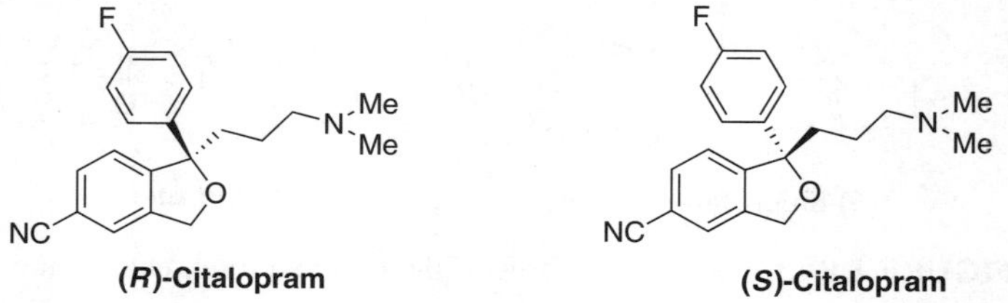

FIGURE 8.10 Chemical structures of the enantiomers of citalopram.

8.3.5 Drugs with One Enantiomer Antagonizing the Effects of the Other Enantiomer

One well-known drug is citalopram, which is a selective serotonin reuptake inhibitor used as an antidepressant. Citalopram (Fig. 8.10) is a racemic mixture, and its (*S*)-enantiomer is a potent inhibitor of serotonin reuptake while the (*R*)-isomer actually reduces the activities of the (*S*)-isomer both in vitro [47–49] and in vivo [50,51]. How can this be explained pharmacologically? Studies have demonstrated that the (*S*)- and (*R*)-enantiomers bind to different sites on the serotonin transporter. The binding of the (*S*)-enantiomer inhibits the reuptake activity, while the binding of (*R*)-enantiomer to an allosteric site (a different site from the normal ligand binding site with the ability to affect the receptor activity) reduces the binding of the (*S*)-enantiomer [47,49]. Since the inhibition of serotonin reuptake activity is the intended purpose of the drug, the (*S*)-enantiomer should be a better drug compared with the racemates. In fact, the (*S*)-enantiomer (marketed as escitalopram) has been developed as a new and better antidepressant [52].

8.3.6 Unique Chiral Drugs

There are other unique chiral drugs that do not belong to any of the above-discussed categories. One example is omeprazole, a drug used for the treatment of gastroesophageal reflex diseases, and once the world's best-selling drug [53]. As a racemic mixture, omeprazole is a pro-drug as both the (*R*)- and (*S*)-enantiomers are converted in vivo in an acidic environment to the nonchiral active form (Fig. 8.11) that potently inhibits the proton pump (H^+/K^+-ATPase) in the gastric secretory membrane of the parietal cell [53]. Interestingly, the (*R*)- and (*S*)-enantiomers are metabolized differently in humans, and there is some variability

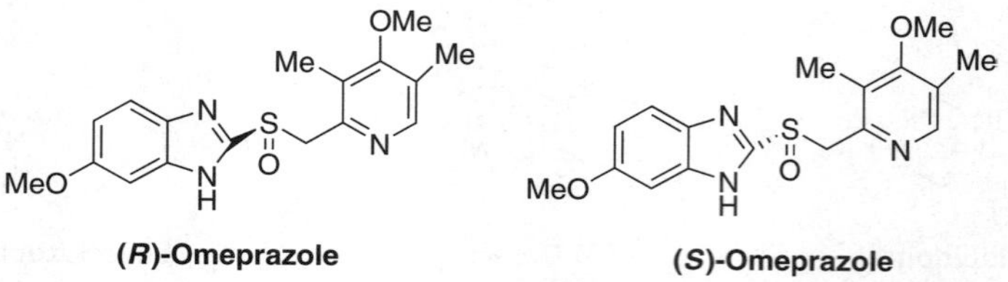

FIGURE 8.11 Chemical structures of the enantiomers of omeprazole.

between different populations of patients [54]. This variation of metabolism may contribute to the difficulty of dosing and variable efficacy of omeprazole [53,54]. Thus the (S)-enantiomer of omeprazole, marketed as esomeprazole, was developed to reduce the interindividual variability. Indeed, several studies have shown that esomeprazole has a quicker onset of action and improved efficacy compared with the racemic mixture omeprazole [53,55–58]. However, these claims of superiority by esomeprazole have been disputed [59,60].

8.3.7 Development Strategies for Chiral Drugs

With the cost of developing a drug exceeding US $1 billion [61] and the pipeline dwindling, drug companies are under increasing pressure to streamline the drug development process and increase the success rate through the clinical study phases. In dealing with the chirality of drugs, two strategies are applicable. To develop new chemical identities, it is important to separate the enantiomers and fully understand the pharmacological properties of each enantiomer. This will allow the identification of active enantiomers, which is then selected to enter into clinical studies. Enantiomers may provide several benefits versus racemates, for example, a possible improved safety margin through increasing receptor selectivity and potency while reducing off-target adverse effects. The dose may be reduced because the inactive enantiomer is eliminated. Other considerations include pharmacokinetic improvement, decreased interindividual variability, and decreased potential for drug-drug interactions, as discussed in Chapter 9. The technology to synthesize and separate racemates has advanced significantly in recent years, making the development of single enantiomers technically and economically possible. In addition, drug regulators are increasingly demanding pharmacokinetic, pharmacodynamic, and toxicological information on the enantiomers of chiral drugs [16,62–65]. Thus it is generally agreed that enantiomers should be developed for new drugs whenever possible, unless the racemic mixture has apparent advantages or the obstacles to separation and manufacture of enantiomers are technically insurmountable and economically unreasonable.

Another strategy applies to the existing racemic or diastereomer mixture drugs already on the market by identifying the active enantiomer and developing it as a new drug. This procedure, referred as chiral switch, has become quite common and is often successful [19,37,46,62,64–67]. Omeprazole to esomeprazole and albuterol to levalbuterol are just two of the successful switches.

8.4 SUMMARY

Pharmacology deals with the effects of drugs on the body. The first step for a drug to produce a biological effect requires the specific and high-affinity binding of the drug to a receptor, which then activates a second messenger system to transduce the signal from cell surface to intracellular organelles such as the nuclei and to amplify from a simple receptor binding to a complex cellular response. Chiral properties of the chemical compounds play an important role in the determination of specific binding and thus pharmacological actions of the drug. The current drug discovery and development trend is to select and develop enantiomers that may provide improved selectivity and potency, and reduced adverse events.

REFERENCES

1. Wang, W., Wang, E. Q., Balthasar, J. P. (2008). Monoclonal antibody pharmacokinetics and pharmacodynamics. *Clin. Pharmacol. Ther*. **84**, 548–558.

2. Mould, D. R., Sweeney, K. R. (2007). The pharmacokinetics and pharmacodynamics of monoclonal antibodies—mechanistic modeling applied to drug development. *Curr. Opin. Drug. Discov. Devel*. **10**, 84–96.

3. Lobo, E. D., Hansen, R. J., Balthasar, J. P. (2004). Antibody pharmacokinetics and pharmacodynamics. *J. Pharm. Sci*. **93**, 2645–2668.

4. Luttrell, L. M. (2006). Transmembrane signaling by G protein-coupled receptors. *Methods Mol. Biol*. **332**, 3–49.

5. Luttrell, L. M. (2008). Reviews in molecular biology and biotechnology: transmembrane signaling by G protein-coupled receptors. *Mol. Biotechnol*. **39**, 239–264.

6. Tibes, R., Trent, J., Kurzrock, R. (2005). Tyrosine kinase inhibitors and the down of molecular cancer therapeutics. *Annu. Rev. Pharmacol. Toxicol*. **45**, 357–384.

7. Alexander, S. P., Mathie, A., Peters, J. A. (2008). Guide to Receptors and Channels (GRAC), 3rd edition. *Br. J. Pharmacol*. **153** Suppl 2, S1–S209.

8. Lohse, M. J. (2010). Dimerization in GPCR mobility and signaling. *Curr. Opin. Pharmacol*. **10**, 53–58.

9. Schulte, G., Levy, F. O. (2007). Novel aspects of G-protein-coupled receptor signalling—different ways to achieve specificity. *Acta Physiol. (Oxf)* **190**, 33–38.

10. Kenakin, T. (2003). Ligand-selective receptor conformations revisited: the promise and the problem. *Trends Pharmacol. Sci*. **24**, 346–354.

11. Rovira, X., Pin, J. P., Giraldo, J. (2010). The asymmetric/symmetric activation of GPCR dimers as a possible mechanistic rationale for multiple signalling pathways. *Trends Pharmacol. Sci*. **31**, 15–21.

12. Luttrell, L. M., Gesty-Palmer, D. (2010). Beyond desensitization: physiological relevance of arrestin-dependent signaling. *Pharmacol. Rev*. 62: 305–330.

13. Gainetdinov, R. R., Premont, R. T., Bohn, L. M., Lefkowitz, R. J., Caron, M. G. (2004). Desensitization of G-protein-coupled receptors and neuronal functions. *Annu. Rev. Neurosci*. **27**, 107–144.

14. Ferguson, S. S. G. (2001). Evolving concepts in G protein-coupled receptor endocytosis: the role in receptor desensitization and signaling. *Pharmacol. Rev*. **53**, 1–24.

15. Nunez, M. C., Garcia-Rubino, M. E., Conejo-Garcia, A., Cruz-Lopez, O., Kimatral, M., Gallo, M. A., Espinosa, A. Campos, J. M. (2009). Homochiral drugs: a demanding tendency of the pharmaceutical industry. *Curr. Med. Chem*. **16**, 2064–2074.

16. Shimazawa, R., Nagai, N., Toyoshima, S., Okuda, H. (2008). Present state of new chiral drug development and review in Japan. *J. Health Sci*. **54**, 23–29.

17. Ranade, V. V., Somberg, J. C. (2005). Chiral cardiovascular drugs: an overview. *Am. J. Ther*. **12**, 439–459.

18. Siebert, C. D., Hansicke, A., Nagel, T. (2008). Stereochemical comparison of nebivolol with other beta-blockers. *Chirality* **20**, 103–109.

19. Francotte, E., Lindner, W. (2006). *Chirality in Drug Research*. Weinheim: Wiley-VCH.

20. Hutt, A. J., Tan, S. C. (1996). Drug chirality and its clinical significance. *Drugs* **52** Suppl 5, 1–12.

21. Lane, R. M., Baker, G. B. (1999). Chirality and drugs used in psychiatry: nice to know or need to know? *Cell. Mol. Neurobiol*. **19**, 355–372.

22. Leonov, A., Bielory, L. (2007). Chirality in ocular agents. *Curr. Opin. Allergy Clin. Immunol*. **7**, 418–423.

23. Mather, L. E., Edwards, S. R. (1998). Chirality in anaesthesia—ropivacaine, ketamine and thiopentone. *Curr. Opin. Anaesthesiol*. **11**, 383–390.

24. Mather, L. E. (2005). Stereochemistry in anaesthetic and analgesic drugs. *Minerva Anesthesiol*. **71**, 507–516.

25. Nau, C., Strichartz, G. R. (2002). Drug chirality in anesthesia. *Anesthesiology* **97**, 497–502.

26. Reddy, I. K., Kommuru, T. R., Zaghloul, A. A., Khan, M. A. (2000). Chirality and its implications in transdermal drug development. *Crit. Rev. Ther. Drug Carrier Syst*. **17**, 285–325.

27. Garg, R., Yusuf, S. (1995). Overview of randomized trials of angiotensin-converting enzyme inhibitors on mortality and morbidity in patients with heart failure. *JAMA* **273**, 1450–1456.

28. Chirumamilla, R. R., Marchant, R., Nigam, P. (2001). Captopril and its synthesis from chiral intermediates. *J. Chem. Tech. Biotech*. **76**, 123–127.

29. Li, J. J. (2009). *Triumph of the Heart: The Story of Statins*. Oxford: Oxford University Press.

30. Sadowitz, B., Maier, K. G., Gahtan, V. (2010). Basic science review: Statin therapy—Part I: The pleiotropic effects of statins in cardiovascular disease. *Vasc. Endovasc. Surg*. **44**, 241–251.

31. Roth, B. D. (2002). The discovery and development of atorvastatin, a potent novel hypolipidemic agent. *Prog. Med. Chem*. **40**, 1–22.

32. Smallwood, J. K., Robertson, D. W., Steinberg, M. I. (1989). Electrophysiological effects of flecainide enantiomers in canine Purkinje fibres. *Naunyn Schmiedebergs Arch. Pharmacol*. **339**, 625–629.

33. Banitt, E. H., Schmid, J. R., Newmark, R. A. (1986). Resolution of flecainide acetate, *N*-(2-piperidylmethyl)-2,5-bis(2,2,2-trifluoroethoxy)benzamide acetate, and antiarrhythmic properties of the enantiomers. *J. Med. Chem*. **29**, 299–302.

34. Wong, D. T., Fuller, R. W., Robertson, D. W. (1990). Fluoxetine and its two enantiomers as selective serotonin uptake inhibitors. *Acta Pharm. Nord*. **2**, 171–180.

35. Robertson, D. W., Krushinski, J. H., Fuller, R. W., Leander, J. D. (1988). Absolute configurations and pharmacological activities of the optical isomers of fluoxetine, a selective serotonin-uptake inhibitor. *J. Med. Chem*. **31**, 1412–1417.

36. Penning, T. D., Zhu, G. D., Gandhi, V. B., Gong, J. C., Liu, X. S., Shi, Y., Klinghofer, V., Johnson, E. F., Donawho, C. K., Frost, D. J., Bontcheva-Diaz, V., Bouska, J. J., Osterling, D. J., Olson, A. M., Marsh, K. C., Luo, Y., Giranda, V. L. (2009). Discovery

of the poly(ADP-ribose) polymerase (PARP) inhibitor 2-[(R)-2-methylpyrrolidin-2-yl]-1H-benzimidazole-4-carboxamide (ABT-888) for the treatment of cancer. *J. Med. Chem*. **52**, 514–523.

37. Agranat, I., Caner, H., Caldwell, J. (2002). Putting chirality to work: the strategy of chiral switches. *Nat. Rev. Drug Discov*. **1**, 753–768.

38. Hardikar, M. S. (2008). Chiral non-steroidal anti-inflammatory drugs—a review. *J. Indian Med. Assoc*. **106**, 615–618, 622, 624.

39. Chen, C. S., Shieh, W. R., Lu, P. H., Harriman, S., Chen, C. Y. (1991). Metabolic stereoisomeric inversion of ibuprofen in mammals. *Biochim. Biophys. Acta Prot. Struct. Mol. Enzymol*. **1078**, 411–417.

40. Tracy, T. S., Hall, S. D. (1992). Metabolic inversion of (*R*)-ibuprofen. Epimerization and hydrolysis of ibuprofenyl-coenzyme A. *Drug Metab. Dispos*. **20**: 322–327.

41. Nguyen, L. A., He, H., Pham-Huy, C. (2006). Chiral drugs. An overview. *Int. J. Biomed. Sci*. **2**, 85–100.

42. Cotzias, G. C., Papavasiliou, P. S., Gellene, R. (1969). Modification of Parkinsonism—chronic treatment with L-dopa. *N. Engl. J. Med*. **280**, 337–345.

43. Cotzias, G. C., Van Woert, M. H., Schiffer, L. M. (1967). Aromatic amino acids and modification of parkinsonism. *N. Engl. J. Med*. **276**, 374–379.

44. Hutt, A. J., O'Grady, J. (1996). Drug chirality: a consideration of the significance of the stereochemistry of antimicrobial agents. *J. Antimicrob. Chemother*. **37**, 7–32.

45. Davis, R., Bryson, H. M. (1994). Levofloxacin. A review of its antibacterial activity, pharmacokinetics and therapeutic efficacy. *Drugs* **47**, 677–700.

46. Agranat, I., Wainschtein, S. R. (2010). The strategy of enantiomer patents of drugs. *Drug Discov. Today* **15**, 163–170.

47. Plenge, P., Gether, U., Rasmussen, S. G. (2007). Allosteric effects of R- and S-citalopram on the human 5-HT transporter: evidence for distinct high- and low-affinity binding sites. *Eur. J. Pharmacol*. **567**, 1–9.

48. Hyttel, J., Bogeso, K. P., Perregaard, J., Sanchez, C. (1992). The pharmacological effect of citalopram residues in the (S)-(+)-enantiomer. *J. Neural Transm. Gen. Sect*. **88**, 157–160.

49. Chen, F., Larsen, M. B., Sanchez, C., Wiborg O. (2005). The S-enantiomer of R,S-citalopram, increases inhibitor binding to the human serotonin transporter by an allosteric mechanism. Comparison with other serotonin transporter inhibitors. *Eur. Neuropsychopharmacol*. **15**, 193–198.

50. Mork, A., Kreilgaard, M., Sanchez, C. (2003). The R-enantiomer of citalopram counteracts escitalopram-induced increase in extracellular 5-HT in the frontal cortex of freely moving rats. *Neuropharmacology* **45**, 167–173.

51. Sanchez, C., Bogeso, K. P., Ebert, B., Reines, E. H., Braestrup, C. (2004). Escitalopram versus citalopram: the surprising role of the R-enantiomer. *Psychopharmacology (Berl)* **174**, 163–176.

52. Montgomery, S. A., Moller, H. J. (2009). Is the significant superiority of escitalopram compared with other antidepressants clinically relevant? *Int. Clin. Psychopharmacol*. **24**: 111–118.

53. Olbe, L., Carlsson, E., Lindberg, P. (2003). A proton-pump inhibitor expedition: the case histories of omeprazole and esomeprazole. *Nat. Rev. Drug Discov*. **2**: 132–139.

54. Andersson, T. (1996). Pharmacokinetics, metabolism and interactions of acid pump inhibitors. Focus on omeprazole, lansoprazole and pantoprazole. *Clin. Pharmacokinet.* **31**, 9–28.

55. Dent, J. (2003). Review article: pharmacology of esomeprazole and comparisons with omeprazole. *Aliment. Pharmacol. Ther.* **17** Suppl 1, 5–9.

56. Kahrilas, P. J., Falk, G. W., Johnson, D. A., Schmitt, C., Collins, D. W., Whipple, J., D'Amico, D., Hamelin, B., Joelsson, B. (2000). Esomeprazole improves healing and symptom resolution as compared with omeprazole in reflux oesophagitis patients: a randomized controlled trial. The Esomeprazole Study Investigators. *Aliment. Pharmacol. Ther.* **14**, 1249–1258.

57. Rohss, K., Lind, T., Wilder-Smith, C. (2004). Esomeprazole 40 mg provides more effective intragastric acid control than lansoprazole 30 mg, omeprazole 20 mg, pantoprazole 40mg and rabeprazole 20 mg in patients with gastro-oesophageal reflux symptoms. *Eur. J. Clin. Pharmacol.* **60**, 531–539.

58. Pai, V., Pai, N. (2007). Recent advances in chirally pure proton pump inhibitors. *J. Indian Med. Assoc.* **105**, 469–470, 472, 474.

59. Chiba, N. (2005). Esomeprazole was not better than omeprazole for resolving heartburn in endoscopy-negative reflux disease. *ACP J Club* **142**, 6.

60. Therapeutics Initiative. Do single stereoisomer drugs provide value? *Therapeut. Lett.* June-Sept. 2002, Issue **45**.

61. Adams, C. P., Brantner, V. V. (2010). Spending on new drug development1. *Health Econ.* **19**, 130–141.

62. Agrawal, Y. K., Bhatt, H. G., Raval, H. G., Oza, P. M., Gogoi, P. J. (2007). Chirality—a new era of therapeutics. *Mini Rev. Med. Chem.* **7**, 451–460.

63. Batra, S., Seth, M., Bhaduri, A. P. (1993). Chirality and future drug design. *Prog. Drug Res.* **41**, 191–248.

64. Burke, D., Henderson, D. J. (2002). Chirality: a blueprint for the future. *Br. J. Anaesth.* **88**, 563–576.

65. Caner, H., Groner, E., Levy, L., Agranat, I. (2004). Trends in the development of chiral drugs. *Drug Discov. Today* **9**, 105–110.

66. Blin, O. (2004). Chiral switch: towards a better benefit-risk ratio? *Therapie* **59**, 625–628.

67. Patil, P. A., Kothekar, M. A. (2006). Development of safer molecules through chirality. *Indian J. Med. Sci.* **60**, 427–437.

CHAPTER 9

PHARMACOKINETICS OF CHIRAL DRUGS

HANQING DONG
OSI Pharmaceuticals, A Wholly Owned Subsidary of Astellas US,
Farmingdale, NY

XIAOCHUAN GUO and ZENGBIAO LI
Drumetix Laboratories, LLC, Greensboro, NC

In the current pharmaceutical industry, the failure of drug candidates in clinical trials is commonly caused by the presence of toxicity and/or a lack of efficacy, which are often related to inferior pharmacokinetic properties [1]. It is also well known that chiral drugs often display stereoselectivity in their pharmacokinetics and disposition, especially in the process of metabolism [2]. Therefore, this

Chiral Drugs: Chemistry and Biological Action, First Edition. Edited by Guo-Qiang Lin, Qi-Dong You and Jie-Fei Cheng.
© 2011 John Wiley & Sons, Inc. Published 2011 by John Wiley & Sons, Inc.

chapter presents various aspects of the pharmacokinetics of chiral drugs, including stereoselectivity in the absorption, distribution, metabolism, and excretion (ADME) processes, as well as stereoselective drug-drug interactions.

9.1 BASIC CONCEPTS IN PHARMACOKINETICS

Pharmacokinetics describes how the body handles a specific drug after its administration. Pharmacokinetics usually studies the extent and rate of absorption (the process of a drug entering the body), distribution (the dispersion or dissemination of a drug throughout fluids and tissues of the body), metabolism (the irreversible biotransformation of parent drugs into their metabolites), and excretion (the elimination of drug-related species from the body) [3]. The ADME processes are crucial for drug actions and can assume equal relevance to the actual pharmacological effects of the drugs at their target sites [4]. In pharmacokinetic studies, drug levels in various parts of the body (blood, various tissues and organs, and excreta) are measured as a function of time after dosing. A typical profile of drug plasma concentration versus time, after an intravascular or extravascular administration, is shown in Figure 9.1.

Several useful parameters, including total plasma clearance (CL, the elimination of a drug from body by the processes of metabolism and excretion), volume of distribution (Vd), elimination half-life ($t_{1/2}$), and area under the curve (AUC) after an intravascular administration, can be calculated from the plasma concentration vs. time curve. Likewise, maximum concentration (C_{max}), time to reach C_{max} (t_{max}), $t_{1/2}$, and AUC can also be determined after an extravascular administration. Another important parameter in pharmacokinetics is bioavailability (F), which represents the extent to which a drug becomes available in the general circulation, that is, the fraction of the administered dose reaching the systemic circulation. F is equal to 1 when a drug is administered intravascularly (e.g., intravenously), while F is usually less than 1 when a drug is administered extravascularly (e.g., orally). In general, the bioavailability of a drug depends on its

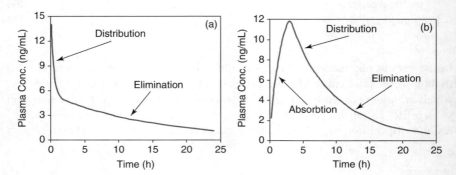

FIGURE 9.1 Representative profiles of plasma concentration vs. time after intravascular (a) and extravascular (b) administrations.

absorption and first-pass metabolism before the drug molecules reach the systemic circulation. If the doses in both intravenous (IV) and oral (PO) routes are equal, F can be calculated by the following equation: $F = AUC_{PO}/AUC_{IV}$, where AUC_{PO} and AUC_{IV} represent the areas under the concentration vs. time curves after oral and IV administrations, respectively. In the case of different dosages in both routes and assuming linear pharmacokinetics for the drug, F can be calculated by the modified equation $F = (AUC_{PO}/AUC_{IV}) \times (Dose_{IV}/Dose_{PO})$.

9.2   STEREOSELECTIVITY IN ADME PROPERTIES OF CHIRAL DRUGS

9.2.1   Absorption

Absorption is the phase in which a drug enters the bloodstream from the administration site. When a drug is dosed via intravascular routes (e.g., intraarterial or intravenous), it bypasses the absorption phase as the drug is injected directly into the bloodstream. The common extravascular administration sites usually include the gut (orally), connective tissues (subcutaneously), muscles (intramuscularly), skin, eyes, nose, and various mucosa.

A drug can be absorbed by passive diffusion and/or with the aid of active transporters. Passive diffusion across cellular membranes is dominated by lipophilicity, molecular weight, and the extent of ionization at physiological pH. As a spontaneous process, passive diffusion does not require cellular energy as the molecules move simply from areas of higher concentrations to areas of lower concentrations. In general, there is little or no enantiomer-specific difference. In contrast, active transport is a mediated process of molecules moving across a biological membrane against a concentration gradient. As a mediated process, active transport utilizes energy and can discriminate between enantiomers. In other words, both the amount and the rate of absorption for the two enantiomers of a drug may be different if their interactions with the responsible transporter(s) are stereoselective.

The L-isomer of dopa (**1**) is used to treat Parkinson disease and dopamine-responsive dystonia (Fig. 9.2). This drug is the metabolic precursor of the neurotransmitters dopamine, norepinephrine (noradrenaline), and epinephrine (adrenaline). There is evidence indicating the existence of an active transport mechanism responsible for the intestinal absorption of L-dopa. Some in vitro mucosal uptake experiments have demonstrated that L-dopa can be accumulated against a concentration gradient and its absorption can be inhibited under

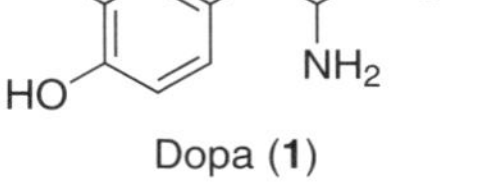

Dopa (**1**)

FIGURE 9.2   Structure of dopa (**1**).

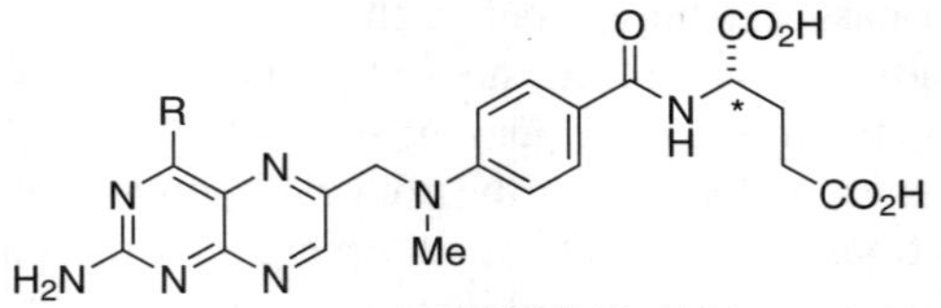

MTX (2): R = NH$_2$

FA (3):   R = OH

FIGURE 9.3   Structures of MTX (2) and FA (3).

anoxic conditions (characterized by the absence of oxygen supply to an organ or a tissue) and by the presence of other amino acids. It is further observed that L-dopa is more readily absorbed than D-dopa. This is in accordance with the stereospecificity of an intestinal amino acid transport system, where only the L-isomers of most naturally occurring amino acids are absorbed by this mechanism [5].

Methotrexate (MTX or L-amethopterin, 2), an analog of folic acid (FA, 3), is used clinically as an antineoplastic and antirheumatic agent (Fig. 9.3). This drug acts by inhibiting the metabolism of FA. Both MTX and FA are absorbed by the small intestine in a process mediated by the folate transporter system. In vitro experiments using rabbit intestinal brush-border membrane vesicles revealed a significant difference between MTX and its enantiomer (D-amethopterin) in the absorption process. MTX and FA mutually inhibit their uptake in a competitive manner, with K_i values of approximately 1 μM, whereas D-amethopterin inhibits the uptake of FA competitively with a K_i value of 60 μM. In other words, the affinity of the D-isomer for the folate transporter is approximately 60-fold weaker than that of the L-isomer. When both enantiomers were orally administered to humans, the plasma concentration of MTX was much higher than that of D-amethopterin, with the C_{max} and AUC of MTX being approximately 30- and 40-fold greater, respectively, than those of D-amethopterin. The significant differences in C_{max} and AUC of the two enantiomers are most likely due to the stereoselectivity in the intestinal absorption process [6].

9.2.2   Distribution

Distribution is the phase in which a drug is reversibly transferred from the central compartment (usually plasma) to peripheral compartments (tissues). This process includes two successive stages: dilution of the absorbed drug in blood and distribution of the drug to tissues. As for a chiral drug, the binding of its two enantiomers with plasma and tissue proteins is often stereoselective because of the chiral nature of proteins, resulting in a different plasma concentration vs. time profile for each enantiomer.

Volume of distribution (Vd) is a pharmacological term used to quantify the distribution of a drug between plasma and the rest of the body after administration. It is defined as the theoretical volume of fluid required to dissolve the amount of the drug in the body at a concentration equal to that observed in the plasma, and is equal to the total amount of the drug in the body divided by the drug concentration in the plasma. Vd is not a true physiological fluid or tissue space, but it does indicate how a drug distributes throughout the body, depending on its physicochemical properties (e.g., lipophilicity, solubility, charge, size). For highly tissue-bound drugs, plasma concentrations will be low and Vd will be high. A low Vd indicates that a drug mainly remains in the plasma instead of distributing into tissues and organs. It should be noted that Vd refers to the initial volume of distribution before the establishment of equilibrium. There is another parameter, volume of distribution at steady state (Vd$_{SS}$), which is determined based on the plasma drug concentration at equilibrium. Because the calculation uses drug concentrations at different times, the values of Vd and Vd$_{SS}$ can be very different when a drug is distributed slowly. Table 9.1 lists the Vd$_{SS}$ values of several chiral drugs (Fig. 9.4) following IV administration as racemates into humans [7–17].

In most cases, drug distribution is mainly determined by protein binding in plasma and tissues. The general equation for such binding in plasma at reversible equilibrium, [Drug] + [Protein] $\leftrightarrow$ [Drug-Protein], can be described as $K_a = C_b/[C_u \bullet (C_p - C_b)]$ where K_a is the association constant, C_p the protein concentration, C_b the concentration of protein-bound drug, and C_u the concentration of unbound (free) drug. The total plasma drug concentration is C, which is equal to $C_b + C_u$. Based on the above equation, the free fraction of drug in plasma ($f_u = C_u/C$) can be generated as $f_u = 1/[1 + K_a \bullet (C_p - C_b)]$. In the cases that the protein concentrations in plasma are much higher than C_b (i.e., $C_p \gg C_b$), the equation can be simplified as $f_u = 1/[1 + K_a \bullet C_p]$. In

TABLE 9.1 Stereoselectivity-Related Volume of Distribution at Steady State (Vd$_{SS}$)

| Drug | Vd$_{SS}$ (l/kg) | | Ratio (*R/S*) | Ref. |
	(*R*)-	(*S*)-		
Disopyramide (**4**)	59.8[a]	32.8[a]	1.82	7
Ketorolac (**5**)	0.129	0.282	0.46	8
Mepivacaine (**6**)	243[a]	155[a]	1.57	9
Methadone (**7**)	497[a]	289[a]	1.72	10
Pimobendan (**8**)	2.34 (−)[b]	1.74 (+)[b]	1.34 (−/+)[b]	11
Reboxetine (**9**)	0.39 (*R, R*)	0.92 (*S, S*)	0.42	12
Rolipram (**10**)	0.52	0.82	0.63	13
Salbutamol (**11**)	2.00	1.77	1.13	14
Thiopental (**12**)	139[a]	114[a]	1.22	15
Tocainide (**13**)	136[a]	134[a]	1.02	16
Verapamil (**14**)	2.74	6.42	0.43	17

[a]cited as l, not l/kg.
[b]*R/S* configuration unassigned.

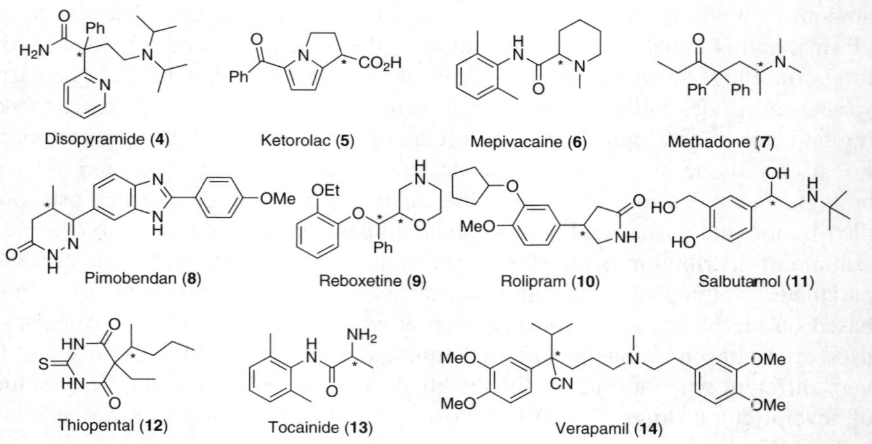

Disopyramide (4) Ketorolac (5) Mepivacaine (6) Methadone (7)

Pimobendan (8) Reboxetine (9) Rolipram (10) Salbutamol (11)

Thiopental (12) Tocainide (13) Verapamil (14)

FIGURE 9.4 Structures of chiral drugs **4–14**.

TABLE 9.2 Stereoselectivity-Related Plasma Protein Binding

Drug	Species	f_u (%)		Ratio (R/S)	Ref.
		(R)-	(S)-		
Bupivacaine (**15**)	Human	6.6	4.5	1.47	18
Carvedilol (**16**)	Rat	0.68	1.04	0.65	19
Etodolac (**17**)	Human	0.29	0.72	0.40	20
	Rat	0.39	1.16	0.34	21
Gallopamil (**18**)	Human	4.0	5.7	0.70	22
Halofantrine (**19**)	Rat	0.037	0.039	0.95	23
Hydroxychloroquine (**20**)	Human	65.9	47.3	1.39	24
Ketorolac (**5**)	Human	0.45	0.61	0.74	25
Propafenone (**21**)	Human	3.9	2.5	1.56	26
	Rat	7.4	9.7	0.76	
Talinolol (**22**)	Human	21.0	21.8	0.96	27
Thiamylal (**23**)	Human	11.8	6.7	1.76	28
Warfarin (**24**)	Human	5.52	4.12	1.34	29
	Rat	1.31	0.84	1.56	

other words, higher protein concentration (C_p) or higher affinity of a drug with protein (K_a) will translate into less free drug in plasma.

Two enantiomers of a chiral drug usually have different binding affinities with plasma proteins. Table 9.2 lists the stereoselectivity of several chiral drugs (Fig. 9.5) when binding to plasma proteins from different species [18–29].

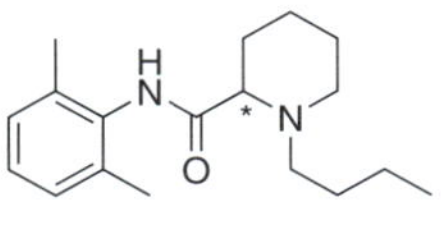

FIGURE 9.5 Structures of chiral drugs **15–24**.

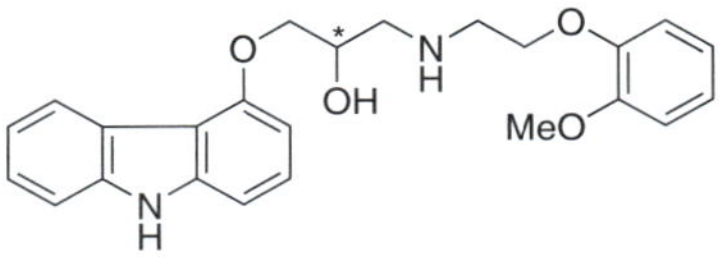

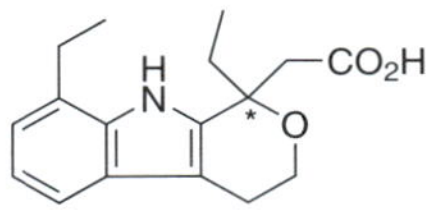

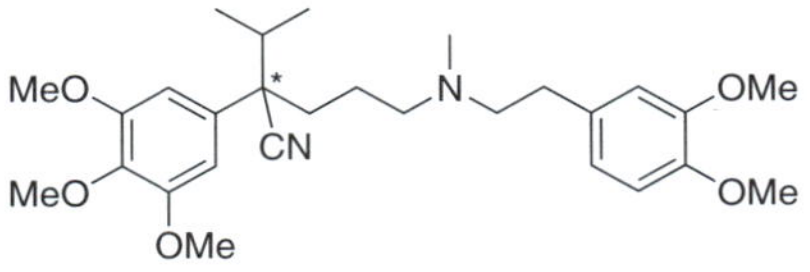

The selectivity in protein binding of two enantiomers can vary from species to species. As seen in Table 9.2, the free fractions of the nonsteroidal antiinflammatory drug etodolac (**17**) increased from human to rat for both enantiomers. However, the R/S ratios were different, being 0.40 in human and 0.34 in rat. In the case of warfarin (**24**), the free fractions of both enantiomers decreased from human to rat, with R/S ratios of 1.34 in human and 1.56 in rat. Interestingly, the binding preference for (R)- and (S)-enantiomers sometimes can be reversed from species to species. For example, the (R)-enantiomer of the antiarrhythmic drug propafenone (**21**) has a higher free fraction in human plasma, while the (S)-enantiomer has higher free fraction in rat plasma.

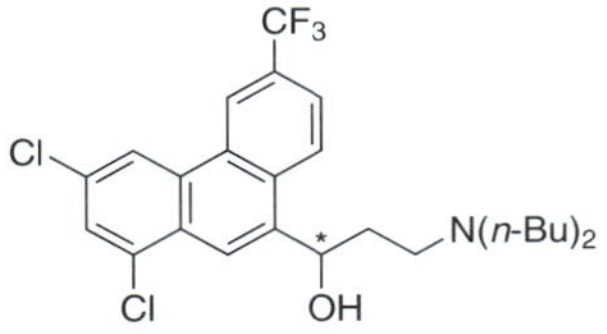

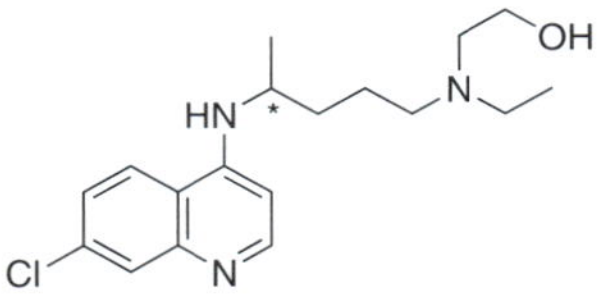

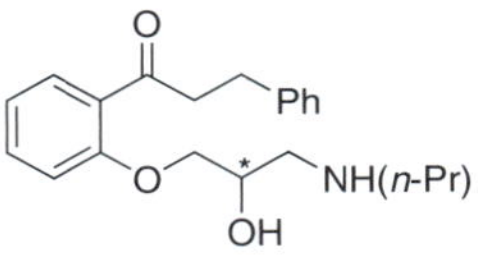

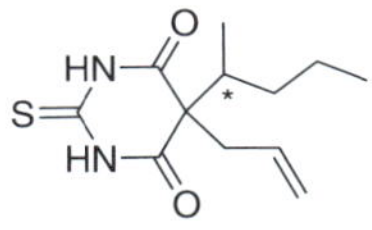

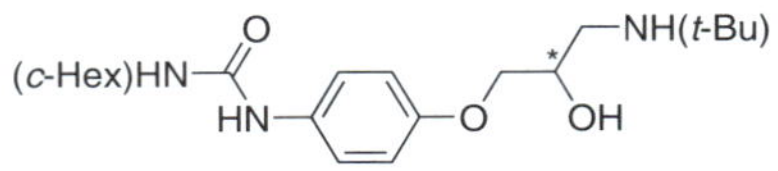

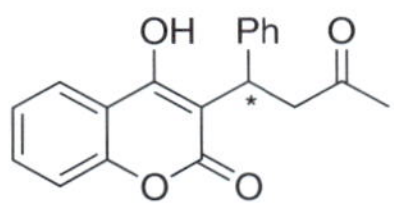

Serum albumin and α_1-acid glycoprotein are the two major drug-binding plasma proteins. Acidic drugs tend to bind serum albumin, while basic drugs prefer to bind α_1-acid glycoprotein. The two enantiomers of a chiral drug may have different selectivity when binding with different serum proteins. Propranolol (**25**) (Fig. 9.6) is a basic drug and has stronger binding affinity for α_1-acid glycoprotein than for serum albumin. The free fractions of (R)- and (S)-enantiomers of the drug in binding with human α_1-acid glycoprotein were 16.2% and 12.7%,

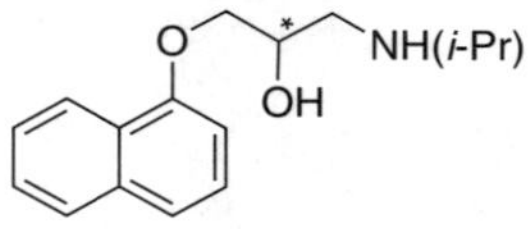

Propranolol (**25**)

FIGURE 9.6   Structure of propranolol (**25**).

respectively. Interestingly, the selectivity was reversed when binding with human serum albumin, with corresponding free fractions of (*R*)- and (*S*)-enantiomers increasing to 60.7% and 64.9%, respectively. In experiments with human plasma with all the drug-binding proteins present, the binding affinity showed a stereo-selectivity similar to that observed for α_1-acid glycoprotein, with free fractions of 25.3% and 22.0% for (*R*)- and (*S*)-enantiomers, respectively. This is consistent with the predominant binding of this basic drug to α_1-acid glycoprotein in plasma [30].

Enantiomers of a chiral drug may also exhibit stereoselectivity in binding with different lipoprotein fractions of plasma. Brocks et al. studied the distribution of halofantrine (**19**) enantiomers in different lipoprotein fractions of human, dog, and rat plasmas [31]. Their results showed that the (*R*)-enantiomer was preferentially associated with lipoprotein-rich fractions of plasma and the (*S*)-enantiomer was more prevalent in the lipoprotein-deficient fractions, although the selectivity ratios varied from species to species to some extent [31].

In addition to plasma protein binding, another important factor affecting drug distribution is tissue protein binding. Likewise, two enantiomers of a chiral drug can also have different binding affinities for tissue proteins. Brocks et al. studied the stereoselectivity of etodolac (**17**) enantiomers in different rat tissues, using an equilibrium dialysis assay [32]. While the concentrations of both enantiomers were found to be greater in plasma than in tissues, the stereoselectivties in plasma and various tissues were different, with *S/R* concentration ratios in plasma, liver, kidney, heart, fat, and brain of 0.31, 1.14, 1.01, 0.77, 0.78, and 0.45, respectively. Further analysis of some tissues revealed that the *S/R* ratios of unbound enantiomers were 1.15, 1.23, 2.14, and 3.69 in brain, liver, heart, and kidney, respectively. This illustrates that the (*R*)-isomer has higher binding affinity for these tissues than the (*S*)-isomer, with the highest degree of stereoselectivity in heart and kidney [32].

Carvedilol (**16**) is a lipophilic β-adrenoceptor antagonist with vasodilating properties. In rats, the plasma protein binding of the drug is >98% for plasma and >96% for serum albumin solution. The *S/R* concentration ratio of unbound drug was 1.53 for plasma but 0.78 for serum albumin, showing an inverse selectivity. The drug concentrations in tissues were always higher than those in plasma, with *S/R* concentration ratios of 2.3 in liver tissue and 1.3–1.4 in other tissues, while the *S/R* ratio in plasma was 0.6. Such inverse *S/R* ratios in tissues and plasma led to significantly different tissue-to-plasma ratios for (*R*)- and (*S*)-enantiomers [tissue-to-plasma ratios at steady state: lung, (*S*) 76, (*R*) 34; liver, (*S*) 21, (*R*) 5;

and kidney, (S) 8, (R) 3]. The concentrations of both enantiomers in blood cells were lower than those in plasma. The S/R concentration ratio in blood cells was 1.7, inverse of that in plasma as well [19].

It is also important to note that dosing a single enantiomer sometimes results in different pharmacokinetics compared to dosing a racemate (refer to Section 9.3 on drug-drug interactions for details). For example, when individual enantiomers of the antiarrhythmic drug disopyramide (**4**) were separately dosed intravenously in humans, there was no difference in the plasma clearance, renal clearance, half-life, and volume of distribution between the two enantiomers, as calculated from either total or unbound drug concentrations. However, when a racemic form of the drug was dosed, the D-isomer had lower plasma clearance and renal clearance, a longer half-life, and a smaller apparent volume of distribution than the L-isomer. Such discrimination was probably caused by the difference in concentration-dependent plasma protein binding of the two enantiomers, with D-disopyramide being more avidly bound to plasma proteins at lower concentrations than L-disopyramide [7].

9.2.3 Metabolism

Drug metabolism is the phase in which a drug is subjected to various biotransformation reactions. Typically, drug metabolism produces metabolites of lower lipophilicity relative to parent drugs by adding ionizable or hydrophilic groups. Biotransformation reactions can be divided into two categories: phase I reactions (functionalization) and phase II reactions (conjugation). Phase I reactions involve the generation of a functional group or the modification of an existing group by oxidation, reduction, or hydrolysis. Phase II reactions involve the coupling of a drug or its metabolite(s) to an endogenous conjugating agent (e.g., glucuronic acid, glutathione, sulfate moiety, acetyl group).

The stereoselectivity in the metabolic transformations of drugs can be substrate oriented and/or product oriented [33]. The former refers to the preferential metabolism of one enantiomer over the other, while the latter refers to the preferential formation of one particular enantiomer from a single prochiral substrate in the metabolic process. When two enantiomers of a chiral drug interact with a metabolizing enzyme (a chiral macromolecule), the formation of a pair of diastereomeric complexes may yield chiral discrimination as a result of different affinities and/or reactivities at the binding step and/or the catalytic step. Therefore, the products (metabolites) formed in this process may vary to some extent, resulting in stereoselectivity in metabolism. On the other hand, when a prochiral drug binds to an enzyme, it forms a chiral substrate-enzyme complex, as in a general catalytic asymmetric organic reaction, providing stereoselectivity in the formation of chiral metabolites.

Cytochrome P450 (CYP) enzymes represent a large and diverse superfamily of heme-thiolate proteins. They are named after the UV/VIS spectral absorbance peak of their Fe(II)-carbon monoxide-bound species near 450 nm. The CYP superfamily exists in all domains of life and is believed to have originated from an ancestral gene over 3 billion years ago [34]. CYP enzymes are classified in terms of families, subfamilies, and individual members according to a systematic alphanumeric designation based on homology in amino acid sequences. They can catalyze a broad range of metabolic reactions. These can be classified into four major categories: 1) hydroxylation, in which a hydrogen atom is replaced by a hydroxyl group; 2) epoxidation of substrates with a carbon-carbon double bond; 3) heteroatom oxidations on nitrogen or sulfur; and 4) reduction, which takes place under limited oxygen conditions. It should be noted that dealkylation reactions of some alkyl-substituted heteroatoms (i.e., O, S, and N) are actually due to the hydroxylation of the carbon atom adjacent to the heteroatom followed by the release of a carbonyl group. Another characteristic of CYP enzymes is their lack of high substrate specificity, meaning that almost any organic molecule can reach their active sites. There are about 500 different CYP isozymes responsible for most phase I metabolic reactions by means of an oxidative process. An inherently high degree of chemical reactivity, together with a lack of high substrate specificity, would suggest that CYP enzymes are not particularly stereoselective. In contrast to expectations, the CYP enzymes actually tend to show significant stereoselectivity on chiral and prochiral substrates (both substrate oriented and product oriented). This can be explained by two dominant structural features, a Fe-porphyrin-oxygen complex and an apoprotein associated with a specific enzyme. Both are common to all isozymes of CYP and are independent of the source. The Fe-porphyrin-oxygen complex functions to transfer an oxygen atom to the substrate via the reactive oxenoid intermediate, while the apoprotein forms the architecture of the active site and governs which part of the substrate can gain the necessary access. As a result, they control both regioselectivity and stereoselectivity in metabolism [35].

Table 9.3 shows the substrate stereoselectivity of several chiral drugs (Fig. 9.7) in terms of their regio- and stereoselectivity of various CYP-mediated biotransformation reactions [36–42].

The anticoagulant drug warfarin (**24**) has been used in the clinic for more than 40 years, with the pharmacological activity mainly residing with the (S)-isomer. The regio- and stereoselectivity of warfarin by CYP-mediated metabolism has been well explored with recombinantly expressed CYP enzymes. The predominant route of metabolism for the active enantiomer is by means of 7-hydroxylation of the coumarin ring, which is primarily catalyzed by CYP2C9. The metabolism of the (R)-enantiomer is primarily by CYP1A2 to 6- and 8-hydroxyl warfarin, and also by CYP3A4 to 10-hydroxyl warfarin. The molecular basis of preferential 7-hydroxylation of (S)-warfarin by human CYP2C9 has been studied in detail with crystal structures of human CYP2C9, both ligand free and in complex with (S)-warfarin [43].

A nonchiral compound can become chiral by metabolism of enantiotopic groups at a prochiral center. Such examples include the reduction of a ketone or the oxidation of "CH_2" to a chiral alcohol and the oxidation of the nitrogen in amines or the sulfur in sulfides to chiral N-oxides or sulfoxides. Risperidone (**31**), an antipsychotic agent, is mainly metabolized by means of 9-hydroxylation

TABLE 9.3   Regio- and Stereoselectivity in Terms of CYP-Mediated Metabolism

Drug	Pathway	Enzyme	Selectivity	Ref.
Acenocoumarol (**26**)	6-Hydroxylation	CYP2C9	$S/R = 37^a$	36
	7-Hydroxylation	CYP2C9	$S/R = 31^a$	
	8-Hydroxylation	CYP2C9	$S/R = 17^a$	
Cibenzoline (**27**)	p-Hydroxylation	CYP2D6	$R >> S^c$	37
Disopyramide (**4**)	Mono-N-Dealkylation	CYP3A3	$S/R = 1.4^b$	38
		CYP3A4	$S/R = 2.2^b$	
Fluoxetine (**28**)	N-Dealkylation	CYP2C9	$R/S = 5^a$	39
		CYP2D6	$R/S = 1.3^a$	
Ifosfamide (**29**)	4-Hydroxylation	CYP3A4	$R/S = 2.7^b$	40
		CYP2B6	$R/S = 0.09^b$	
	N^2-Dechloroethylation	CYP3A4	$R/S = 1.4^b$	
		CYP2B6	$R/S = 0.16^b$	
	N^3-Dechloroethylation	CYP3A4	$R/S = 7.4^b$	
		CYP2B6	$R/S = 0.08^b$	
Omeprazole (**30**)	Hydroxylation	CYP2C19	$R/S = 20^a$	41
	Sulfone formation	CYP3A4	$S/R = 10^a$	
	5-O-Demethylation	CYP2C19	$S/R = 11^a$	
Warfarin (**24**)	7-Hydroxylation	CYP2C9	$S >> R^{b,d}$	42
	6-Hydroxylation	CYP1A2	$R >> S^{b,d}$	
	8-Hydroxylation	CYP1A2	$R >> S^{b,d}$	
	10-Hydroxylation	CYP3A4	$R >> S^{b,d}$	

[a] Based on intrinsic clearance.
[b] based on rate of metabolite formation.
[c] No p-hydroxyl metabolite formed from the S-isomer.
[d] At least 5-fold (exact ratios not calculated).

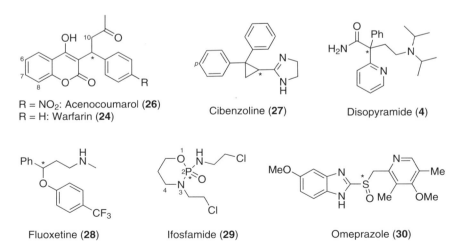

R = NO₂: Acenocoumarol (**26**)
R = H: Warfarin (**24**)

Cibenzoline (**27**)

Disopyramide (**4**)

Fluoxetine (**28**)

Ifosfamide (**29**)

Omeprazole (**30**)

FIGURE 9.7   Structures of chiral drugs **26–30**.

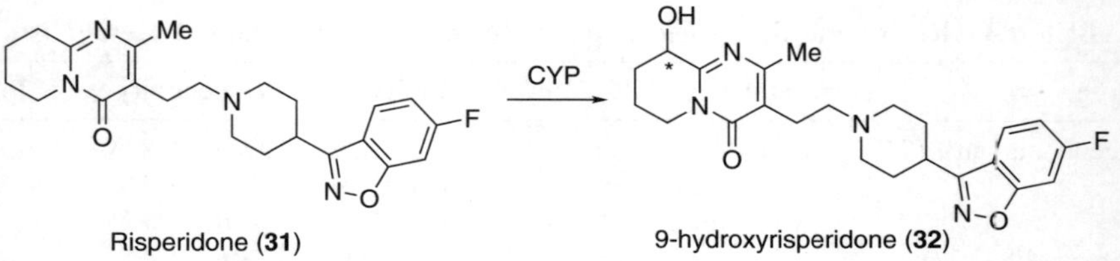

FIGURE 9.8   CYP-mediated 9-hydroxylation of risperidone (**31**).

in human liver, producing two enantiomers, (R)- and (S)-9-hydroxyrisperidone (**32**) (Fig. 9.8). Interestingly, the formation of these chiral metabolites from the prochiral risperidone (**31**) is mediated by different CYP enzymes. CYP2D6 predominately promoted the (S)-9-hydroxylation of risperidone, the major metabolic pathway in human liver, whereas CYP3A4- and CYP3A5-mediated formation of (S)-9-hydroxyrisperidone was almost four times lower than that of the (R)-isomer [44].

A nonchiral drug can also become chiral by metabolism at a site remote from the prochiral center. For example, phenytoin (**33**), a widely prescribed anticonvulsant, is a prochiral compound that is eliminated in humans almost entirely by oxidative metabolism to 5-(4-hydroxyphenyl)-5-phenylhydantoin (**34**), existing as (R)- and (S)-enantiomers (Fig. 9.9). Levy et al. studied the stereoselective metabolism of phenytoin by cDNA-expressed CYP2C9 and CYP2C19. Recombinant CYP2C9 was highly (S)-enantioselective (S/R ratio = 43–44), whereas recombinant CYP2C19 exhibited low enantioselectivity (S/R ratio = 1.2–1.3). The S/R ratios of metabolites were independent of substrate concentration for both enzymes [45].

Many chiral drugs undergo metabolic conversion in which a second chiral center is introduced, thus producing diastereomers. This is exemplified by the metabolism of hexobarbital (**35**) in rat liver microsomes (Fig. 9.10). (S)-hexobarbital was metabolized 1.5 times faster than its (R)-enantiomer, and a fairly high stereoselective hydroxylation of both enantiomers was observed. The ratios of the resulting hydroxyl metabolites (3'S)-OH-hexobarbital to (3'R)-OH-isomer from (S)-hexobarbital and (R)-hexobarbital were 4.6:1 and 5.8:1, respectively [46].

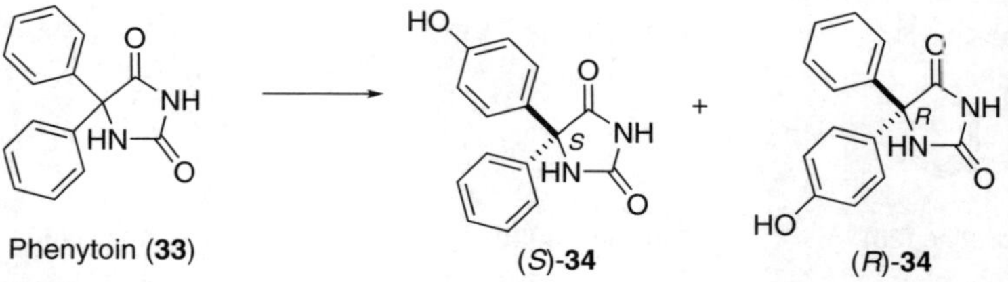

FIGURE 9.9   CYP-mediated hydroxylation of phenytoin (**33**).

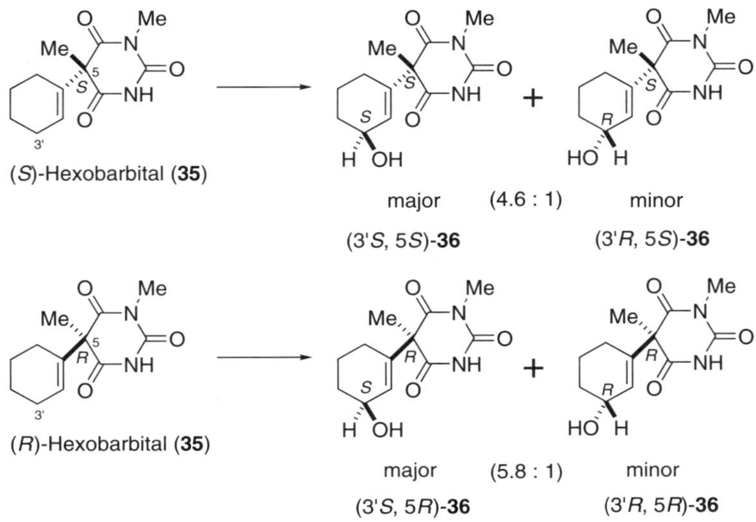

FIGURE 9.10 Stereoselective metabolism of (*S*)- and (*R*)-hexobarbital (**35**).

The structural characteristics of the apoprotein of a specific CYP enzyme dictate the enantiomeric discrimination associated with the biotransformation of chiral substances, so the stereoselectivity of metabolic reactions can be viewed as a characteristic physical property of each enzyme. The ratio of relative rates for the formation of a specific metabolite from two enantiomers of a chiral drug by a pure CYP enzyme should give a powerful index for categorizing and identifying that isozyme. A more powerful index could be the ratio provided by a prochiral substrate, since the architecture of the active site is the only physical feature in the process that can lead to differences in rates of formation of two enantiomers from a prochiral substrate [35]. Several mammalian CYP isozymes, including CYP2A6, CYP2B4, CYP2C5, CYP2C8, CYP2C9, and CYP3A4, have been resolved crystallographically [47]. This structural information, as well as the use of site-directed mutagenesis in the study of CYP structure-function relationships [48], is pivotal in understanding the interactions between substrates and specific CYP enzymes. With recent developments in computational software that facilitates the estimation of binding affinity, it is anticipated that the routes of metabolism can be predicted in the near future.

Another important aspect of drug metabolism is conjugation, which is energy-consuming biosynthesis involving the linkage of the drug or its metabolite(s) to an endogenous conjugating agent (endocon) to give a characteristic product known as a conjugate. The endocons can be divided into two groups, achiral and chiral. The former include methyl, acetyl, sulfate, etc. Chiral endocons, which are quantitatively more significant, include glucuronic acid, glutathione, etc., derived from the chiral pool with fixed configurations. There are two possible types of stereochemistry regarding the conjugation reactions of chiral drugs: substrate-oriented when they are conjugated at different rates with an achiral endocon and substrate/product-oriented when coupled with a chiral endocon to form pairs of diastereomeric conjugates.

Uridine 5′-diphosphate glucuronosyltransferases (EC 2.4.1.17; UGTs) are the major vertebrate phase II conjugative metabolism enzymes, which are not only abundant in liver but also present in other tissues, including lungs, kidneys, gastrointestinal tract, mammary gland, and prostate. UGTs catalyze the covalent addition of a glucuronic acid moiety from the uridine 5′-diphosphate glucuronic acid (UDPGA, **37**) donor ligand to nucleophilic functional groups (such as alcohols, phenols, thiols, carboxylic acids, and amines) present on a range of structurally diverse acceptor substrates (Fig. 9.11). The conjugative adducts, called glucuronides, are often biologically inactive and more hydrophilic than the parent drugs so that they can be readily excreted from the body through urine or bile [49].

Propranolol (**25**) is a nonselective β-adrenergic blocking agent used as a racemic mixture in the treatment of hypertension, cardiac arrhythmias, and angina pectoris. This drug is metabolized through various pathways,

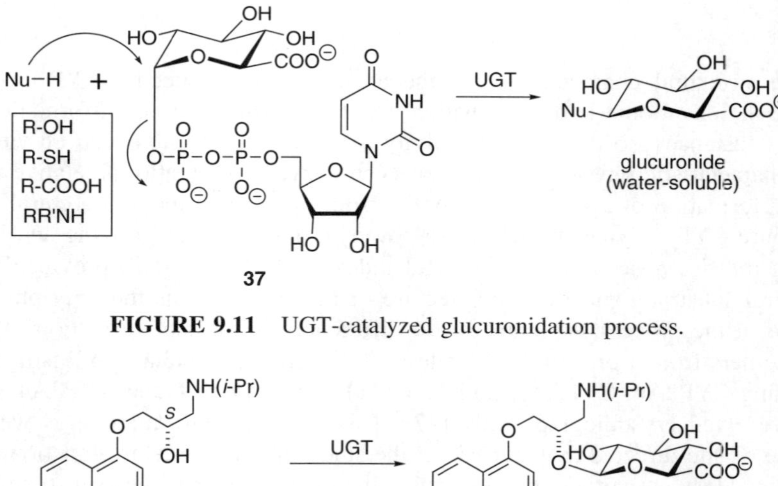

FIGURE 9.11 UGT-catalyzed glucuronidation process.

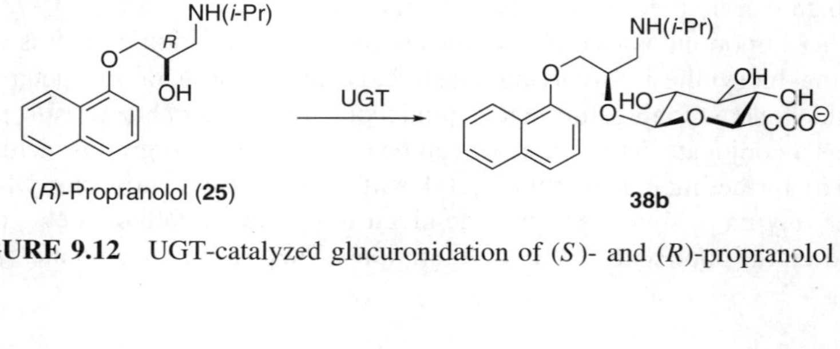

FIGURE 9.12 UGT-catalyzed glucuronidation of (S)- and (R)-propranolol (**25**).

including conjugation and side chain/ring oxidation. Conjugative metabolism via glucuronidation produces two diastereomeric O-glucuronides **38a** and **38b** in humans, as well as in dog and rat (Fig. 9.12). Finel et al. screened 15 recombinant human UGTs for the glucuronidation of propranolol (**25**) and found that UGT1A9 and UGT1A10 displayed high and, surprisingly, opposite stereoselectivity in the metabolism of this drug. (S)-propranolol was glucuronidated by UGT1A9, expressed in liver, much faster than the (R)-enantiomer, with reaction velocities of 21.6 pmol/mg/min and 2.0 pmol/mg/min, respectively. In contrast, UGT1A10, expressed in intestine, exhibited the reverse enantiomer preference, with reaction velocities of 8.7 pmol/mg/min and 36.7 pmol/mg/min for (S)- and (R)-propranolol, respectively. The results from recombinant UGTs are qualitatively consistent with the data from human liver microsomes and human intestinal microsomes, that is, more (S)-propranolol glucuronide **38a** was generated in human liver microsomes, whereas more (R)-propranolol glucuronide **38b** was produced in human intestinal microsomes [50].

Sulfotransferases (SULTs) are another class of phase II metabolism enzymes, catalyzing the sulfate conjugation of a wide variety of endogenous compounds as well as drugs and xenobiotics. SULTs are expressed not only in liver but also in kidneys, stomach, colon, and small intestine. Sulfation involves the transfer of a sulfate group from 3′-phosphoadenosine 5′-phosphosulfate (PAPS, **39**) to the nucleophilic functional groups in drug substrates or their metabolites (Fig. 9.13), generally resulting in the abolition of biological activity [51].

The metabolism of the β_2-receptor agonist drugs, for example, salbutamol **11** (also named albuterol), in humans occurs almost exclusively by sulfate conjugation of the phenolic hydroxyl groups (Fig. 9.14). Walle et al. utilized the human hepatoma cell line HepG2, which expresses the highly active M form of human

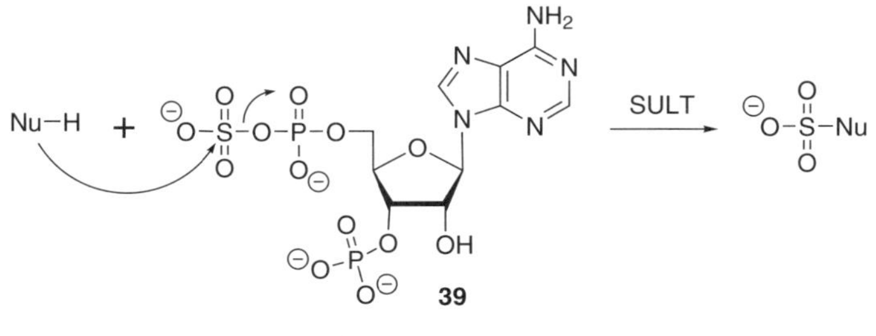

FIGURE 9.13 SULT-catalyzed sulfation process.

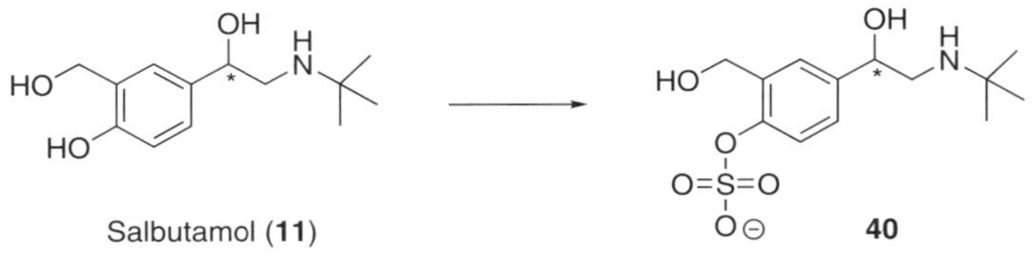

FIGURE 9.14 Sulfation of salbutamol (**11**).

phenolsulfotransferase (PST), in the biosynthesis of sulfate conjugate of salbutamol and assigned the position of the resulting sulfate with liquid secondary ion mass spectrometry MS/MS. A clear preference for sulfation was observed for the (R)-enantiomer. Kinetic experiments with individual enantiomers gave apparent K_m values of 115 μM and 528 μM for (R)- and (S)-salbutamol, respectively, with an apparent V_{max} value 1.7 times higher for (R)-salbutamol. Thus the efficiency of sulfation (V_{max}/K_m) was 7.8 fold higher for (R)-salbutamol [52]. Experiments using the recombinant human M form of PST for salbutamol showed a similar trend, with apparent K_m values of 93 μM and 919 μM for the (R)- and (S)-enantiomers, respectively, and comparable V_{max} values [53].

Fenoterol (**41**) is also a β_2 agonist drug used as a racemate of the (R, R)- and (S, S)-enantiomers (Fig. 9.15). This compound has three potential phenolic hydroxyl groups for sulfate conjugation. Walle et al. applied the human HepG2 cell line, together with HPLC isolation and FAB/MS/MS, to determine the sites for sulfation. It was observed that the sulfation occurred either at the 4-hydroxyl group of one phenyl ring or one of the two 3',5'-hydroxyl groups of another phenyl ring. The stereoselective sulfation was further studied with recombinant human M-PST (monoamine form of the phenolsulfotransferases) and P-PST (phenol form of the phenolsulfotransferases). M-PST preferentially sulfated the 4-hydroxyl group and showed significant stereoselectivity between the enantiomers. The formation of this sulfate conjugate was about sixfold higher for (R, R)-enantiomer than for the (S, S)-enantiomer. In contrast, P-PST exclusively sulfated one of the 3',5'-hydroxyl groups, with slight preference for the (S, S)-enantiomer [54].

Some chiral drugs experience a chiral inversion during the metabolic process, which means the switch from one enantiomeric configuration to the other. There are two kinds of chiral inversion: unidirectional and bidirectional.

Unidirectional inversion is usually facilitated by the presence of certain proteins or enzymes. 2-Aryl-propionic acid derivatives (the "profens"), which are nonsteroidal antiinflammatory drugs (NSAIDs) used in the treatment of pain and inflammation, have been a focus in the field of chiral inversion. For these agents with a chiral center adjacent to the carboxyl group, only the (S)-enantiomer has analgesic and antiinflammatory effects, with the (R)-enantiomer being weakly active or inactive in in vitro assays. However, the activities of two enantiomers of many profens are essentially indistinguishable in the in vivo setting, owing to the unidirectional chiral inversion of the inactive (R)-enantiomer to the active

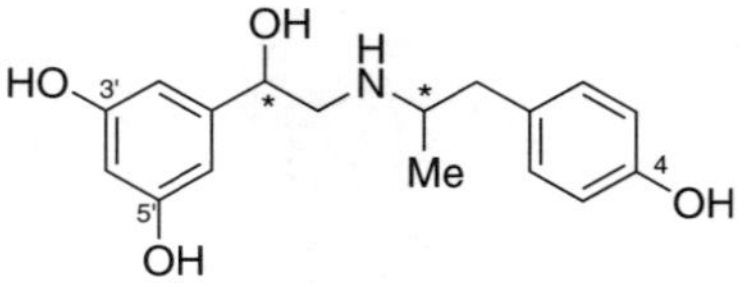

Fenoterol (**41**)

FIGURE 9.15 Structure of fenoterol (**41**).

(S)-enantiomer. For example, ibuprofen (**42**) was developed as an antirheumatic drug in the 1960s as a racemate, with the (S)-enantiomer being about 160 times more potent than the (R)-enantiomer in inhibiting prostaglandin synthesis in vitro. In the body, (R)-ibuprofen can undergo rapid and substantial chiral inversion into the active (S)-enantiomer but not vice versa. This interesting phenomenon represents a metabolic activation of the (R)-enantiomer [55]. When racemic ibuprofen is administered, the internal exposure is principally (S)-ibuprofen, with little (R)-enantiomer present in the circulation. Therefore, racemic ibuprofen and (S)-ibuprofen can be viewed to be essentially bioequivalent, but the use of (S)-ibuprofen reduces variability in configuration inversion. The strategy of switching ibuprofen from a racemate to its chiral form was quite successful in its clinical application. So far, (S)-ibuprofen has been clinically used in Germany, Austria, and Switzerland [56].

The mechanism of enzymatic inversion of (R)-ibuprofen (**42**) is shown in Figure 9.16 [57]. In the first step, the carboxylic acid is converted into an intermediary thioester **43a** by a microsomal or mitochondrial acyl-CoA ligase. This enzyme, usually the long-chain fatty acid ligase (EC 6.2.1.2.3), catalyzes the incorporation of CoA-SH to the (R)-enantiomer only, that is, in an enantiospecific manner, dictating the unidirectional character of this inversion process. The formed thioester **43a** is then converted to the opposite configuration by an epimerase [58], and finally the resultant thioester **43b** is hydrolyzed by a hydrolase, releasing (S)-ibuprofen. The therapeutic consequence of the chiral inversion of ibuprofen is that the administered dose of racemate actually provides greater than the theoretical amount of active component present in the formulation. This has an impact on the calculation of pharmacokinetic parameters, such as the clearance and volume of distribution of the enantiomers. Moreover, the formation of CoA thioesters of profens may induce significant toxicological complications by the (R)-enantiomer due to potential lipid incorporation.

It should be pointed out that such chiral inversion may be species dependent (Fig. 9.17). In a human study with an oral dose (25 mg/kg) of (R)-flurbiprofen (**44**), the (S)-isomer was not detected in plasma or urine, indicating the absence

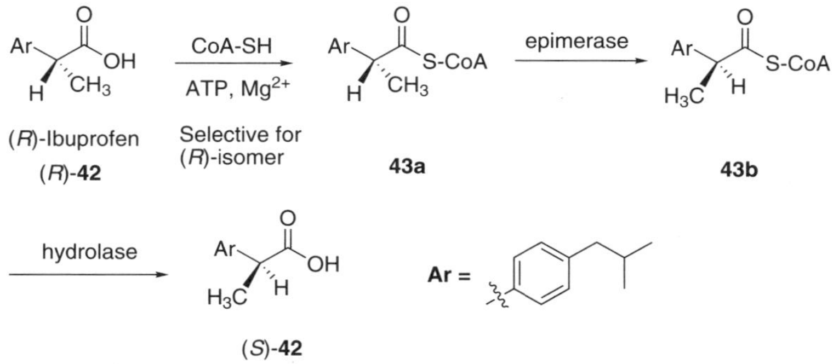

FIGURE 9.16   Chiral inversion of (R)-ibuprofen (**42**).

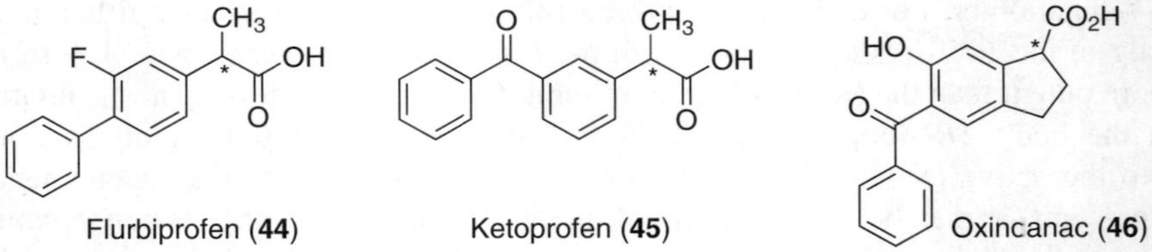

Flurbiprofen (**44**)   Ketoprofen (**45**)   Oxincanac (**46**)

FIGURE 9.17 Structures of flurbiprofen (**44**), ketoprofen (**45**), and oxindanac (**46**).

of chiral inversion. Similar results were observed in a pharmacokinetic study with horses. Although the mean plasma enantiomeric R/S ratio was 54.6:45.4 after the IV dosage of 0.5 mg/kg of racemic flurbiprofen, the antipode was not detected in plasma after IV administration of either (R)- or (S)-flurbiprofen. However, in rat (R)-flurbiprofen showed some degree of chiral inversion. After IV administration (5 mg/kg) of (R)-flurbiprofen to the rat, both enantiomers were found in plasma, with AUC (S):AUC (R) = 0.10–0.16 [59].

Another chiral NSAID drug, ketoprofen (**45**), also showed species-dependent chiral inversion. When racemic ketoprofen was intravenously administered into the rat at 10 mg/kg, a pseudoequilibrium was rapidly attained. The plasma concentrations were always greater for (S)-ketoprofen than for (R)-ketoprofen. On average, the AUCs of the (S)-enantiomer were approximately 11-fold greater than those for the (R)-enantiomer. This significant difference in AUCs was mainly due to the chiral inversion of (R)-ketoprofen to (S)-ketoprofen [60]. In humans, the pharmacokinetics of ketoprofen (**45**) are quite different. When either racemic ketoprofen (50 mg/kg dose), (S)-ketoprofen (25 mg/kg dose), or (R)-ketoprofen (25 mg/kg dose) were ingested, the mean AUCs of the (S)-enantiomer were determined to be 4.87, 4.90, and 0.65 mg·h/l, respectively, while the mean AUCs of the (R)-isomer were determined to be 4.65, 0, and 4.74 mg·h/l, respectively, suggesting that (R)-ketoprofen could undergo very limited unidirectional inversion to its (S)-form in humans [61]. Interestingly, a bidirectional inversion of ketoprofen (**45**) was observed in CD-1 mice. The pharmacokinetic study was conducted in groups of CD-1 mice dosed intraperitoneally with (R)-, (S)-, and racemic ketoprofen at doses of 5, 5, and 10 mg/kg, respectively. After the racemic dose, no significant stereoselectivity was observed for the peak plasma concentration C_{max} [7.53 vs. 6.07 μg/ml for (S)- and (R)-isomers, respectively]. When the enantiomers were dosed separately, the administered enantiomer presented a significantly greater C_{max} than its antipode [8.67 vs. 3.18 μg/ml for the (S)- and (R)-isomers, respectively after dosing (S)-ketoprofen and 2.11 vs. 3.22 μg/ml for the (S)- and (R)-isomers, respectively, after dosing (R)-ketoprofen]. The presence of a significant amount of the antipodes suggested that a bidirectional chiral inversion of ketoprofen in CD-1 mice took place [62]. It should be pointed out that the (S) to (R) chiral inversion is very rare for NSAID drugs, but has been observed as shown above, as well as for 2-phenylpropionic acid and oxindanac (**46**) in dogs [63].

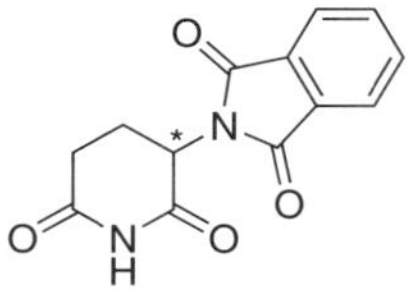

Thalidomide (**47**)

FIGURE 9.18 Structure of thalidomide (**47**).

Thalidomide (**47**), a former racemic sedative withdrawn from the market in the 1960s because of its severe teratogenic effects, is known to undergo a bidirectional chiral inversion/racemization (Fig. 9.18). Such an inversion was observed in aqueous media and was catalyzed by serum albumin. When both enantiomers were incubated in aqueous phosphate buffer (pH = 7.4) at 37 °C, the half-lives of the racemization, calculated from the graphs of the racemization kinetics, were essentially the same, with $t_{1/2}$ = 289 and 261 min for (*R*)- and (*S*)-thalidomide, respectively. Citric plasma from human and rabbit both significantly accelerated the racemization. In human citric plasma $t_{1/2}$ were 11.5 and 8.3 min, while in rabbit citric plasma $t_{1/2}$ were 9.3 and 6.5 min for (*R*)- and (*S*)-thalidomide, respectively. The racemization process of (*S*)-thalidomide was faster than that of (*R*)-thalidomide in both species [64]. Eriksson et al. further characterized the in vitro inversion and degradation of (*R*)- and (*S*)-thalidomide in heparinized human blood from four healthy volunteers. The mean rate constants for chiral inversion of (*R*)- and (*S*)-thalidomide at 37 °C were 0.30 and 0.31 h^{-1}, respectively, while rate constants for degradation were 0.17 and 0.18 h^{-1}, respectively. In the in vivo setting, the interconversion of both enantiomers was faster than under in vitro conditions, with the (*R*)-enantiomer predominating at equilibrium. The mean rate constants of chiral inversion were 0.17 h^{-1} for the (*R*) to (*S*) inversion, and 0.12 h^{-1} for the (*S*) to (*R*) inversion. The elimination of (*R*)-thalidomide was threefold faster than that of (*S*)-thalidomide, with rate constants of 0.079 and 0.24 h^{-1}, respectively [65].

9.2.4 Excretion

Excretion is the phase in which a drug and its metabolite(s) are eliminated from the body. Kidneys and liver are the main organs involved in this process. In general, hydrophilic compounds are excreted more rapidly than lipophilic compounds.

Renal excretion includes three separate processes in kidneys: glomerular filtration, active tubular excretion, and passive tubular reabsorption. Glomerular filtration is a simple filtration of plasma through the pores of the glomeruli, whose sizes allow only the unbound free drug and metabolites to be eliminated. The amount of compound cleared in this process is dependent on the renal blood flow. Although this process is considered to be nonstereoselective for unbound drugs, enantiomers may have different rates of filtration because of their different binding affinities with plasma proteins. Active tubular excretion is the process by which ionized compounds are excreted from tubules with specific transporters (P-glycoprotein, multidrug resistance-associated proteins, and organic anion and cation transporters) and energy supply. Because enantiomers usually have different binding affinities with transporters, this process is mostly stereoselective for chiral compounds [66]. Plasma pH, a factor influencing the degree of ionization, plays a prominent role in this process. Passive tubular reabsorption is the process by which nonionized compounds return to capillaries via a favorable concentration gradient, thus it is usually considered to be nonstereoselective. This process involves mostly the more lipophilic parent compounds, and the elimination depends on the urine pH as it determines the percentage of the reabsorbed nonionized form. Total renal elimination of a drug and its metabolites is the sum of these three processes and therefore is normally a stereoselective event overall for a chiral drug.

E-10-hydroxynortriptyline (E-10-OH-NT, **49**) is the major metabolite of the tricyclic antidepressant nortriptyline (NT, **48**) (Fig. 9.19). E-10-OH-NT is about 50% as potent as NT in inhibiting the neuronal uptake of noradrenaline in vitro, but has much lower affinity for muscarinic receptors in vitro and fewer anticholinergic side effects in vivo. In healthy men receiving a single oral dose of 75 mg racemic E-10-OH-NT, the urinary excretion rate of unconjugated (−)-E-10-OH-NT was higher than that of (+)-E-10-OH-NT, but the trend was reversed for the corresponding glucuronides. The recovery of unconjugated (−)-E-10-OH-NT was much higher than that of (+)-E-10-OH-NT, with 35.8% and 14.4% for (−)- and (+)-enantiomers, respectively. As for their glucuronides in urine, a significantly higher proportion was recovered for the (+)-isomer (64.4%) than for the (−)-isomer (35.3%). The total oral plasma clearance and the metabolic clearance by glucuronidation was twofold higher for (+)-E-10-OH-NT (2.72 l/kg/h) than for (−)-E-10-OH-NT (1.26 l/kg/h). Renal clearance of both enantiomers exceeded the glomerular filtration rate, with 0.44 l/kg/h for (+)-E-10-OH-NT and 0.57 l/kg/h for (−)-E-10-OH-NT, suggesting that active tubular secretion was involved in this process with a moderate stereoselectivity preference for (−)-E-10-OH-NT [67].

Oxprenolol (**50**) is a β-adrenergic blocking agent marketed as a racemic mixture, with the (S)-enantiomer 10–35 times more active than the (R)-enantiomer (Fig. 9.20). This drug is extensively metabolized in humans, and more than 50%

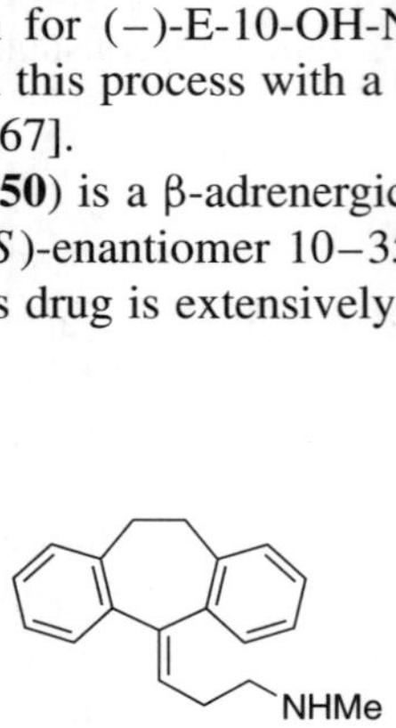

Nortriptyline (NT, **48**) 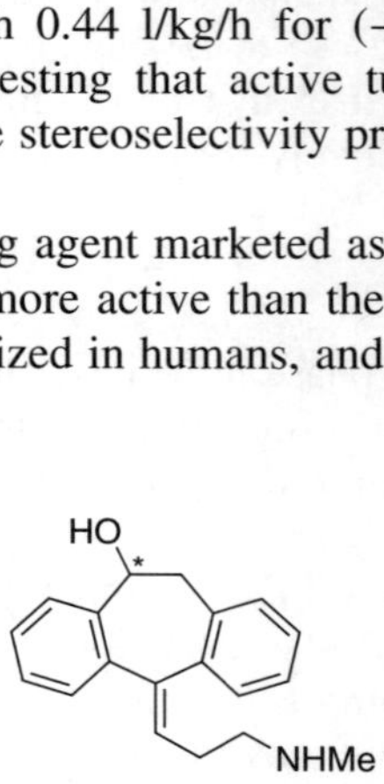   E-10-OH-NT (**49**)

FIGURE 9.19 Structures of NT (**48**) and E-10-OH-NT (**49**).

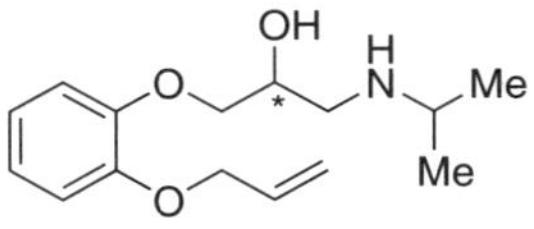

Oxprenolol (**50**)

FIGURE 9.20   Structure of oxprenolol (**50**).

of the dose is excreted in urine as the O-glucuronide of the parent drug. In healthy men receiving a single 80-mg oral dose of racemic oxprenolol hydrochloride salt, there was only a small difference in disposition between (R)- and (S)-oxprenolol, whereas a marked difference in disposition between the O-glucuronides was noted. The renal clearances of (R)- and (S)-oxprenolol were 11.1 and 7.6 ml/min, respectively. However, the renal clearance of (R)-oxprenolol glucuronide (CL = 171.7 ml/min) was about three times higher than that of (S)-oxprenolol glucuronide (CL = 49.0 ml/min). The renal clearance of (R)-oxprenolol glucuronide was larger than the glomerular filtration rate, suggesting the involvement of active tubular secretion. In contrast, the renal clearance of (S)-oxprenolol glucuronide was smaller than the glomerular filtration rate, probably because of higher plasma binding of the enantiomer. The free percentages of (S)- and (R)-glucuronides in plasma were 51.0% and 59.7%, respectively [68].

Hepatic elimination of a drug and its metabolites involves three main processes in the liver: uptake, metabolism, and biliary secretion. Hepatocytes usually excrete unchanged drugs or metabolites by active processes involving transporters. After biliary secretion, the excreted compounds reach the duodenum, where they can be excreted from the body via feces, partially reabsorbed through enterohepatic cycling, or metabolized by the intestinal flora (e.g., deconjugation of phase II metabolites).

Carvedilol (**16**) is a racemic antihypertensive drug with both β-blocking and vasodilating properties. ^{14}C-labeled (R)-/unlabeled (S)-carvedilol and ^{14}C-labeled (S)-/unlabeled (R)-carvedilol mixtures were administered orally and intravascularly to bile duct-cannulated rats. The biliary excretion of each enantiomer and the composition of metabolites were measured by chiral HPLC with a radiodetector and a UV detector. As for the parent drug, oral administration produced no enantiomeric difference in biliary excretion, with radioactivity of 41.4% and 41.5% for the (R)- and (S)-enantiomers, respectively, whereas intravascular administration resulted in a slight enantiomeric difference, with radioactivity of 43.7% and 40.0% for the (R)- and (S)-enantiomers, respectively. As for the metabolites in the bile, significant enantioselectivity was observed for both 1-hydroxy and 8-hydroxy O-glucuronides, interestingly with opposite preference (Fig. 9.21). The S/R ratios of 1-hydroxy O-glucuronides were 0.59 (**51a:51b**) for oral dosing and 0.43 (**51a:51b**) for IV dosing, while the S/R ratios of 8-hydroxy O-glucuronides were 3.29 (**52a:52b**) and 2.63 (**52a:52b**) for oral and IV dosing, respectively. Because the hydroxyl metabolites are rapidly biotransformed to O-glucuronides that are excreted predominantly into the bile, such differences of the hydroxy

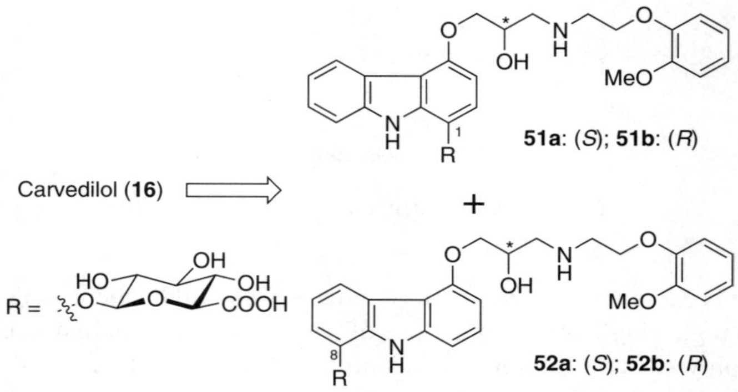

Carvedilol (16)

51a: (*S*); 51b: (*R*)

+

52a: (*S*); 52b: (*R*)

FIGURE 9.21   Hydroxy *O*-glucuronides of carvedilol (**16**).

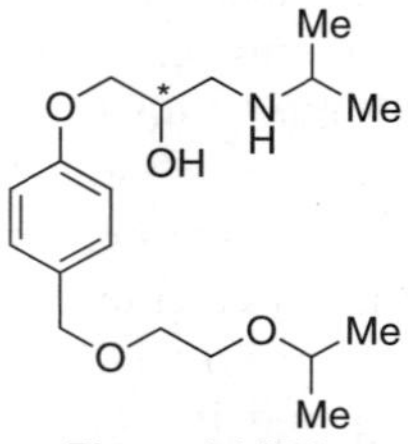

Bisoprolol (**53**)

FIGURE 9.22   Structure of bisoprolol (**53**).

O-glucuronides in the bile most likely arise from selective oxidative metabolism in the liver [69].

It should be noted that a drug and its metabolites may have multiple elimination pathways. In this case, the stereoselective appearance in the excreta may not necessarily reflect stereoselectivity in the secretion process itself. For example, bisoprolol (**53**) is a highly β_1-selective adrenoceptor blocking agent marketed as a racemic mixture, although the activity of the (*S*)-enantiomer is about 30–80 times higher than that of the (*R*)-enantiomer (Fig. 9.22). Bisoprolol is eliminated by two pathways, hepatic metabolism and renal excretion. When racemic bisoprolol was orally administered to beagle dogs, the cumulative recovery of (*S*)-bisoprolol in urine was significantly larger than that of (*R*)-bisoprolol based on the extrapolation to time infinity, with *S*/*R* ratio of 1.29. However, no significant difference was observed for the renal clearance of both enantiomers, with $CL_r = 25.3$ and 26.7 ml/min for (*S*)-and (*R*)-bisoprolol, respectively. This can be explained by the difference in their metabolic clearances, with $CL_m = 38.5$ and 67.1 ml/min for (*S*)-and (*R*)-bisoprolol, respectively [70]. Another example is that (*S*)- and (*R*)-etodolac (**17**) glucuronides in rat bile appeared to be stereoselective (*S*/*R* = 3:1). However, this was not due to the stereoselectivity in the secretion process; instead it likely arose from the process of glucuronide formation [21].

9.3 STEREOSELECTIVE DRUG-DRUG INTERACTIONS OF CHIRAL DRUGS

It has been generally accepted that it is drug metabolism that introduces the greatest degree of stereoselectivity into drug disposition, so the pharmacokinetics of two enantiomers of a chiral drug can be affected by a coadministered drug through inhibition or induction of metabolic enzymes. In other words, the plasma concentration of one enantiomer can display a clearly disproportional change, compared with the other enantiomer, when a coadministered drug interferes with CYP activity [71,72].

A classic example is the coadministration of warfarin (**24**) with cimetidine (**54**) or sulfinpyrazone (**55**) (Fig. 9.23). In the CYP-mediated stereoselective metabolism of warfarin, the (*R*)-enantiomer is preferentially hydroxylated at the 6- and 8-positions via CYP1A2, while the (*S*)-enantiomer is primarily hydroxylated at the 7-position via CYP2C9. In a human trial of eight volunteers, oral coadministration of cimetidine, an inhibitor of CYP1A2, caused a significant increase in AUC and $t_{1/2}$ for the (*R*)-enantiomer. The AUC increased from 96.1 to 126.5 μg·h/ml and $t_{1/2}$ increased from 47.8 to 57.8 h. The influence on the active (*S*)-warfarin was almost negligible, with AUC from 80.0 to 81.1 μg·h/ml and $t_{1/2}$ from 38.4 to 37.7 h [73]. In contrast, coadministration of sulfinpyrazone, a competitive inhibitor for the 7-hydroxylation of (*S*)-warfarin mediated by CYP2C9, decreased the clearance of active (*S*)-warfarin (from 2.0 to 1.2 ml/h/kg) and accordingly increased its hypoprothrombinemia effect [74].

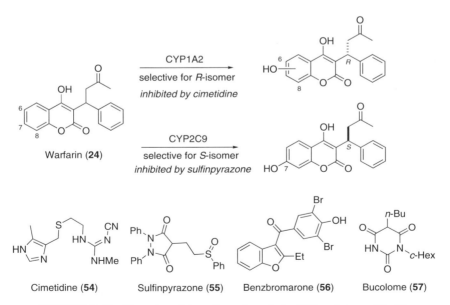

FIGURE 9.23 Drug-drug interaction of warfarin (**24**) and drugs **54–57**.

Similarly, benzbromarone (**56**) and bucolome (**57**) could also decrease the clearance of (S)-warfarin, thus exhibiting augmented anticoagulant effect [75].

Enzyme induction from a coadministered drug can also affect the stereoselectivity in the metabolism of a chiral drug. Hexobarbital (**35**) is a chiral drug with the (R)-enantiomer being the eutomer (bioactive enantiomer). The major enzyme for its metabolism is CYP2C19, which is known to be inducible by rifampicin. Hexobarbital racemate (single 500-mg oral dose) was coadministered with rifampicin (600 mg daily for 14 days) to six young and six old healthy male volunteers. For the young volunteers, the clearance of (R)-hexobarbital increased by 74-fold (from 15.6 to 1146.7 ml/min/kg), while the clearance of the (S)-enantiomer increased by only 6-fold (from 1.9 to 11.9 ml/min/kg). In this case, the therapeutic efficacy would be diminished to a larger extent than anticipated in young patients because of the enhanced metabolism of the active (R)-enantiomer by CYP2C19 induction. The differential induction of selective metabolism was altered by age. In the case of the old volunteers, the induction of (R)-hexobarbital clearance was significantly lower compared with the young volunteers, with only a 19-fold increase (from 10.3 to 199.9 ml/min/kg), while the induction of the (S)-enantiomer clearance was still of the same order of magnitude (from 1.8 to 10.7 ml/min/kg) [76].

Cigarette smoking, a known CYP1A2 inducer, can also alter drug disposition. Mexiletine (**58**) is an oral class Ib antiarrhythmic agent used as a racemic mixture in the treatment of ventricular arrhythmias (Fig. 9.24). Mexiletine is extensively metabolized in humans, with <10% of the dose being excreted unchanged in urine. Clinical studies suggest that CYP1A2 is involved in the metabolism. Nineteen healthy men (10 smokers and 9 nonsmokers) participated in a study to evaluate the effects of ciprofloxacin (**59**), a known inhibitor of CYP1A2, on the stereoselective disposition of mexiletine. The total clearances of (R)- and (S)-mexiletine (37 and 39 l/h) were significantly higher for smokers compared with nonsmokers (26 and 24 l/h), with a 42% and 63% increase, respectively. Coadministration of ciprofloxacin decreased the total clearances of (R)- and (S)-mexiletine for both smokers and nonsmokers. A greater extent of decrease was observed for (S)-mexiletine than for (R)-mexiletine in both smokers [by 3 l/h and 5 l/h for (R)- and (S)-enantiomers, respectively] and nonsmokers [by 2 l/h and 5 l/h for (R)- and (S)- enantiomers, respectively]. Consequently, the mean R/S ratio of mexiletine clearance was modified by the coadministration of ciprofloxacin in

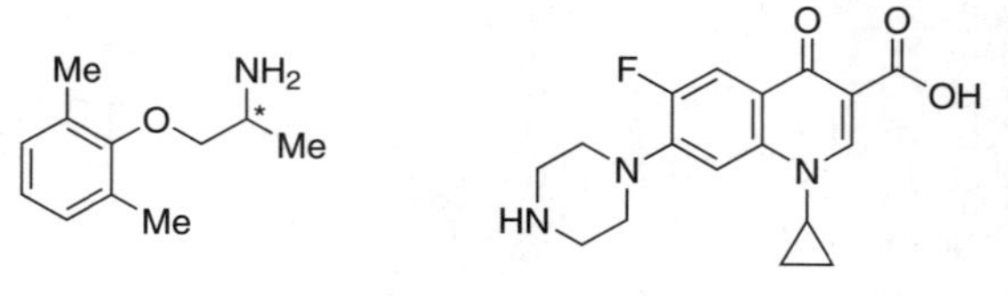

Mexiletine (**58**)   Ciprofloxacin (**59**)

FIGURE 9.24 Structures of mexiletine (**58**) and ciprofloxacin (**59**).

both smokers (from 0.93 to 1.02) and nonsmokers (from 1.13 to 1.31). However, the magnitude of drug-drug interaction due to the concomitant administration of mexiletine and ciprofloxacin in patients was considered of no clinical significance [77].

Administration of a racemic form of a chiral drug could be considered as giving two different drugs at the same time. The enantiomer-enantiomer interactions usually result from the competition for the same metabolic pathway(s) mediated by the same enzyme(s). In other words, the same enzyme(s) can mediate the biotransformation of two enantiomers of a chiral drug at different rates. For example, ifosfamide (**29**) is a nitrogen mustard alkylating agent used in the treatment of cancer. The pharmacokinetics were studied after the IV administration of individual enantiomers or pseudoracemates **29a** and **29b** to male Sprague-Dawley rats (Fig. 9.25). After the separate administration of both enantiomers, the pharmacokinetic parameters for (R)- and (S)-ifosfamide were not significantly different. However, after the administration of pseudoracemates, the R/S concentration ratio was initially 0.96 at 5 min after dosing and gradually decreased to 0.60 at 2 h after dosing. Multiple parameters for (R)- and (S)-ifosfamide also showed significant changes compared with those following the administration of individual enantiomers, especially for (S)-ifosfamide. The half-life ($t_{1/2}$) was prolonged from 29.6 min to 34.3 min for the (R)-enantiomer and from 32.6 min to 41.8 min for the (S)-enantiomer. The mean residence time was also prolonged, from 42.7 min to 49.4 min for the (R)-enantiomer and from 47.1 min to 60.3 min for the (S)-enantiomer. The AUC$_{0-inf}$ had a greater increase for the (S)-enantiomer (from 5,110 to 6,259 μM·min) than that for the (R)-enantiomer (from 4,259 to 4,853 μM·min). The total clearance for the (R)-enantiomer had almost no change, from 18.0 to 17.9 ml/min/kg, while a significant decrease was observed for the (S)-enantiomer, from 16.8 to 14.0 ml/min/kg. However, no enantioselectivity in Vd$_{SS}$, renal clearance, and blood cell partitioning was observed regardless of the two enantiomers being administered separately or in combination as a racemate. The results suggested that each enantiomer inhibited the metabolism of its antipode in a competitive manner, with (S)-ifosfamide being affected to a larger extent. Such a hypothesis was supported by the results from in vitro metabolism and inhibition experiments [78].

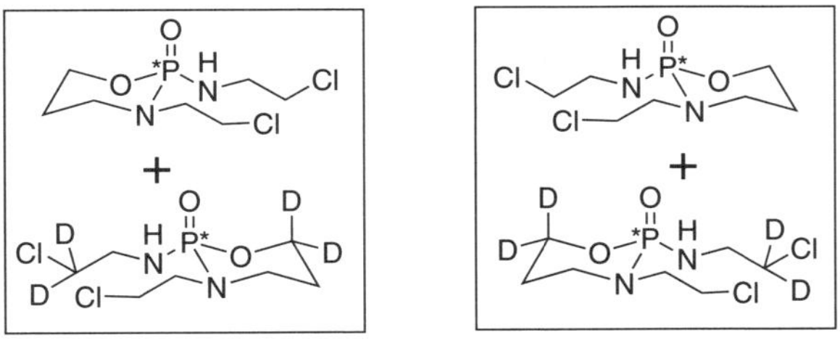

Pseudo-racemate **29a**
(R)-/d4-(S)-Ifosfamide

Pseudo-racemate **29b**
(S)-/d4-(R)-Ifosfamide

FIGURE 9.25 Composition of pseudoracemates **29a** and **29b**.

Propafenone (**21**) is an antiarrhythmic drug marketed as a racemic mixture. Both enantiomers are equally potent sodium channel blockers. In addition to the effect on sodium conductance, the drug has β-adrenoreceptor-blocking properties, which are mediated by the (S)-enantiomer. The major oxidative metabolism is CYP2D6-mediated 5-hydroxylation. In vitro incubation experiments using human liver microsomes for individual enantiomers showed enantioselectivity for 5-hydroxylation, with $V_{max} = 10.2$ and 5.5 pmol/μg/h and $K_m = 5.3$ and 3.0 μM for (S)- and (R)-propafenone, respectively. However, no substrate selectivity was observed after incubation of pseudoracemate [(R)-/d4-(S)-propafenone] (Fig. 9.26). The rates of 5-hydroxylation of both enantiomers were decreased, compared with the data obtained after the incubation of individual enantiomers [$V_{max} = 3.1$ and 3.3 pmol/μg/h for d4-(S)-propafenone and (R)-propafenone, respectively]. This result was explained by an enantiomer-enantiomer interaction. Further experiments revealed the K_i values of 2.9 μM for the inhibition of 5-hydroxylation of d4-(S)-propafenone by (R)-propafenone and 5.2 μM for inhibition of 5-hydroxylation of (R)-propafenone by d4-(S)-propafenone, which suggested that (R)-propafenone was a more potent inhibitor than the (S)-enantiomer with respect to the CYP2D6-mediated 5-hydroxylation [79].

The enantiomer-enantiomer interaction of propafenone was also observed in the human in vivo setting. Seven healthy male volunteers participated in a single-blind, randomized study to evaluate the individual enantiomers and the racemate (150 mg, every 6 hours for 4 days). The clearances after the separate administration of both enantiomers were 2,521 and 1,279 ml/min for (S)- and (R)-propafenone, respectively. The clearance for (R)-propafenone, after the administration of the racemate, had almost no change (1,460 ml/min), but the clearance for (S)-propafenone considerably decreased to 920 ml/min. This reduced clearance resulted in significantly higher plasma concentration of (S)-propafenone after administration of 150 mg of racemate than what would be predicted based on pharmacokinetic data obtained after administration of 150 mg of (S)-propafenone. The steady-state plasma concentration after administration of 150 mg of (S)-propafenone averaged 196 ng/ml. Thus one would predict the mean plasma concentration after the administration of 150 mg of racemate, which consists of only 75 mg of the β-blocking (S)-enantiomer, to be no higher than 100 ng/ml. However, the observed average plasma concentration was 224 ng/ml. As mentioned above, the β-blocking effects of propafenone are mediated

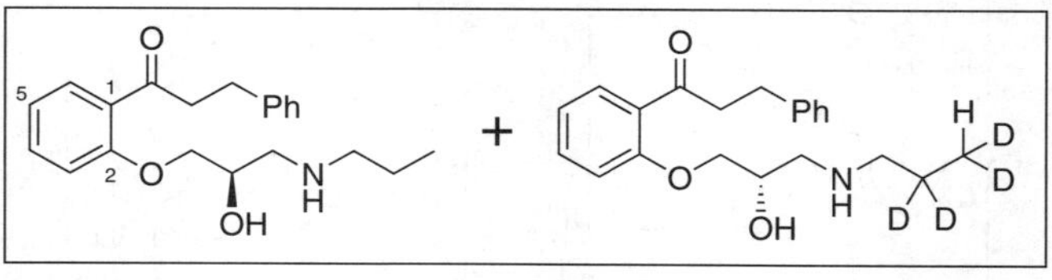

Pseudo-racemate **21**

(R)-/d4-(S)-Propafenone

FIGURE 9.26 Composition of pseudoracemate **21**.

exclusively by the (S)-enantiomer. The increased plasma concentration of (S)-propafenone as a consequence of the enantiomer-enantiomer interaction would lead to enhanced blockade. In fact, the maximum exercise heart rate (ΔHR_{max}), an index of β-blockade, was significantly altered compared with subjects dosed with placebo. ΔHR_{max} was -4.3 beats/min during administration of (S)-propafenone but increased to -8.8 beats/min during administration of the racemate, in which only one-half was (S)-propafenone. (R)-propafenone had no effect on ΔHR_{max} compared with placebo ($\Delta HR_{max} = -1.8$ beats/min). This pharmacodynamic effect showed excellent correlation with the pharmacokinetic results [80].

9.4 CONCLUDING REMARKS

Two enantiomers of a chiral drug frequently display different pharmacokinetic profiles in terms of absorption, distribution, metabolism, and excretion. Other factors, such as route of administration, age and sex of subjects, disease states and lifestyle, as well as genetic polymorphism of CYP enzymes, can also have different influences on each enantiomer in the pharmacokinetic process [2a]. The administration of racemic mixtures can be regarded as dosing a fixed combination of two drugs, that is, the two enantiomers in 1:1 ratio. So far, many pharmacokinetic studies on chiral drugs have been performed with nonstereospecific assays. The resulting data in this circumstance could be highly misleading, especially when attempting to link the plasma drug concentration and the related pharmacological effects. Therefore, the development of stereospecific analytical methodologies to detect both drug and metabolite concentrations in certain biological media is an essential requirement for the determination of pharmacokinetic parameters, especially for chiral drugs with significant differences between two enantiomers in efficacy and/or toxicity. Various chiral separation techniques using HPLC, gas chromatography, supercritical fluid chromatography, and capillary electrophoresis have been developed for enantiomeric resolution. The measurement of each enantiomer in in vitro and in vivo biological experiments will greatly help the understanding of pharmacokinetic and pharmacodynamic relationships of chiral drugs [2c, 81].

REFERENCES

1. Kola, I., Landis, J. (2004). Can the pharmaceutical industry reduce attrition rates? *Nat. Rev. Drug Discov.* **3**, 711–715.
2. (a) Brocks, D. R. (2006). Drug disposition in three dimensions: an update on stereoselectivity in pharmacokinetics. *Biopharm. Drug Dispos.* **27**, 387–406; (b) Lu, H. (2007). Stereoselectivity in drug metabolism. *Exp. Opin. Drug Metab. Toxicol.* **3**, 149–158; (c) Campo, V. L., Bernardes, L. S. C., Carvalho, I. (2009). Stereoselectivity in drug metabolism: molecular mechanisms and analytical methods. *Curr. Drug Metab.* **10**, 188–205.

3. Taylor, J. B., Triggle, D. J. (2007). *Comprehensive Medicinal Chemistry II*, Volume 5, ADME-Tox approaches. Oxford: Elsevier.

4. Burke, D., Henderson, D. J. (2002). Chirality: a blueprint for the future. *Br. J. Anaesth.*, **88**, 563–576.

5. Wade, D. N., Mearrick, P. T., Morris, J. L. (1973). Active transport of L-dopa in the intestine. *Nature* **242**, 463–465.

6. (a) Itoh, T., Ono, K., Koido, K. -I., Li, Y. -H., Yamada, H. (2001). Stereoselectivity of the folate transporter in rabbit small intestine: studies with amethopterin enantiomers. *Chirality* **13**, 164–169; (b) Hendel, J., Brodthagen, H. (1984). Entero-hepatic cycling of methotrexate estimated by use of the D-isomer as a reference marker. *Eur. J. Clin. Pharmacol.* **26**, 103–107.

7. Giacomini, K. M., Nelson, W. L., Pershe, R. A., Valdivieso, L., Turner-Tamiyasu, K., Blaschke, T. F. (1986). In vivo interaction of the enantiomers of disopyramide in human subjects. *J. Pharmacokinet. Biopharm.* **14**, 335–356.

8. Hamunen, K., Maunuksela, E. -L., Sarvela, J., Bullingham, R. E. S., Olkkola, K. T. (1999). Stereoselective pharmacokinetics of ketorolac in children, adolescents and adults. *Acta Anaesthesiol. Scand.* **43**, 1041–1046.

9. Vree, T. B., Beumer, E. M. C., Lagerwerf, A. J., Simon, M. A. M., Gielen, M. J. M. (1992). Clinical pharmacokinetics of R(+)- and S(−)-mepivacaine after high doses of racemic mepivacaine with epinephrine in the combined psoas compartment/sciatic nerve block. *Anesth. Analg.* **75**, 75–80.

10. Kristensen, K., Blemmer, T., Angelo, H. R., Christrup, L. L., Drenck, N. E., Rasmussen, S. N., Sjøgren, P. (1996). Stereoselective pharmacokinetics of methadone in chronic pain patients. *Ther. Drug Monit.* **18**, 221–227.

11. Chu, K.-M., Shieh, S.-M., Hu, O. Y.-P. (1995). Plasma and red blood cell pharmacokinetics of pimobendan enantiomers in healthy Chinese. *Eur. J. Clin. Pharmacol.* **47**, 537–542.

12. Fleishaker, J. C., Mucci, M., Pellizzoni, C., Poggesi, I. (1999). Absolute bioavailability of reboxetine enantiomers and effect of gender on pharmacokinetics. *Biopharm. Drug Dispos.* **20**, 53–57.

13. Krause, W., Kühne, G., Sauerbrey, N. (1990). Pharmacokinetics of (+)-rolipram and (-)-rolipram in healthy volunteers. *Eur. J. Clin. Pharmacol.* **38**, 71–75.

14. Boulton, D. W., Fawcett, J. P. (1996). Enantioselective disposition of salbutamol in man following oral and intravenous administration. *Br. J. Clin. Pharmacol.* **41**, 35–40.

15. Nguyen, K. T., Stephens, D. P., McLeish, M. J., Crankshaw, D. P., Morgan, D. J. (1996). Pharmacokinetics of thiopental and pentobarbital enantiomers after intravenous administration of racemic thiopental. *Anesth. Analg.* **83**, 552–558.

16. Thomson, A. H., Murdoch, G., Pottage, A., Kelman, A. W., Whiting, B., Hillis, W. S. (1986). The pharmacokinetics of R- and S-tocainide in patients with acute ventricular arrhythmias. *Br. J. Clin. Pharmacol.* **21**, 149–154.

17. Eichelbaum, M., Mikus, G., Vogelgesang, B. (1984). Pharmacokinetics of (+)-, (-)-and (+/-)-verapamil after intravenous administration. *Br. J. Clin. Pharmacol.* **17**, 453–458.

18. Burm, A. G. L., van der Meer, A. D., van Kleef, J. W., Zeijlmans, P. W. M., Groen, K. (1994). Pharmacokinetics of the enantiomers of bupivacaine following intravenous administration of the racemate. *Br. J. Clin. Pharmacol.* **38**, 125–129.

19. Stahl, E., Mutschler, E., Baumgartner, U., Spahn-Langguth, H. (1993). Carvedilol stereopharmacokinetics in rats: affinities to blood constituents and tissues. *Arch. Pharm. (Weinheim)* **326**, 529–533.

20. Brocks, D. R., Jamali, F., Russell, A. S. (1991). Stereoselective disposition of etodolac enantiomers in synovial fluid. *J. Clin. Pharmacol*. **31**, 741–746.

21. Brocks, D. R., Jamali, F. (1990). The pharmacokinetics of etodolac enantiomers in the rat: lack of pharmacokinetic interaction between enantiomers. *Drug Metab. Dispos*. **18**, 471–475.

22. Gross, A. S., Eser, C., Mikus, G., Eichelbaum, M. (1993). Enantioselective gallopamil protein binding. *Chirality* **5**, 414–418.

23. Brocks, D. R. (2002). Stereoselective halofantrine and desbutylhalofantrine disposition in the rat: Cardiac and plasma concentrations and plasma protein binding. *Biopharm. Drug Dispos*. **23**, 9–15.

24. Brocks, D. R., Skeith, K. J., Johnston, C., Emamibafrani, J., Davis, P., Russell, A. S., Jamali, F. (1994). Hematologic disposition of hydroxychloroquine enantiomers. *J. Clin. Pharmacol*. **34**, 1088–1097.

25. Vakily, M., Corrigan, B., Jamali, F. (1995). The problem of racemization in the stereospecific assay and pharmacokinetic evaluation of ketorolac in human and rats. *Pharm. Res*. **12**, 1652–1657.

26. Mehvar, R. (1991). Apparent stereoselectivity in propafenone uptake by human and rat erythrocytes. *Biopharm. Drug Dispos*. **12**, 299–310.

27. Zschiesche, M., Lemma, G. L., Klebingat, K. -J., Franke, G., Terhaag, B., Hoffmann, A., Gramatté, T., Kroemer, H. K., Siegmund, W. (2002). Stereoselective disposition of talinolol in man. *J. Pharm. Sci*. **91**, 303–311.

28. Sueyasu, M., Fujito, K., Shuto, H., Mizokoshi, T., Kataoka, Y., Oishi, R. (2000). Protein binding and the metabolism of thiamylal enantiomers in vitro. *Anesth. Analg*. **91**, 736–740.

29. Yacobi, A., Levy, G. (1977). Protein binding of warfarin enantiomers in serum of humans and rats. *J. Pharmacokinetics Biopharm*. **5**, 123–131.

30. Mehvar, R., Brocks, D. R. (2001). Stereospecific pharmacokinetics and pharmacodynamics of β-adrenergic blockers in humans. *J. Pharm. Pharm. Sci*. **4**, 185–200.

31. Brocks, D. R., Ramaswamy, M., MacInnes, A. I., Wasan, K. M. (2000). The stereoselective distribution of halofantrine enantiomers within human, dog and rat plasma lipoproteins. *Pharm. Res*. **17**, 427–431.

32. Brocks, D. R., Jamali, F. (1991). Enantioselective pharmacokinetics of etodolac in the rat: tissue distribution, tissue binding, and in vitro metabolism. *J. Pharm. Sci*. **80**, 1058–1061.

33. Caldwell, J. (1995). Stereochemical determinants of the nature and consequences of drug metabolism. *J. Chromatogr. A*. **694**, 39–48.

34. (a) Soucek, P., Gut, I. (1992). Cytochromes P-450 in rats: structures, functions, properties and relevant human forms. *Xenobiotica* **22**, 83–103; (b) Danielson, P. B. (2002). The cytochrome P450 superfamily: biochemistry, evolution and drug metabolism in humans. *Curr. Drug Metab*. **3**, 561–597.

35. Trager, W. F. (1989). Stereochemistry of cytochrome P-450 reactions. *Drug Metab. Rev*. **20**, 489–496.

36. Thijssen, H. H., Flinois, J. P., Beaune, P. H. (2000). Cytochrome P4502C9 is the principal catalyst of racemic acenocoumarol hydroxylation reactions in human liver microsomes. *Drug Metab. Dispos.* **28**, 1284–1290.

37. Niwa, T., Shiraga, T., Mitani, Y., Terakawa, M., Tokuma, Y., Kagayama, A. (2000). Stereoselective metabolism of cibenzoline, an antiarrhythmic drug, by human and rat liver microsomes: possible involvement of CYP2D and CYP3A. *Drug Metab. Dispos.* **28**, 1128–1134.

38. Echizen, H., Tanizaki, M., Tatsuno, J., Chiba, K., Berwick, T., Tani, M., Gonzalez, F. J., Ishizaki, T. (2000). Identification of CYP3A4 as the enzyme involved in the mono-N-dealkylation of disopyramide enantiomers in humans. *Drug Metab. Dispos.* **28**, 937–944.

39. Margolis, J. M., O'Donnell, J. P., Mankowski, D. C., Ekins, S., Obach, R. S. (2000). (R)-, (S)-, and racemic fluoxetine N-demethylation by human cytochrome P450 enzymes. *Drug Metab. Dispos.* **28**, 1187–1191.

40. Lu, H., Wang, J. J., Chan, K. K., Philip, P. A. (2006). Stereoselectivity in metabolism of ifosfamide by CYP3A4 and CYP2B6. *Xenobiotica* **36**, 367–385.

41. Äbelö, A., Andersson, T. B., Antonsson, M., Naudot, A. K., Skanberg, I., Weidolf, L. (2000). Stereoselective metabolism of omeprazole by human cytochrome P450 enzymes. *Drug Metab. Dispos.* **28**, 966–972.

42. Rettie, A. E., Korzekwa, K. R., Kunze, K. L., Lawrence, R. F., Eddy, A. C., Aoyama, T., Gelboin, H. V., Gonzalez, F. J., Trager, W. F. (1992). Hydroxylation of warfarin by human cDNA-expressed cytochrome P-450: a role for P-4502C9 in the etiology of (S)-warfarin–drug interactions. *Chem. Res. Toxicol.* **5**, 54–59.

43. Williams, P. A., Cosme, J., Ward, A., Angove, H. C., Vinkovic, D. M., Jhoti, H. (2003). Crystal structure of human cytochrome P4502C9 with bound warfarin. *Nature* **424**, 464–468.

44. Yasui-Furukori, N., Hidestrand, M., Spina, E., Facciola, G., Scordo, M. G., Tybring, G. (2001). Different enantioselective 9-hydroxylation of risperidone by the two human CYP2D6 and CYP3A4 enzymes. *Drug Metab. Dispos.* **29**, 1263–1268.

45. Bajpai, M., Roskos, L. K., Shen, D. D., Levy, R. H. (1996). Roles of cytochrome P4502C9 and cytochrome P4502C19 in the stereoselective metabolism of phenytoin to its major metabolite. *Drug Metab. Dispos.* **24**, 1401–1403.

46. Miyano, K., Fujii, Y., Toki, S. (1980). Stereoselective hydroxylation of hexobarbital enantiomers by rat liver microsomes. *Drug Metab. Dispos.* **8**, 104–110.

47. (a) Lewis, D. F. V., Ito, Y., Goldfarb, P. S. (2006). Cytochrome P450 structures and their substrate interactions. *Drug Devel. Res.* **66**, 19–24; (b) Lewis, D. F. V., Ito, Y., Goldfarb, P. S. (2006). Investigating human P450s involved in drug metabolism via homology with high-resolution P450 crystal structures of the CYP2C subfamily. *Curr. Drug Metab.* **7**, 589–598.

48. Domanski, T. L., Halpert, J. R. (2001). Analysis of mammalian cytochrome P450 structure and function by site-directed mutagenesis. *Curr. Drug Metab.* **2**, 117–137.

49. (a) Tukey, R. H., Strassburg, C. P. (2000). Human UDP-glucuronosyltransferases: metabolism, expression and disease. *Annu. Rev. Pharmacol. Toxicol.* **40**, 581–616; (b) King, C. D., Rios, G. R., Green, M. D., Tephly, T. R. (2000). UDP-glucuronosyltransferases. *Curr. Drug Metab.* **1**, 143–161.

50. Sten, T., Qvisen, S., Uutela, P., Luukkanen, L., Kostiainen, R., Finel, M. (2006). Prominent but reverse stereoselectivity in propranolol glucuronidation by human UDP-glucuronosyltransferases 1A9 and 1A10. *Drug Metab. Dispos.* **34**, 1488–1494.

51. (a) Blanchard, R. L., Freimuth, R. R., Buck, J., Weinshilboum, R. M., Coughtrie, M. W. H. (2004). A proposed nomenclature system for the cytosolic sulfotransferase (SULT) superfamily. *Pharmacogenetics* **14**, 199–211; (b) Wang, J., Falany, J. L., Falany, C. N. (1998). Expression and characterization of a novel thyroid hormone-sulfating form of cytosolic sulfotransferase from human liver. *Mol. Pharmacol*. **53**, 274–282; (c) Her, C., Kaur, G. P., Athwal, R. S., Weinshilboum, R. M. (1997). Human sulfotransferase SULT1C1: cDNA cloning, tissue-specific expression, and chromosomal localization. *Genomics* **41**, 467–470.

52. Walle, T., Walle, U. K., Thornburg, K. R., Schey, K. L. (1993). Stereoselective sulfation of albuterol in humans: biosynthesis of the sulfate conjugate by HEP-G2 cells. *Drug Metab. Dispos*. **21**, 76–80.

53. Hartman, A. P., Wilson, A. A., Wilson, H. M. (1998). Enantioselective sulfation of β_2-receptor agonists by the human intestine and the recombinant M-form phenolsulfotransferase. *Chirality* **10**, 800–803.

54. Wilson, A. A., Wang, J., Koch, P., Walle, T. (1997). Stereoselective sulphate conjugation of fenoterol by human phenolsulphotransferases. *Xenobiotica* **27**, 1147–1154.

55. Davies, N. M. (1998). Clinical pharmacokinetics of ibuprofen: the first 30 years. *Clin. Pharmacokinet*. **34**, 101–154.

56. Agranat, I., Caner, H., Caldwell, J. (2002). Putting chirality to work: the strategy of chiral switches. *Nat. Rev.: Drug Discov*. **1**, 753–768.

57. Landoni, M. F., Soraci, A. (2001). Pharmacology of chiral compounds: 2-arylpropionic acid derivatives. *Curr. Drug Metab*. **2**, 37–51.

58. Tanner, M. E. (2002). Understanding nature's strategies for enzyme-catalyzed racemization and epimerization. *Acc. Chem. Res*. **35**, 237–246.

59. (a) Jamali, F., Berry, B. W., Tehrani, M. R., Russell, A. S. (1988). Stereoselective pharmacokinetics of flurbiprofen in humans and rats. *J. Pharm. Sci*. **77**, 666–669; (b) Berry, B. W., Jamali, F. (1989). Enantiomeric interaction of flurbiprofen in the rat. *J. Pharm. Sci*. **78**, 632–634; (c) Soraci, A. L., Tapia, O., Garcia, J. (2005). Pharmacokinetics and synovial fluid concentrations of flurbiprofen enantiomers in horses: chiral inversion. *J. Vet. Pharmacol. Ther*. **28**, 65–70.

60. Foster, R. T., Jamali, F. (1988). Stereoselective pharmacokinetics of ketoprofen in the rat: influence of route of administration. *Drug Metab. Dispos*. **16**, 623–626.

61. Jamali, F., Russell, A. S., Foster, R. T., Lemko, C. (1990). Ketoprofen pharmacokinetics in humans: evidence of enantiomeric inversion and lack of interaction. *J. Pharm. Sci*. **79**, 460–461.

62. Jamali, F., Lovlin, R., Aberg, G. (1997). Bi-directional chiral inversion of ketoprofen in CD-1 mice. *Chirality* **9**, 29–31.

63. (a) Tanaka, Y., Shimomura, Y., Hirota, T., Nozaki, A., Ebata, M., Takasaki, W., Shigehara, E., Hayashi, R., Caldwell, J. (1992). Formation of glycine conjugate and (−)-(R)-enantiomer from (+)-(S)-2-phenylpropionic acid suggesting the formation of the CoA thioester intermediate of (+)-(S)-enantiomer in dogs. *Chirality* **4**, 342–348; (b) King, J. N., Mauron, C., LeGoff, C., Hauffe, S. (1994). Bidirectional chiral inversion of the enantiomers of the nonsteroidal antiinflammatory drug oxindanac in dogs. *Chirality* **6**, 460–466.

64. Knoche, B., Blaschke, G. (1994). Investigations on the in vitro racemization of thalidomide by high-performance liquid chromatography. *J. Chromatogr. A* **666**, 235–240.

65. (a) Eriksson, T., Bjorkman, S., Roth, B., Fyge, A., Hoglund, P. (1995). Stereospecific determination, chiral inversion in vitro and pharmacokinetics in humans of the enantiomers of thalidomide. *Chirality* **7**, 44–52; (b) Eriksson, T., Bjorkman, S., Roth, B., Fyge, A., Hoglund, P. (1998). Enantiomers of thalidomide: blood distribution and the influence of serum albumin on chiral inversion and hydrolysis. *Chirality* **10**, 223–228.

66. (a) Moaddel, R., Patel, S., Jozwiak, K., Yamaguchi, R., Ho, P. C., Wainer, I. W. (2005). Enantioselective binding to the human organic cation transporter-1 (hOCT1) determined using an immobilized hOCT1 liquid chromatographic stationary phase. *Chirality* **17**, 501–506; (b) Moaddel, R., Yamaguchi, R., Ho, P. C., Patel, S., Hsu, C. P., Subrahmanyam, V., Wainer, I. W. (2005). Development and characterization of an immobilized human organic cation transporter based liquid chromatographic stationary phase. *J. Chromatogr. B* **818**, 263–268; (c) Ott, R. J., Giacomini, K. M. (1993). Stereoselective interactions of organic cations with the organic cation transporter in OK cells. *Pharm. Res.* **10**, 1169–1173; (d) Zhang, L., Schaner, M. E., Giacomini, K. M. (1998). Functional characterization of an organic cation transporter (hOCT1) in a transiently transfected human cell line (HeLa). *J. Pharmacol. Exp. Ther.* **286**, 354–361.

67. Dahlpuustinen, M. L., Perry, T. L., Dumont, E., Bahr, C., Nordin, C., Bertilsson, L. (1989). Stereoselective disposition of racemic E-10-hydroxynortriptyline in human beings. *Clin. Pharmacol. Ther.* **45**, 650–656.

68. Laethem, M. E., Lefebvre, R. A., Belpaire, F. M., Vanhoe, H. L., Bogaert, M. G. (1995). Stereoselective pharmacokinetics of oxprenolol and its glucuronides in humans. *Clin. Pharmacol. Ther.* **57**, 419–424.

69. Fujimaki, M., Shintani, S., Hakusui, H. (1991). Stereoselective metabolism of carvedilol in the rat: use of enantiomerically radiolabeled pseudoracemates. *Drug Metab. Dispos.* **19**, 749–753.

70. Horikiri, Y., Suzuki, T., Mizobe, M. (1997). Stereoselective pharmacokinetics of bisoprolol after intravenous and oral administration in beagle dogs. *J. Pharm. Sci.* **86**, 560–564.

71. Gibaldi, M. (1993). Stereoselective and isozyme-selective drug interactions. *Chirality* **5**, 407–413.

72. Kroemer, H. K., Fromm, M. F., Eichelbaum, M. (1996). Stereoselectivity in drug metabolism and action: Effects of enzyme inhibition and induction. *Ther. Drug Monit.* **18**, 388–392.

73. Choonara, I. A., Cholerton, S., Haynes, B. P., Breckenridge, A. M., Park, B. K. (1986). Stereoselective interaction between the R-enantiomer of warfarin and cimetidine. *Br. J. Clin. Pharmacol.* **21**, 271–277.

74. Toon, S., Low, L. K., Gibaldi, M., Trager, W. F., O'Reilly, R. A., Motley, C. H., Goulart, D. A. (1986). The warfarin-sulfinpyrazone interaction: stereochemical considerations. *Clin. Pharmacol. Ther.* **39**, 15–24.

75. (a) Takahashi, H., Sato, T., Shimoyama, Y., Shioda, N., Shimizu, T., Kubo, S., Tamura, N., Tainaka, H., Yasumori, T., Echizen, H., Echizen, H. (1999). Potentiation of anticoagulant effect of warfarin caused by enantioselective metabolic inhibition by the uricosuric agent benzbromarone. *Clin. Pharmacol. Ther.* **66**, 569–581; (b) Takahashi, H., Kashima, T., Kimura, S. (1999). Pharmacokinetic interaction between warfarin and a uricosuric agent, bucolome: Application of in vitro approaches to predicting in vivo reduction of (*S*)-warfarin clearance. *Drug Metab. Dispos.* **27**,

1179–1186; (c) Matsumoto, K., Ishida, S., Ueno, K., Hashimoto, H., Takada, M., Tanaka, K., Kamakura, S., Miyatake, K., Shibakawa, M. (2001). The stereoselective effects of bucolome on the pharmacokinetics and pharmacodynamics of racemic warfarin. *J. Clin. Pharmacol.* **41**, 459–464.

76. Smith, D. A., Chandler, M. H. H., Shedlofsky, S. I., Wedlund, P. J., Blouin, R. A. (1991). Age-dependent stereoselective increase in the oral clearance of hexobarbitone isomers caused by rifampicin. *Br. J. Clin. Pharmacol.* **32**, 735–739.

77. Labbe, L., Robitaille, N. M., Lefez, C., Potvin, D., Gilbert, M., O'Hara, G., Turgeon, J. (2004). Effects of ciprofloxacin on the stereoselective disposition of mexiletine in man. *Ther. Drug Monit.* **26**, 492–498.

78. Lu, H., Wang, J. J., Chan, K. K. (2006). Enantiomer-enantiomer interaction of ifosfamide in the rat. *Xenobiotica* **36**, 535–549.

79. Kroemer, H. K., Fischer, C., Meese, C. O., Eichelbaum, M. (1991). Enantiomer/enantiomer interaction of (*S*)- and (*R*)-propafenone for cytochrome P450IID6-catalyzed 5-hydroxylation: in vitro evaluation of the mechanism. *Mol. Pharmacol.* **40**, 135–142.

80. (a) Kroemer, H. K., Fromm, M. F., Buhl, K., Terefe, H., Blaschke, G., Eichelbaum, M. (1994). An enantiomer-enantiomer interaction of (*S*)- and (*R*-propafenone modifies the effect of racemic drug therapy. *Circulation* **89**, 2396–2400; (b) Li, G., Gong, P. L., Qiu, J., Zeng, F. D., Klotz, U. (1998). Stereoselective steady state disposition and action of propafenone in Chinese subjects. *Br. J. Clin. Pharmacol.* **46**, 441–445.

81. Nguyen, L. A., He, H., Pham-Huy, C. (2006). Chiral drugs: an overview. *Int. J. Biomed. Sci.* **2**, 85.

CHAPTER 10

TOXICOLOGY OF CHIRAL DRUGS

GUANG YANG
GlaxoSmithKline, R&D China, Shanghai, China

HAI-ZHI BU
3D BioOptima Co. Ltd, Suzhou, China

10.1 STEREOCHEMISTRY AND STEREOPHARMACOLOGY

The presence of one chiral center in a molecule gives rise to a pair of enantiomers as an object and its mirror image (Fig. 10.1). An equimolar mixture of both enantiomers is known as a racemate. Among all the synthetic, semisynthetic, and natural small-molecule drugs currently in use, more than half contain one or more chiral centers or centers of unsaturation [1–4].

Enantiomeric drugs have two structurally similar forms that can behave very differently in biological systems because of their different shapes in

Chiral Drugs: Chemistry and Biological Action, First Edition. Edited by Guo-Qiang Lin, Qi-Dong You and Jie-Fei Cheng.
© 2011 John Wiley & Sons, Inc. Published 2011 by John Wiley & Sons, Inc.

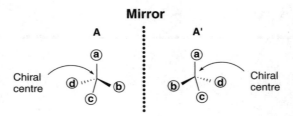

FIGURE 10.1   A chiral molecule (A) and its mirror image (A′) in a tetrahedral configuration. Letters in circles (a, b, c, and d) represent four different groups attached to a chiral center (e.g., a carbon atom).

three-dimensional space as well as the homochiral nature of biological systems. Humans and other living organisms are primarily composed of chiral molecules such as L-amino acids, D-carbohydrates, etc. The macromolecules, for example, proteins and DNAs, adopt unique secondary structures with right-handed α-helices and β-DNA configurations. The tertiary structures of proteins, receptors, and enzymes create stereospecific three-dimensional environments (domains or molecular pockets) for binding, catalysis, and stabilization. Furthermore, during normal growth and development of the body, special rotation patterns and left-right "mirror" symmetry are evolved and established for cardiac and gastrointestinal systems. Thus chiral drugs are expected to be stereoselective in all stages of disposition processes, such as absorption, metabolism, distribution, protein/receptor binding and transportation, enzyme catalysis, and so on, leading to complex pharmacological and toxicological behaviors. One isomer could display the desired pharmacokinetic and pharmacodynamic properties and therapeutic index, while the other may be less optimal or the antipode.

The enantiomer of a chiral drug possessing therapeutic activity of interest is termed a eutomer (bioactive enantiomer), while that with lower affinity or less activity is a distomer. The "eudismic ratio" measures the pharmacological potency ratio of the eutomer to the distomer, or the stereoselectivity of a chiral drug [5]. It is noted that the eudismic designation refers to one biological action only, and for a dual-action drug the eutomer for one activity may be the distomer for the other.

In this chapter, three categories of chiral drugs in terms of their enantiomeric toxicological profiles are summarized and compared.

10.2   TOXICOLOGY OF CHIRAL DRUGS

Mechanisms of drug-induced toxicities are usually complicated, as multiple contributing factors may be associated. There are generally five classes of drug toxicities, including on-target mechanism-based toxicities, hypersensitivity and immunological reactions, off-target pharmacology, biological activation to toxic metabolites, and idiosyncratic toxicities [6]. The underlying mechanisms can be more or less understood by two general concepts: "metabolic activation" [7,8] and "covalent binding" (to tissue proteins) [9,10]. Drugs or chemical substances can be "activated" or further converted to reactive metabolites via metabolic transformation in a living system. Pharmacologically active metabolites usually show pharmacological effects similar to those of their parent drugs in the target tissues [11], but they may exhibit unrelated and undesired activities at other sites, leading to clinical toxicities. Covalent modification of the cellular or tissue proteins in normal homeostasis by bioactive drug molecules and their reactive metabolites can provide a key measure of toxic effects. Other important toxicity indexes of pharmaceutical molecules include receptor-mediated signaling, DNA/protein damage repair, cell proliferation, and immune responses. In general, drug toxicities occur within one of three defined contexts: drug overdoses, drug-drug interactions, and adverse effects in therapeutic doses.

Stereospecific pharmacological and toxicological effects have been demonstrated in various therapeutic agents including antibiotic [12], cardiovascular [13,14], chemotherapeutic [15], psychotropic [16], pulmonary [17], and rheumatic [18,19] drugs. Chirality in drugs, as discussed above, may have significant impacts on those processes, depending on the nature of the interaction with chiral macromolecules in the body, for example, drug efflux system, plasma protein binding, drug metabolism, and receptor (target) interaction. For example, the expression of P-glycoprotein (PgP) of the membrane efflux system appears to be enantiospecifically regulated up and down by R- and S-cetirizine, respectively [20]. Only the L-enantiomer of mefloquine blocks the uptake of cyclosporine and vinblastine via stereoselective inhibition of the membrane PgP [21,22]. Human albumin preferentially binds to enantiomeric S-ketoprofen [23], R-warfarin [24,25], and S-(+)-chloroquine [26,27]. Plasma $\alpha 1$-acid glycoprotein shows stereospecific affinity for R-(−)-disopyramide, S-(−)-verapamil, and R-(+)-propranolol [28]. In metabolism, stereospecificity, as a feature of metabolic enzymes and pathways, is responsible for the majority of the differences observed in enantioselective drug disposition [29]. The combined stereospecific effects of each process can result in marked pharmacokinetic and pharmacodynamic differences between enantiomers of chiral drugs, likely leading to stereoselective toxicities.

The unique three-dimensional structures of DNAs, when interacting with certain chiral molecules, may cause stereospecific carcinogenicity and mutagenicity, as illustrated in the cytochrome P450-mediated epoxide transformations [30–33]. Aflatoxin B1 (AFB1) requires bioactivation to AFB1-8,9-epoxide for carcinogenicity, and glutathione-S-transferase (GST)-catalyzed conjugation of activated AFB1 with glutathione (GSH) is a critical determinant of susceptibility to the mycotoxin (Scheme 10.1) [30]. With the use of a rabbit liver microsomal system to generate both AFB1 exo- and endo-epoxide isomers, it was found that the rabbit liver cytosolic GSTs catalyzed formation of both AFB1 exo- and endo-epoxide-GSH conjugates, whereas pulmonary cytosolic GSTs catalyzed only the formation of the exo-stereoisomer at detectable levels. Despite a preference for conjugating the more mutagenic AFB1 exo-epoxide isomer, the relatively low

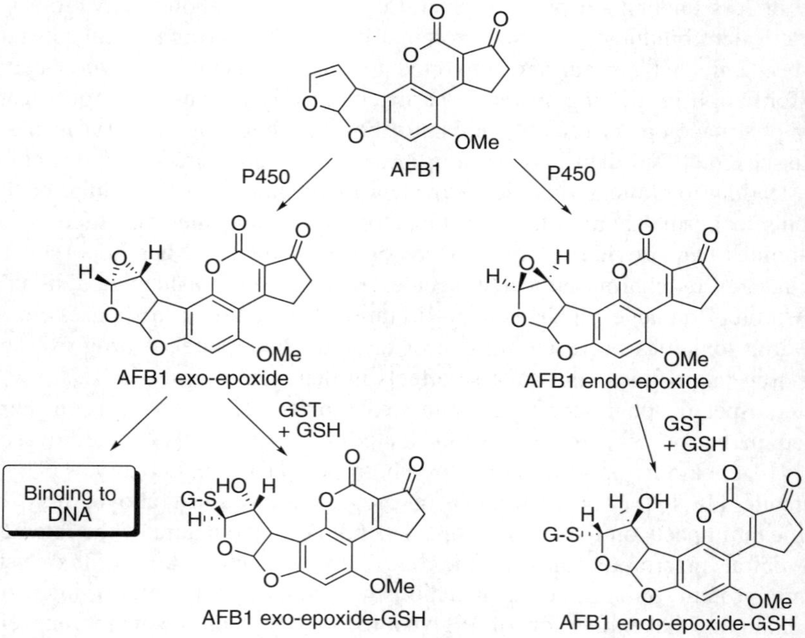

SCHEME 10.1 Glutathione S-transferase (GST)-catalyzed conjugation of AFB1 exo- and endoepoxide with glutathione (GSH).

capacity for GST-catalyzed detoxification of bioactivated AFB1 in lung may be an important factor in the susceptibility of the lung to AFB1 toxicity [30]. The two enantiomers of styrene-7,8-oxide and various thioether metabolites of racemic styrene 7,8-oxide were tested for their direct mutagenicity in *Salmonella typhimurium* TA100 [31]. The mutagenicity data suggest that the R-enantiomer is more mutagenic than the S-enantiomer, with the racemic mixture intermediate between the two. The thioether metabolites were not mutagenic. The difference in the mutagenicities of enantiomers probably resulted from a stereoselective process in the *Salmonella* tester strain. At the present time it is not clear whether the rate-limiting reaction is the interaction of the enantiomers with DNAs or some other cellular components [31]. Benzo[α]pyrene (B[α]P), a known environmental pollutant and tobacco smoke carcinogen, is metabolically activated to highly tumorigenic B[α]P diol epoxide derivatives (Fig. 10.2) that predominantly form N^2-guanine adducts in cellular DNA. Although nucleotide excision repair (NER) is an important cellular defense mechanism, the molecular basis of recognition of these bulky lesions is poorly understood. To investigate the effects of DNA adduct structure on NER, three stereoisomeric and conformationally different B[α]P-N^2-dG lesions were site-specifically incorporated into identical 135-mer duplexes and their response to purified NER factors was investigated [33]. By permanganate footprinting assay, the NER lesion recognition factor XPC/HR23B exhibits, in each case, remarkably different patterns of helix opening that is also

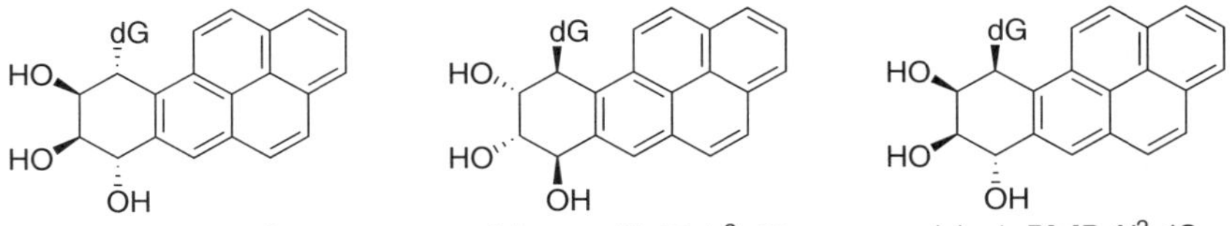

(−)-*trans*-B[a]P-N^2-dG   (+)-*trans*-B[a]P-N^2-dG   (+)-*cis*-B[a]P-N^2-dG

FIGURE 10.2 Structures of stereoisomeric (+)-*cis*-, (+)-*trans*-, and (−)-*trans*-B[α]P-N^2-dG adducts.

markedly distinct in the case of an intrastrand cross-linked cisplatin adduct. The different extents of helix distortions, as well as differences in the overall binding of XPC/HR23B to double-stranded DNA containing either of the three stereoisomeric B[α]P-N^2-dG lesions, are correlated with dual incisions catalyzed by a reconstituted incision system of six purified NER factors, and by the full NER apparatus in cell-free nuclear extracts [33].

In the same context of drug toxicity, the toxicological properties in a pair of enantiomers can be similar or entirely opposite. They can reside in the eutomer, the distomer, or the racemate mixture of a chiral drug. Chiral inversion, a biotransformation process converting one stereoisomer to its antipode with no other change in the molecule (Scheme 10.2), also occurs frequently in vivo [34]. This adds a new dimension of uncertainty and complication in drug toxicity

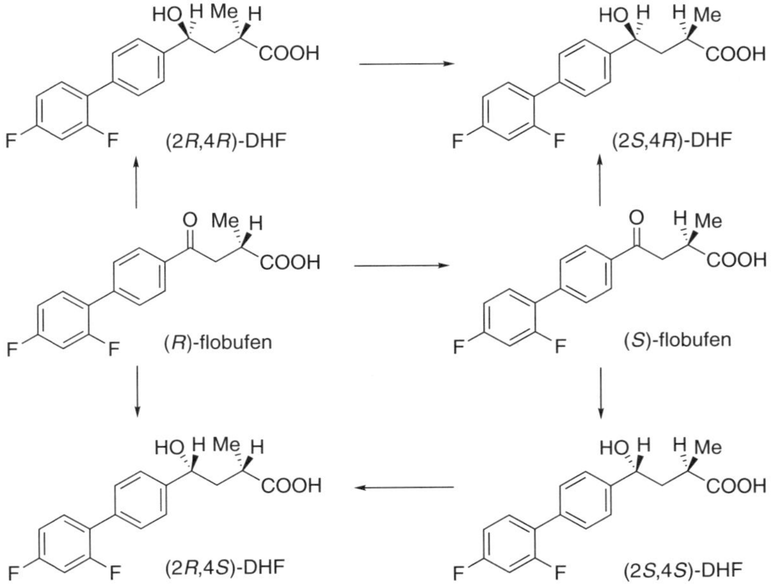

SCHEME 10.2 Chiral inversion of flobufen enantiomers and 4-dihydroflobufen diastereoisomers (DHF) observed in primary culture of hepatocytes in humans. Respective arrows indicate the main direction of inversion.

determination. Thalidomide, a medication for anxiety and sleeping disorder, represents a well-known, classical example of complexity and challenge in terms of stereoselective drug toxicity. The tragic consequences of thalidomide therapy in the early 1960s were originally believed to be the result of the teratogenic potential of the S-(−)-distomer [35]. (A teratogenic fetus is a fetus with deficient, redundant, misplaced, or grossly misshapen parts.) Over 2,000 cases of serious birth defects in children born of women who took thalidomide during pregnancy were reported. It is apparent that the case of S-thalidomide is not the only reason for using the appropriate animal models for the evaluation of teratogenic effects [36–38] and investigation of the chiral inversion properties of thalidomide [39]. The teratogenic potencies of the enantiomers of 2-(2,6-dioxopiperidine-3-yl)-phthalimidine (EM-12), a teratogenic thalidomide analog, were investigated in *Callithrix jacchus*, a primate very sensitive to the teratogenic action of this thalidomide analog [37]. The results indicate that the S-(−)-form of EM-12 is clearly more teratogenic than the R-(+)-form. However, the interpretation of the findings becomes difficult, since both enantiomers racemize in vivo with appreciable rates. Therefore, it cannot be concluded as yet that the R-(+)-form lacks all teratogenic potential [37]. In human blood, mean rate constants of chiral inversion of (+)-(R)-thalidomide and (−)-(S)-thalidomide were 0.30 and 0.31 h^{-1}, respectively, while rate constants of degradation were 0.17 and 0.18 h^{-1}, respectively, at 37°C [39]. There was rapid interconversion in vivo in humans, with the (+)-(R)-enantiomer predominating at equilibrium. The pharmacokinetics of (+)-(R)- and (−)-(S)-thalidomide could be characterized by means of two one-compartment models connected by rate constants for chiral inversion (Fig. 10.3). Mean rate constants for in vivo inversion were 0.17 h^{-1} (R to S) and 0.12 h^{-1} (S to R) and for elimination 0.079 h^{-1} (R) and 0.24 h^{-1} (S), that is, a considerably faster rate of elimination of the (−)-(S)-enantiomer. Putative differences in therapeutic or adverse effects between (+)-(R)- and (−)-(S)-thalidomide would be abolished to a large extent by rapid interconversion in vivo [39].

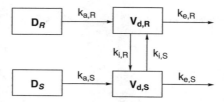

FIGURE 10.3 Model used to characterize in vivo pharmacokinetics of (+)-(R)- and (−)-(S)-thalidomide after oral administration of either enantiomer or racemate. Abbreviations: D_R and D_S, presystemic depots of (+)-(R)- and/or (−)-(S)-thalidomide; $k_{a,R}$ and $k_{a,S}$, rate constants of absorption; $k_{i,R}$ and $k_{i,S}$, rate constants of chiral inversion; $k_{e,R}$ and $k_{e,S}$, rate constants of elimination; $V_{d,R}$ and $V_{d,S}$, compartmental distribution volumes of the respective enantiomers.

In the following sections, each category of stereospecific toxicity of chiral drugs is discussed in more detail.

10.3 TOXICITY OF DISTOMERS

Interactions of both enantiomers of a chiral drug may differ at the active sites through which pharmacological action is mediated. Actions and levels of activity of the stereoisomers in vivo may also differ. All the pharmacological activity may reside in a single enantiomer, whereas several possibilities exist for the other enantiomer. It may be inactive, have a qualitatively different effect or an antagonistic effect, or produce greater toxicity. Two isomers may have nearly identical qualitative pharmacological activity, qualitatively similar pharmacological activity but quantitatively different potency, or qualitatively different pharmacological activity. It should be noted that many chiral drugs frequently have their toxicities residing mostly in the pharmacologically inactive enantiomers (i.e., distomers). For such chiral drugs, pure enantiomeric forms (i.e., eutomers) rather than their racemic forms usually show clear advantages in terms of therapeutic index. For example, L-dopa is an effective medicine in the treatment of Parkinson disease, but L-dopa is an antipode that shows a number of adverse effects, including nausea, vomiting, anorexia, involuntary movements, and granulocytopenia [40,41]. Racemic tetramisole, a former nematocide, has multiple toxic effects such as vertigo, headache, vomiting, abdominal pain, etc., which appear to be mainly associated with its d-isomer. In fact, levamisole, the active l-form of tetramisole, is used as an anthelmintic in the treatment of many nematode infestations, particularly in veterinary applications. It is also an immunomodulator as an adjunct with fluorouracil to make it work better against the colon cancer after surgical resection of the primary tumor.

Ofloxacin, a quinolone antibiotic drug targeting DNA gyrase (bacterial topoisomerase II) [42], exhibits stereoselective antimicrobial activity against both gram-positive and gram-negative bacteria with eudismic ratios of 8–128 for the S-(−)-enantiomer [43,44]. The eutomer S-(−)-ofloxacin shows much greater selectivity (6.7-fold over the R-(+)-isomer) against the mammalian topoisomerase II [42,45]. In addition, ofloxacin also demonstrates stereoselective distribution, metabolism, and renal excretion, which appear to be species dependent as well [46–48]. Ofloxacin can cause neurotoxicity that is believed to reside in its dextrorotary form because of these stereospecific effects [49]. For this reason, levofloxacin, the active S-enantiomer of ofloxacin, has been developed and marketed as a broad-spectrum antibacterial agent.

The S-(+)-enantiomer of ketamine, an anesthetic agent, shows much greater analgesic and anesthetic potency (4 times more potent) than its R-antipode in both animals and humans [50,51], whereas the postanesthetic reactions, for example, hallucinations, agitation, and restlessness, are almost exclusively associated with the distomeric R-(−)-form [51,52]. Penicillamine, a metabolite of penicillin although it has no antibiotic properties, is clinically used as a copper chelator in the treatment of copper poisoning and Wilson disease. The *l*- and *d*-forms of penicillamine have a eudismic ratio of one, but with differential toxicities. The *l*-enantiomer, not the *d*-isomer, has been known to possess serious adverse effects since 1948, including growth inhibition [53], seizures [53], impaired L-amino acid intestinal absorption [54], optic neuritis [18,55], nephrotic syndrome [56], and pyridoxine antagonism [57]. In addition, a greater mutagenic potency of the *l*-distomer, compared to the *d*-eutomer, was also reported [58]. The *Salmonella typhimurium* strain TA92 was found to be more sensitive than TA100 to the mutagenic action of L-cysteine. This strain allowed the unambiguous realization of a (weak) mutagenic effect of L-cysteine even in the absence of mammalian enzyme preparations. D-Cysteine did not show mutagenicity under any experimental conditions. However, it was strongly bacteriotoxic. On the other hand, both enantiomers of penicillamine exerted clear mutagenic effects. Vigabatrin is an antiepileptic drug that inhibits the catabolism of γ-aminobutyric acid (GABA) by irreversibly inhibiting GABA transaminase by its *S*-(+)-eutomer [59]. It is an analog of GABA, but it is not a receptor agonist. The *R*-(−)-distomer, on the other hand, does not show any inhibitory activity of GABA. Moreover the *R*-(−)-antipode appears to be highly toxic [60].

10.4 TOXICITY OF EUTOMERS AND RACEMATES

Pharmacological and toxicological effects can coexist in the same enantiomer or both. Citalopram, an antidepressant, is a racemic drug and a selective serotonin reuptake inhibitor. Based on the inhibition of 5-HT uptake in vitro and the potentiation of 1-5-HTP in vivo, the pharmacological activities of citalopram and its metabolites (*N*-desmethylcitalopram and *N*-didesmethylcitalopram) were found to reside in the *S*-(+)-enantiomers [61]. In the 5-HT uptake test, eudismic ratios were estimated to be 167 and 6.6 for the enantiomers of citalopram and *N*-demethylcitalopram, respectively. The pharmacological profiles of the eutomers of citalopram and *N*-demethylcitalopram very much resemble those of their respective racemates [61]. Escitalopram, the active *S*-(+)-eutomer of citalopram, at a dose of 10 mg/day, is shown to be at least as effective as the racemic citalopram at a dose of 40 mg/day [62]. Meanwhile, escitalopram demonstrates a faster onset of action and an improved tolerability profile. However, there is a wide range of inter- and intrapatient variabilities in escitalopram concentrations in serum [63]. As a serotonergic agent, escitalopram occasionally causes a variety of adverse reactions, such as hyponatremia, serotonin syndrome, and restless leg syndrome [64−70]. In addition, both the *S*-enantiomer and the racemic citalopram show a similar QT prolongation and rare cases of torsade de pointes (Forest Pharmaceuticals, Inc., 2009). In conclusion, the clinical superiority of escitalopram remains to be demonstrated.

Salbutamol (albuterol) is a short-acting β_2-adrenergic receptor agonist used for the relief of bronchospasm in conditions such as asthma and chronic obstructive pulmonary disease. Racemic albuterol is composed of an equimolar mixture of stereoisomers. For asthma therapy, (R)-albuterol is the eutomer and (S)-albuterol is the distomer. By interacting with β_2-adrenoceptors, (R)-albuterol has bronchodilating, bronchoprotective, and antiedematous properties and inhibits activation of mast cells and eosinophils. In contrast, (S)-albuterol does not activate β_2-adrenoceptors and does not modify activation of β_2-adrenoceptors by (R)-albuterol, so that for many years the distomer was presumed to be biologically inert. Recently, it has been established that regular and excessive use of racemic albuterol induces paradoxical reactions in some subjects with asthma. Because such effects cannot be accounted for by activation of β_2-adrenoceptors, the pharmacological profile of (S)-albuterol has been more carefully defined. (S)-albuterol has distinctive pharmacological properties that are unrelated to activation of β_2-adrenoceptors. Thus (S)-albuterol intensifies bronchoconstrictor responses of sensitized guinea pigs and induces hypersensitivity of asthmatic airways; it also promotes the activation of human eosinophils in vitro [71]. These actions of (S)-albuterol may explain why racemic albuterol can intensify allergic bronchospasm and promote eosinophil activation in asthmatic airways. The capacity of (S)-albuterol to elevate intracellular Ca^{2+} would account for the paradox because this action will oppose, or even nullify, the consequences of adenylyl cyclase activation by (R)-albuterol. Because (S)-albuterol is metabolized more slowly than (R)-albuterol and is retained preferentially within the airways, paradoxical effects become more prominent during regular and excessive use of racemic albuterol. Because (S)-albuterol has detrimental effects in asthmatic airways, levalbuterol [homochiral (R)-albuterol] should have advantages over racemic albuterol in therapy for asthma. Both enantiomers of albuterol may cause hyperkalemia and have been implicated in eosinophil activation and proinflammatory properties [17]. Even though inhalation of (R)-albuterol produces significantly greater bronchodilatation than the equivalent dose of the racemate in humans [72], the clinical superiority of the eutomer is under further debating [73,74].

Cyclophosphamide (also known as cytophosphane) is a nitrogen mustard alkylating agent from the oxazophorines group [75,76]. The main effect of cyclophosphamide is due to its metabolite phosphoramide mustard. This metabolite is only formed in cells that have low levels of aldehyde dehydrogenase (ALDH). Phosphoramide mustard forms DNA cross-links between (interstrand cross-linkages) and within (intrastrand cross-linkages) DNA strands at guanine N-7 positions. This leads to cell death. The main use of cyclophosphamide is together with other chemotherapy agents in the treatment of lymphomas, some forms of leukemia, and some solid tumors [75,76]. It is a chemotherapy drug that works by slowing or stopping cell growth. Cyclophosphamide also decreases the immune system's response to various diseases and conditions. Therefore, it has been used in various nonneoplastic autoimmune diseases where disease-modifying antirheumatic drugs have become ineffective. Cyclophosphamide is converted by mixed-function oxidase enzymes in the liver to active metabolites. The main active metabolite is 4-hydroxycyclophosphamide, which exists in equilibrium with its tautomer, aldophosphamide. Most of the aldophosphamide is oxidized by ALDH to make carboxyphosphamide. A small proportion of aldophosphamide is converted into phosphoramide mustard and acrolein. Acrolein is toxic to the bladder epithelium and can lead to hemorrhagic cystitis. This can be prevented through the use of aggressive hydration and/or mesna (sodium 2-mercaptoethane sulfonate), which is a sulfhydryl donor and binds acrolein. Apparently, the two enantiomers of the drug exert similar toxicities and side effects [75,76]. Many people taking cyclophosphamide do not have serious side effects. Side effects include chemotherapy-induced nausea and vomiting, bone marrow suppression, stomachache, diarrhea, darkening of the skin/nails, alopecia (hair loss) or thinning of hair, changes in color and texture of the hair, and lethargy. Hemorrhagic cystitis is a frequent complication, but this is prevented by adequate fluid intake and mesna. In addition, cyclophosphamide is itself carcinogenic, potentially causing transitional cell carcinoma of the bladder as a long-term complication. It can lower the body's ability to fight an infection. It can cause temporary or (rarely) permanent sterility.

10.5   SIGNIFICANCE OF CHIRAL INVERSION

The enantiomeric composition of a chiral drug and/or its chiral metabolite(s) can be changed in biological systems by enzyme(s) mediating the process of chiral inversion, in which one stereoisomer is converted into its antipode with no other changes in the molecule [77]. This inversion process is usually both species- and substrate dependent [34]. The significance of chiral inversion to drug activity and toxicity has been summarized and discussed in a series of review articles [34,78], as shown in Table 10.1. Two general mechanisms of chiral inversion have been observed, including 1) formation of a covalent intermediate followed by a stepwise enzyme-catalyzed transformation and 2) interaction of two opposing metabolic processes, such as oxidation and reduction [34]. Both mechanisms operate unidirectionally or bidirectionally, which alters the enantiomeric ratio R/S in a chiral drug and thus the toxicological profiles.

The 2-arylpropionates (APAs) drugs, a subset of nonsteroidal antiinflammatory drugs including ibuprofen, ketoprofen, fenprofen, benoxaprophen, etc., are the best-studied chiral molecules in terms of their biochemical inversion in biological systems. The therapeutic efficacy of the APA profens is achieved via the stereoselective inhibition of cyclooxygenases (COXs), key enzymes in prostaglandin biosynthesis, by the active S-(+)-eutomers. Because of their high blood-brain barrier (BBB) penetration (through diffusion), these APA drugs can also exert antipyretic and analgesic functions in the brain [79]. The unidirectional chiral inversion of R-(−)-distomers to S-(+)-eutomers is a key pathway of the metabolism of the APA drugs in vivo [80–83]. Only the inactive R-(−)-antipodes can undergo chiral inversion in the body by hepatic enzymes and not vice versa [84–86]. The hepatic chiral inversion process is also species- and substrate dependent. For instance, R-(−)-ibuprofen is inverted in all species tested, whereas R-(−)-ketoprofen is inverted in rats but not in humans [57]. The

TABLE 10.1   Harmful effects of racemization of optically active drugs

Optically Active Drugs	Therapeutic Group	Species	Harmful Effects
(−)-Thalidomide	Immunomodulatory & antiangiogenic	Human	Malformation of embryos in pregnant woman
(−)-Ibuprofen	Antianalgesic	Human	Inactive
D-DOPA	Parkinson disease	Human	Inactive
L-Sucrose	Sweetening agent	Human	Nonmetabolized
D-Ribose sugar	Sugar	Human	Less therapeutics
L-Penicillamine	Antiarthritic	Human	Toxic
(+)-Warfarin	Anticoagulant	Human	Inactive
L-Peptide (V13KD)	Antimicrobial	*Candida albicans*	Trypsin proteolysis
S-Albuterol	Adrenergic	Rat	Partially active
(+)-Clausenamide	Synaptic transducer	Rat	Low synaptic transmission
S-Propranolol	β-Blockers	Human	Toxic
R-Ketoprofen	Antiallodynic	Human	Inactive
R-(−)-Vigabatrin	Antiepileptic	Human	Highly toxic
S-Tiaprofenic acid	NSAID	Human	Inactive
N-Alkylated dihydro-pyridine-(−)-AC 394	Antitumor & antimetastatic	Human	Inactive
(+)-Tramadol	Analgesic	Human	Nausea & vomiting
R-CC-4047	Immunomodulatory	Human	Inactive
(+)-Thyroxine	Hormone	Human	Inactive
S-Flurbiprofen	NSAID	Human	Inactive
(−)-Ketamine	Anesthetic	Rat	Inactive
(+)-Methadone	Analgesic	Human	Inactive
(+)-Morphine	Analgesic	Human	Inactive
(−)-Tetramisole	Anthelmintic	Human	Inactive
S-2-[2,6-dioxo-piperidine-3-yl]-phthalimidine	Sedative	Monkey	Teratogenic
R-Fenoprofen	NSAID	Rat	Inactive
(−)-Fluoxetine	Antidepressant	Human	Inactive
(+)-Verapamil	Calcium channel blocker	Human	Inactive

NSAID: Nonsteroidal anti-inflammatory drug.

molecular mechanism of the profen chiral inversion is stepwise, involving the initial stereoselective formation of a covalent acyl-CoA thioester adduct between profen and coenzyme A (CoA) through long-chain acyl-CoA synthetase catalysis [82,87–89]. The acyl-CoA intermediates formed during the chiral inversion are of toxicological significance since they are potential modulators of lipid metabolism. It has been shown that the acyl-CoA thioester inhibits the fatty acid oxidation in mitochondria, leading to microvesicular steatosis [90]. In addition, the acyl-CoA

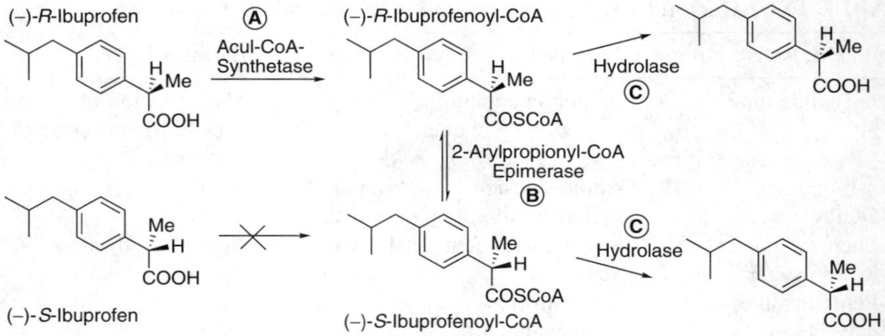

SCHEME 10.3 Proposed mechanism of chiral inversion of 2-arylpropionic acids.

intermediate can be incorporated into adipose tissue triglycerides, altering plasma membrane and second messenger mechanisms [91]. Nevertheless, since R-(−)-ibuprofen is readily inverted in the body, the chirally pure R-(−)-ibuprofen as a pro-drug is suggested to be a safer alternative to the available racemic ibuprofen [92]. Clinically, R-(−)-ibuprofen is shown to contribute to the therapeutic effects not only via chiral inversion to S-(+)-eutomer but also via direct COX2 inhibition by its thioester intermediate, R-(−)-ibuprofenyl-CoA [93]. In addition, R-(−)-ibuprofen may be the active ingredient contributing to the temperature-lowering and analgesic effects of the racemic ibuprofen in brain, because of the fact that the lipophilic (R)-ibuprofenyl-CoA can penetrate the BBB membrane more easily than the (S)-ibuprofen and the localization and expression of the 2-arylpropionyl-CoA epimerase in the brain (Scheme 10.3) [94,95].

Thalidomide, a former racemic sedative, was withdrawn from the market in the 1960s due to its severe teratogenic effects (phocomelia, amelia). The S-(−)-distomer of thalidomide was originally believed to be responsible for the toxic teratogenic effects based on animal model studies using Smith–Magenis syndrome mice [35]. When more sensitive animal teratogen models (e.g., New Zealand White rabbits) were used, however, the enantiomers of thalidomide exhibited equivalent teratogen potential [36,38]. Furthermore, a thalidomide derivative, EM12, displayed the similar stereo equivalent teratogenic potency in the nonhuman primate *Callithrix jacchus* [37]. Thalidomide and its derivatives have also been demonstrated to undergo facile bidirectional chiral inversion (racemization) both in aqueous solution and in vivo (Scheme 10.4) [96–100]. Thus exposure to either enantiomer of thalidomide would present a risk of fetal harm. Obviously, it is difficult to attribute the observed pharmacological and toxicological effects to a single enantiomer of thalidomide, not to mention the effects of its numerous chiral and achiral metabolites. Thalidomide and its derivatives apparently work by potently inhibiting the tumor necrosis factor-α pathways. There is currently a renewed interest in restricted thalidomide therapies for erythema nodosum leprosum, aptosis, Behcet syndrome, and some autoimmune diseases because of its remarkable immunomodulatory [101], antiangiogenic, and antiinflammatory effects [102].

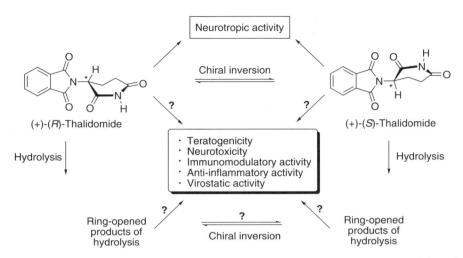

SCHEME 10.4 Interplay of biological activities, metabolism, and stereoselectivity of thalidomide enantiomers and their ring-opened metabolites.

10.6 DISCUSSION

The stereoselective and stereospecific features of biological systems contribute to the chiral differentiation of pharmacologically active molecules. Enantiomers of chiral drugs may differentiate in the processes of absorption, distribution, metabolism, and excretion, and may have distinct disease indications and states, routes of administration, genetic susceptibility, and drug-drug interactions. The existing interconversion and biotransformation processes further complicate the elucidation of each enantiomer's action. No generalizations can be drawn for drug enantiomers since they exhibit a wide range of variations in their toxicological effects either structurally or mechanistically. A better understanding of the contributing factors of chirality through chiral separation and testing may greatly aid in drug development by identifying chirality-related properties at earlier stages.

REFERENCES

1. Millership, J. S., Fitzpatrick, A. (1993). Commonly used chiral drugs—a survey. *Chirality* **5**, 573–576.
2. Caner, H., Groner, E., Levy, L., Angranat I. (2004). Trends in the development of chiral drugs. *Drug Discov. Today* **9**, 105–110.
3. Nguyen, L. A., He, H., and Pham-Huy, C. (2006). Chiral drugs. An overview. *Int. J. Biomed. Sci.* **2**, 85–100.
4. Cahn, R. S., Ingold, C., Prelog, V. (1966). Specification of molecular chirality. *Angew. Chem. Int. Ed.* **5**, 385–415.

5. Lehmann, P. A. F., DeMiranda, J. F. R., Ariens, E. J. (1976). In: *Progress in Drug Research*, *vol. 20* (Junker, E, Ed.), Basel: Birkhäuser, p. 101–142.

6. Liebler, D. C., Guengerich, F. P. (2007) Elucidating mechanisms of drug-induced toxicity. *Nat. Rev. Drug Discov.* **4**, 410–420.

7. Mitchell, J. R., Jollow, D. J., Potter, W. Z., Davis, D. C., Gillette, J. R., Brodie, B. B. (1973). Acetaminophen-induced hepatic necrosis. 1. Role of drug-metabolism. *J. Pharmacol. Exp. Ther.* **187**, 185–194.

8. Jollow, D. J., Mitchell, J. R., Potter, W. Z., Davis, D. C., Gillette, J. R., Brodie, B. B. (1973). Acetaminophen-induced hepatic necrosis. 2. Role of covalent binding in vivo. *J. Pharmacol. Exp. Ther.* **187**, 195–202.

9. Miller, J. A. (1970). Carcinogenesis by chemicals—an overview. *Cancer Res.* **30**, 559–576.

10. Miller, J. A. (1994). Brief history of chemical carcinogenesis. *Cancer Lett.* **83**, 9–14.

11. Fura, A., Shu, Y.Z., Zhu, M., Hanson, R. L., Roongta, V., Humphreys, W. G. (2004). Discovering drugs through biological transformation: role of pharmacologically active metabolites in drug discovery. *J. Med. Chem.* **47**, 4339–4351.

12. Hutt, A. J., O'Grady, J. (1996). Drug chirality: a consideration of the significance of the stereochemistry of antimicrobial agents. *J. Antimicrob. Chemother.* **37**, 7–32.

13. Mehvar, R., Brocks, D. R., Vakily, M. (2002). Impact of stereoselectivity on the pharmacokinetics and pharmacodynamics of antiarrhythmic drugs. *Clin. Pharmacokinet.* **41**, 533–558.

14. Ranade, V. V., Somberg, J. C. (2005) Chiral cardiovascular drugs: an overview. *Am. J. Ther.* **12**, 439–459.

15. Wainer, I. W., Granvil, C. P. (1993). Stereoselective separations of chiral anticancer drugs and their application to pharmacodynamic and pharmacokinetics studies. *Ther. Drug Monit.* **15**, 570–575.

16. Baker, G. B., Prior, T. I. (2002). Stereochemistry and drug efficacy and development: relevance of chirality to antidepressant and antipsychotic drugs. *Ann. Med.* **34**, 537–543.

17. Vakily, M., Mehvar, R., Brocks, D. (2002). Stereoselective pharmacokinetics and pharmacodynamics of anti-asthma agents. *Ann. Pharmacother.* **36**, 693–701.

18. Kean, W. F., Lock, C. J., Howard-Lock, H. E. (1991). Chirality in antirheumatic drugs. *Lancet* **338**, 1565–1568.

19. Williams, K. M. (1990). Enantiomers in arthritic disorders. *Pharmacol. Ther.* **46**, 273–295.

20. Shen, S., He, Y., Zeng, S. (2007). Stereoselective regulation of MDR1 expression in Caco-2 cells by cetirizine enantiomers. *Chirality* **19**, 485–490.

21. Lu, L., Leonessa, F., Baynham, M. T., Clarke, R., Gimenez, F., Pham, Y. T., Roux, F., Wainer, I. W. (2001). The enantioselective binding of mefloquine enantiomers to P-glycoprotein determined using an immobilized- P-glycoprotein liquid chromatographic stationary phase. *Pharm. Res.* **18**, 1327–1330.

22. Pham, Y. T., Regina, A., Farinotti, R., Couraud, P., Wainer, I. W., Roux, F., Gimenez, F. (2000). Interactions of racemic mefloquine and its enantiomers with P-glycoprotein in an immortalized rat brain capillary endothelial cell line GPNT. *Biochim. Biophys. Acta* **1524**, 212–219.

23. Pistolozzi, M., Bertucci, C. (2008). Species-dependent stereoselective drug binding to albumin: a circular dichroism study. *Chirality* **20**, 552–558.

24. Fitos, I., Visy, J., Kardos, J. (2002). Stereoselective kinetics of warfarin binding to human serum albumin: effect of an allosteric interaction. *Chirality* **14**, 442–448.

25. Tatsumi, A., Kadobayashi, M., Iwakawa, S. (2007). Effect of ethanol on the binding of warfarin enantiomers to human serum albumin. *Biol. Pharm. Bull.* **30**, 826–829.

26. Augustijns, P., Verbeke, N. (1993). Stereoselective pharmacokinetic properties of chloroquine and de-ethyl-chloroquine in humans. *Clin. Pharmacokinet.* **24**, 259–269.

27. Oforiadjei, D., Ericsson, O., Lindstrom, B., Sjoqvist, F. (1986). Protein-binding of chloroquine enantiomers and disethylchloroquine. *Br. J. Clin. Pharmacol.* **22**, 356–358.

28. Hanada, K., Ohta, T., Hirai, M., Arai, M., and Ogata, H. (2000). Enantioselective binding of propranolol, dispyramide, and verapamil to human α_1-acid glycoprotein. *J. Pharm. Sci.* **89**, 751–757.

29. Caldwell, J., Winter, S. M., Hutt, A. J. (1988). The pharmacological and toxicological significance of the stereochemistry of drug disposition. *Xenobiotica* **18**, Suppl. 1, 59–70.

30. Stewart, R. K., Serabjit-Singh, C. J., Massey, T. E. (1996). Glutathione *S*-transferase-catalyzed conjugation of bioactivated B-1 in rabbit lung and liver. *Toxicol. Appl. Pharmacol.* **140**, 499–507.

31. Pagano, D. A., Yagen, B., Hernandez, O., Bend, J. R., Zeiger, E. (1982). Mutagenicity of (R)- and (S)-styrene 7,8-oxide and the intermediary mercapturic acid metabolites formed from styrene 7,8-oxide. *Environ. Mutagen.* **4**, 575–584.

32. Carlson, G. P., Turner, M., Mantick, N. A. (2006). Effects of styrene and styrene oxide on glutathione-related antioxidant enzymes. *Toxicology* **227**, 217–226.

33. Mocquet, V., Kropachev, K., Kolbanovskiy, M., Kolbanovskiy, A., Tapias, A., Cai, Y., Broyde, S., Geacintov, N. E., Egly, J. M. (2007). The human DNA repair factor XPC-HR23B distinguishes stereoisomeric benzo[a]pyrenyl-DNA lesions. *EMBO J.* **26**, 2923–2932.

34. Wsol, V., Skalova, L., Szotakova, B. (2004). Chiral inversion of drugs: coincidence or principle? *Curr. Drug Metab.* **5**, 517–533.

35. Blaschke, G., Kraft, H. P., Fickentscher, K., Kohler, F. (1979). Chromatographic-separation of racemic thalidomic and teratogenic activity of its enantiomers. *Arzneimittel Forsch.* **29**, 1640–1642.

36. Scott, W. J., Fradkin, R., Wilson, J. G. (1977). Non-confirmation of thalidomide induced teratogenesis in rats and mice. *Teratology* **16**, 333–335.

37. Heger, W., Klug, S., Schmahl, H. J., Nau, H., Merker, H. J., Neubert, D. (1988). Embryotoxic effects of thalidomide derivatives on the non-human primate *Callithrix jacchus*. 3. Teratogenic potency of the EM-12 enantiomers. *Arch. Toxicol.* **62**, 205–208.

38. Fabro, S., Smith, R. L. Williams, R. T. (1967). Toxicity and teratogenicity of optical isomers of thalidomide. *Nature* **215**, 296.

39. Eriksson, T., Bjorkman, S. B., Roth, A., Fyge, P., Hoglund, P. (1995). Stereospecific determination, chiral inversion in-vitro and pharmacokinetics in humans of the enantiomer of thalidomide. *Chirality* **7**, 44–52.

40. Cotzias, G. C., Van Woert, M. H., and Schiffer, L. M. (1967). Aromatic amino acid and modification of parkinsonism. *N. Engl. J. Med*. **276**, 374–379.

41. Cotzias, G. C., Papavasiliou, P. S., Gellene, R. (1969). Modification of parkinsonism—chronic treatment with L-DOPA. *N. Engl. J. Med*. **280**, 337–345.

42. Sato, K., Hoshino, K., Tanaka, M., Hayakawa, I., Osada, Y., Nishino, T. et al. (1993). In: *Molecular Biology of DNA Topoisomerases and its Application to Chemotherapy* (Andoh, T., Ikeda, H., Oguro, M., Eds). Boca Raton, FL: CRC Press, p. 167–176.

43. Hayakawa, I., Atarashi, S., Yokohama, S., Imamura, M., Sakano, K. I., Furukawa, M. (1986). Synthesis and antibacterial activities of optically-active ofloxacin. *Antimicrob. Agents Chemother*. **29**, 163–164.

44. Atarashi, S., Yokohama, S., Yamazaki, K. I., Sakano, K. I., Imamura, M., Hayakawa, I. (1987). Synthesis and antibacterial activities of optically-active ofloxacin and its fluoromethyl derivative. *Chem. Pharmaceut. Bull*. **35**, 1896–1902.

45. Hoshino, K., Sato, K., Akahane, K., Yoshida, A., Hayakawa, I., Sato, M., Une, T., Osada, Y. (1991). Significance of the methyl-group on the oxazine ring of ofloxacin derivatives in the inhibition of bacterial and mammalian type-II topoisomerases. *Antimicrob. Agents Chemother*. **35**, 302–312.

46. Okazaki, O., Kurata, T., Tachizawa, H. (1989). Stereoselective metabolic disposition of enantiomers of ofloxacin in rats. *Xenobiotica* **19**, 419–429.

47. Okazaki, O., Kurata, T., Hakusui, H., Tachizawa, H. (1991). Stereoselective glucuronidation of ofloxacin in rat-liver microsomes. *Drug Metab. Dispos*. **19**, 376–380.

48. Okazaki, O., Kurata, T., Hakusui, H., Tachizawa, H. (1992). Species-related stereoselective disposition of ofloxacin in the rat, dog and monkey. *Xenobiotica* **22**, 439–450.

49. Akahane, K., Tsutomi, Y., Kimura, K., Kitano, K. (1994). Levofloxacin, an optical isomer of ofloxacin, has attenuated epileptogenic activity in mice and inhibitory potency in GABA receptor-binding. *Chemotherapy (Basel)* **40**, 412–417.

50. Marietta, M. P., Way, W. L., Trevor, A. J. (1977). Pharmacology of ketamine enantiomorphs in rat. *J. Pharmacol. Exp. Ther*. **202**, 157–165.

51. White, P. F., Ham, J., Way, W. L., Trevor, A. J. (1980). Pharmacology of ketamine isomers in surgical patients. *Anesthesiology* **52**, 231–239.

52. Kohrs, R., Durieux, M. E. (1998). Ketamine: teaching an old drug new tricks. *Anesth. Analg*. **87**, 1186–1193.

53. Wilson, J. E., Du Vigneaud, V. (1948). L-Penicillamine as a metabolic antagonist. *Science* **107**, 653.

54. Wass, M., Evered, D. F. (1970). Transport of penicillamine across mucosa of rat small intestine in vitro. *Biochem. Pharmacol*. **19**, 1287–1295.

55. Tu, J., Blackwell, R. Q., Lee, P. F. (1963). DL-Penicillamine as a cause of optic axial neuritis. *JAMA* **185**, 83–86.

56. Sternlieb, I. (1966). Pencillamine and nephritic syndrome-results in patients with hepatolenticular degeneration. *JAMA* **198**, 1311–1312.

57. Williams, K. M. (1990). Enantiomers in arthritic disorders. *Pharmacol. Ther*. **46**, 273–295.

58. Glatt, H., Oesch, F. (1985). Mutagenicity of cysteine and penicillamine and its enantiomeric selectivity. *Biochem. Pharmacol*. **34**, 3725–3728.

59. Gidal, B. E., Privitera, M. D., Sheth, R. D., Gilman, J. T. (1999). Vigabatrin: a novel therapy for seizure disorders. *Ann. Pharmacother.* **33**, 1277–1286.

60. Sheean, G., Schramm, T., Anderson, D. S., Eadie, M. J. (1992). Vigabatrin—plasma enantiomer concentrations and clinical effects. *Clin. Exp. Neurol.* **29**, 107–116.

61. Hyttel, J., Bogeso, K. P., Perregaard, J., Sanchez, C. (1992). The pharmacological effect of citalopram presides in the (S)-(+)-enantiomer. *J. Neural Trans.* **88**, 157–160.

62. Burke, W. J., Gergel, I., Bose, A. (2002). Fixed-dose trial of the single isomer SSRI escitalopram in depressed outpatients. *J. Clin. Psychiatry* **63**, 331–336.

63. Reis, M., Cherma, M. D., Carlsson, B., Bengtsson, F., and Task Force for TDM of Escitalopram in Sweden. (2007). Therapeutic drug monitoring of escitalopram in an outpatient setting. *Ther. Drug Monit.* **29**, 758–766.

64. Covyeou, J. A., Jackson, C. W. (2007). Hyponatremia associated with escitalopram. *N. Engl. J. Med.* **356**, 94–95.

65. Grover, S., Biswas, P., Bhateja, G., Kulhara, P. (2007). Escitalopram-associated hyponatremia. *Psychiatry Clin. Neurosci.* **61**, 132–133.

66. Huska, M. T., Catalano, G., Catalano, M. C. (2007). Serotonin syndrome associated with the use of escitalopram. *CNS Spectr.* **12**, 270–274.

67. Nahshoni, E., Weizman, A., Shefet, D., Pik, N. (2004). A case of hyponatremia associated with escitalopram. *J. Clin. Psychiatry* **65**, 1722.

68. Olsen, D., Dart, R. C., and Robinett, M. (2004) Severe serotonin syndrome from escitalopram over-dose. *J. Toxicol. Clin. Toxicol.* **42**, 744–745.

69. Page, R. L., Ruscin, J. M., Bainbridge, J. L., Brieke, A. A. (2008). Restless leg syndrome induced by escitalopram: case report and review of the literature. *Pharmacotherapy* **28**, 271–280.

70. Vari, G., Beckson, M. (2007). Escitalopram-associated serotonin toxicity. *J. Clin. Psychopharmacol.* **27**, 229–230.

71. Page, C. P., Morley, J. (1999). Contrasting properties of albuterol stereoisomers. *J. Allergy Clin. Immunol.* **104**, S31–S41.

72. Nelson, H. S. (1999). Clinical experience with levalbuterol. *J. Allergy Clin. Immunol.* **104**, S77–S84.

73. Anonymous. (2006). A levalbuterol metered-dose inhaler (Xopenex HFA) for asthma. Is the R-isomer alone better than racemic albuterol? *Med. Lett. Drugs Ther.* **48**, 21–22 and 24.

74. Boulton, D. W., Fawcett, J. P. (1997). Pharmacokinetics and pharmacodynanics of single oral doses of albuterol and its enantiomers in humans. *Clin. Pharmacol. Ther.* **62**, 138–144.

75. Eichelbaum, M. (1995). Side effects and toxic reactions of chiral drugs: a clinical perspective. *Arch. Toxicol. Suppl.* **17**, 514–521.

76. Scott, A. K. (1993). Stereoisomers and drug toxicity—The value of single stereoisomer therapy. *Drug Safety* **8**, 149–159.

77. Caldwell, J., Hutt, A. J., Fournel-Gigleux, S. (1988). The metabolic chiral inversion and dispositional enantioselectivity of the 2-arylpropionic acids and their biological consequences. *Biochem. Pharmacol.* **37**, 105–114.

78. Ali, I., Gupta, V. K., Aboul-Enein, H. Y., Singh, P., Sharma, B. (2007). Role of racemization in optically active drugs development. *Chirality* **19**, 453–463.

79. Bannwarth, B., Lapicque, F., Pehourcq, F., Gillet, P., Schaeverbeke, T., Laborde, C. Dehais, J., Gaucher, A., Netter, P. (1995). Stereoselective disposition of ibuprofen enantiomers in human cerebrospinal-fluid. *Br. J. Clin. Pharmacol.* **40**, 266–269.

80. Adams, S. S., Bresloff, P., Mason, C. G. (1976). Pharmacological differences between optical isomers of ibuprofen—Evidence for metabolic inversion of (−)-isomer. *J. Pharm. Pharmacol.* **28**, 256–257.

81. Kaiser, D. G., Van-Giessen, G. J., Reischer, R. J., Wechter, W. J. (1976). Isomeric inversion of ibuprofen (R)-enantiomer in humans. *J. Pharm. Sci.* **65**, 269–273.

82. Wechter, W. J., Loughhead, D. G., Reischer, R. J., Van-Giessen, G. J., Kaiser, D. G. (1974). Enzymatic inversion at saturated carbon—Nature and mechanism of inversion of R(−)-p-iso-butyl hydratropic acid. *Biochem. Biophys. Res. Commun.* **61**, 833–837.

83. Reichel, C., Brugger, R., Bang, H., Geisslinger, G., Brune, K. (1997). Molecular cloning and expression of a 2-arylpropionyl-coenzyme A epimerase: a key enzyme in the inversion metabolism of ibuprofen. *Mol. Pharmacol.* **51**, 576–582.

84. Landoni, M. F., Soraci, A. (2001). Pharmacology of chiral compounds: 2-aryl-propionic acid derivatives. *Curr. Drug Metab.* **2**, 37–51.

85. Marzo, A., Heftman, E. (2002). Enantioselective analytical methods in pharmacokinetics with specific reference to genetic polymorphic metabolism. *J. Biochem. Biophys. Methods* **54**, 57–70.

86. Knihinicki, R. D., Day, R. O., Williams, K. M. (1991). Chiral inversion of 2-arylpropionic acid nonsteroidal anti-inflammatory drugs. 2. Racemization and hydrolysis of (R)-ibuprofen-CoA and (S)-ibuprofen-CoA thioesters. *Biochem. Pharmacol.* **42**, 1905–1911.

87. Nakamura, Y., Yamaguchi, T., Takahashi, S., Hashimoto, S., Iwatani, K., Nakagawa, Y. (1981). Optical isomerization mechanism of (R)-hydratropic acid derivatives. *J. Pharmacobio-Dyn.* **4**, S1.

88. Williams, K. M., and Day, R. O. (1985) Stereoselective disposition—basis for variability in response to NSAIDs. *Agents Actions Suppl.* **17**, 119–126.

89. Tracy, T. S., Hall, S. D. (1992). Metabolic inversion of (R)-ibuprofen-epimerization and hydrolysis of ibuprofenyl-coenzyme-A. *Drug Metab. Dispos.* **20**, 322–327.

90. Freneaux, E., Fromenty, B. A., Berson, A., Labbe, G., Degott, C., Letteron, P., Larrey, D., Pessayre, D. (1990). Stereoselective and nonstereoselective effects of ibuprofen enantiomers on mitochondrial beta-oxidation of fatty-acids. *J. Pharmacol. Exp. Ther.* **255**, 529–535.

91. Williams, K., Day, R., Knihinicki, R., Duffield, A. (1986). The stereoselective uptake of ibuprofen enantiomers into adipose-tissue. *Biochem. Pharmacol.* **35**, 3403–3405.

92. Knihinicki, R. D., Day, R. O., Graham, G. G., Williams, K. M. (1990). Stereoselective disposition of ibuprofen and flurbiprofen in rats. *Chirality* **2**, 134–140.

93. Neupert, W., Brugger, R., Euchenhofer, C., Brune, K., Geisslinger, G. (1997). Effects of ibuprofen enantiomers and its coenzyme A thioesters on human prostaglandin endoperoxide synthases. *Br. J. Pharmacol.* **122**, 487–492.

94. Reichel, C., Brugger, R., Bang, H., Geisslinger, G., Brune, K. (1997). Molecular cloning and expression of a 2-arylpropionyl-coenzyme A epimerase: a key enzyme in the inversion metabolism of ibuprofen. *Mol. Pharmacol.* **51**, 576–582.

95. Evans, A. M. (1996). Pharmacodynamics and pharmacokinetics of the profens: enantioselectivity clinical implications, and special reference to S(+)-ibuprofen. *J. Clin. Pharmacol*. **36**, 7S–15S.

96. Reist, M., Carrupt, P. A., Francotte, E., Testa, B. (1998). Chiral inversion and hydrolysis of thalidomide: mechanisms and catalysis by bases and serum albumin, and chiral stability of teratogenic metabolites. *Chem. Res. Toxicol*. **11**, 1521–1528.

97. Eriksson, T., Bjorkman, S., Hoglund, P. (2001). Clinical pharmacology of thalidomide. *Eur. J. Clin. Pharmacol*. **57**, 365–376.

98. Schmahl, H. J., Nau, H., Neubert, D. (1988). The enantiomers of the teratogenic thalidomide analog em-12. 1. Chiral inversion and plasma pharmacokinetics in the marmoset monkey. *Arch. Toxicol*. **62**, 200–204.

99. Teo, S. K., Chen, Y., Muller, G. W., Chen, R. S., Thomas, S. D., Stirling, D. I., Chandula, R. S. (2003). Chiral inversion of the second generation IMiD™ CC-4047 (ACTIMID™) in human plasma and phosphate-buffered saline. *Chirality* **15**, 348–351.

100. Yamada, T., Okada, T., Sakaguchi, K., Ohfune, Y., Ueki, H., Soloshonok, V. A. (2006). Efficient asymmetric synthesis of novel 4-substituted and configurationally stable analogues of thalidomide. *Org. Lett*. **8**, 5625–5628.

101. Pham-Huy, C., Galons, H., Voisin, J., Zhu, J. R., Righenzi, S., Warnet, J. M., Clande, J. R., Duc, H. T. (1997). In vitro and in vivo immunosuppressive potential of thalidomide and its derivative N-hydroxythalidomide, alone and in combination with cyclosporine A. *Int. J. Immunopharmacol*. **19**, 289–296.

102. Davies, N. M., and Teng, X. V. (2003) Importance of chirality in drug therapy and pharmacy practice: implications for psychiatry. *Adv. Pharm*. **1**, 242–252.

CHAPTER 11

REPRESENTATIVE CHIRAL DRUGS

JIANGQIN SUN
Otsuka Shanghai Research Institute, Pudong New District, Shanghai, China

DINGGUO LIU
Pfizer, San Diego, CA

ZHIMIN WANG
Sundia MedTech Company Ltd., Pudong New District, Shanghai, China

Chiral Drugs: Chemistry and Biological Action, First Edition. Edited by Guo-Qiang Lin, Qi-Dong You and Jie-Fei Cheng.
© 2011 John Wiley & Sons, Inc. Published 2011 by John Wiley & Sons, Inc.

11.1 INTRODUCTION

Since the FDA issued its guidelines governing the development of chiral drugs in 1992, chirality in drug molecules has been a major factor in drug development. In fact, racemic substances as new drugs have nearly disappeared. Thus the focus of a large number of pharmaceutical companies is on the development of chiral drugs as single enantiomers because of their pharmaceutical market importance.

A number of single-enantiomer drugs based on big-selling racemic drugs were rapidly developed and launched in the late 1990s. For example, AstraZeneca developed a single-enantiomer version of its bestseller Losec, or omeprazole, to produce Nexium, or esomeprazole, the third biggest-selling medicine in the world. Sepracor developed levalbuterol, a single-enantiomer version of albuterol, to avoid the side effects of the racemic product. Other commercially important products include (S)-ibuprofan, (S)-naproxen, (S)-ibuprofen, and (R)-fluoxetine. By applying "chiral switch", or single enantiomers of previously approved racemates, a pharmaceutical company can extend a product's life cycle.

In the past decades, approximately 80% of the small-molecule drugs approved by the FDA were chiral, of which nearly 75% were approved as single-enantiomer products. Single-enantiomer drug sales show a continuous growth worldwide and many of the top selling drugs are marketed as single enantiomers.

To date, the growth in single-enantiomer products has produced several blockbuster compounds such as Pfizer's Lipitor (atorvastatin) and Zoloft (sertraline); Amgen and Johnson & Johnson's Epogen (epoetin alfa); Merck & Co.'s Zocor (simvastatin) and Singulair (montelukast); Sanofi-Aventis' and Bristol-Myers Squibb's Plavix (clopidogrel); GlaxoSmithKline's Advair (fluticasone and salmeterol); Genentech and Roche's Rituxan (rituximab); AstraZeneca's Nexium (esomeprazole); and Novartis' Diovan (valsartan).

Currently, chiral formations of small molecules are an essential part of the discovery and development of new medicines. Here we provide a summary of some major chiral drugs.

11.2 REPRESENTATIVE CHIRAL DRUGS

11.2.1 Moxifloxacin

Trade name:	Avelox, Avalox, Avelon, and Vigamox
Manufacturer:	Bayer/Shionogi/Schering-Plough
Country in which first launched:	US
Year of introduction:	1999
Molecular weight:	437.90
CAS registration number:	186826-86-8
Structure:	

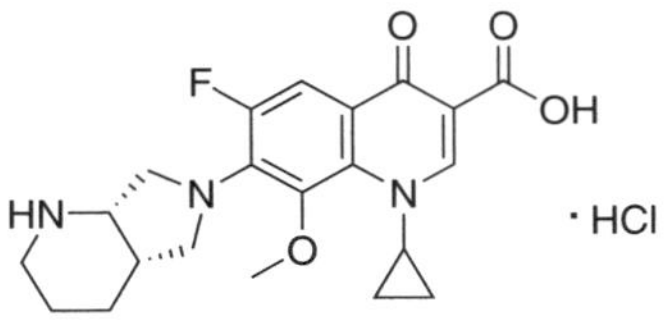

FIGURE 11.1   Structure of moxifloxacin (**1**).

Moxifloxacin (**1**, Fig. 11.1) is the fourth-generation fluoroquinolone antibiotic for the treatment of certain bacterial infections. It is a broad-spectrum antibiotic and is active against both gram-positive and gram-negative bacteria. It functions by inhibiting DNA gyrase, a type II topoisomerase, and topoisomerase IV, enzymes necessary to unwind the bacterial DNA chain, thereby inhibiting cell division [1]. In 1999, the US FDA approved moxifloxacin for use in the United States to treat severe and life threatening bacterial infections, and since then it has been marketed by Bayer worldwide under the brand names Avelox, Avalox, and Avelon for oral administration. The drug is indicated for the treatment of acute bacterial sinusitis, acute bacterial exacerbation of chronic bronchitis (ABECB), community-acquired pneumonia (CAP), complicated intra-abdominal infections, and skin and skin structure infections (SSSIs). However, serious questions were raised within the FDA regarding the safety and efficacy of moxifloxacin. Of particular concern is the fact that moxifloxacin may induce cardiac arrhythmias. Additionally, moxifloxacin interacts with a number of other drugs, as well as a number of herbal and natural supplements [2]. Resistance to moxifloxacin and other fluoroquinolones may evolve rapidly, even during a course of

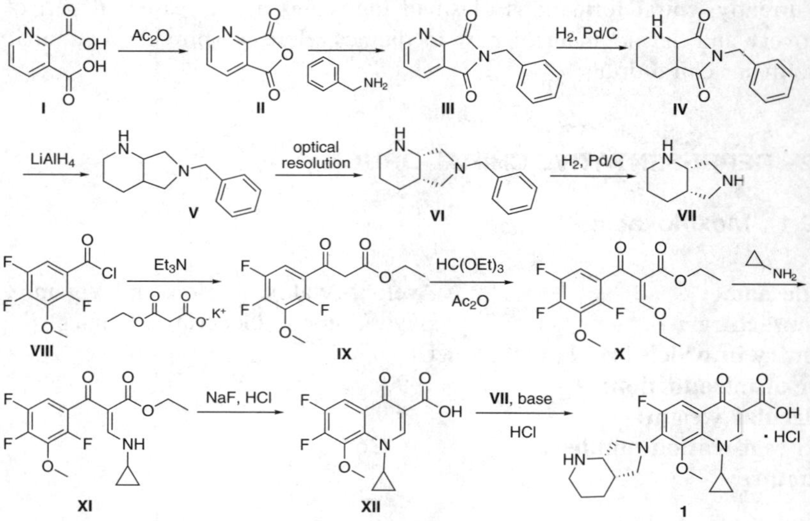

SCHEME 11.1 Synthesis of moxifloxacin (**1**).

treatment. Because of its safety concerns and ever-increasing bacterial resistance, moxifloxacin is considered a drug of last resort when other antibiotics have failed.

There is one synthetic route disclosed in a conference abstract [3]. Starting with pyridine-2,3-dicarboxylic acid (**I**) (Scheme 11.1), the corresponding anhydride (**II**) is obtained through treatment with acetic anhydride. Compound **II** is condensed with benzylamine to afford the benzylimide (**III**). Hydrogenation of **III** over a catalytic amount of Pd/C yielded 8-benzyl-2,8-diazabicyclo[4.3.0]nonane-7,9-dione (**IV**), which is further hydrogenated with LiAlH$_4$ to give (*cis*-8-benzyl-2,8-diazabicyclo [4.3.0]nonane (**V**). The optical resolution of **V** with tartaric acid afforded *cis*-(*S,S*)-isomer (**VI**) D-(−)-tartrate salt, which is converted into desirable enantiomerically pure (*S,S*)-8-benzyl-2,8-diazabicyclo[4.3.0]nonane (**VI**). Hydrogenation of **VI** over Pd/C affords (*S,S*)-2,8-diazabicyclo[4.3.0]nonane (**VII**). The other segment, 1-cyclopropyl-6,7-difluoro-8-methoxy-4-oxo-1,4-dihydroquinoline-3-carboxylic acid (**XII**), is prepared from 2,4,5-trifluoro-3-methoxybenzoyl chloride (**VIII**) in four steps. Thus reaction of **VIII** with malonic acid monoethyl ester monopotassium salt and triethylamine affords 2-(2,4,5-trifluoro-3-methoxybenzoyl) acetic acid ethyl ester (**IX**), which is condensed with triethyl orthoformate to yield the corresponding ethoxymethylene derivative (**X**). Compound **X** is treated with cyclopropylamine to give the cyclopropylaminomethylene derivative (**XI**), which is cyclized to **XII** in the presence of NaF in DMF. Finally, **XII** is condensed with amine **VII** in basic medium followed by acidified with HCl to give moxifloxacin (**1**).

11.2.2 Oseltamivir Phosphate

Trade name:	Tamiflu
Manufacturer:	Roche/Chugai/Gilead
Country in which first launched:	US
Year of introduction:	1999
Molecular weight:	410.40
CAS registration number:	204255-11-8
Structure:	

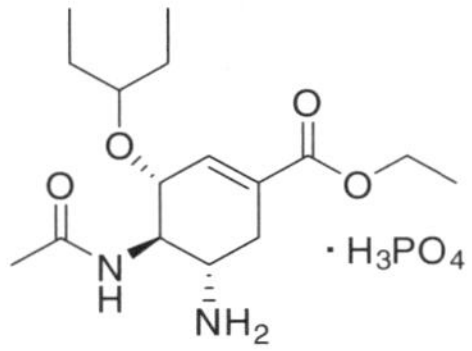

FIGURE 11.2 Structure of oseltamivir phosphate (**2**).

Oseltamivir (**2**, Fig. 11.2) is an inhibitor of influenza neuraminidase, serving as a competitive inhibitor toward sialic acid, which is found on the surface proteins of normal host cells. By blocking the activity of the neuraminidase, it can prevent the release of new copies from infected cells and slow the spread of nonresistant strains of the influenza virus between cells in the body [4]. Oseltamivir was approved in 1999 by FDA in the US for the treatment and prophylaxis of influenza virus A and B infection. Oseltamivir is the first orally active neuraminidase inhibitor and was developed by Gilead Sciences as a pro-drug. It is currently marketed by Roche under the trade name Tamiflu [5]. Oseltamivir is used to treat flu symptoms caused by influenza virus in patients who have had symptoms for less than 2 days. It may also be given to prevent influenza in people who may have been exposed but do not yet have symptoms. However, Oseltamivir is not intended to treat the common cold and should not be used in place of getting a yearly flu shot. The usual adult dosage for treatment of influenza is 75 mg twice daily for 5 days, beginning within 2 days of the appearance of symptoms and with decreased doses for children and patients with renal impairment. Common adverse drug reactions (ADRs) associated with oseltamivir therapy (occurring in over 1% of clinical trial participants) include nausea, vomiting, diarrhea, abdominal pain, and headache. As with other antivirals, there are concerns about increased resistance to oseltamivir with its widespread use.

Several groups have published their work toward the total synthesis of oseltamivir [5–9]. The original commercial process involved in shikimic acid is described in Chapter 6. The synthetic route developed by Shibasaki's group is described in Scheme 11.2. (−)-Quinic acid (**I**) is treated with 2,2-dimethoxypropane (**II**) in the presence of p-toluenesulfonic acid in refluxing acetone to give the protected lactone (**III**), which is ring-opened with sodium ethoxide in ethanol to generate the ethyl ester (**IV**). Mesylation

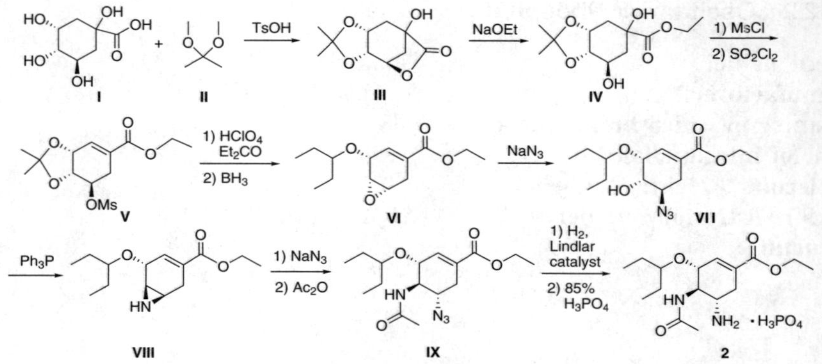

SCHEME 11.2   Shibasaki's synthesis of oseltamivir (**2**).

of (**IV**) followed by dehydration with SO_2Cl_2 in dichloromethane affords the cyclohexenecarboxylate (**V**). Reaction of **V** with 3-pentanone and $HClO_4$ is followed by ring opening of ketal with borane methyl sulfide complex to yield the 3-pentyl ether, which is further converted to the epoxide (**VI**) by treatment with $KHCO_3$ in hot ethanol. Epoxide (**VI**) is reacted with sodium azide and ammonium chloride in ethanol/water, resulting in the azido alcohol (**VII**). The cyclization of **VII** with Ph_3P in refluxing THF/CH_3CN or trimethylphosphine in anhydrous acetonitrile provides aziridine (**VIII**). The aziridine is opened by sodium azide in hot DMF, and the resulting free amine is further acetylated with acetic anhydride to yield the azidoacetamide (**IX**). Finally, hydrogenation of (**IX**) over Lindlar catalyst or over Raney Ni in ethanol is followed by treatment with 85% phosphoric acid to accomplish the synthesis of target compound (**2**).

The synthetic approach developed by Corey shown in Scheme 11.3 does not involve the usage of either shikimic acid or azide. Trifluoroethyl acrylate (**I**) is reacted with butadiene in the presence of the S-proline-derived catalyst (**II**) to give the intermediate (**III**) with enantioselectivity >97%. Ammonolysis of the intermediate (**III**) followed by iodolactamization affords the lactam (**IV**). **IV** is converted to the intermediate (**V**) effeciently with the use of DBU and NBS. The lactam (**V**) is treated with Cs_2CO_3 and reacted with NBA to give the cyclohexene derivative (**VI**). **VI** is converted into the desirable product in three steps. That is, the cyclization of **VI** gives an intermediary N-acetylaziridine with KHMDS, followed by ring opening of the resulting aziridine by 3-pentanol in the presence of a catalytic amount of cupric triflate and deprotection of Boc group with H_3PO_4 to give oseltamivir.

Two similar routes using organocatalysis published recently by Hayashi's [8] and Ma's [9] groups are shown in Scheme 11.4. Alkoxyaldehyde (**I**) is reacted with different nitroethene derivatives using pyrrolidine-based organocatalysts (**II**) and (**III**) to give the intermediates (**IV**) and (**V**), respectively. The chiral intermediates (**IV**) and (**V**) are reacted first with vinylphosphonate and then with p-methylbenzenethiol to give the ring products (**VI**) and (**VII**), respectively, which are both converted to oseltamivir through some known reactions.

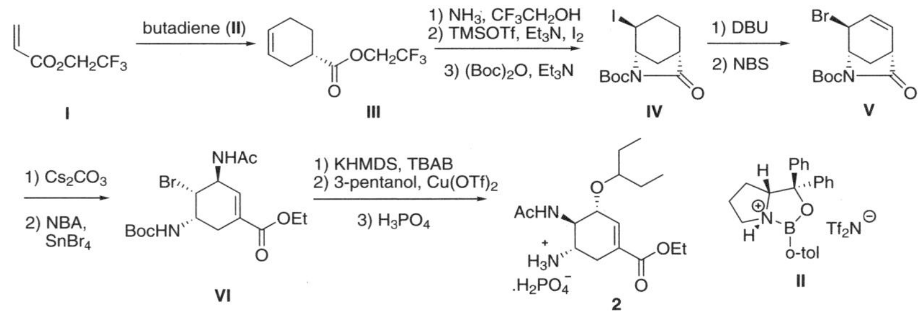

SCHEME 11.3 Corey's synthesis of oseltamivir (**2**).

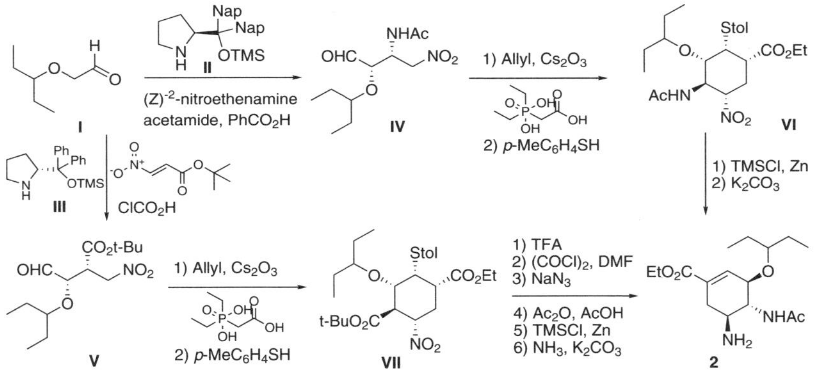

SCHEME 11.4 Synthesis of oseltamivir (**2**) by Hayashi and Ma.

11.2.3 Indinavir Sulfate

Trade name:	Crixivan
Manufacturer:	Merck
Country in which first launched:	US
Year of introduction:	1996
Molecular weight:	711.88
CAS registration number:	157810-81-6
Structure:	

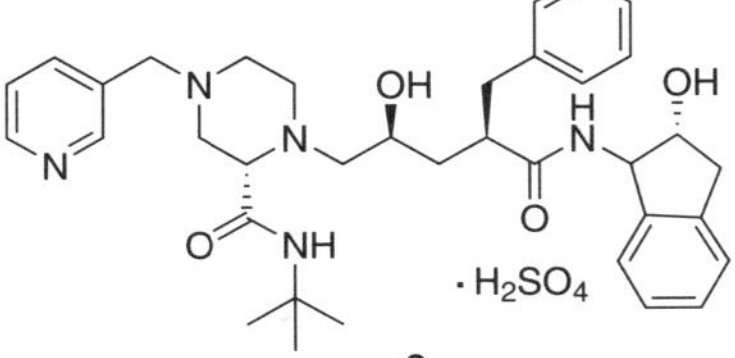

FIGURE 11.3 Structure of indinavir sulfate (**3**).

Indinavir (Fig. 11.3) is a potent inhibitor of the HIV protease enzyme that is critical to the replication of HIV virus and is marketed to treat HIV infection as a single agent or in combination with other anti-HIV drugs. Indinavir was developed by Merck and approved by FDA in March 1996. It was more powerful than previously introduced antiretroviral drugs. Combination therapy with dual NRTIs set the standard for treatment of HIV/AIDS. Unfortunately, indinavir wears off quickly after dosing and therefore requires frequent dosing (3 times a day) in order to prevent HIV from developing drug-resistant mutations including resistances to other protease inhibitors [10]. There are restrictions on what sorts of food may be eaten concurrently with the drug. Indinavir was demonstrated in clinical trials to be safe and effective when combined with other drugs, and it is suggested that indinavir, combined with Norvir, is an effective option for people for whom a protease inhibitor-based drug regimen has failed in the past [11]. Indinavir forms crystals in the urine, which occurs in as many as 40% of people taking this drug. To reduce the risk of kidney stone formation and other side effects, it is advised that people taking indinavir drink at least six 8-ounce glasses of water a day.

Several procedures have been disclosed for the synthesis of indinavir [12]. Shown in Scheme 11.5 is the route published by Merck [13,14]. Commencing with cis-(1S,2R)-indanediol (I), it is treated with concentrated H_2SO_4 in acetonitrile to give cis-(1S,2R)-1-aminoindan-2-ol (II), which is subjected to acetylation with 3-phenylpropionyl chloride (III), followed by cyclization with isopropenyl

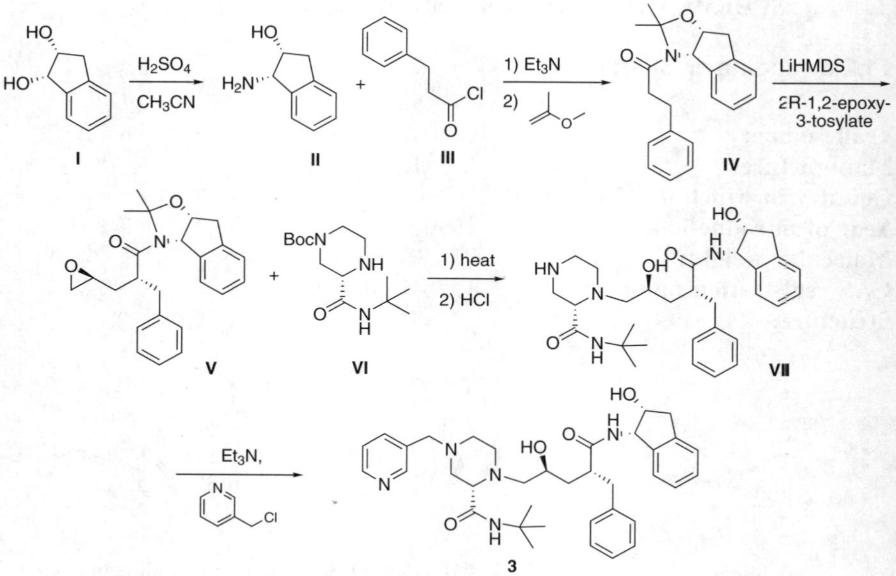

SCHEME 11.5   Synthesis of indinavir (**3**).

methyl ether to yield the acetonide amide (**IV**). The α-alkylation of amide (**IV**) with (*S*)-(+)-glycidyl *p*-toluenesulfonate in the presence of lithium hexamethyldisilylamide (LiHMDS) affords the chiral epoxide (**V**). The epoxide is opened with 4-(*tert*-butoxycarbonyl)-*N*-*tert*-butylpiperazine-2-(*S*)-carboxamide (**VI**) in refluxing isopropyl acetate followed by deprotection with aqueous HCl to give the dihydroxydiamide (**VII**). Finally, (**VII**) is condensed with 3-(chloromethyl)-pyridine in the presence of triethylamine in DMF to provide indinavir.

11.2.4 Ertapenem Sodium

Trade name:	Invanz
Manufacturer:	Merck
Country in which first launched:	US
Year of introduction:	2001
Molecular weight:	497.50
CAS registration number:	153773-82-1
Structure:	

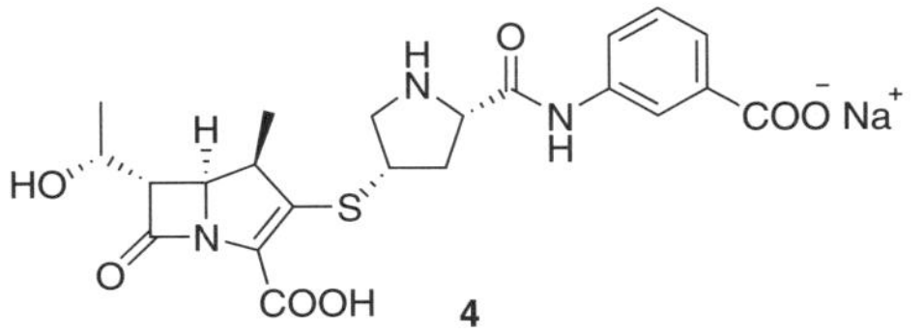

FIGURE 11.4 Structure of ertapenem sodium (**4**).

Ertapenem (**4**, Fig. 11.4) is a carbapenem antibiotic marketed as a first-line treatment for cephalosporin-resistant gram-negative infections, including complicated intra-abdominal, skin and skin-structure, urinary tract, and acute pelvic infections [15]. Structurally, it is very similar to meropenem, which has a 1-β-methyl group. In 2001, Ertapenem was approved in the US and marketed by Merck under the trademark of Invanz. Ertapenem was demonstrated to be clinically effective against gram-negative bacteria but lacking activity against MRSA, ampicillin-resistant enterococci, *Pseudomonas aeruginosa*, or *Acinetobacter* species. Additionally, Ertapenem showed useful activity against anaerobic bacteria in clinical trials. It should not be used as empirical treatment for hospital-acquired infections because of its lack of activity against *Pseudomonas aeruginosa*. In practice, it is reserved primarily for use against extended-spectrum β-lactamase (ESBL)-producing and high-level Ampc-producing gram-negative bacteria. Ertapenem is dosed as 1 g given by intravenous injection over 30 minutes, or 1g diluted with 3.2 ml of 1% lidocaine given intramuscularly. Acquired resistance to ertapenem is usually mediated by upregulation of efflux mechanisms and by the selection of porin-deficient mutants, and organisms that produce a metallo-β-lactamase are innately immune to ertapenem. Few adverse effects of ertapenem were observed,

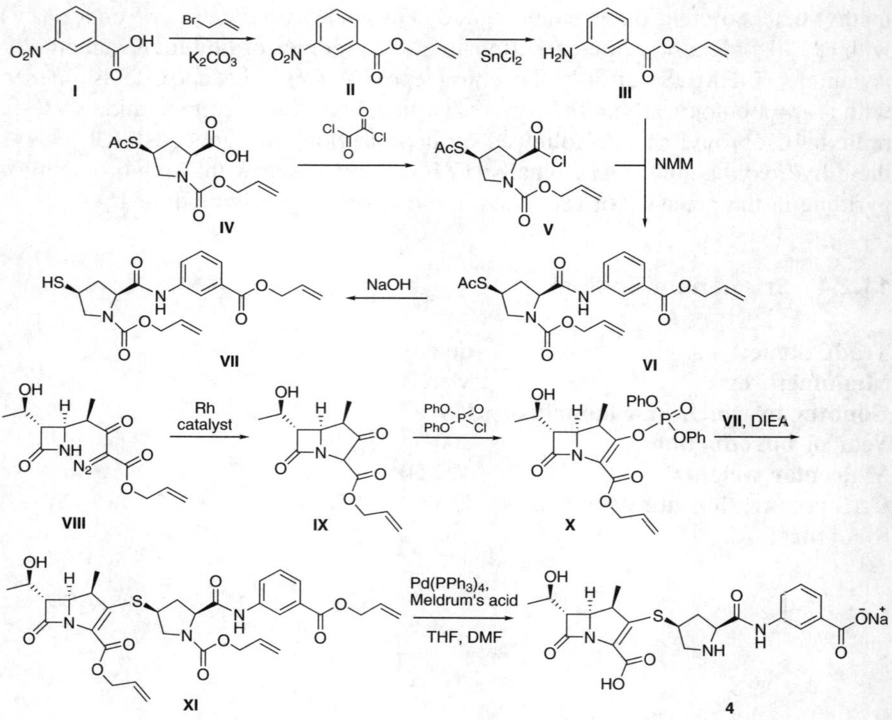

SCHEME 11.6 Synthesis of ertapenem sodium (**4**).

and the only absolute contraindication is a previous anaphylactic reaction to ertapenem or other β-lactam antibiotics.

There is one synthetic route disclosed in Merck's patent [16,17]. As illustrated in Scheme 11.6, esterification of 3-nitrobenzoic acid (**I**) with allyl bromide in the presence of K_2CO_3 in DMF provides the corresponding allyl ester (**II**), which is reduced with $SnCl_2$ in refluxing ethyl acetate to yield the 3-aminobenzoate (**III**). The condensation of **III** with the pyrrolidinecarbonyl chloride (**V**), which is obtained by reaction of the corresponding carboxylic acid (**IV**) with oxalyl chloride, in the presence of NMM in CH_2Cl_2 affords the expected amide (**VI**), which is further deacetylated with NaOH in allyl alcohol to give sulfanylpyrrolidine (**VII**). The cyclization of 2-diazo-4-[3-(1-hydroxyethyl)-4-oxoazetidin-2-yl]-3-oxopentanoic acid allyl ester (**VIII**) is achieved in the presence of rhodium octanoate to yield the 2-oxocarbapenam derivative (**IX**), which by reaction with chlorophosphoric acid diphenyl ester and DIEA in acetonitrile affords the enol phosphate (**X**). Condensation of **X** with the sulfanylpyrrolidine (**VII**) under basic conditions affords the corresponding adduct MK-0826 diallyl ester (**XI**), which is finally deprotected with Pd(PPh$_3$)$_4$ in the presence of Meldrum's acid and further converted to the sodium salt (**4**).

11.2.5 Rivastigmine Tartrate

Trade name:	Exelon
Manufacturer:	Novartis
Country in which first launched:	US
Year of introduction:	2007
Molecular weight:	400.43
CAS registration number:	129101-54-8
Structure:	

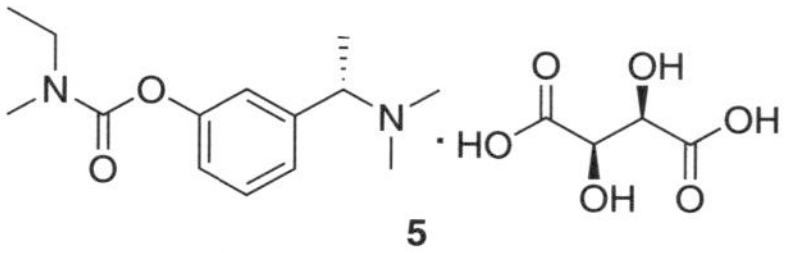

FIGURE 11.5 Structure of rivastigmine tartrate (**5**).

Rivastigmine tartrate (**5**, Fig. 11.5) is a cholinesterase inhibitor. By blocking the cholinesterase enzyme from breaking down acetylcholine, it increases both the level and duration of action of the neurotransmitter acetylcholine. In 2007, the U.S. FDA approved both rivastigmine capsules and rivastigmine patch for the treatment of mild to moderate dementia of the Alzheimer's type and for mild to moderate dementia related to Parkinson disease. Rivastigmine has been used by more than 6 million patients worldwide and has demonstrated significant efficacy in improving the cognitive (thinking and memory), functional (activities of daily living), and behavioral problems that are commonly associated with Alzheimer [18] and Parkinson [19] disease dementias. In particular, rivastigmine appears to have marked effects in patients showing a more aggressive course of disease, such as those with a younger age of onset or a poor nutritional status, or those experiencing symptoms such as delusions or hallucinations. Rivastigmine may cause some common side effects such as nausea and vomiting. The rivastigmine patch provides a clinical effect comparable to a common 12 mg/day capsule, but with three times fewer reports of nausea and vomiting.

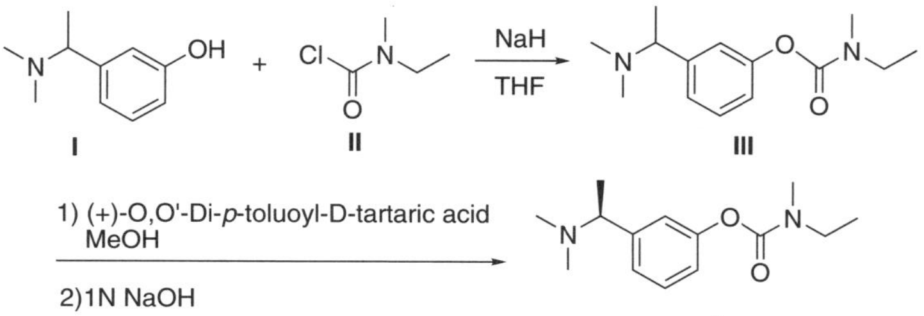

SCHEME 11.7 Synthesis of rivastigmine (**5**).

The synthesis of rivastigmine was developed by Enz et al. and published in both patents [20] and literature [21]. As shown in Scheme 11.7, reaction of 3-[1-(dimethylamino) ethyl] phenol (**I**) with *N*-ethyl-*N*-methylcarbamoyl chloride (**II**) and sodium hydride in THF provides racemic *N*-ethyl-*N*-methylcarbamic acid 3-[1-(dimethylamino)ethyl]phenyl ester (**III**), which is optically resolved by recrystallization with (+)-*O,O'*-di-*p*-toluoyl-D-tartaric acid in methanol followed by treatment with base to afford rivastigmine.

11.2.6   Cisatracurium Besilate

Trade name:	Nimbex/Nimbium
Manufacturer:	Abbott/GlaxoSmithKline
Country in which first launched:	US
Year of introduction:	1995
Molecular weight:	1243.48
CAS registration number:	96946-42-8
Structure:	

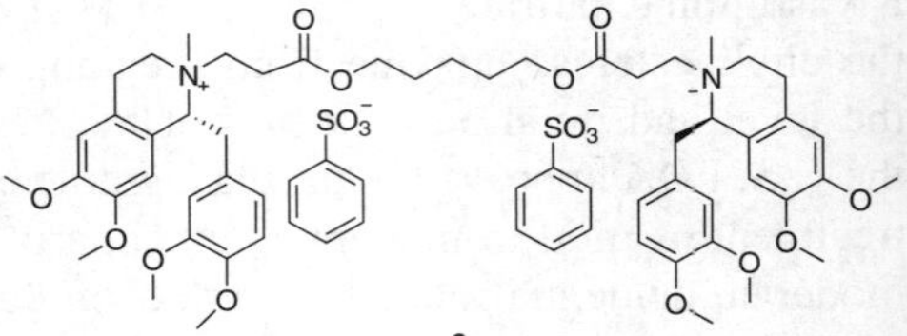

FIGURE 11.6   Structure of cisatracurium besilate (**6**).

Cisatracurium besilate (besylate) (**6**, Fig. 11.6) is a nondepolarizing neuromuscular blocking agent. It inhibits neuromuscular transmission by competing with acetylcholine for the cholinergic receptors at the motor end plate, thereby antagonizing the action of acetylcholine. Cisatracurium was codeveloped by Abbott and GlaxoSmithKline and received approval from the US FDA in 1995 as an adjunct to general anesthesia to facilitate endotracheal intubation and to induce skeletal muscle relaxation in surgical patients [22]. Cisatracurium besilate is the *R-cis*, *R'-cis* isomer of atracurium besilate and is approximately three-fold more potent than the mixture of isomers that constitute the parent drug. Cisatracurium has an intermediate onset of action and is therefore not recommended for use in rapid-sequence endotracheal intubation. Eighty percent of a dose is metabolized via Hofmann elimination, and the active metabolites of cisatracurium contain less laudanosine (which causes hypotension, central nervous system excitement, and seizures) than atracurium.

The synthesis of cisatracurium besilate commences with 1,2,3,4-tetrahydropapaverine (**IV**) [23] (Scheme 11.8). The optical resolution of **IV** with *N*-acetyl-L-leucine yields (*R*)-tetrahydropapaverine (**V**), which is condensed with 1,5-pentamethylene diacrylate (**III**) [obtained by esterification of 1,5-pentanediol

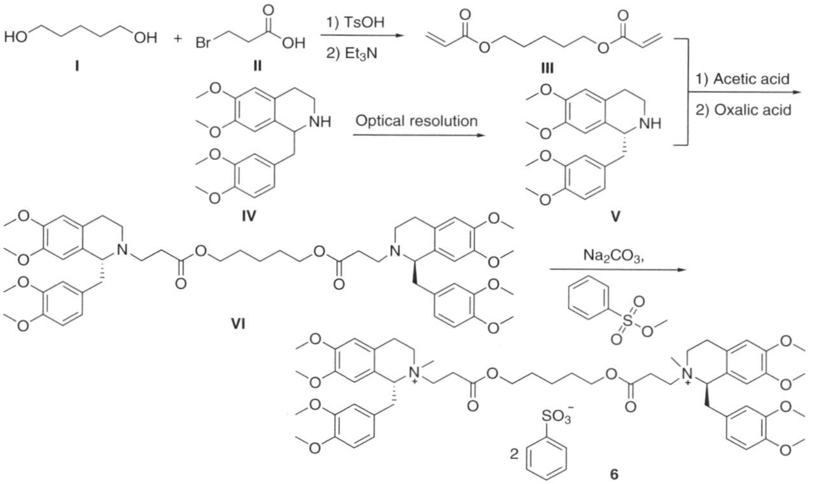

SCHEME 11.8   Synthesis of cistracurium besilate (**6**).

(**I**) with 3-bromopropionic acid in the presence of p-toluenesulfonic acid followed by dehydrobromination with triethylamine] in hot glacial acetic acid and treated with oxalic acid to afford the bis-tetrahydropapaverine derivative (**VI**). Finally, this compound is treated with aqueous Na_2CO_3 to eliminate the oxalic acid and then treated with methyl benzenesulfonate at room temperature. The resulting product is a 58:34:6 mixture of the (1R-cis, 1'R-cis)-, (1R-cis, 1'R-$trans$)- and (1R-$trans$, 1'R-$trans$)-isomers, which is separated by column chromatography over silica gel using an 80:20:5 mixture of dichloromethane methanol and methanesulfonic acid.

11.2.7   Darifenacin

Trade name:	Enablex, Emselex
Originator:	Pfizer
Country in which first introduced:	Germany
Year of introduction:	2005
Molecular weight:	426.55
CAS registration number:	133099-04-4
Structure:	

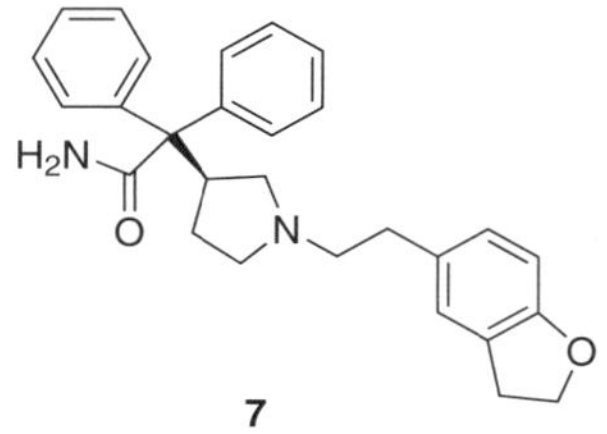

FIGURE 11.7   Structure of darifenacin (**7**).

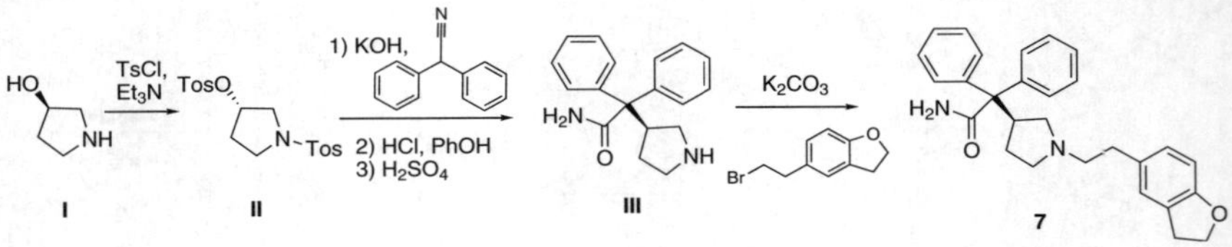

SCHEME 11.9   Synthesis of darifenacin (**7**).

Darifenacin (**7**, Fig. 11.7), a novel muscarinic M3 receptor selective antagonist, was launched for the treatment of urinary incontinence and overactive bladder. The overactivity of the detrusor muscle will lead to overactive bladder symptoms, and the mAChR M3 subtype in the detrusor is responsible for voiding contractions. Darifenacin selectively antagonizes the activity of the muscarinic M3 receptor to lower the overactivity of the detrusor muscle. Darifenacin has a higher level of M3 selectivity than the previously marketed antimuscarinic agents. In addition, darifenacin demonstrates greater effect on tissues in which the predominant receptor type is M3 rather than M1 or M2. In animal models, it shows greater selectivity for inhibition of detrusor contraction over salivation or tachycardia. The main adverse effects of darifenacin are dry mouth, constipation, dyspepsia, and urinary tract infection [24].

There is a synthetic route published in reference articles [25–28]. Starting from 3-(*R*)-hydroxypyrrolidine (**I**) (Scheme 11.9), it is tosylated on the nitrogen atom and through Mitsunobu reaction to introduce a tosyloxy group and finish the stereochemical inversion of the 3-position. The resulting intermediate (**II**) is anionic alkylated with diphenylacetonitrile, cleavage of the *N*-tosyl protecting group, and hydrolysis of the cyano group to give the intermediate (**III**). The intermediate (**III**) is reacted with 5-(2-bromoethyl)-2, 3-dihydrobenzofuran by an SN$_2$ reaction to give darifenacin.

11.2.8   Ramelteon

Trade name: Rozerem
Originator: Takeda
Country in which first introduced: US
Year of introduction: 2005
Molecular weight: 259.34
CAS registration number: 196597-26-9
Structure:

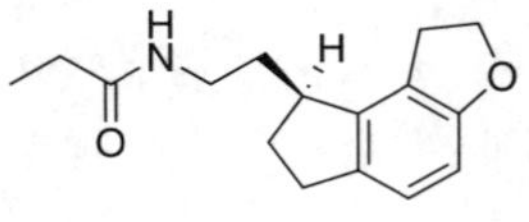

8

FIGURE 11.8   Structure of ramelteon (**8**).

Ramelteon (**8**, Fig. 11.8), an agonist of the melatonin receptor, was launched for the treatment of insomnia. It has high selectivity for the MT1 and MT2 subtypes, which are responsible for the maintenance of circadian rhythms, over the MT3 subtype, responsible for other melatonin functions. It has little affinity not only to the GABA receptor but also to the neurotransmitter, dopaminergic, opiate, and benzodiazepine receptors, which means an improved drug safety profile devoid of many side effects. By substitution of the methoxy group of melatonin with a stereoconstrained furan ring, the oxygen is fixed in a way that is favorable for selective binding to the MT1 receptor. Furthermore, the (*S*)-configuration of the side chain also contributes to the selectivity for MT1 ($K_i = 13.8$ pM) over MT3 ($K_i = 2.6$ mM). The adverse effects from clinical studies were somnolence, dizziness, nausea, fatigue, headache, and insomnia [29–31].

The synthesis of ramelteon is as follows (Scheme 11.10) [32,33]: 6-Methoxyindanone (**I**) is reacted with diethyl cyanomethyl phosphonate to give the unsaturated nitrile which is then reduced and acylated with propionyl chloride to give the intermediate (**II**). The intermediate **II** is asymmetrically hydrogenated with a ruthenium-BINAP catalyst to give the enantiomerically pure (*S*)-isomer (**III**). The (*S*)-isomer (**III**) is brominated at the C-5 position, demethylated, and allylated to yield the intermediate (**IV**). The intermediate (**IV**) is converted to compound (**V**) by Claisen rearrangement. The subsequent ozonolysis of the vinyl moiety, hydrogenation, conversion of the primary alcohol to the mesylate, and finally ring closure give ramelteon (**8**).

Ramelteon can also be synthesized as follows (Scheme 11.11) [34,35]: 6-Hydroxy-1-indanone (**I**) is reacted with allyl bromide to give an *O*-allyl indanone, which was converted to intermediate (**II**) by Claisen rearrangement. The subsequent ozonolysis of the vinyl moiety, hydrogenation, conversion of the primary alcohol to the tosylate, and finally ring closure give the intermediate (**III**). The intermediate (**III**) is reacted with benzyl *P,P*-diethylphosphonoacetate and the subsequent hydrolysis and hydrogenation give the intermediate (**IV**), which gives the key intermediate (**V**) by resolution with (*R*)-α-methyl-benzylamine. The intermediate (**V**) is reacted with $SOCl_2$ and NH_3H_2O to give an amide, which is reduced to amine. Finally, the amine is reacted with propionic anhydride in pyridine to give ramelteon.

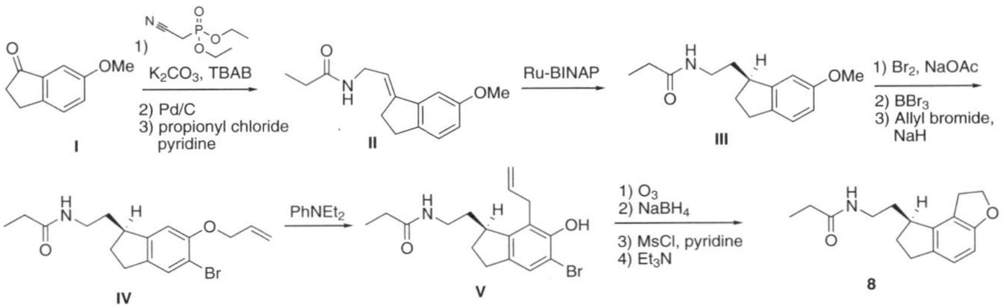

SCHEME 11.10   Yamashita's synthesis of ramelteon (**8**).

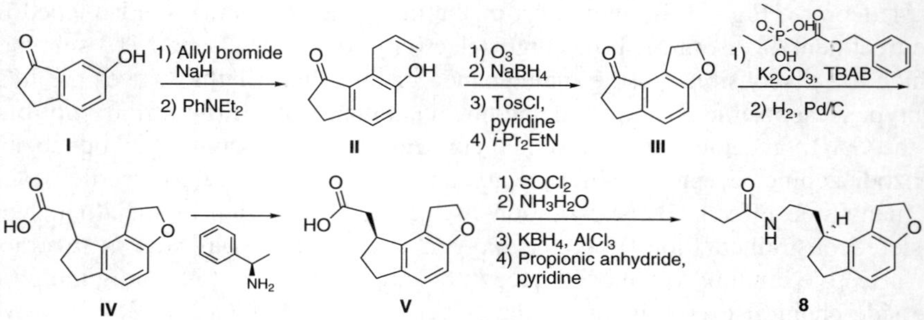

SCHEME 11.11 Process for preparing ramelteon (**8**).

11.2.9 Ivabradine

Trade name:	Procoralan
Originator:	Servier
Country in which first introduced:	UK
Year of introduction:	2006
Molecular weight:	505.05
CAS registration number:	148849-67-6
Structure:	

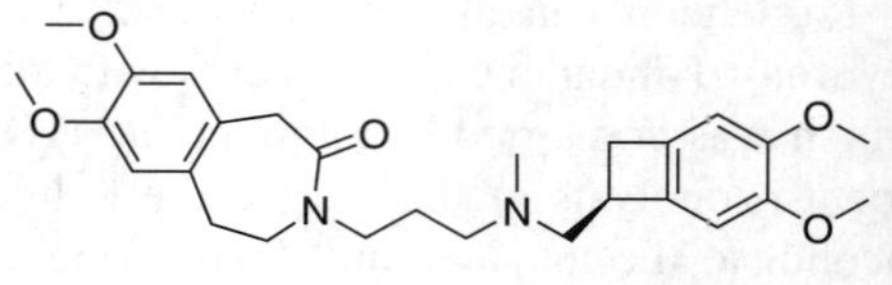

FIGURE 11.9 Structure of ivabradine (**9**).

Ivabradine (**9**, Fig. 11.9), a specific inhibitor of I(f) current through its contact with f-channels on the intracellular side of the plasma membrane, is used for the treatment of angina. Ivabradine is a new heart rate-reducing compound that acts specifically on the sinoatrial (SA) node. It reduces the speed of diastolic depolarization and decreases heart rate. It has been approved for the treatment of chronic stable angina and provides a viable alternative to patients with a contraindication or intolerance of β-blockers. Evaluation is also under way for the potential treatment of ischemic heart disease. The most common adverse effects are visual disturbances, headache, and dizziness [36–38].

The preparation of ivabradine is as follows (Scheme 11.12) [39–41]: Starting with 4,5-dimethoxybenzocyclobutane-1-carbonitrile (**I**), it is reduced with borane in THF and then acylated with ethyl chloroformate. The resulting carbamate is treated with LiAlH₄ to provide racemic N-(4,5-dimethoxybenzocyclobutan-1-yl-N-methylamine (**II**). The desired (S)-enantiomer (**III**) is achieved by optical

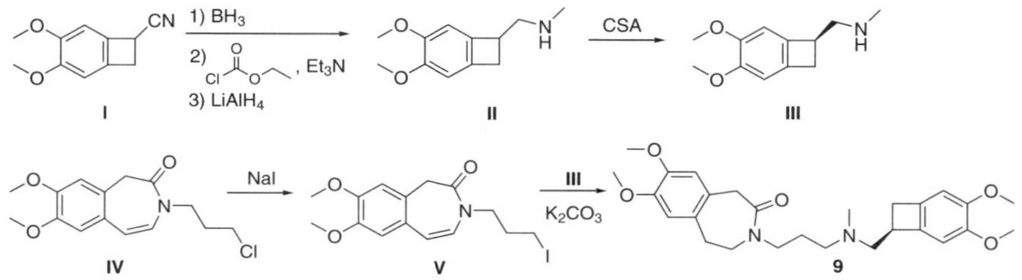

SCHEME 11.12   Synthesis of ivabradine (**9**).

resolution of (**II**) with CSA. 3-(3-Chloropropyl)-7,8-dimethoxy-2,3-dihydro-1H-3-benzazepin-2-one (**IV**) is converted into its corresponding iodide (**V**) by reaction with NaI. The intermediate (**III**) is then condensed with the terminal iodide (**V**). The double bond of the final intermediate (**VI**) is reduced with hydrogen over Pd(OH)$_2$ in AcOH to generate ivabradine. The active drug component is isolated as its hydrochloride salt.

11.2.10   Sitagliptin

Trade name:	Januvia
Originator:	Merck
Country in which first introduced:	Mexico
Year of introduction:	2006
Molecular weight:	407.31
CAS registration number:	486460-32-6
Structure:	

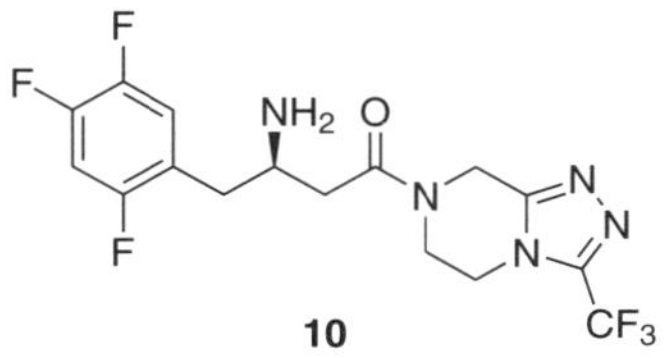

FIGURE 11.10   Structure of sitagliptin (**10**).

Sitagliptin (**10**, Fig. 11.10), a competitive dipeptidyl peptidase-4 (DPP-4) inhibitor, is a first-in-class oral drug launched for the treatment of type 2 diabetes. It acts by slowing the inactivation of incretins, which are endogenous peptides involved in the physiological regulation of glucose homeostasis. By inhibiting DPP-4, sitagliptin increases the concentration and duration of active incretin levels, such as the levels of GLP-1 and GIP, which in turn results in increased insulin release and decreased glucagon levels in a glucose-dependent manner. This drives blood glucose levels toward normal. Sitagliptin is a potent, competitive, reversible inhibitor of DPP-4. The (R)-enantiomer of sitagliptin is more potent than the (S)-enantiomer. In addition, sitagliptin is highly selective for DPP-4 versus other closely related enzymes in the DPP-4 gene family such as fibroblast activation protein-a, DPP-8, and DPP-9. The recommended regimen of sitagliptin for all approved indications is 100 mg once daily with or without food. After oral administration of a single 100-mg dose, sitagliptin exhibits an absolute bioavailability of approximately 87%, with peak plasma concentrations occurring 1−4 h postdose [42,43].

The initial chemical synthesis of sitagliptin is shown as follows (Scheme 11.13) [44]: Starting from 3-oxo-4-(2,4,5- trifluorophenyl)butyric acid methyl ester (**I**), it is first reduced enantioselectively to 3(S)-hydroxy-4-(2,4,5-trifluorophenyl)butyric acid methyl ester with hydrogen over (S)-BINAP-RuCl$_2$ catalyst, followed by hydrolysis to give the corresponding acid (**II**). The intermediate (**II**) is condensed with O-benzylhydroxylamine, cyclized to an azetidinone, and treated with LiOH to give 3(R)-(benzyloxyamino)-4-(2,4,5-trifluorophenyl)butyric acid (**III**). 2-Hydrazinopyrazine (**IV**) is first acylated by trifluoroacetic anhydride to give (**V**), followed by polyphosphoric acid-mediated cyclization to the triazolopyrazine, and subsequent hydrogenation to give 3-trifluoromethyl-5,6,7,8-tetrahydro-[1,2,4]-triazolo[4,3-a]pyrazine (**VI**). The reaction of intermediate (**III**) with (**VI**) and the subsequent hydrogenation provide the desired product sitagliptin.

A more convenient route to synthesize sitagliptin is shown as follows (Scheme 11.14): Starting from 2,4,5-trifluorophenyl acetic acid (**I**), it is

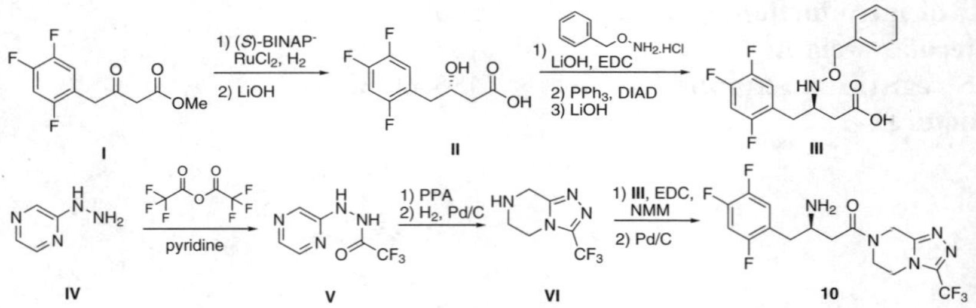

SCHEME 11.13   First-generation synthesis of sitagliptin (**10**).

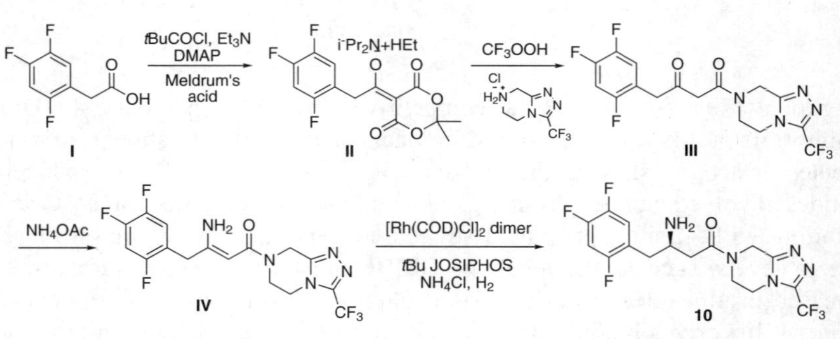

SCHEME 11.14   Second-generation synthesis of sitagliptin (**10**).

activated with pivaloyl chloride, then reacted with Meldrum's acid under the catalysis of DMAP to give intermediate (**II**). The intermediate (**II**) is reacted with 3-trifluoromethyl-5,6,7,8-tetrahydro-[1,2,4]-triazolo[4,3-a]pyrazine to give intermediate (**III**). The intermediate (**III**) is then converted to enamine (**IV**) by reaction with NH_4OAc. It should be emphasized that from (**I**) to (**IV**), this is an easily operated one-pot process. A key enantioselective reduction of (**IV**) with $[Rh(COD)Cl]_2$ dimer as metal precursor and tBu JOSIPHOS as ligand gives the product sitagliptin [45–49].

11.2.11 Ixabepilone

Trade name:	Ixempra
Originator:	BMS
Country in which first introduced:	US
Year of introduction:	2007
Molecular weight:	506.70
CAS registration number:	219989-84-1
Structure:	

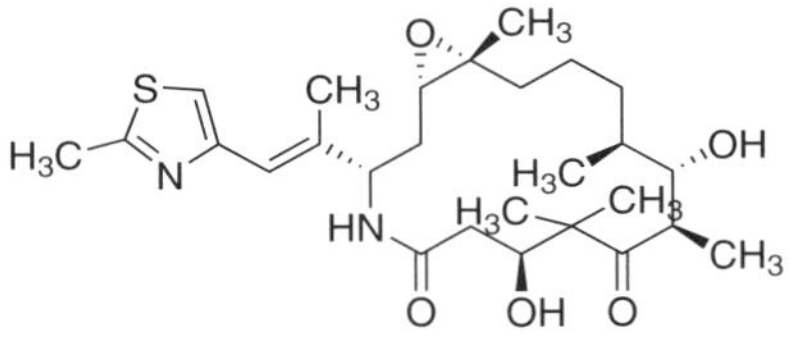

11

FIGURE 11.11 Structure of ixabepilone (**11**).

Ixabepilone (**11**, Fig. 11.11), a semisynthetic analog of epothilone B, was launched for the treatment of metastatic or locally advanced breast cancer. Ixabepilone is the first approved anticancer drug within the epothilone family. Epothilones are novel cytotoxic macrolides found in bacterial fermentation. Their mechanism of action involves binding to and stabilizing microtubules, which results in mitotic arrest and apoptosis. Epothilone B exhibits potent in vitro anticancer activity, but the activity is modest, owing to poor metabolic stability and unfavorable pharmacokinetics. Ixabepilone is a semisynthetic analog of epothilone B, in which the naturally existing lactone moiety is replaced by a lactam. It is indicated for use in combination with capecitabine in patients in whom treatment with anthracycline and taxane has previously failed. It is also approved as monotherapy for the treatment of metastatic or locally advanced breast cancer in patients whose tumors are resistant or refractory to anthracyclines, taxanes, and capecitabine. The most common adverse reactions associated with ixabelipone are peripheral sensory neuropathy, alopecia, fatigue, myalgia, nausea, vomiting, stomatitis, diarrhea, and musculoskeletal pain [50–53].

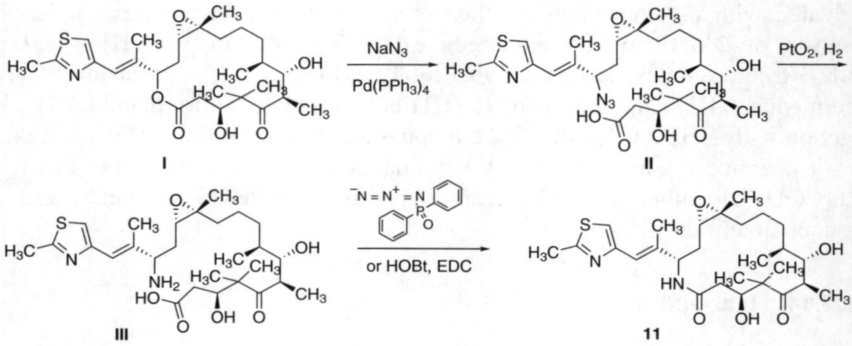

SCHEME 11.15 Synthesis of ixabepilone (**11**).

Ixabepilone is synthesized as follows (Scheme 11.15) [54–57]: Starting from epothilone B (**I**), the lactone group is cleaved with sodium azide and tetrakis(triphenylphosphine) palladium to an azido carboxylic acid intermediate (**II**), which is subsequently reduced to the corresponding amino acid with hydrogen over platinum oxide and cyclized by means of diphenylphosphoryl azide or HOBt and EDC to give the product ixabepilone.

11.2.12 Maraviroc

Trade name:	Selzentry
Originator:	Pfizer
Country in which first introduced:	US
Year of introduction:	2007
Molecular weight:	513.67
CAS registration number:	376348-65-1
Structure:	

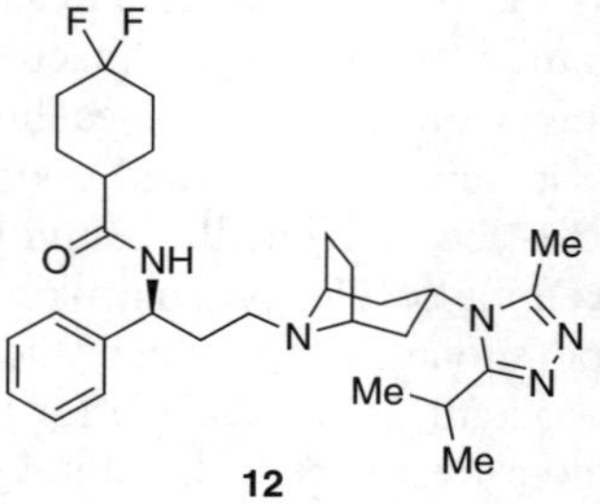

FIGURE 11.12 Structure of maraviroc (**12**).

Maraviroc (**12**, Fig. 11.12), the first CCR5 receptor antagonist, was launched for the treatment of HIV-1. The CCR5 receptor is one of two principal chemokine

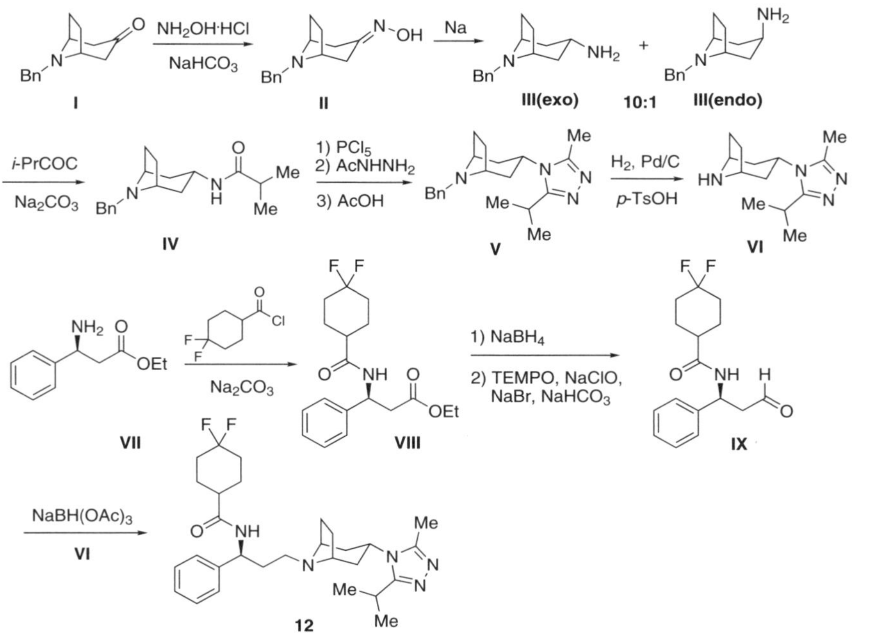

SCHEME 11.16 Synthesis of maraviroc (**12**).

coreceptors for viral entry into the host cell. Maraviroc binding slowly to CCR5 induces its conformational changes, thereby preventing CCR5 binding to the viral gp120 protein and the ultimate CCR5-mediated virus-cell fusion that is necessary for HIV invasion. Maraviroc is approved for use in combination with other antiretroviral drugs in adult patients with R5-tropic HIV-1 infection. With an IC_{50} of 11 nM for inhibition of gp120 binding to CCR5 and enhanced solubility compared to most HIV drugs, maraviroc has a lower pill burden. While there are no contraindications, maraviroc should be used with caution in patients with liver dysfunction, high risk of cardiovascular events, and postural hypotension [58–61].

Maraviroc (**12**) is synthesized as follows (Scheme 11.16): Starting from 8-benzyl-8-azabicyclo[3.2.1]octan-3-one (**I**), it is reacted with hydroxylamine to give an oxime (**II**). The intermediate (**II**) is reduced by Na in toluene to give a 10:1 mixture of *exo-/endo*-isomers (**III**). The intermediate (**III**) is reacted with *i*-PrCOCl to give amide (**IV**), which is crystallized to purge the undesired *endo*-amide isomer. The amide (**IV**) is subsequently cyclized with acetic hydrazide to give 1,2,4-triazole (**V**). The subsequent reduction gives the key intermediate (**VI**). Compound **VII** is reacted with 4,4-difluorocyclohexane carboxylic acid chloride to give the intermediate (**VIII**), which is reduced and oxidized to give the aldehyde (**IX**). The aldehyde (**IX**) is reacted with the intermediate (**VI**) to provide maraviroc [62–67].

11.2.13 Alvimopan

Trade name:	Entereg
Originator:	Lilly
Country in which first introduced:	US
Year of introduction:	2008
Molecular weight:	424.53
CAS registration number:	156053-89-3
Structure:	

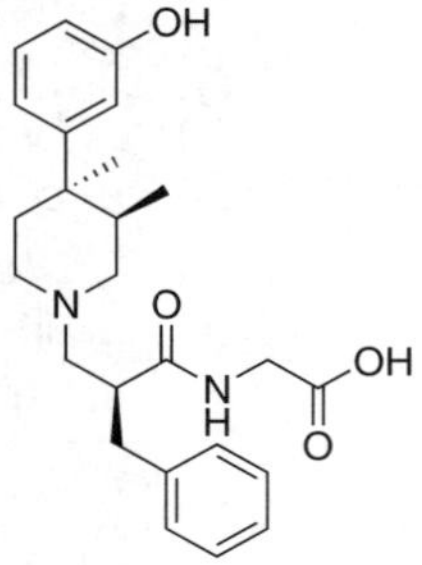

13

FIGURE 11.13 Structure of alvimopan (**13**).

Alvimopan (**13**, Fig. 11.13), a peripherally restricted antagonist of the μ-opioid receptor, is marketed for the oral treatment of postoperative ileus (POI) following bowel resection surgery. Ileus is a complication that affects almost all patients undergoing bowel surgery. It is thought that the activation of μ-opioid receptors in the GI tract by endogenous opioids, as well as by the use of opioid analgesics for the treatment of pain, plays an important role in the pathophysiology of POI. Activation of these peripheral μ-opioid receptors leads to an increase in colonic muscle tone and a decrease in propulsive activity in the GI tract. Hence, administration of opioids for postoperative pain relief can prolong POI. The adverse effect of opioids on the GI tract can be reversed with previously marketed μ-opioid receptor antagonists. However, these antagonists are also active in the CNS, which results in the inhibition of analgesic effects of the opioids as well. By contrast, the zwitterionic character and the high polarity of alvimopan restrict its penetration through the blood-brain barrier, thereby limiting its activity to the periphery. The most common adverse events associated with alvimopan treatment are delayed micturition, anemia, hypokalemia, dyspepsia, urinary retention, and back pain [68,69].

Alvimopan (**13**) can be synthesized as follows (Scheme 11.17): Starting from 3-(isopropoxy)bromobenzene (**I**), it is converted to Grignard's reagent and treated with 1,3-dimethyl-4-piperidinone to give the adduct, which is purified by crystallization in heptane to give *cis*-product (**II**). The *cis*-alcohol (**II**) is reacted with ethyl chloroformate, and the resulting amine is resolved with (−)-DTTA, followed by thermal elimination to give the olefin (**III**). The olefin (**III**) is treated with BuLi and then reacted with Me$_2$SO$_4$ to give the methyl product, which is

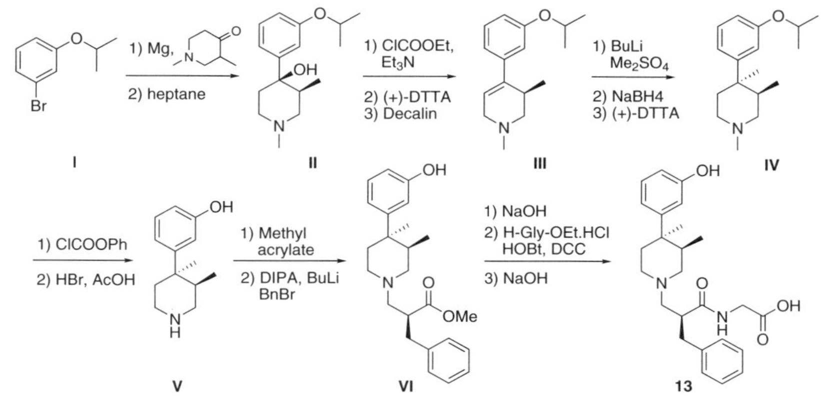

SCHEME 11.17 Synthesis of alvimopan (**13**).

reduced with NaBH₄ and resolved with (+)-DTTA to give the enantiomerically pure intermediate (**IV**). The intermediate (**IV**) is reacted with phenyl chloroformate and then treated with HBr/AcOH to give the amine (**V**). The amine (**V**) is reacted with methyl acrylate to give an adduct that was alkylated with LDA to give the intermediate (**VI**), which was hydrolyzed, condensed with glycine, and again hydrolyzed to give alvimopan [70,71].

11.2.14 Rivaroxaban

Trade name:	Xarelto
Originator:	Bayer
Country in which first introduced:	Canada
Year of introduction:	2008
Molecular weight:	435.88
CAS registration number:	366789-02-8
Structure:	

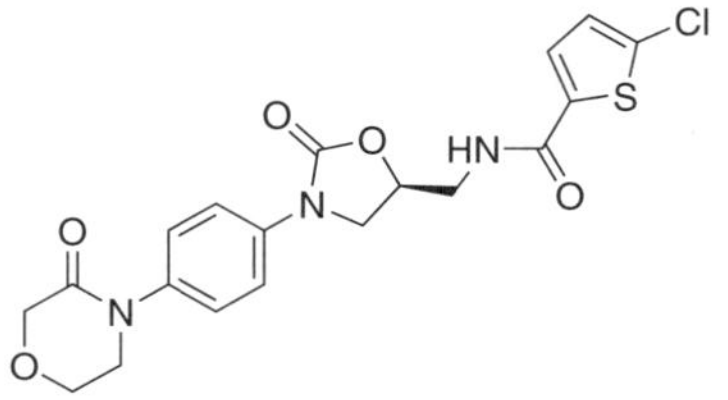

14

FIGURE 11.14 Structure of rivaroxaban (**14**).

Rivaroxaban (**14**, Fig. 11.14), a recently marketed inhibitor of factor Xa (FXa), was launched for the treatment of coagulation and venous thromboembolism. FXa is an attractive target for anticoagulation pathway, as it catalyzes the conversion of prothrombin to thrombin, blocking the burst of thrombin-mediated activation of

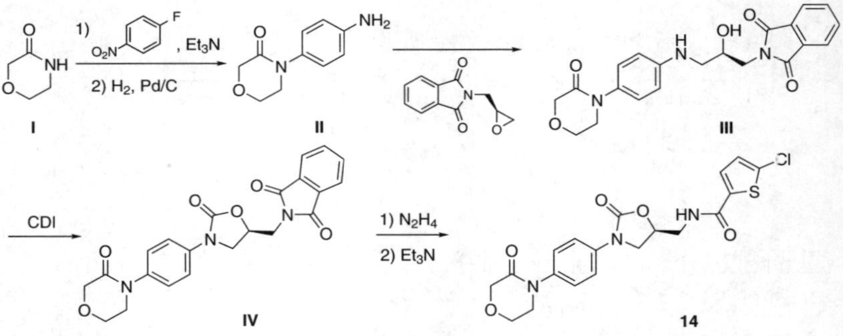

SCHEME 11.18   Synthesis of rivaroxaban (**14**).

coagulation. Rivaroxaban is an oxazolidinone derivative optimized for inhibiting both free FXa and FXa bound in the prothrombinase complex. It is a highly selective direct FXa inhibitor with oral bioavailability and rapid onset of action. It inhibits FXa with high potency and shows good selectivity over other related serine proteases. Regarding safety, owing to its mechanism of action, there is a bleeding risk with treatment with rivaroxaban. Other adverse effects include asthenia, dizziness, headache, and nausea [72–74].

Rivaroxaban is synthesized as follows (Scheme 11.18): Condensation of 3-morpholinone (**I**) with 4-fluoronitrobenzene followed by catalytic hydrogenation provides *N*-(*p*-aminophenyl)morpholinone (**II**). The intermediate (**II**) is reacted with (*S*)-2-(phthalimidomethyl)oxirane to give aminoalcohol adduct (**III**), which is cyclized by CDI to form the central oxazolidinone (**IV**). Deprotection and acylation with 5-chlorothiophene-2-carbonyl chloride afford Rivaroxaban [72,75–77].

11.2.15   Bortezomib

Trade name:	Velcade
Originator:	Millenium
Country in which first introduced:	US
Year of introduction:	2003
Molecular weight:	384.24
CAS registration number:	179324-69-7
Structure:	

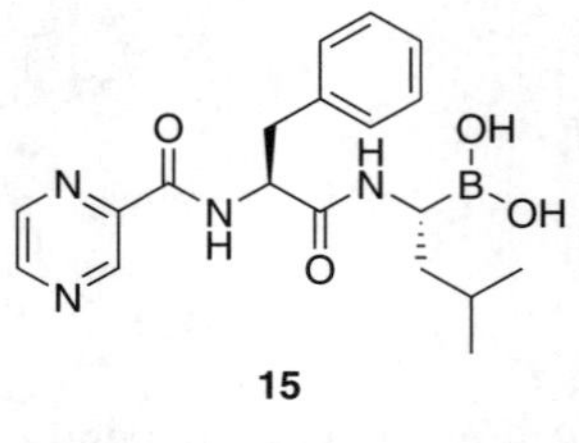

15

FIGURE 11.15   Structure of bortezomib (**15**).

Bortezomib (**15**, Fig. 11.15), a potent ubiquitin proteasome (26S) inhibitor, was launched in the US as a treatment for multiple myeloma. This proteasome is required for the proteolytic degradation of the majority of cellular proteins and is present in all cells. It is required for the control of inflammatory processes, cell cycle regulation, and gene expression. In recent years it has become a novel target in cancer treatment. Inhibition of the proteasome may prevent degradation of proapoptotic factors, permitting activation of programmed cell death in neoplastic cells dependent upon suppression of proapoptotic pathways. Bortezomib is a N-acyl-pseudodipeptidyl boronic acid that is formulated as a mannitol ester. The boronic acid moiety provides some measure of selective proteasome inhibition relative to many other serine proteases. The most common adverse events are gastrointestinal (GI) effects and asthenia. Other adverse effects include peripheral neuropathy, myelosuppression, and shingles [78–81].

There are mainly two routes to synthesize bortezomib, and we introduce the route shown as follows (Scheme 11.19): Starting from N-tert-butanesulfinyl aldimine (**I**), it is borylated under the catalysis of (ICy)CuOt-Bu to give the intermediate (**II**). Removal of the N-sulfinyl group and condensation with L-Boc-Phe-OH gives the intermediate (**III**), which is deprotected and coupled with 2-pyrazinecarboylic acid to give pinacol boronate, which is hydrolyzed to give bortezomib [82–85].

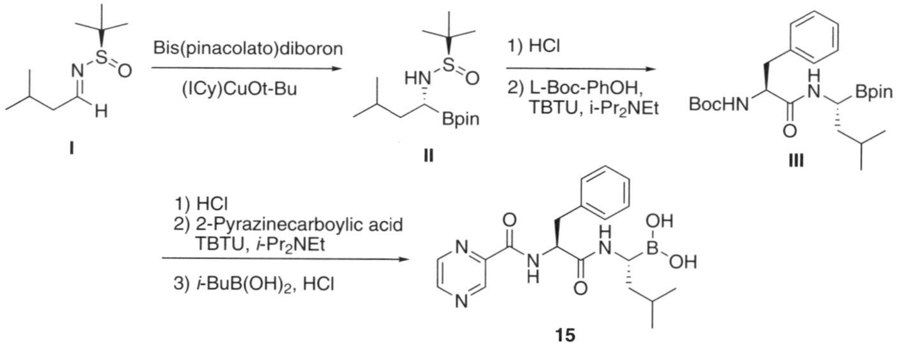

SCHEME 11.19   Synthesis of bortezomib (**15**).

11.2.16 Atorvastatin

Trade name:	Lipitor
Originator:	Warner-Lambert
Country in which first introduced:	UK
Year of introduction:	1997
Molecular weight:	558.64
CAS registration number:	134523-00-5

Structure:

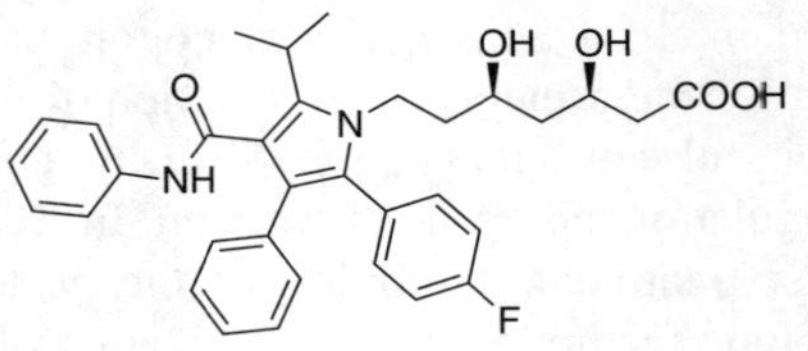

16

FIGURE 11.16   Structure of atorvastatin (**16**).

Atorvastatin (**16**, Fig. 11.16), a competitive inhibitor of HMG-CoA reductase, was launched for lowering blood cholesterol. HMG-CoA reductase is an enzyme found in liver tissue that plays a key role in production of cholesterol in the body. It catalyzes the reduction of 3-hydroxy-3-methylglutaryl-coenzyme A to meval-onate, which is the rate-limiting step in hepatic cholesterol biosynthesis. Inhibition of the enzyme decreases de novo cholesterol synthesis, increasing expression of low-density lipoprotein (LDL) receptors on hepatocytes. This increases LDL uptake by the hepatocytes, decreasing the amount of LDL-cholesterol in the blood. Atorvaststin also stabilizes plaque and prevents strokes through anti-inflammatory and other mechanisms. Atorvastatin has rapid oral absorption, with an approximate time of 1–2 hours to maximum plasma concentration. The most common side effects are weakness, dizziness, chest pain, and insomnia [86–89].

Atorvastatin can be synthesized as follows (Scheme 11.20): Starting from the chiral ester (**I**), it is reacted with diphenylamine acetamide and reduced with Et$_2$BOMe and NaBH$_4$ to give the diol (**II**). The diol (**II**) is protected and reduced with Raney Ni to give the amine (**III**). The intermediate (**IV**) is reacted first with

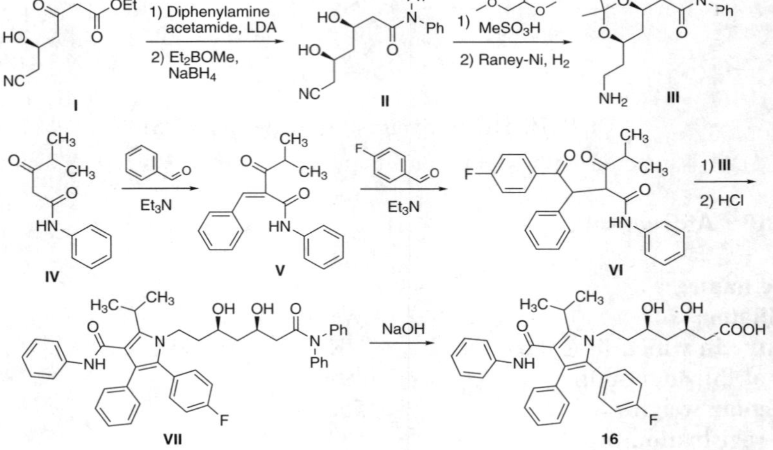

SCHEME 11.20   Synthesis of atorvastatin (**16**).

benzaldehyde and then with 4-fluorobenzaldehyde to give the intermediate (**VI**), which was reacted with the amine (**III**) and deprotected to give atorvastatin [90–93].

11.2.17 Clopidogrel

Trade name:	Plavix
Originator:	Sanofi
Country in which first introduced:	US
Year of introduction:	1998
Molecular weight:	321.82
CAS registration number:	113665-84-2
Structure:	

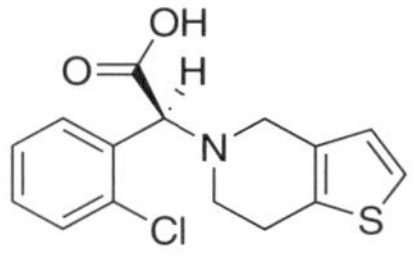

17

FIGURE 11.17 Structure of clopidogrel (**17**).

Clopidogrel (**17**, Fig. 11.17), an oral thienopyridine class antiplatelet agent, is used to inhibit blood clots in coronary artery disease, peripheral vascular disease, and cerebrovascular disease. Clopidogrel is a pro-drug. When it metabolizes in the body, it can specifically and irreversibly inhibit the P2Y12 subtype of ADP receptor, which is important in aggregation of platelets and cross-linking by the protein fibrin. Blockade of this receptor inhibits platelet aggregation by blocking activation of the glycoprotein IIb/IIIa pathway. The IIb/IIIa complex functions as a receptor mainly for fibrinogen and vitronectin but also for fibronectin and von Willebrand factor. Activation of this receptor complex is the "final common pathway" for platelet aggregation and is important in the cross-linking of platelets by fibrin. Adverse effects include hemorrhage, severe neutropenia, and thrombotic thrombocytopenic purpura (TTP) [94–96].

Clopidogrel can be synthesized as follows (Scheme 11.21): Starting from the acid (**I**), it is reacted with MeOH and then resolved or resolved first and then

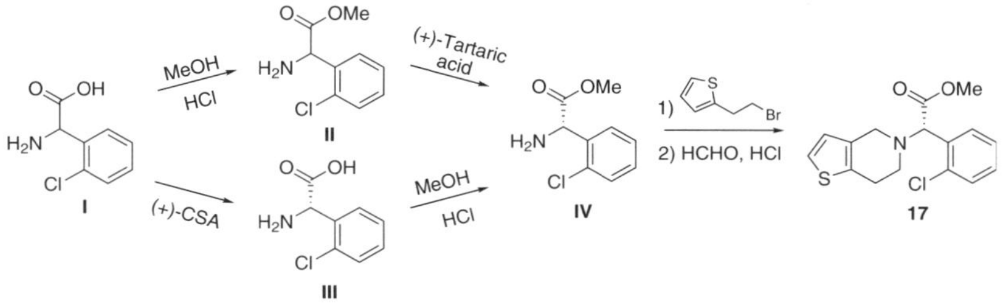

SCHEME 11.21 Synthesis of clopidogrel (**17**).

reacted with MeOH to give the chiral intermediate (**IV**), which is reacted with 2-(2-bromoethyl) thiophene and cyclized to give clopidogrel [97–99].

11.2.18   Esomeprazole

Trade name:	Nexium
Originator:	Astrazenica
Country in which first introduced:	Sweden
Year of introduction:	2001
Molecular weight:	345.42
CAS registration number:	119141-88-7
Structure:	

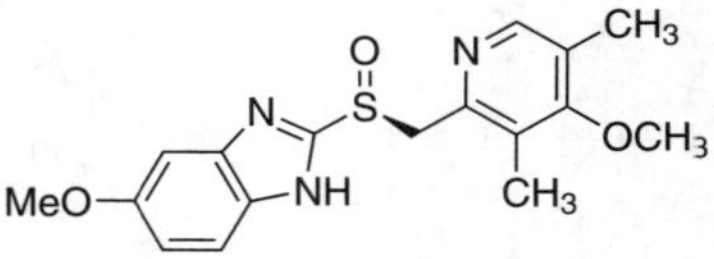

18

FIGURE 11.18   Structure of esomeprazole (**18**).

Esomeprazole (**18**, Fig. 11.18), a proton pump inhibitor, was launched for the treatment of dyspepsia, peptic ulcer disease, gastroesophageal reflux disease, and Zollinger–Ellison syndrome. Esomeprazole is the (*S*)-enantiomer of omeprazole, and it is claimed that the (*S*)-enantiomer has improved efficacy over the racemic mixture of omeprazole. However, this greater efficacy has been disputed, with some claiming it offers no benefit over its older form. Common side effects include headache, diarrhea, nausea, dry mouth, and abdominal pain. More severe side effects are severe allergic reactions, chest pain, dark urine, fast heartbeat, and fever [100–104].

Esomeprazole can be synthesized by either asymmetric synthesis or optical resolution [105–113]. The typical synthesis of esomeprazole is as follows

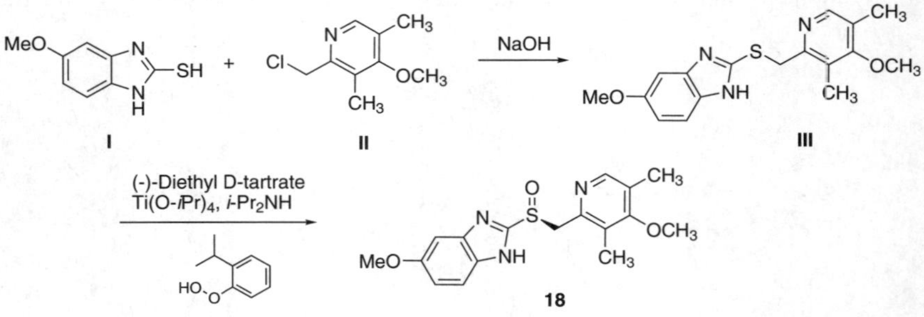

SCHEME 11.22   Synthesis of esomeprazole (**18**).

(Scheme 11.22): Reaction of thiol (**I**) and chloromethyl pyridine derivative (**II**) yields the thioether(**III**), which is enantiomerically oxidized to give esomeprazole.

11.2.19 Montelukast

Trade name:	Singulair
Originator:	Merck Frosst
Country in which first introduced:	US
Year of introduction:	1997
Molecular weight:	586.18
CAS registration number:	158966-92-8
Structure:	

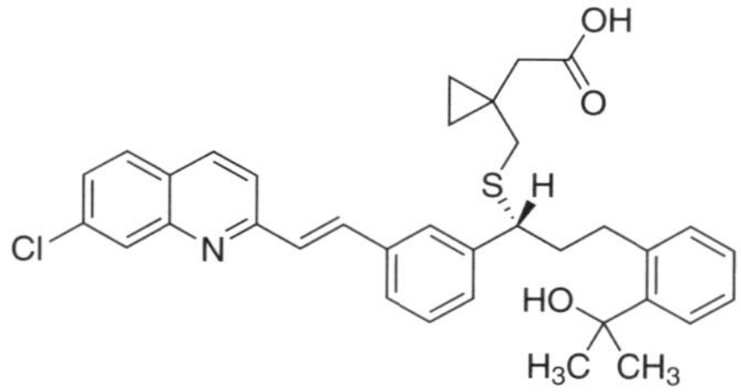

FIGURE 11.19 Structure of montelukast (**19**).

Montelukast (**19**, Fig. 11.19), a CysLT1 receptor antagonist, was launched for the maintenance treatment of asthma and to relieve symptoms of seasonal allergies. It is usually administered orally. Montelukast is an oral antagonist; it blocks the action of leukotriene D4 on the cysteinyl leukotriene receptor CysLT1 in the lungs and bronchial tubes by binding to it. This reduces the bronchoconstriction otherwise caused by the leukotriene, and results in less inflammation.

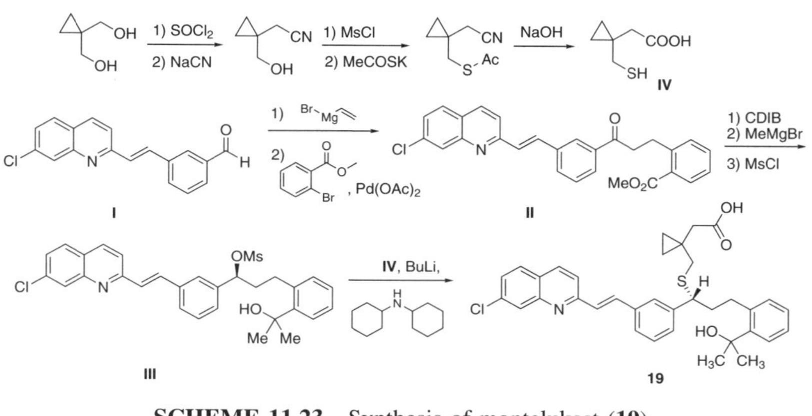

SCHEME 11.23 Synthesis of montelukast (**19**).

Adverse effects include gastrointestinal disturbances, sleep disorders, hypersensitivity reactions, and increased bleeding tendency, aside from many other generic adverse reactions [114,115].

Montelukast can be synthesized as follows (Scheme 11.23): The intermediate (**IV**) can be synthesized through simple reactions. The aldehyde (**I**) is reacted first with vinylmagnesium bromide to give olefin, which was coupled with 2-bromobenzoic acid methyl ester to give the intermediate (**II**). The intermediate (**II**) is first reduced with CDIB, then reacted with MeMgBr, and finally reacted with MsCl to give the intermediate (**III**). The intermediate (**III**) is reacted with thiol (**IV**) to give montelukast [116–119].

11.2.20 Rosuvastatin

Trade name: Crestor
Originator: Shionogi
Country in which first introduced: The Netherlands
Year of introduction: 2003
Molecular weight: 481.54
CAS registration number: 287714-41-4
Structure:

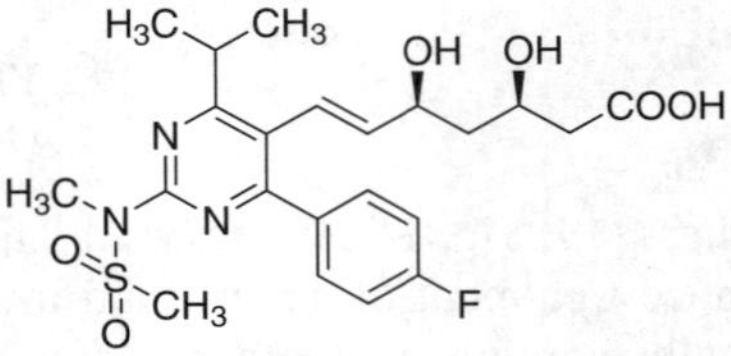

FIGURE 11.20 Structure of rosuvastatin (**20**).

Rosuvastatin (**20**, Fig. 11.20), a competitive inhibitor of HMG-CoA reductase, was launched to treat high cholesterol and related conditions, and to prevent cardiovascular disease. HMG-CoA reductase is an enzyme found in liver tissue that plays a key role in production of cholesterol in the body. It catalyzes the reduction of 3-hydroxy-3-methylglutaryl-coenzyme A to mevalonate, which is the rate-limiting step in hepatic cholesterol biosynthesis. Inhibition of the enzyme decreases de novo cholesterol synthesis, increasing expression of low-density lipoprotein receptors on hepatocytes. The most common side effects are weakness, dizziness, and insomnia. As with all statins, there is a concern of rhabdomyolysis, a severe undesired side effect, as well as a kidney toxicity warning [120–124].

The synthesis of rosuvastatin is as follows (Scheme 11.24): Starting from ethyl 2-isobutylacetate (**I**), it is reacted with 4-fluorobenzaldehyde, oxidized with DDQ, and then cyclized with S-methylthiourea to give the intermediate (**II**).The intermediate (**II**) is oxidized with mCPBA, reacted with MeNH₂, and then reacted

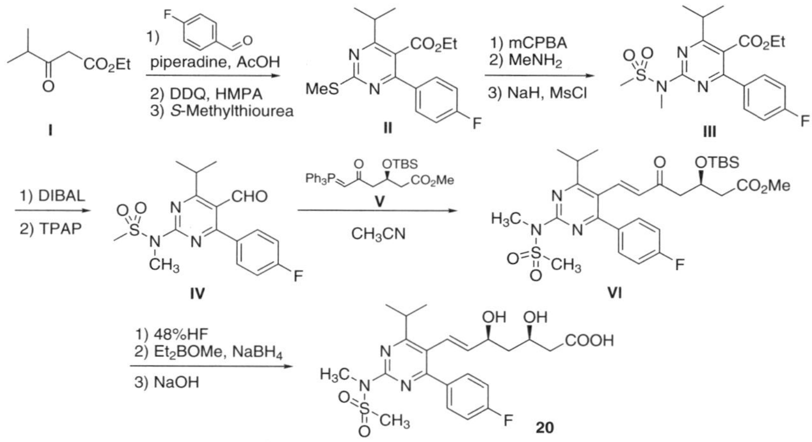

SCHEME 11.24   Synthesis of rosuvastatin (**20**).

with MsCl to give the intermediate (**III**). The intermediate (**III**) is reduced with DIBAL-H and oxidized with TPAP to give the aldehyde (**IV**), which is reacted with the intermediate (**V**) to give the adduct (**VI**). The adduct (**VI**) is reduced and deprotected to give rosuvastatin [125–128].

11.2.21   Ezetimibe

Trade name:	Ezetrol
Originator:	Schering-Plough
Country in which first introduced:	Germany
Year of introduction:	2002
Molecular weight:	409.43
CAS registration number:	163222-33-1
Structure:	

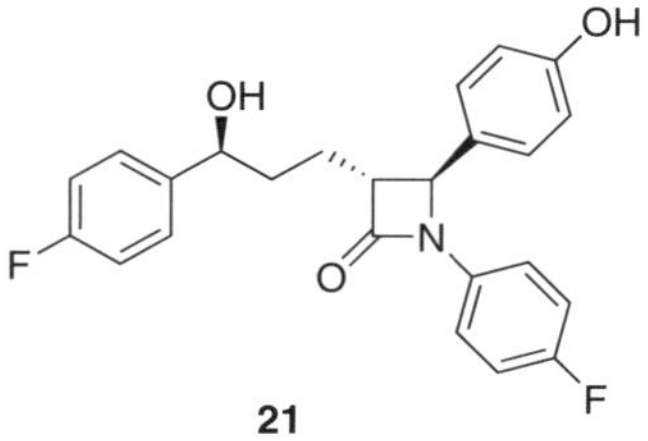

FIGURE 11.21   Structure of ezetimibe (**21**).

Ezetimibe (**21**, Fig. 11.21) is a drug that lowers cholesterol. It acts by localizing at the brush border of the small intestine and decreasing the absorption of cholesterol in the intestine. Specifically, it appears to bind to a critical mediator of cholesterol absorption, the Niemann-Pick C1-Like 1 (NPC1L1) protein in gastrointestinal

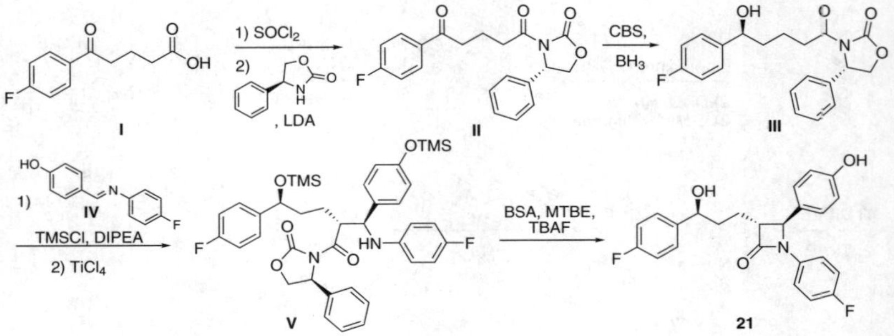

SCHEME 11.25   Synthesis of ezetimibe (**21**).

tract epithelial cells as well as in hepatocytes. In addition to this direct effect, decreased cholesterol absorption leads to an upregulation of LDL receptors on the surface of cells and an increased LDL-cholesterol uptake into cells, thus decreasing levels of LDL in the blood plasma that contribute to atherosclerosis and cardiovascular events. The common side effects include headache and/or diarrhea [129–133].

Ezetimibe can be synthesized as follows (Scheme 11.25): Starting from the acid (**I**), it is reacted with (*S*)-(+)-4-phenyl-2-oxazolidinone to give the intermediate (**II**), which is reduced with CBS and BH$_3$ to give the chiral alcohol (**III**). The chiral alcohol (**III**) is first reacted with the intermediate (**IV**) and TMSCl and then treated with TiCl$_4$ to give the intermediate (**V**), which is cyclized with BSA and deprotected with TBAF to give ezetimibe [134–138].

11.2.22   Gemcitabine

Trade name:	Gemzar
Originator:	Lilly
Country in which first introduced:	US
Year of introduction:	1995
Molecular weight:	263.2
CAS registration number:	95058-81-4
Structure:	

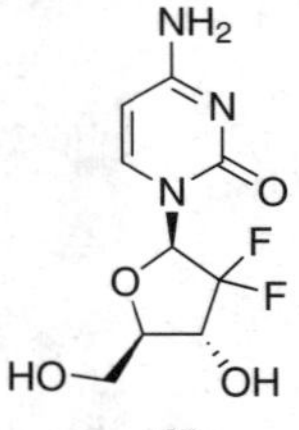

FIGURE 11.22   Structure of gemcitabine (**22**).

Gemcitabine (**22**, Fig. 11.22) is a nucleoside analog used as chemotherapy. As with fluorouracil and other analogs of pyrimidines, the triphosphate analog of gemcitabine replaces one of the building blocks of nucleic acids, in this case cytidine, during DNA replication. The process arrests tumor growth, as only one additional nucleoside can be attached to the "faulty" nucleoside, resulting in apoptosis. Another target of gemcitabine is the enzyme ribonucleotide reductase (RNR). The diphosphate analog binds to RNR active site and inactivates the enzyme irreversibly. Once RNR is inhibited, the cell cannot produce the deoxyribonucleotides required for DNA replication and repair, and cell apoptosis is induced. Gemcitabine is used in various carcinomas: non-small-cell lung cancer, pancreatic cancer, bladder cancer, and breast cancer. It is being investigated for use in esophageal cancer, and is used experimentally in lymphomas and various other tumor types. Gemcitabine represents an advance in pancreatic cancer care. The common side effects are flulike symptoms (muscle pain, fever, headache, chills, fatigue), vomiting, poor appetite, and skin rash [139–141].

The synthesis of gemcitabine is shown as follows (Scheme 11.26): Starting from difluorolatone (**I**), it is reacted with cinnamoyl chloride and recrystallization to give the intermediate (**II**). The intermediate (**II**) is reduced with LTBA and then reacted with $MeSO_2Cl$ to give the intermediate (**III**), which is reacted with N-acetyl cytosine and treated with Na_2CO_3 to give the intermediate (**IV**). The intermediate (**IV**) is deprotected with NH_3 and HCl to give gemcitabine [142–147].

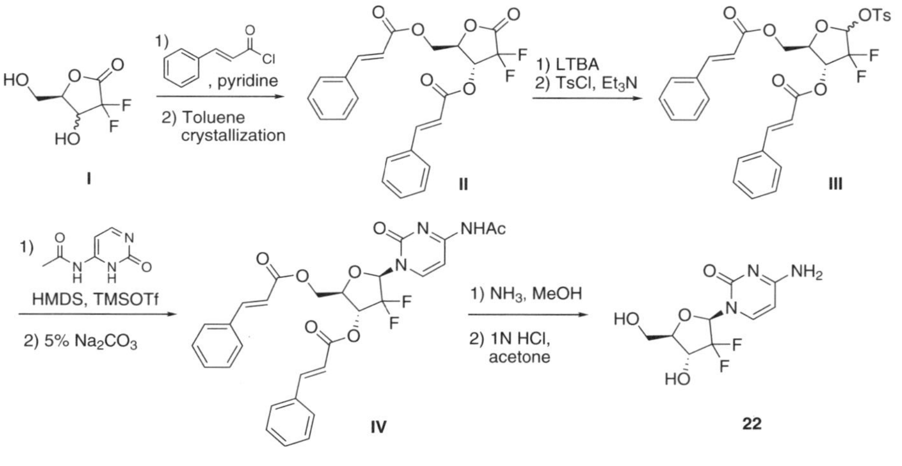

SCHEME 11.26 Synthesis of gemcitabine (**22**).

11.2.23   Latanoprost

Trade name: Xalatan
Originator: Columbia University
Country in which first introduced: US
Year of introduction: 1997
Molecular weight: 432.59
CAS registration number: 130209-82-4
Structure:

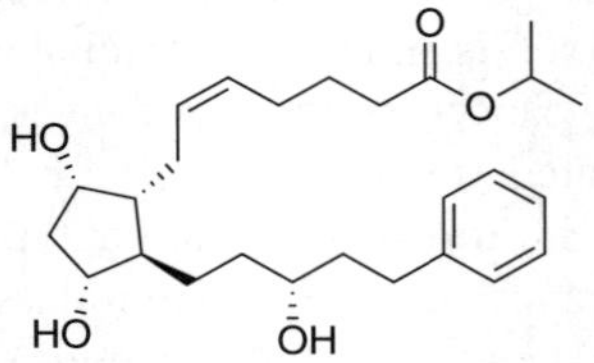

FIGURE 11.23   Structure of latanoprost (**23**).

Latanoprost (**23**, Fig. 11.23) is a prostaglandin analog used for controlling the progression of glaucoma or ocular hypertension by reducing intraocular pressure. It works by increasing the outflow of aqueous fluid from the eyes. The common side effects include reddening of the eyes, blurred vision, and eyelid redness [148–150].

Latanoprost can be synthesized as follows (Scheme 11.27): Starting from the benzoyl protected Corey lactone (**I**), it is converted to aldehyde and reacted with a dimethyl (2-oxo-4-phenylbutyl) phosphonate to give the olefin (**II**). The olefin (**II**)

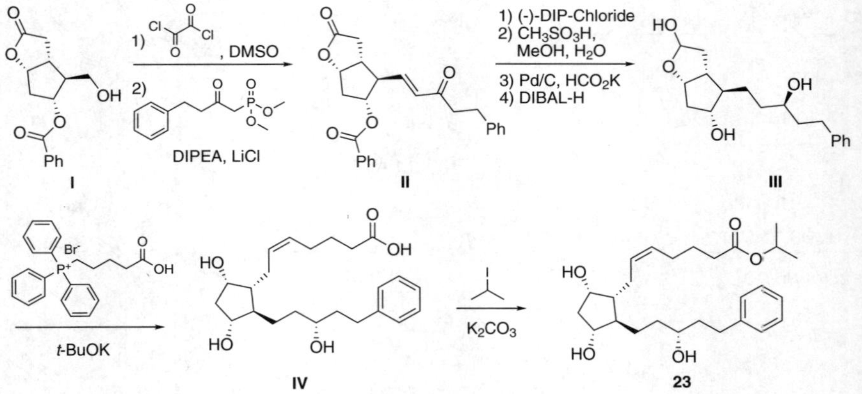

SCHEME 11.27   Synthesis of latanoprost (**23**).

is enantiomerically reduced by (−)-DIP-chloride, deprotected, and reduced first by Pd/C and then by DIBAL-H to give the intermediate (**III**). The intermediate (**III**) is reacted first with (4-carboxybutyl) triphenylphosphonium bromide and then with isopropyliodide to give latanoprost [151–155].

11.2.24 Levofloxacin

Trade name:	Levaquin
Originator:	Daiichi
Country in which first introduced:	US
Year of introduction:	1996
Molecular weight:	361.37
CAS registration number:	100986-85-4
Structure:	

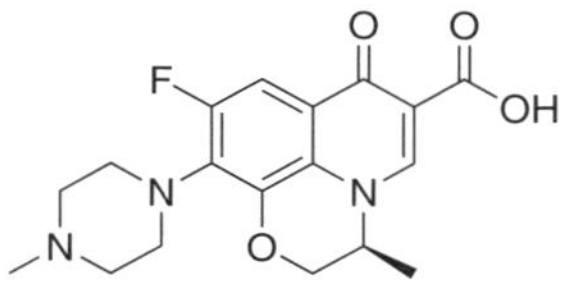

24

FIGURE 11.24   Structure of levofloxacin (**24**).

Levofloxacin (**24**, Fig. 11.24) is a synthetic chemotherapeutic antibiotic of the fluoroquinolone drug class and is used to treat common to severe bacterial infections. Levofloxacin is a (*S*)-enantiomer of ofloxacin and is more effective than the latter. Levofloxacin is a broad-spectrum antibiotic that is active against both gram-positive and gram-negative bacteria. It functions by inhibiting DNA gyrase, a type II topoisomerase, and topoisomerase IV, which is an enzyme necessary to separate replicated DNA, thus inhibiting cell division. Levofloxacin is rapidly and, in essence, completely absorbed after oral administration. The serious adverse effects that may occur as a result of levofloxacin therapy include irreversible peripheral neuropathy, spontaneous tendon rupture and tendonitis, toxic epidermal necrolysis, and *Clostridium difficile*-associated disease (CDAD) [156–162].

There are many routes to synthesize levofloxacin, and we introduce the route shown as follows (Scheme 11.28): Starting from 2,3,4,5-trifluorobenzoic acid (**I**), it is converted to the intermediate (**II**) using the common method in fluoroquinolone synthesis. The intermediate (**II**) is reacted with (S)-(+)-2-amino-1-propanol to give the chiral amino alcohol (**III**), which is treated first with NaH and then KOH to give the cyclized product (**V**). The intermediate (**V**) is reacted with *N*-methylpiperazine to give levofloxacin [163–167].

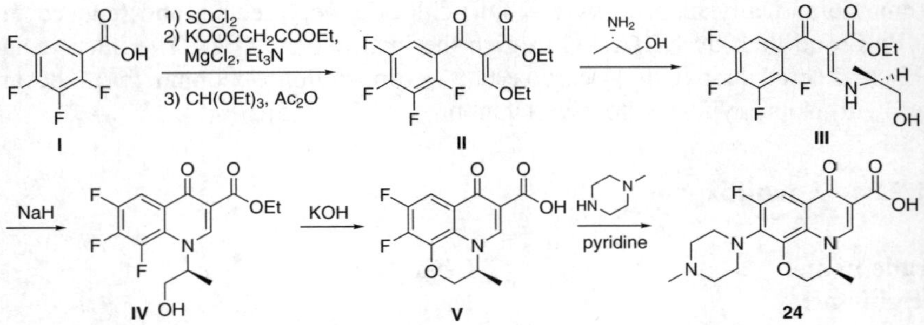

SCHEME 11.28   Synthesis of levofloxacin (**24**).

11.2.25   Efavirenz

Trade name: Sustiva
Originator: Dupont Merck Pharmaceutical Co.
Country in which first introduced: US
Year of introduction: 1998
Molecular weight: 315.67
CAS registration number: 154598-52-4
Structure:

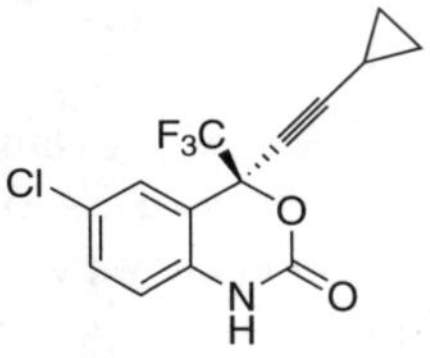

FIGURE 11.25   Structure of efavirenz (**25**).

Efavirenz (**25**, Fig. 11.25) is a nonnucleoside reverse transcriptase inhibitor and is used as part of highly active antiretroviral therapy for the treatment of human immunodeficiency virus (HIV) type 1. The virus reverse transcriptase enzyme is an essential enzyme to transcribe viral RNA into DNA. Efavirenz is not effective against HIV-2, as the pocket of the HIV-2 reverse transcriptase has a different structure. Efavirenz is used to treat HIV infection. It is never used alone and is always given in combination with other drugs. Psychiatric symptoms, including insomnia, confusion, memory loss, and depression, are common, and more serious symptoms such as psychosis may occur in patients with compromised liver or kidney function [168–170].

Efavirenz can be synthesized as follows (Scheme 11.29): Starting from *p*-chloroaniline (**I**), it is reacted with pivaloyl chloride to give the amide (**II**), which is treated with *n*-BuLi and reacted with ethyl trifluoroacetate to give the

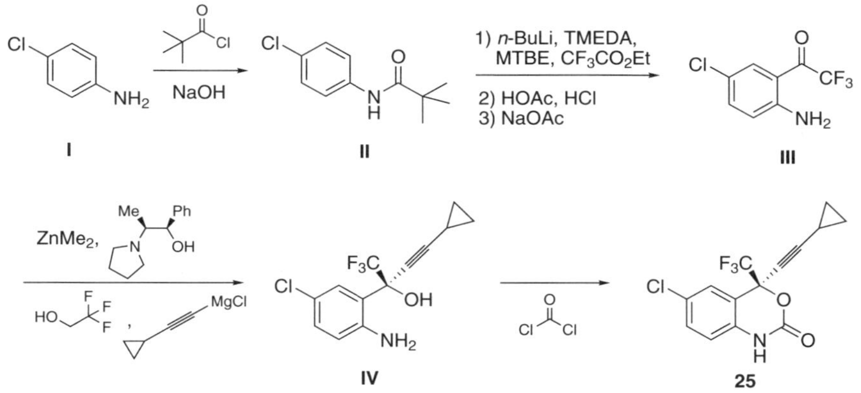

SCHEME 11.29   Synthesis of efavirenz (**25**).

intermediate (**III**). Chloromagnesium acetylide is treated with ZnMe$_2$ and reacted with the intermediate (**III**) in the presence of (1*R*,2*S*)-*N*-pyrrolidinylnorephedrine and 2,2,2-trifluoroethanol to give the adduct (**IV**). The adduct (**IV**) is cyclized by phosgene to give efavirenz [171–174].

REFERENCES

1. Robinson, M. J., Martin, B. A., Gootz, T. D., McGuirk, P. R., Osheroff, N. (1992). Effects of novel fluoroquinolones on the catalytic activities of eukaryotic topoisomerase II: Influence of the C-8 fluorine group. *Antimicrob. Agents Chemother*. **36**, 751–756.

2. *Drug card for Moxifloxacin*. Canada: DrugBank.

3. Petersen, U., Bremm, K. D., Dalhoff, A., Endermann, R., Heilmann, W., Krebs, A., Schenke T. (1996). Synthesis and in vitro activity of BAY 12-8039, a new 8-methoxy-quinolone. *36th Interscience Conference Antimicrobial Agents Chemotherapy* (Sept 15–18, New Orleans) Abstract F1.

4. Roche Laboratories, Inc. Tamiflu product information.

5. Shibasaki, M., Kanai, M. (2008). Synthetic strategies for oseltamivir phosphate. *Eur. J. Org. Chem*. **11**, 1839–1850.

6. Rohloff, J. C., Kent, K. M., Postich, M. J., Becker, M. W., Chapman, H. H., Kelly, D. E., Lew, W., Louie, M. S., McGee, L. R., Prisbe, E. J., Schultze, L. M., Yu, R. H., Zhang, L. J. (1998). Practical total synthesis of the anti-influenza drug GS-4104. *J. Org. Chem*. **63**, 4545–4550.

7. Yeung, Y. Y., Hong, S., Corey, E. J. (2006). A short enantioselective pathway for the synthesis of the anti-influenza neuramidase inhibitor oseltamivir from 1, 3-butadiene and acrylic acid. *J. Am. Chem. Soc*. **128**, 6310–6311.

8. Ishikawa, H., Suzuki, T., Hayashi, Y. (2009). High-yielding synthesis of the anti-influenza neuramidase inhibitor (−)-oseltamivir by three "one-pot" operations. *Angew. Chem. Int. Ed*. **48**, 1304–1307.

9. Zhu, S. L., Yu, S. Y., Wang, Y., Ma, D. W. (2010). Organocatalytic Michael addition of aldehydes to protected 2-amino-1-nitroethenes: the practical syntheses of oseltamivir (Tamiflu) and substituted 3-aminopyrrolidines. *Angew. Chem. Int. Ed.* **49**, 4656–4660.

10. Danner, S. A., Carr, A., Leonard, J. M., Lehman, L. M., Gudiol, F., Gonzales, J., Raventos, A., Rubio, R., Bouza, E., Pintado, V., Aguado, A. G., Delomas, J. G., Delgado, R., Borleffs, J. C. C., Hsu, A., Valdes, J. M., Boucher, C. A. B., Copper, D. A., Gimeno, C., Clotet, B., Tor, J., Ferrer, E., Martinez, P. L., Moreno, S., Zancada, G., Alcami, J., Noriega, A. R., Pulido, F., Glassman, H. N. (1995). A short-term study of the safety, pharmacokinetics and efficacy of ritonavir, an inhibitor of HIV-1 protease. *N. Engl. J. Med.* **333**, 1528–1533.

11. Campo, R. E. (2000). Efficacy of indinavir/ritonavir-based regimens among patients with prior protease inhibitor failure. *3rd International Workshop on Salvage Therapy for HIV Infection*. Chicago. April 12–14, Abstract and poster presentation 7.

12. Askin, D. (1998). The synthesis of indinavir and other clinically useful HIV-1 protease inhibitors. *Curr. Opin. Drug Discov. Dev.* **1**, 338–348

13. Askin, D., Eng, K. K., Rossen, K., Purick, R. M., Wells, K. M., Volante, R. P., Reider, P. J. (1994). Highly diastereoselective reaction of a chiral, non-racemic amide enolate with (S)-glycidyl tosylate. Synthesis of the orally active HIV-1 protease inhibitor L-735524. *Tetrahedron Lett.* **35**, 673–676.

14. Dorsey, B. D., Levin, R. B., McDaniel, S. L., Vacca, J. P., Guare, J. P. Darke, P. L., Zugay, J. A., Emini, E. A., Schleif, W. A., Quintero, J. C., Lin, J. H., Chen, I. W., Holloway, M. K., Fitzgerald, P. M. D., Axel, M. G., Ostovic, D., Anderson, P. S., Huff, J. R. (1994). L-735524: The Design of a potent and orally bioavailable HIV protease inhibitor. *J. Med. Chem.* **37**, 3443–3451.

15. Livermore, D. M., Mushtaq, S., Warner, M. (2005). Selectivity of ertapenem for *Pseudomonas aeruginosa* mutants cross-resistant to other carbapenems. *J. Antimicrob. Chemother.* **55**, 306–311.

16. Betts, M. J., Davies, G. M., Swain, M. L. (1993). Carbapenems containing a carboxy substituted phenyl group, processes for their preparation, intermediates and use as antibiotics. EP 0579826, JP 1994506704, US 5478820, US 5652233, US 5856321, WO 9318078.

17. Williams, J. M., Brands, K. M. J., Skerlj, R., Houghton, P. (1999). Process for synthesizing carbapenem antibiotics. WO 9945010.

18. Birks, J., Iakovidou, V., Tsolaki, M. (2000). Rivastigmine for Alzheimer's disease. *Cochrane Database Sys. Rev.* **2**, CD001191.

19. Emre, M., Aarsland, D., Albanese, A. (2004). Rivastigmine for dementia associated with Parkinson' disease. *N. Engl. J. Med.* **351**, 2509–2518.

20. Enz, A. (Sandoz Patent GmbH). (1988). Phenylcarbamate. AU 8812554, BE 1001467, CH 675720, DE 3805744, FR 2611707, GB 2203040, JP 88238054.

21. Amstutz, R., Enz, A., Marzi, M., Boelsterli, J., Walkinshaw, M. (1990). Cyclic phenyl-carbamates of the myotin-type and their action on acetylcholinesterase. *Helv. Chim. Acta.* **73**, 739–753.

22. Esmaoglu, A., Akin, A., Mizrak, A., Turk, Y., Boyaci, A. (2006). Addition of cisatracurium to lidocaine for intravenous regional anesthesia. *J. Clin. Anesth.* **18**, 194–197.

23. Hill, G. L., Turner, D. A. (1992). Neuromuscular blocking agents. CH 683427, EP 539470, FR 2665791, GB 2260763, JP 93508648, WO 9200965.

24. Matsumoto, Y., Miyazato, M., Furuta, A., Torimoto, K., Hirao, Y., Chancellor, M. B., Yoshimura, N. (2010). Differential roles of M2 and M3 muscarinic receptor subtypes in modulation of bladder afferent activity in rats. *Urology* **75**, 862–867.

25. Tyagi, O. D., Ray, P. C., Chauhan, Y. K., Rao, K. B., Reddy, N. M., Reddy, D. S. P. (2009). Improved process for producing darifenacin. WO 2009125430.

26. Ramakrishnan, A., Kapkoti, G. S., Hashmi, A. M., Sahoo, A. (2009). Novel process for the preparation of (3SM-[2-(2,3-dihydro-5-benzofuranyl)ethyl]-α,α-diphenyl-3-pyrrolidineacetamide hydrobromide. WO 2009125426.

27. Lin, R., Deng, Q. H., Zhou, M., Li, Z. C., Zeng, G. C. (2009). Study on the synthesis of (3S)-1-[2-(2,3-dihydro-5-benzofuranyl)ethyl]-α,α-diphenyl-3-pyrrolidineacetamide hydrobromide (darifenacin hydrobromide). *Huaxue Shijie* **50**, 100–103.

28. Liu, C. H., Yu, X., Yuan, Z. D. (2007). Synthesis of (3S)-1-[2-(2,3-dihydro-5-benzofuranyl)ethyl]-α,α-diphenyl-3-pyrrolidineacetamide. *Zhongguo Yiyao Gongye Zazhi* **38**, 825–827.

29. Owen, R. T. (2006). Ramelteon: profile of a new-promoting medication. *Drugs Today* **42**, 255–263.

30. Miyamoto, M., Nishikawa, H., Doken, Y., Hirai, K., Uchikawa, O., Ohkawa, S. (2004). The sleep-promoting action of ramelteon (TAK-375) in freely moving cats. *Sleep* **27**, 1319–1325.

31. Zammit, G., Erman, M., Wang-Weigand, S., Sainati, S., Zhang, J., Roth, T. (2007). Evaluation of the efficacy and safety of ramelteon in subjects with chronic insomnia. *J. Clin. Sleep Med.* **3**, 495–504.

32. Yamashita, M., Yamano, T. (2009). Synthesis of melatonin receptor agonist Ramelteon via Rh-catalyzed asymmetric hydrogenation of an allylamine. *Chem. Lett.* **38**, 100–101.

33. Uchikawa, O., Fukatsu, K., Tokunoh, R., Kawada, M., Matsumoto, K., Imai, Y., Hinuma, S., Kato, K., Nishikawa, H., Hirai, K., Miyamoto, M., Ohkawa, S. (2002). Synthesis of a novel series of tricyclic indan derivatives as melatonin receptor agonists. *J. Medicinal Chem.* **45**, 4222–4239.

34. Bhanu, M. N., Sinha, C., Aher, B., Bandal, A., Parab, A. (2010). Process for the preparation of ramelteon. WO 2010055481.

35. Camps Garcia, P., Masllorens Llinas, E. (2009). Process for preparing ramelteon. WO 2009106966.

36. Sulfi, S., Timmis, A. D. (2006). Ivabradine—the first selective sinus node I(f) channel inhibitor in the treatment of stable angina. *Int. J. Clin. Pract.* **60L**, 222–228.

37. Tardif, J. C., Ford, I., Tendera, M., Bourassa, M. G. Fox K. (2005). Efficacy of ivabradine, a new selective I(f) inhibitor, compared with atenolol in patients with chronic stable angina. *Eur. Heart. J.* **26**, 2529–2536.

38. Fox, K., Ford, I., Steg, P. G., Tendera, M., Ferrari, R. (2008). Ivabradine for patients with stable coronary artery disease and left-ventricular systolic dysfunction: a randomized, double-blind, placebo-controlled trial. *Lancet* **372**, 807–816.

39. Bose, P., Siripalli, U. B. R., Kandadai, A. S. (2010). Process for preparation of ivabradine via resolution of N-[[3,4-dimethoxybicyclo[4.2.0]octa-1,3,5-trien-7-yl]methyl]-N-methylamine. WO 2010072409.

40. Luo, J. Z., Yan, Y. M., Tu, Y. R. (2008). Process for preparation of ivabradine hydrochloride and its stable crystal form. WO 2008125006.

41. Dwivedi, S. D., Kumar, R., Patel, S. T., Shah, A. P. C. (2008). Process for preparation of ivabradine hydrochloride. WO 2008065681.

42. Herman, G. A., Stevens, C., Van Dyck, K., Bergman, A., Yi, B., De Smet, M., Snyder, K. Hilliard, D., Tanen, M., Tanaka, W., Wang, A. Q., Zeng, W., Musson, D., Winchell, G., Davies, M. J., Ramael, S., Gottesdiener K. M., Wagner, J. A. (2005). Pharmacokinetics and pharmacodynamics of sitagliptin, an inhibitor of dipeptidyl peptidase IV, in healthy subjects: results from two randomized, double-blind, placebo-controlled studies with single oral doses. *Clin. Pharmacol. Ther.* **78**, 675–688.

43. Herman, G. A., Bergman, A., Liu, F., Stevens, C., Wang, A. Q., Zeng, W., Chen, L., Snyder, K., Hilliard, D., Tanen, M., Tanaka, W., Meehan, A G., Lasseter, K., Dilzer, S., Blum, R., Wagner, J. A. (2006). Pharmacokinetics and pharmacodynamic effects of the oral DPP-4 inhibitor sitagliptin in middle-aged obese subjects. *J. Clin. Pharmacol.* **46**, 876–886.

44. Hansen, K. B., Balsells, J., Dreher, S., Hsiao, Y., Kubryk, M., Palucki, M., Rivera, N., Steinhuebel, D., Armstrong, J. D., Askin, D., Grabowski, E. J. J. (2005). First generation process for the preparation of the DPP-IV inhibitor sitagliptin. *Org. Process Res. Dev.* **9**, 634–663.

45. Wu, S., Yu, B., Wang, Y. J., Delice, A., Zhu, J. Y. (2010). Process and intermediates for the preparation of N-acylated-4-aryl beta-amino acid derivatives. WO 2010078440

46. Kothari, H. M., Dave, M. G., Pandey, B., Shukla, B. S. (2010). Improved process for preparation of (2*R*)-4-oxo-4-[3-(trifluoromethyl)-5,6-dihydro[1,2,4]-triazolo[4,3-a]pyrazin-7(8h)-yl]-1-(2,4,5-trifluorophenyl)butan-2-amine and new impurities in preparation thereof. WO 2010032264.

47. Steinhuebel, D., Sun, Y. K., Matsumura, K., Sayo, N., Saito, T. (2009). Direct asymmetric reductive amination. *J. Am. Chem. Soc.* **131**, 11316–11317.

48. Hansen, K. B., Hsiao, Y., Xu, F., Rivera, N., Clausen, A., Kubryk, M., Krska, S., Rosner, T., Simmons, B., Balsells, J., Ikemoto, N., Sun, Y. K., Spindler, F., Malan, C., Grabowski, E. J. J., Armstrong, J. D. (2009). Highly efficient asymmetric synthesis of sitagliptin. *J. Am. Chem. Soc.* **131**, 8798–8804.

49. Clausen, A. M., Dziadul, B., Cappuccio, K. L., Kaba, M., Starbuck, C., Hsiao, Y., Dowling, T. M. (2006). Identification of ammonium chloride as an effective promoter of the asymmetric hydrogenation of a β-enamine amide. *Org. Process Res. Dev.* **10**, 723–726.

50. Lee, F. Y. F, Borzilleri, R., Fairchild, C. R., Kamath, A., Smykla, R., Kramer, R., Vite, G. (2008). Preclinical discovery of ixabepilone, a highly active antineoplastic agent. *Cancer Chemother. Pharmacol.* **63**, 157–166.

51. Goodin, S. (2008). Ixabepilone: a novel microtubule-stabilizing agent for the treatment of metastatic breast cancer. *Am. J. Health Syst. Pharm.* **65**, 2017–2026.

52. Thomas, E. S., Gomez, H. L., Li, R. K., Chung, H. C., Pein, L. E., Chan, V. F., Jassem, J., Pivot, X. B., Klimovsky J. V., de Mendoza, F. H., Xu, B. H., Campone, M., Lerzo, G. L., Peck, R. A., Mukhopadhyay, P., Vahdat, L T., Roche, H. H. (2007). Ixabepilone plus capecitabine for metastatic breast cancer progressing after anthracycline and taxane treatment. *J. Clin. Oncol.* **25**, 5210–5217.

53. Aghajanian, C., Burris, H. A., Jones, S., Spriggs, D. R., Cohen, M. B., Peck, R., Sabbatini, P., Hensley, M. L., Greco, F. A., Dupont, J., O'Connor, O. A. (2007). Phase I study of the novel epothilone analog ixabepilone (BMS-247550) in patients with advanced solid tumors and lymphomas. *J. Clin. Oncol.* **25**, 1082–1088.

54. Li, W. S., Thornton, J. E., Guo, Z. R., Swaminathan, S. (2002). A process for the preparation of epothilone analogs and intermediates. WO 2002060904.

55. Danishefsky, S. J., Lee, C. B., Chappell, M., Stachel, S., Chou, T. C. (2001). Synthesis of epothilones, intermediates and analogs for use in treatment of cancers with multidrug resistant phenotype. WO 2001064650.

56. Kim, S. H., Johnson, J. A. (2000). A process for the reduction of oxiranyl epothilones to olefinic epothilones. WO 2000071521.

57. Borzilleri, R. M., Zheng, X. P., Schmidt, R. J., Johnson, J. A., Kim, S. H., DiMarco, J. D., Fairchild, C. R., Gougoutas, J. Z., Lee, F. Y. F., Long, B. H., Vite, G. D. (2000). A novel application of a Pd(0)-catalyzed nucleophilic substitution reaction to the regio- and stereoselective synthesis of lactam analogues of the epothilone natural products. *J. Am. Chem. Soc.* **122**, 8890–8897.

58. Abel, S., Russell, D., Whitlock, L. A., Ridgway, C. E., Nedderman, A. N., Walker, D. K. (2008). Assessment of the absorption, metabolism and absolute bioavailability of maraviroc in healthy male subjects. *Br. J. Clin. Pharmacol.* **65**, 60–67.

59. Abel, S., Back, D. J., Vourvahis, M. (2009). Maraviroc: pharmacokinetics and drug interactions. *Antiviral Ther.* **14**, 607–618.

60. Emmelkamp, J. M., Rockstroh, J. K. (2007). CCR5 antagonist: comparison of efficacy, side effects, pharmacokinetics and interactions—review of the literature. *Eur. J. Med. Res.* **12**, 409–417.

61. Stephenson, J. (2007). Researchers buoyed by novel HIV drugs: will expand drug arsenal against resistant virus. *JAMA* **297**, 1535–1536.

62. Gant, T. G., Sarshar, S. (2008). 8-Azabicyclo[3.2.1]octane-base compounds, preparation and utility of CCR5 inhibitors and use in the treatment of infections. WO 2008146605.

63. Haycock-Lewandowski, S. J., Wilder, A., Ahman, J. (2008). Development of a bulk enabling route to Maraviroc (UK-427,857), a CCR-5 receptor antagonist. *Org. Process Res. Dev.* **12**, 1094–1103.

64. Ahman, J., Birch, M., Haycock-Lewandowski, S. J., Long, J., Wilder, A. (2008). Process research and scale-up of a commercializable route to Maraviroc (UK-427,857), a CCR-5 receptor antagonist. *Org. Process Res. Dev.* **12**, 1104–1113.

65. Tung, R. (2008). Preparation of deuterated triazolyl tropane derivatives as CCR5 receptor inhibitors. WO 2008063600.

66. Lou, S., Moquist, P. N., Schaus, S. E. (2007). Asymmetric allylboration of acyl imines catalyzed by chiral diols. *J. Am. Chem. Soc.* **129**, 15398–15404.

67. Price, D. A., Gayton, S., Selby, M. D., Ahman, J., Haycock-Lewandowski, S., Stammen, B. L., Warren, A. (2005). Initial synthesis of UK-427,857 (Maraviroc). *Tetrahedron Lett.* **46**, 5005–5007.

68. Neary, P., Delaney, P. (2005). Alvimopan. *Expert. Opin. Investig. Drugs* **14**, 479–488

69. Schmidt, W. K. (2001). Alvimopan (ADL 8-2698) is a novel peripheral opioid antagonist. *Am. J. Surg.* **182**, 27S–38S.

70. Werner, J. A., Cerlone, L. R., Frank, S. A., Ward, J. A., Labib, P., Tharp-Taylor, R. W., Ryan, C. W. (1996). Synthesis of *trans*-3, 4-dimethyl-4-(3-hydroxyphenyl) piperidine opioid antagonist: application of the *cis*-thermal elimination of carbonates to alkaloid synthesis. *J. Org. Chem*. **61**, 587–597.

71. Buehler, J. D. (2007). Compositions containing opioid antagonists. WO 2007047935.

72. Roehrig S., Straub A., Pohlmann J., Lampe, T., Pernerstorfer, J., Schlemmer, K. H., Reinemer, P., Perzborn, E. (2005). Discovery of the novel antithrombotic agent 5-chloro-*N*-({(5S)-2-oxo-3-[4-(3-oxomorpholin-4-yl)phenyl]-1,3-oxazolidin-5-yl}methyl)thiophene-2-carboxamide (BAY59-7939): an oral direct factor Xa inhibitor. *J. Medicinal Chem*. **48**, 5900–5908.

73. Eriksson B. I., Borris L. C., Dahl O.E., Haas, S., Huisman, M. V., Kakkar, A. K., Muehlhofer, E., Dierig, C., Misselwitz, F., Kalebo, P. (2006). A once-daily, oral, direct Factor Xa inhibitor, rivaroxaban (BAY 59-7939), for thrombopropylaxis after total hip replacement. *Circulation* **114**, 2374–2381.

74. Kakkar, A. K., Brenner, B., Dahl, O.E., Eriksson, B. I., Mouret, P. Muntz, J., Soglian, A. G., Pap, A. F., Misselwitz, F., Hass, S. (2008). Extended duration rivaroxaban versus short-term enoxaparin for the prevention of venous thromboembolism after total hip arthroplasty: a double-blind, randomized controlled trial. *Lancet* **372**, 31–39.

75. Masse, C. E. (2009). Substituted oxazolidinone derivatives. WO 2009023233.

76. Straub, A., Lampe, T., Pohlmann, J., Roehrig, S., Perzborn, E., Schlemmer, K., Pernerstorfer, J. Substituted oxozolidinones and their use in the field of blood coagulation. US 7157456.

77. Berwe, M., Thomas, C., Rehse, J., Grotjohann, D. (2005). Preparation of rivaroxaban. WO 2005068456.

78. Adams, J., Kauffman, M. (2004). Development of the proteasome inhibitor Velcade (Bortezomib). *Cancer Invest*. **22**, 304–311.

79. Bonvini, P., Zorzi, E., Basso, G., Rosolen, A. (2007). Bortezomib-mediated 26S proteasome inhibition causes cell-cycle arrest and induces apoptosis in CD30+ anaplastic large cell lymphoma. *Leukemia* **21**, 838–842.

80. Voorhees, P. M., Dees, E. C., O'Neil, B., Orlowski, R. Z. (2003). The proteasome as a target for cancer therapy. *Clin. Cancer Res*. **9**, 6316–6325.

81. Anargyrou, K., Dimopoulos, M. A., Sezer, O., Terpos, E. (2008). Novel antimyeloma agents and angiogenesis. *Leuk. Lymphoma* **49**, 677–689.

82. Zhu, Y. Q., Zhao, X., Zhu, X. R., Wu, G., Li, Y. J., Ma, Y. H., Yuan, Y. X., Yang, J., Hu, Y., Ai, L., Gao, Q. Z. (2009). Design, synthesis, biological evaluation, and structure-activity relationship (SAR) discussion of dipeptidyl boronate proteasome inhibitors, Part I: comprehensive understanding of the SAR of -amino acid boronates. *J. Medicinal Chem*. **52**, 4192–4199.

83. Palle, R. V., Kadaboina, R., Murki, V., Manda, A., Gunda, N., Pulla, R. R., Hanmanthu, M., Mopidevi, N. N., Ramdoss, S. K. (2009). Preparation of bortezomib in crystalline form. WO 2009036281.

84. Janca, M., Dobrovolny, P. (2009). Preparation of bortezomib, a boronic acid dipeptide. WO 2009004350.

85. Beenen, M. A., An, C., Ellman, J. A. (2008). Asymmetric copper-catalyzed synthesis of α-amino boronate esters from *N*-tert-butanesulfinyl aldimines. *J. Am. Chem. Soc*. **130**, 6910–6911.

86. Nissen, S. E., Nicholls, S. J., Sipahi, I., Libby, P., Raichlen, J. S., Ballantyne, C. M., Davignon, J., Erbel, R., Fruchart, J. C., Tardif, J. C., Schoenhagen, P., Crowe, T., Cain, V., Wolski, K., Goormastic, M., Tuzcu, E. M. (2006). Effect of very high-intensity statin therapy on regression of coronary atherosclerosis: the ASTEROID trial. *JAMA* **295**, 1556–1565.

87. Sever, P. S., Dahlöf, B., Poulter, N. R., Wedel, H., Beevers, G., Caulfield, M., Collins, R., Kjeldsen, S. E., Kristinsson, A., Mclnnes, G. T., Mehlsen, J., Nieminen, M., O'Brien, E., Ostergren, J. (2003). Prevention of coronary and stroke events with atorvastatin in hypertensive patients who have average or lower-than-average cholesterol concentrations, in the Anglo-Scandinavian Cardiac Outcomes Trial—Lipid Lowering Arm (ASCOT-LLA): a multicentre randomised controlled trial trial. *Lancet* **361**, 1149–1158.

88. Colhoun, H. M., Betteridge, D. J., Durrington, P. N., Hitman, G. A., Nell, H. A. M., Livingstone, S. J., Thomason, M. J., Mackness, M. I., Charlton-Menys, V., Fuller, J. H. (2004). Primary prevention of cardiovascular disease with atorvastatin in type 2 diabetes in the Collaborative Atorvastatin Diabetes Study (CARDS): multicentre randomised placebo-controlled trial. *Lancet* **364**, 685–696.

89. Ghirlanda, G., Oradei, A., Manto, A., Lippa, S., Uccioli, L., Caputo, S., Greco, A., Littarru, G. (1993). Evidence of plasma CoQ10-lowering effect by HMG-CoA reductase inhibitors: a double-blind, placebo-controlled study. *J. Clin. Pharmacol.* **33**, 226–229.

90. Dwivedi, S. D., Patel, D. J., Vinchhi, K. M., Rupapara, M. L. (2010). Process for the preparation of amorphous atorvastatin calcium via saponification of atorvastatin tert-butyl ester. US 2010190999.

91. Pai, G. G., Nanda, K., Chaudhari, N. P., Anjaneyulu, A., Ghogare, B. N. (2009). Process for preparation of 2-[2-(4-fluorophenyl)-2-oxo-1-phenylethyl]-4-methyl-3-oxopentanoic acid phenylamide. WO 2009144736.

92. Oren, J., Dolitzky, B., Harel, Z., Perlman, N., Lidor-Hadas, R. (2004). An improved method of synthesis of 3, 5-dihydroxy-7-pyrrol-1-yl heptanoic acids (atorvastatin derivatives). WO 2004046105.

93. Bulter, D. E., Le, T. V., Nanninga, T. N. (1994). Process for trans-6-[2-(substituted—pyrrol-1-yl)alkyl]pyran-2-one-inhibitors of cholesterol synthesis. US 5298627.

94. Chan, F. K., Ching, J. Y., Hung, L. C., Wong, V. W., Leung, V. K., Kung, N. N. (2005). Clopidogrel versus aspirin and esomeprazole to prevent recurrent ulcer bleeding. *N. Engl. J. Med.* **352**, 238–244.

95. Ho, P. M., Maddox, T. M., Wang, L., Fihn, S. D., Jesse, R. L., Peterson, E. D., Rumsfeld, J. S. (2009). Risk of adverse outcomes associated with concomitant use of clopidogrel and proton pump inhibitors following acute coronary syndrome. *JAMA*, **301**, 937–944.

96. Collet, J. P., Hulot J. S., Pena A., Villard E., Esteve J. B., Silvain J., Payot L., Brugier D., Cayla, G., Beygul, F., Bensimon, G., Funck-Brentano, C., Montalescot, G. (2009). Cytochrome P450 2C19 polymorphism in young patients treated with clopidogrel after myocardial infarction: a cohort study. *Lancet* **373**, 309–317.

97. Simonic, I., Benkic, P., Zupet, R., Smrkolj, M., Stukelj, M. (2008). Process for the synthesis of clopidogrel and new forms of pharmaceutically acceptable salts thereof. WO 2008034912.

98. Wang, L. X., Shen, J. F., Tang, Y., Chen, Y., Wang, W., Cai, Z. G., Du, Z. J. (2007). Synthetic improvements in the preparation of clopidogrel. *Org. Process Res. Dev.* **11**, 487–489.

99. Descamps, M., Radisson, J. (1992). Process for the preparation of an *N*-phenylacetic derivative of tetrahydrothieno [3.2.c] pyridine and its intermediate of synthesis. EP 446569.

100. Yang, Y. X., Lewis, J. D., Epstein, S., Metz, D. C. (2006). Long-term proton pump inhibitor therapy and risk of hip fracture. *JAMA* **296**, 2947–2953.

101. Shoshana, J., Herzig, M. D., Michael, D., Howell, M. D. (2009). Acid-suppressive medication use and the risk for hospital-acquired pneumonia. *JAMA* **301**, 2120–2128.

102. Stedman, C. A., Barclay, M. L. (2000). Review article: comparison of the pharmacokinetics, acid suppression and efficacy of proton pump inhibitors. *Aliment. Pharmacol. Ther.* **14**, 963–978.

103. Norgard, N. B., Mathews, K. D., Wall, G. C. (2009). Drug-drug interaction between clopidogrel and the proton pump inhibitors. *Ann. Pharmacother.* **43**, 1266–1274.

104. Röhss, K., Hasselgren, G., Hedenström, H. (2002). Effect of esomeprazole 40mgvs omeprazole 40mg on 24-hour intragastric pH in patients with symptoms of gastroesophageal reflux disease. *Dig. Dis. Sci.* **47**, 954–958.

105. Stepankova, H., Zezula, J., Hajicek, J., Kral, V. (2010). Manufacture of pure or enriched esomeprazole comprazole comprises oxidizing 5-methoxy-2-(4-methoxy-3,5-dimethyl-pyridin-2-yl methylsulfanyl)-1H-benzoimidazole with a hydroperoxide on a catalyst constituted by a chiral metallic complex. WO 2010091652

106. Parthasaradhi Reddy, B., Rathnakar Reddy, K., Raji Reddy, R., Muralidhara Reddy, D. (2009). A process for preparation of enantiomerically pure esomeprazole. WO 2009040825.

107. Jiang, B., Zhao, X. L., Dong, J. J., Wang, W. J. (2009). Catalytic asymmetric oxidation of heteroaromatic sulfides with tert-butyl hydroperoxide catalyzed by a titanium complex with a new chiral 1,2-diphenylethane-1,2-diol ligand. *Eur. J. Org. Chem.* **7**, 987–991.

108. Lindberg, P., Weidolf, L. (1999). Method for the treatment of gastric acid-related diseases and production of medication using (-) enantiomer of omperazole. US5877192.

109. Domingo Coto, A., Comely, A., Verdaguer Espaulella, X., Rafecas Jane, L. (2007). A process for the preparation of the (S)-enantiomer of omeprazole. WO 2007074099

110. Thennati, R., Rehani, R. B., Soni, R. R., Chhabada, V. C., Patel, V. M. (2003). Optically active substituted pyridinylmethylsulfinyl-benzimidazoles and salts. WO 2003089408.

111. Cotton, H., Elebring, T., Larsson, M., Li, L., Sorensen, H., Von Unge, S. (2000). Asymmetric synthesis of esomeprazole. *Tetrahedron Asymmetry* **11**, 3819–3825.

112. Holt, R., Lindberg, P., Reeve, C., Taylor, S. (1996). Enantioselective preparation of pharmaceutically active sulfoxides by biooxidation. WO 9617076

113. Von Unge, S. (1997). A process for the optical purification of enantiomerically enriched benzimidazole derivatives. WO 9702261.

114. Jones, T. R., Labelle, M., Belley, M., Champion, E., Charette, L., Evans, J., Fordhutchinson, A. W., Ganthier, J. Y., lord, A., Masson, P., Mcauliffe, M., Mcfarlane, C. S., Metters, K. M., Pickett, C., Piechuta, H., Rochette, C., Radger, I. W., Sawyer,

N., Young, R. N., Zamboni, R., Abraham, W. M. (1995). Pharmacology of Montelukast sodium, a potent and selective leukotriene D4 receptor antagonist. *Can. J. Physiol Pharmacol*. **73**, 191–201.

115. Cheng, H., Leff, J. A., Amin, R., Gertz, B. J., DeSmet, M., Noonan, N, Rogers, J. D., Malbecq, W., Meisner, D., Somers, G. (1996). Pharmacokinetics, bioavailability, and safety of Montelukast sodium (MK-0476) in healthy males and females. *Pharm. Res*. **13**, 445–452.

116. Ray, P. C., Reddy, G. M., Tirmalaiah, G., Rammohan, K. (2010). Novel intermediates for producing [*R*-(E)]-1-[[[1-[3-[2-(7-chloro-quinolinyl)ethenyl]phenyl]-3-[2-(1-hydroxy-methylethyl)phenyl]propyl]thio]methyl]cyclopropaneacetic acid, monosodium salt and process thereof. WO 2010064257.

117. Nakka, K. C. S., Kashyap, T., Singh, J. (2010). An improved process for the preparation of montelukast sodium and its intermediates. WO 2010064109.

118. Halama, A., Jirman, J., Bouskova, O., Gibala, P., Jarrah, K. (2010). Improved process for the preparation of montelukast: development of an efficient synthesis, identification of critical impurities and degradants. *Org. Process Res. Dev*. **14**, 425–431

119. Halama, A., Jirman, J. (2008). A method for the preparation of montelukast. WO 2008083635.

120. Nissen, S. E., Nicholls, S. J., Sipahi, I., Libby, P., Raichlen, J. S., Ballantyne, C. M., Davignon, J., Erbel, R., Fruchart, J. C., Tardlf, J. C., Schoenhagen, P., Crowe, T., Cain, V., Wolski, K., Goormastic, M., Tuzcu, E. M. (2006). Effect of very high-intensity statin therapy on regression of coronary atherosclerosis: the asteroid trial. *JAMA* **295**, 1556–1565.

121. Ridker, P. M., Danielson, E., Fonseca, F. A. H., Genest, J., Gotto, A. M., Kastelein, J. J. P., Koenig, W., Libby, P., Lorenzatti, A. J., macFadyen, J. G., Nordestgaard, B. G., Shepherd, J., Willerson, J. T., Glynn, R. J. (2008). Rosuvastatin to prevent vascular events in men and women with elevated C-reactive protein. *N. Engl. J. Med*. **359**, 2195–2207.

122. Jones, P. H., Davidson, M.H., Stein, E. A., Bays, H. E., McKenney, J. M., Miller, E., Cain, V. A., Blasetto, J. W. (2003). Comparison of the efficacy and safety of rosuvastatin versus atorvastatin, simvastatin, and pravastatin across doses (STELLAR Trial). *Am. J. Cardiol*. **92**, 152–160.

123. McKillop, T. (2003). The statin wars. *Lancet* **362**, 1498.

124. McTaggart, F., Buckett, L., Davidson, R., Holdgate, G., McCormick, A., Schneck, D., Smith, G., Warwick, M. (2001). Preclinical and clinical pharmacology of Rosuvastatin, a new 3-hydroxy-3-methylglutaryl coenzyme A reductase inhibitor. *Am. J. Cardiol*. **87**, 28B–32B.

125. Benkic, P., Bevk, D., Lenarsic, R., Zupancic, S., Vajs, A., Jakse, D. (2010). Process for preparation of rosuvastatin. WO 2010081861.

126. Volk, B., Vago, P., Simig, G., Tompe, P., Barkoczy, J., Mezei, T., Bartha, F., Ruzsics, G., Karasz, A., Kiraly, I., Nagy, K. (2009). Process for preparation of rosuvastatin. WO 2009047576.

127. Butters, M., Lenger, S. R., Murray, P. M., Snape, E. W. (2006). Process for preparing rosuvastatin. WO 2006067456.

128. Watanabe, M., Koike, H., Ishiba, T., Okada, T. Seo, S., Hirai, K. (1997). Synthesis and biological activity of methanesulfonamide pyrimidine- and N-methanesulfonyl pyrrole substituted 3,5-dihydroxy-6-heptenoates, a novel series of HMG-CoA reductase inhibitors. *Bioorg. Med. Chem*. **5**, 437–444.

129. Mitka M. (2008). Cholesterol drug controversy continues. *JAMA* **299**, 2266–2266.

130. Taylor, A. J., Villnes, T. C., Stanek, E. J., Devine, P. J., Griffen, L., Miller, M., Weissman, N. J., Turco, M. (2009). Extended-release niacin or ezetimibe and carotid intima-media thickness. *N. Engl. J. Med.* **361**, 2113–2122.

131. Garcia-Calvo M., Lisnock J., Bull H. G., Hawes B. E., Burnett D. A., Braun M. P., Crona, J. H., Davis, H. R., Dean, D. C., Detmers, P. A., Graziano, M. P., Hughes, M., MacIntyre, D. E., Ogawa, A., O'Neill, K. A., Lyer, S. P. N., Shevell, D. E., Smith, M. M., Tang, Y. S., Makarewicz, A. M., Ujjainwalla, F., Altmann, S. W., Chapman, K. T., Thornberry, N. A. (2005). The target of ezetimibe is Niemann-Pick C1-Like 1 (NPC1L1). *Proc. Natl. Acad. Sci. U. S. A.* **102**, 8132–8137.

132. Kastelein, J. J. P., Akdim, F., Stroes, E. S. G., Zwinderman, A. H., Bots, M. L., Stalenhoef, A. F. H., Visseren, F. L. J., Sijbrands, E. J. G., Trip, M. D., Stein, E. A., Gandet, D., Duivenvoorden, R., Veltri, E. P., Marais, A. D., de Groot, E. (2008). Simvastatin with or without ezetimibe in familial hypercholesterolemia. *N. Engl. J. Med.* **358**, 1431–1443.

133. Brown, B. G., Taylor, A. J. (2005). Does ENHANCE diminish confidence in lowering LDL or in ezetimibe? *N. Engl. J. Med.* **358**, 1504–1547.

134. Vaccaro, W. (1998). Carboxy-substituted 2-azetidinones as cholesterol absorption inhibitors. *Bioorg. Medicinal Chem. Lett.* **8**, 319–322.

135. Rosenblum, S. B., Huynh, T., Afonso, A., Davis, H. R. (2000). Synthesis of 3-arylpropenyl, 3-arylpropynyl and 3-arylpropyl 2-azetidinones as cholesterol absorption inhibitor: application of the palladium catalyzed arylation of alkenes and alkynes. *Tetrahedron* **56**, 5735–5742.

136. Kim, G. J., Kim, C. H., Chang, J. Y., Kim, N. D., Chang, Y. K., Lee, G. S. (2010). Method of preparing ezetimibe and intermediates using cyclization, asymmetric reduction and hydrolysis as key steps. WO 2010071358.

137. Chidambaram Venkateswaran, S., Sarin Gurdeep, S., Gupta, P., Wadhwa, L. (2009). Process for preparing ezetimibe using novel allyl intermediates. WO 2009157019.

138. Sasikala, C. H. V. A., Padi, P. R., Sunkara, V., Ramayya, P., Dubey, P. K., Uppala, V. B.R., Praveen, C. (2009). An improved and scalable process for the synthesis of ezetimibe: an antihypercholesterolemia drug. *Org. Process Res. Dev.* **13**, 907–910.

139. Cerqueira, N. M., Fernandes, P. A., Ramos, M. J. (2007). Understanding ribonucleotide reductase inactivation by gemcitabine. *Chemistry Eur. J.* **13**, 8507–8515.

140. Oettle, H., Post, S., Neuhaus, P., Gellert, K., Langrehr, J., Ridwelski, K., Schramm, H., Fahlke, J., Zuelke, C., Burkart, C., Gutberiet, K., Kettner, E., Schmalenberg, H., Weigang-Koehler, Bechstein, W. O., Niedergethmann, M., Schmidt-Wolf, I., Roll, L., Doerken, B., Riess, H. (2007). Adjuvant chemotherapy with gemcitabine vs. observation in patients undergoing curative-intent resection of pancreatic cancer: a randomized controlled trial. *JAMA* **297**, 267–277.

141. Von Der Maase, H., Hansen, S. W., Roberts, J. T., Dogliotti, L., Oliver, T., Moore, M. J., Bodrogi, I., Albers, P., Knuth, A., Lippert, C. M., Kerbrat, P., Rovira, P. S., Wersall, P., Cleall, S. P., Rocychowdhury, D. F., Tomlin, I., Visseren-Grul, C. M., Conte, P. F. (2000). Gemcitabine and cisplatin versus methotrexate, vinblastine, doxorubicin, and cisplatin in advanced or metastatic bladder cancer: results of a large, randomized, multinational, multicenter, phase III study. *J. Clin. Oncol.* **18**, 3068–3077.

142. Vorbrueggen, H. (2009). Method for producing gemcitabine hydrochloride and intermediate products. WO 2009135455.

143. Park, S. J., Oh, C. R., Kim, Y. D. (2008). Process for preparing of 2′-deoxy-2′2′-difluorocytidine. WO 2008117955.

144. Wang, Y. H., Wang, X. H., Qiu, Z. B. (2007). Synthesis of gemcitabine hydrochloride. *Zhongguo Yiyao Gongye Zazhi* **38**, 249–251.

145. Jiang, X. R., Li, J. F., Zhang, R. X., Zhu, Y., Shen, J. S. (2008). An improved preparation process for gemcitabine. *Org. Process Res. Dev.* **12**, 888–891.

146. Palle, V. R. A., Nariyam, S. M., Murki, V., Waghmare, A. A., Mundhada, V. N. (2007). Preparation of Gemcitabine. WO 2007117760.

147. Polturi, R. B., Venkata, S. H., Betini, R. (2005). An improved process for the manufacture of high pure gemcitabine hydrochloride. WO 2005095430.

148. Ishikawa, H., Yoshitomi, T., Mashimo, K., Nakanishi, M., Shimizu, K. (2002). Pharmacological effects of latanoprost, prostaglandin E2, and F2alpha on isolated rabbit ciliary artery. *Graefes Arch. Clin. Exp. Ophthalmol.* **240**, 120–125.

149. Patel, S. S., Spencer, C. M. (1996). Latanoprost. A review of its pharmacological properties, clinical efficacy and tolerability in the management of primary open-angle glaucoma and ocular hypertension. *Drugs Aging* **9**, 363–378.

150. Amano, S., Nakai, Y., Ko, A., Inoue, K., Wakakura, M. (2008). A case of keratoconus progression associated with the use of topical latanoprost. *Jpn. J. Ophthalmol.* **52**, 334–336.

151. Albert, M., Berger, A., De Souza, D., Knepper, K., Sturm, H. (2008). Improved process for the production of prostaglandins and prostaglandin analogs. EP 2143712

152. Yao, C. H., Yang, C. M., Chao, H. H., Lee, G. Ray. (2009). Method for preparing prostaglandin F analogues. US 2009259066.

153. Martynow, J. G., Jozwik, J., Szelejewski, W., Achmatowicz, O., Kutner, A., Wisniewski, K., Winiarski, J., Zegrocka-Stendel, O., Golebiewski, P. (2007). A new synthetic approach to high-purity (15R)-latanoprost. *Eur. J. Org. Chem.* **4**, 689–703.

154. Obadalova R, Pilarcik, T., Slavikova, M. F., Hajicek, J. (2005). Synthesis of latanoprost diastereoisomers. *Chirality* **17(Suppl.)**, S109–S113.

155. Greenwood, A. K., McHattie, D., Thompson, D. G., Clissold, D. W. (2002). Process for the preparation of prostaglandins and analogs thereof. WO 2002096898

156. Morrissey, I., Hoshino, K., Sato, K., Yoshida, A., Hayakawa, I., Bures, M.G., Shen, L. L. (1996). Mechanism of differential activities of ofloxacin enantiomers. *Antimicrob. Agents Chemother.* **40**, 1775–1784.

157. North, D. S., Fish, D. N., Redington, J. J. (1998). Levofloxacin, a second-generation fluoroquinolone. *Pharmacotherapy* **18**, 915–935.

158. Davis, R., Bryson, H. M. (1994). Levofloxacin: a review of its antibacterial activity, pharmacokinetics and therapeutic efficacy. *Drugs* **47**, 677–700.

159. De Sarro, A., De Sarro, G. (2001). Adverse reactions to fluoroquinolones, an overview on mechanistic aspects. *Curr. Med. Chem.* **8**, 371–384.

160. Gopal Rao, G., Mahankali Rao, C. S., Starke, I. (2003). *Clostridium difficile*-associated diarrhoea in patients with community-acquired lower respiratory infection being treated with levofloxacin compared with beta-lactam-based therapy. *J. Antimicrob. Chemother.* **51**, 697–701.

161. Suto, M. J., Domagala, J. M., Roland, G. E., Mailloux, G. B., Cohen, M. A. (1992). Fluoroquinolones: relationships between structural variations, mammalian cell cytotoxicity, and antimicrobial activity. *J. Medicinal Chem.* **35**, 4745–4750.

162. Domagala, J. M. (1994). Structure-activity and structure-side-effect relationships for the quinolone antibacterials. *J. Antimicrob. Chemother*. **33**, 685–706.

163. Bao, J. S., Zhang, H. H., Shi, X., Zeng, H. F., Liang, S. S., Yu, X. H. (2008). Synthesis of (3S)-9-fluoro-2,3-dihydro-3-methyl-10-(4-methyl-1-piperazinyl)-7-oxo-7H-pyrido[1,2,3-de]-1,4-benzoxazine-6-carboxylic acid (levofloxacin). *Hecheng Huaxue* **16**, 721–723.

164. Puig Torres, S., Bessa Bellmunt, J. (2008). Process for the preparation of an antibacterial quinolone compound. WO 2008077643.

165. Bower, J. F., Szeto, P., Gallagher, T. (2007). Enantiopure 1, 4-benzoxazines via 1,2-cyclic sulfamidates, synthesis of levofloxacin. *Org. Lett*. **9**, 3283–3286.

166. Adrio, J., Carretero, J. C., Ruano, J. L. G., Pallares, A., Vicioso, M. (1999). An efficient synthesis of ofloxacin and levofloxacin from 3, 4-difluoroaniline. *Heterocycles* **51**, 1563–1572.

167. Atarashi, S., Yokohama, S, Yamazaki, K., Sakano, K., Imamura, M., Hayakawa, I. (1987). Synthesis and antibacterial activities of optically active ofloxacin and its fluoromethyl derivative. *Chem. Pharmaceut. Bull*. **35**, 1896–1902.

168. Ren, J., Bird, L. E., Chamberlain, P. P., Stewart-Jones, G. B., Stuart, D. I., Stammers, D. K. (2002). Structure of HIV-2 reverse transcriptase at 2.35-Å resolution and the mechanism of resistance to non-nucleoside inhibitors. *Proc. Natl. Acad. Sci. U. S. A*. **99**, 14410–14415.

169. Cespedes, M. S., Aberg, J. A. (2006). Neuropsychiatric complications of antiretroviral therapy. *Drug Safety* **29**, 865–874.

170. Röder, C. S., Heinrich, T., Gehrig, A. K., Mikus, G. (2007). Misleading results of screening for illicit drugs during efavirenz treatment. *AIDS* **21**, 1390–1391.

171. Gurjar, M. K., Deshmukh, A. A., Deshmukh, S. S., Mehta, S. R. (2010). Process for preparation of Efavirenz. WO 2010032259.

172. Nicolaou, K. C., Krasovskiy, A., Majumder, U., Trepanier, V. E., Chen, D. Y. K. (2009). New synthetic technologies for the construction of heterocycles and tryptamines. *J. Am. Chem. Soc*. **131**, 3690–3699.

173. Pierce, M. E., Parsons, R. L. Radesca, L. A., Lo. Y. S., Silverman, S., Moore, J. R., Islam, Q., Choudhury, A., Fortunak, J. M. D., Nguyen, D., Luo, C., Morgan, S. J., Davis, W. P., Confalone, P. N., Chen, C. Y., Tillyer, R. D., Frey, L., Tan, L. S., Xu, F., Zhao, D. L., Thompson, A. S., Corley, E. G., Grabowski, E. J. J., Reamer, R., Reider, P. J. (1998). Practical asymmetric synthesis of efavirenz (DMP 266), an HIV-1 reverse transcriptase inhibitor. *J. Org. Chem*. **63**, 8536–8543.

174. Radesca, L. A., Lo, Y. S., Moore, J. R., Pierce, M. E. (1997). Synthesis of HIV-1 reverse transcriptase inhibitor DMP 266. *Synthetic Commun*. **27**, 4373–4384.

INDEX

Chiral Drugs: Chemistry and Biological Action, First Edition. Edited by Guo-Qiang Lin,
Qi-Dong You and Jie-Fei Cheng.
© 2011 John Wiley & Sons, Inc. Published 2011 by John Wiley & Sons, Inc.